Analysis of Categorical Data with R

CHAPMAN & HALL/CRC
Texts in Statistical Science Series

Series Editors
Francesca Dominici, *Harvard School of Public Health, USA*
Julian J. Faraway, *University of Bath, UK*
Martin Tanner, *Northwestern University, USA*
Jim Zidek, *University of British Columbia, Canada*

Statistical Theory: A Concise Introduction
F. Abramovich and Y. Ritov

Practical Multivariate Analysis, Fifth Edition
A. Afifi, S. May, and V.A. Clark

Practical Statistics for Medical Research
D.G. Altman

Interpreting Data: A First Course in Statistics
A.J.B. Anderson

Introduction to Probability with R
K. Baclawski

Linear Algebra and Matrix Analysis for Statistics
S. Banerjee and A. Roy

Analysis of Categorical Data with R
C. R. Bilder and T. M. Loughin

Statistical Methods for SPC and TQM
D. Bissell

Introduction to Probability
J. K. Blitzstein and J. Hwang

Bayesian Methods for Data Analysis, Third Edition
B.P. Carlin and T.A. Louis

Second Edition
R. Caulcutt

The Analysis of Time Series: An Introduction, Sixth Edition
C. Chatfield

Introduction to Multivariate Analysis
C. Chatfield and A.J. Collins

Problem Solving: A Statistician's Guide, Second Edition
C. Chatfield

Statistics for Technology: A Course in Applied Statistics, Third Edition
C. Chatfield

Bayesian Ideas and Data Analysis: An Introduction for Scientists and Statisticians
R. Christensen, W. Johnson, A. Branscum, and T.E. Hanson

Modelling Binary Data, Second Edition
D. Collett

Modelling Survival Data in Medical Research, Second Edition
D. Collett

Introduction to Statistical Methods for Clinical Trials
T.D. Cook and D.L. DeMets

Applied Statistics: Principles and Examples
D.R. Cox and E.J. Snell

Multivariate Survival Analysis and Competing Risks
M. Crowder

Statistical Analysis of Reliability Data
M.J. Crowder, A.C. Kimber, T.J. Sweeting, and R.L. Smith

An Introduction to Generalized Linear Models, Third Edition
A.J. Dobson and A.G. Barnett

Nonlinear Time Series: Theory, Methods, and Applications with R Examples
R. Douc, E. Moulines, and D.S. Stoffer

Introduction to Optimization Methods and Their Applications in Statistics
B.S. Everitt

Extending the Linear Model with R: Generalized Linear, Mixed Effects and Nonparametric Regression Models
J.J. Faraway

Linear Models with R, Second Edition
J.J. Faraway

A Course in Large Sample Theory
T.S. Ferguson

Multivariate Statistics: A Practical Approach
B. Flury and H. Riedwyl

Readings in Decision Analysis
S. French

Markov Chain Monte Carlo: Stochastic Simulation for Bayesian Inference, Second Edition
D. Gamerman and H.F. Lopes

Bayesian Data Analysis, Third Edition
A. Gelman, J.B. Carlin, H.S. Stern, D.B. Dunson, A. Vehtari, and D.B. Rubin

Multivariate Analysis of Variance and Repeated Measures: A Practical Approach for Behavioural Scientists
D.J. Hand and C.C. Taylor

Practical Data Analysis for Designed Practical Longitudinal Data Analysis
D.J. Hand and M. Crowder

Logistic Regression Models
J.M. Hilbe

Richly Parameterized Linear Models: Additive, Time Series, and Spatial Models Using Random Effects
J.S. Hodges

Statistics for Epidemiology
N.P. Jewell

Stochastic Processes: An Introduction, Second Edition
P.W. Jones and P. Smith

The Theory of Linear Models
B. Jørgensen

Principles of Uncertainty
J.B. Kadane

Graphics for Statistics and Data Analysis with R
K.J. Keen

Mathematical Statistics
K. Knight

Introduction to Multivariate Analysis: Linear and Nonlinear Modeling
S. Konishi

Nonparametric Methods in Statistics with SAS Applications
O. Korosteleva

Modeling and Analysis of Stochastic Systems, Second Edition
V.G. Kulkarni

Exercises and Solutions in Biostatistical Theory
L.L. Kupper, B.H. Neelon, and S.M. O'Brien

Exercises and Solutions in Statistical Theory
L.L. Kupper, B.H. Neelon, and S.M. O'Brien

Design and Analysis of Experiments with SAS
J. Lawson

A Course in Categorical Data Analysis
T. Leonard

Statistics for Accountants
S. Letchford

Introduction to the Theory of Statistical Inference
H. Liero and S. Zwanzig

Statistical Theory, Fourth Edition
B.W. Lindgren

Stationary Stochastic Processes: Theory and Applications
G. Lindgren

The BUGS Book: A Practical Introduction to Bayesian Analysis
D. Lunn, C. Jackson, N. Best, A. Thomas, and D. Spiegelhalter

Introduction to General and Generalized Linear Models
H. Madsen and P. Thyregod

Time Series Analysis
H. Madsen

Pólya Urn Models
H. Mahmoud

Randomization, Bootstrap and Monte Carlo Methods in Biology, Third Edition
B.F.J. Manly

Introduction to Randomized Controlled Clinical Trials, Second Edition
J.N.S. Matthews

Statistical Methods in Agriculture and Experimental Biology, Second Edition
R. Mead, R.N. Curnow, and A.M. Hasted

Statistics in Engineering: A Practical Approach
A.V. Metcalfe

Statistical Inference: An Integrated Approach, Second Edition
H.S. Migon, D. Gamerman, and F. Louzada

Beyond ANOVA: Basics of Applied Statistics
R.G. Miller, Jr.

A Primer on Linear Models
J.F. Monahan

Applied Stochastic Modelling, Second Edition
B.J.T. Morgan

Elements of Simulation
B.J.T. Morgan

Probability: Methods and Measurement
A. O'Hagan

Introduction to Statistical Limit Theory
A.M. Polansky

Applied Bayesian Forecasting and Time Series Analysis
A. Pole, M. West, and J. Harrison

Statistics in Research and Development, Time Series: Modeling, Computation, and Inference
R. Prado and M. West

Introduction to Statistical Process Control
P. Qiu

Sampling Methodologies with Applications
P.S.R.S. Rao

A First Course in Linear Model Theory
N. Ravishanker and D.K. Dey

Essential Statistics, Fourth Edition
D.A.G. Rees

Stochastic Modeling and Mathematical Statistics: A Text for Statisticians and Quantitative
F.J. Samaniego

Statistical Methods for Spatial Data Analysis
O. Schabenberger and C.A. Gotway

Bayesian Networks: With Examples in R
M. Scutari and J.-B. Denis

Large Sample Methods in Statistics
P.K. Sen and J. da Motta Singer

Decision Analysis: A Bayesian Approach
J.Q. Smith

Analysis of Failure and Survival Data
P.J. Smith

Applied Statistics: Handbook of GENSTAT Analyses
E.J. Snell and H. Simpson

Applied Nonparametric Statistical Methods, Fourth Edition
P. Sprent and N.C. Smeeton

Data Driven Statistical Methods
P. Sprent

Generalized Linear Mixed Models: Modern Concepts, Methods and Applications
W.W. Stroup

Survival Analysis Using S: Analysis of Time-to-Event Data
M. Tableman and J.S. Kim

Applied Categorical and Count Data Analysis
W. Tang, H. He, and X.M. Tu

Elementary Applications of Probability Theory, Second Edition
H.C. Tuckwell

Introduction to Statistical Inference and Its Applications with R
M.W. Trosset

Understanding Advanced Statistical Methods
P.H. Westfall and K.S.S. Henning

Statistical Process Control: Theory and Practice, Third Edition
G.B. Wetherill and D.W. Brown

Generalized Additive Models: An Introduction with R
S. Wood

Epidemiology: Study Design and Data Analysis, Third Edition
M. Woodward

Experiments
B.S. Yandell

Texts in Statistical Science

Analysis of Categorical Data with R

Christopher R. Bilder
University of Nebraska-Lincoln
Lincoln, Nebraska, USA

Thomas M. Loughin
Simon Fraser University
Surrey, British Columbia, Canada

CRC Press
Taylor & Francis Group
Boca Raton London New York

CRC Press is an imprint of the
Taylor & Francis Group an **informa** business

A CHAPMAN & HALL BOOK

CRC Press
Taylor & Francis Group
6000 Broken Sound Parkway NW, Suite 300
Boca Raton, FL 33487-2742

© 2015 by Taylor & Francis Group, LLC
CRC Press is an imprint of Taylor & Francis Group, an Informa business

No claim to original U.S. Government works

ISBN 13: 978-1-4398-5567-6 (hbk)

This book contains information obtained from authentic and highly regarded sources. Reasonable efforts have been made to publish reliable data and information, but the author and publisher cannot assume responsibility for the validity of all materials or the consequences of their use. The authors and publishers have attempted to trace the copyright holders of all material reproduced in this publication and apologize to copyright holders if permission to publish in this form has not been obtained. If any copyright material has not been acknowledged please write and let us know so we may rectify in any future reprint.

Except as permitted under U.S. Copyright Law, no part of this book may be reprinted, reproduced, transmitted, or utilized in any form by any electronic, mechanical, or other means, now known or hereafter invented, including photocopying, microfilming, and recording, or in any information storage or retrieval system, without written permission from the publishers.

For permission to photocopy or use material electronically from this work, please access www.copyright.com (http://www.copyright.com/) or contact the Copyright Clearance Center, Inc. (CCC), 222 Rosewood Drive, Danvers, MA 01923, 978-750-8400. CCC is a not-for-profit organization that provides licenses and registration for a variety of users. For organizations that have been granted a photocopy license by the CCC, a separate system of payment has been arranged.

Trademark Notice: Product or corporate names may be trademarks or registered trademarks, and are used only for identification and explanation without intent to infringe.

Library of Congress Cataloging-in-Publication Data

Bilder, Christopher R., author.
 Analysis of categorical data with R / Christopher R. Bilder, Thomas M. Loughin.
  pages cm. -- (CRC texts in statistical science)
 Includes bibliographical references and index.
 ISBN 978-1-4398-5567-6 (hardback)
 1. Categories (Mathematics)--Data processing. 2. R (Computer program language) I. Loughin, Thomas M. II. Title.

QA169.B55 2014
512'.6202855133--dc23   2014019437

Visit the Taylor & Francis Web site at
http://www.taylorandfrancis.com

and the CRC Press Web site at
http://www.crcpress.com

Contents

Preface		**xi**
1	**Analyzing a binary response, part 1: introduction**	**1**
1.1	One binary variable	1
	1.1.1 Bernoulli and binomial probability distributions	1
	1.1.2 Inference for the probability of success	8
	1.1.3 True confidence levels for confidence intervals	17
1.2	Two binary variables	23
	1.2.1 Notation and model	25
	1.2.2 Confidence intervals for the difference of two probabilities	29
	1.2.3 Test for the difference of two probabilities	34
	1.2.4 Relative risks	37
	1.2.5 Odds ratios	40
	1.2.6 Matched pairs data	43
	1.2.7 Larger contingency tables	47
1.3	Exercises	48
2	**Analyzing a binary response, part 2: regression models**	**61**
2.1	Linear regression models	61
2.2	Logistic regression models	62
	2.2.1 Parameter estimation	64
	2.2.2 Hypothesis tests for regression parameters	76
	2.2.3 Odds ratios	82
	2.2.4 Probability of success	86
	2.2.5 Interactions and transformations for explanatory variables	94
	2.2.6 Categorical explanatory variables	100
	2.2.7 Convergence of parameter estimation	110
	2.2.8 Monte Carlo simulation	116
2.3	Generalized linear models	121
2.4	Exercises	129
3	**Analyzing a multicategory response**	**141**
3.1	Multinomial probability distribution	141
3.2	$I \times J$ contingency tables and inference procedures	143
	3.2.1 One multinomial distribution	143
	3.2.2 I multinomial distributions	144
	3.2.3 Test for independence	146
3.3	Nominal response regression models	152
	3.3.1 Odds ratios	160
	3.3.2 Contingency tables	163
3.4	Ordinal response regression models	170
	3.4.1 Odds ratios	177
	3.4.2 Contingency tables	179

          3.4.3 Non-proportional odds model 182
    3.5 Additional regression models 186
    3.6 Exercises 187

4 Analyzing a count response   195
    4.1 Poisson model for count data 195
          4.1.1 Poisson distribution 195
          4.1.2 Poisson likelihood and inference 198
    4.2 Poisson regression models for count responses 201
          4.2.1 Model for mean: Log link 203
          4.2.2 Parameter estimation and inference 204
          4.2.3 Categorical explanatory variables 211
          4.2.4 Poisson regression for contingency tables: loglinear models 218
          4.2.5 Large loglinear models 222
          4.2.6 Ordinal categorical variables 231
    4.3 Poisson rate regression 240
    4.4 Zero inflation 244
    4.5 Exercises 253

5 Model selection and evaluation   265
    5.1 Variable selection 265
          5.1.1 Overview of variable selection 265
          5.1.2 Model comparison criteria 267
          5.1.3 All-subsets regression 268
          5.1.4 Stepwise variable selection 272
          5.1.5 Modern variable selection methods 277
          5.1.6 Model averaging 281
    5.2 Tools to assess model fit 285
          5.2.1 Residuals 286
          5.2.2 Goodness of fit 292
          5.2.3 Influence 296
          5.2.4 Diagnostics for multicategory response models 301
    5.3 Overdispersion 301
          5.3.1 Causes and implications 302
          5.3.2 Detection 305
          5.3.3 Solutions 306
    5.4 Examples 318
          5.4.1 Logistic regression - placekicking data set 318
          5.4.2 Poisson regression - alcohol consumption data set 329
    5.5 Exercises 345

6 Additional topics   355
    6.1 Binary responses and testing error 355
          6.1.1 Estimating the probability of success 355
          6.1.2 Binary regression models 358
          6.1.3 Other methods 361
    6.2 Exact inference 362
          6.2.1 Fisher's exact test for independence 362
          6.2.2 Permutation test for independence 367
          6.2.3 Exact logistic regression 372
          6.2.4 Additional exact inference procedures 380

	6.3	Categorical data analysis in complex survey designs	380
		6.3.1 The survey sampling paradigm	381
		6.3.2 Overview of analysis approaches	382
		6.3.3 Weighted cell counts	386
		6.3.4 Inference on population proportions	387
		6.3.5 Contingency tables and loglinear models	389
		6.3.6 Logistic regression	400
	6.4	"Choose all that apply" data	404
		6.4.1 Item response table	405
		6.4.2 Testing for marginal independence	405
		6.4.3 Regression modeling	412
	6.5	Mixed models and estimating equations for correlated data	419
		6.5.1 Random effects	420
		6.5.2 Mixed-effects models	422
		6.5.3 Model fitting	425
		6.5.4 Inference	432
		6.5.5 Marginal modeling using generalized estimating equations	438
	6.6	Bayesian methods for categorical data	443
		6.6.1 Estimating a probability of success	444
		6.6.2 Regression models	449
		6.6.3 Alternative computational tools	459
	6.7	Exercises	459
A	**An introduction to R**		**473**
	A.1	Basics	473
	A.2	Functions	475
	A.3	Help	476
	A.4	Using functions on vectors	477
	A.5	Packages	478
	A.6	Program editors	479
		A.6.1 R editor	479
		A.6.2 RStudio	479
		A.6.3 Tinn-R	480
		A.6.4 Other editors	482
	A.7	Regression example	482
		A.7.1 Background	482
		A.7.2 Data summary	483
		A.7.3 Regression modeling	485
		A.7.4 Additional items	491
B	**Likelihood methods**		**495**
	B.1	Introduction	495
		B.1.1 Model and parameters	495
		B.1.2 The role of likelihoods	496
	B.2	Likelihood	496
		B.2.1 Definition	496
		B.2.2 Examples	497
	B.3	Maximum likelihood estimates	498
		B.3.1 Mathematical maximization of the log-likelihood function	500
		B.3.2 Computational maximization of the log-likelihood function	501
		B.3.3 Large-sample properties of the MLE	502

	B.3.4	Variance of the MLE	502
B.4	Functions of parameters		504
	B.4.1	Invariance property of MLEs	504
	B.4.2	Delta method for variances of functions	505
B.5	Inference with MLEs		506
	B.5.1	Tests for parameters	506
	B.5.2	Confidence intervals for parameters	510
	B.5.3	Tests for models	511

Bibliography   **513**

Index   **525**

Preface

We live in a categorical world! From a positive or negative disease diagnosis to choosing all items that apply in a survey, outcomes are frequently organized into categories so that people can more easily make sense of them. However, analyzing data from categorical responses requires specialized techniques beyond those learned in a first or second course in Statistics. We offer this book to help students and researchers learn how to properly analyze categorical data. Unlike other texts on similar topics, our book is a modern account using the vastly popular R software. We use R not only as a data analysis method but also as a learning tool. For example, we use data simulation to help readers understand the underlying assumptions of a procedure and then to evaluate that procedure's performance. We also provide numerous graphical demonstrations of the features and properties of various analysis methods.

The focus of this book is on the analysis of data, rather than on the mathematical development of methods. We offer numerous examples from a wide rage of disciplines—medicine, psychology, sports, ecology, and others—and provide extensive R code and output as we work through the examples. We give detailed advice and guidelines regarding which procedures to use and why to use them. While we treat likelihood methods as a tool, they are not used blindly. For example, we write out likelihood functions and explain how they are maximized. We describe where Wald, likelihood ratio, and score procedures come from. However, except in Appendix B, where we give a general introduction to likelihood methods, we do not frequently emphasize calculus or carry out mathematical analysis in the text. The use of calculus is mostly from a conceptual focus, rather than a mathematical one.

We therefore expect that this book will appeal to all readers with a basic background in regression analysis. At times, a rudimentary understanding of derivatives, integrals, and function maximization would be helpful, as would a very basic understanding of matrices, matrix multiplication, and finding inverses of matrices. However, the important points and application advice can be easily understood without these tools. We expect that advanced undergraduates in statistics and related fields will satisfy these prerequisites. Graduate students in statistics, biostatistics, and related fields will certainly have sufficient background for the book. In addition, many students and researchers outside these disciplines who possess the basic regression background should find this book useful both for its descriptions and motivations of the analysis methods and for its worked examples with R code.

The book does not require any prior experience with R. We provide an introduction to the essential features and functions of R in Appendix A. We also provide introductory details on the use of R in the earlier chapters to help inexperienced R users. Throughout the book as new R functions are needed, their basic features are discussed in the text and their implementation shown with corresponding output. We focus on using R packages that are provided by default with the initial R installation. However, we make frequent use of other R packages when they are significantly better or contain functionality unavailable in the standard R packages. The book contains the code and output as it would appear in the R Console; we make minor modifications at times to the output only to save space within the book. Code provided in the book for plotting is often meant for color display rather than the actual black-and-white display shown in the print and some electronic editions.

The data set files and R programs that are referenced in each example are available from the book's website, http://www.chrisbilder.com/categorical. The programs include code used to create every plot and piece of output that we show. Many of these programs contain code to demonstrate additional features or to perform more detailed and complete analyses than what is presented in the book. We strongly recommend that the book and the website be used in tandem, both for teaching and for individual learning. The website also contains many "extras" that can help readers learn the material. Most importantly, we post videos from one of us teaching a course on the subject. These videos include live, in-class recordings that are synchronized with recordings of a tablet computer screen. Instructors may find these videos useful (as we have) for a blended or flipped classroom setting. Readers outside of a classroom setting may also find these videos especially useful as a substitute for a short-course on the subject.

The first four chapters of the book are organized by type of categorical response variable. Within each of these chapters, we first introduce the measurement type, followed by the basic distributional model that is most commonly used for that type of measurement. We slowly generalize to simple regression structures, followed by multiple regressions including transformations, interactions, and categorical explanatory variables. We conclude each of these chapters with some important special cases. Chapter 5 follows with model building and assessment methods for the response variables in the first four chapters. A final chapter discusses additional topics presented as extensions to the previous chapters. These topics include solutions to problems that are frequently mishandled in practice, such as how to incorporate diagnostic testing error into an analysis, the analysis of data from "choose all that apply" questions, and methods for analyzing data arising under a complex survey sampling design. Many of these topics are broad enough that entire books have been written about them, so our treatment in Chapter 6 is meant to be introductory.

For instructors teaching a one-semester course with the book, we recommend covering most of Chapters 1–5. The topics in Chapter 6 provide supplemental material for readers to learn on their own or to provide an instructor a means to go beyond the basics. In particular, topics from Chapter 6 can make good class projects. This helps students gain experience in teaching themselves extensions to familiar topics, which they will face later in industry or in research.

An extensive set of exercises is provided at the end of each chapter (over 65 pages in all!). The exercises are deliberately variable in scope and subject matter, so that instructors can choose those that meet the particular needs of their own students. For example, some carry out an analysis step by step, while others present a problem and leave the reader to choose and implement a solution. An answer key to the exercises is available for instructors using the book for a course. Details on how to obtain the answer key are available through the book's website.

We could not have written this book without the help and support of many people. First and foremost, we thank our families, and especially our wives, Kimberly and Marie, who put in extra effort on our behalf so that we could reserve time to work on the book. We owe them a huge debt for their support and tolerance, but we are hoping that they will settle for a nice dinner. We thank Rob Calver and his staff at CRC Press for their continued support and encouragement during the writing process. We also thank the hundreds of students who have taken categorical courses from us over the last seventeen years. Their feedback helped us to hone the course material and its presentation to what they are today. We especially thank one of our students, Natalie Koziol, who wrote the MRCV package used in Section 6.4 and made the implementation of those methods available to R users. This book was written in LaTeX through LyX, and we are grateful to the many contributors to these open-source projects. Finally, we need to thank our past and present colleagues and mentors at Iowa State, Kansas State, Oklahoma State, Nebraska, and Simon Fraser Universities who have both supported our development and brought us interesting and challenging problems to work on.

Christopher R. Bilder and Thomas M. Loughin
Lincoln, NE and Surrey, BC

Chapter 1

Analyzing a binary response, part 1: introduction

Yes or no. Success or failure. Death or survival. For or against. Binary responses may be the most prevalent type of categorical data. The purpose of this chapter is to show how to estimate and make inferences about a binary response probability and related quantities. We begin in Section 1.1 by examining a homogeneous population where there is one overall probability to be estimated. We generalize this situation in Section 1.2 to the setting where sampled items come from one of two groups.

Throughout Chapter 1 we emphasize the use of R with detailed code explanations. This is done on purpose because we expect that some readers have little R experience beyond the introduction in Appendix A. Future chapters will still emphasize the use of R, but spend less time explaining code.

1.1 One binary variable

1.1.1 Bernoulli and binomial probability distributions

Almost every statistical analysis begins with some kind of statistical model. A statistical model generally takes the form of a probability distribution that attempts to quantify the uncertainty that comes with observing a new response. The model is intended to represent the unknown phenomenon that governs the observation process. At the same time, the model needs to be convenient to work with mathematically, so that inference procedures such as confidence intervals and hypothesis tests can be developed. Selecting a model is typically a compromise between two competing goals: providing a more detailed approximation to the process that generates the data and providing inference procedures that are easy to use.

In the case of binary responses, the natural model is the Bernoulli distribution. Let Y denote a Bernoulli random variable with outcomes of 0 and 1. Typically, we will say $Y = 1$ is a success and $Y = 0$ is a failure. For example, a success would be a basketball free throw attempt that is good or an individual who is cured of a disease by a new drug; a failure would be a free throw attempt that is missed or an individual who is not cured. We denote the probability of success as $P(Y = 1) = \pi$ and the corresponding probability of failure as $P(Y = 0) = 1 - \pi$. The Bernoulli probability mass function (PMF) for Y combines these two expressions into one formula:

$$P(Y = y) = \pi^y (1 - \pi)^{1-y}$$

for $y = 0$ or 1, where we use the standard convention that a capital letter Y denotes the random variable and the lowercase letter y denotes a possible value of Y. The expected value of Y is $E(Y) = \pi$, and the variance of Y is $Var(Y) = \pi(1 - \pi)$.

Often, one observes multiple Bernoulli random variable responses through repeated sampling or *trials* in identical settings. This leads to defining separate random variables for each trial, $Y_1, \ldots, Y_n$, where n is the number of trials. If all trials are identical and independent, we can treat $W = \sum_{i=1}^{n} Y_i$ as a binomial random variable with PMF of

$$P(W = w) = \binom{n}{w} \pi^w (1 - \pi)^{n-w} \tag{1.1}$$

for $w = 0, \ldots, n$. The combination function $\binom{n}{w} = n!/[w!(n-w)!]$ counts the number of ways w successes and $n-w$ failures can be ordered. The expected value of W is $E(W) = n\pi$, and the variance of W is $Var(W) = n\pi(1-\pi)$. Notice that the Bernoulli distribution is a special case of the binomial distribution when $n = 1$.

We next show how R can be used to examine properties of the binomial distribution.

Example: Binomial distribution in R (Binomial.R)

The purpose of this example is to calculate simple probabilities using a binomial distribution and to show how these calculations are performed in R. We will be very basic with our use of R in this example. If you find its use difficult, we recommend reading Appendix A before proceeding further.

Consider a binomial random variable counting the number of successes from an experiment that is repeated $n = 5$ times, and suppose that there is a probability of success of $\pi = 0.6$. For example, suppose an individual has this success rate in a particular card game or shooting a basketball into a goal from a specific location. We can calculate the probability of each number of successes, $w = 0, 1, 2, 3, 4, 5$, using Equation 1.1. For example, the probability of 1 success out of 5 trials is

$$P(W = 1) = \binom{5}{1} 0.6^1 (1 - 0.6)^{5-1} = \frac{5!}{1!4!} 0.6^1 0.4^4 = 0.0768.$$

This calculation is performed in R using the `dbinom()` function:

```
> dbinom(x = 1, size = 5, prob = 0.6)
[1] 0.0768
```

Within the function, the `x` argument denotes the observed value of the binomial random variable (what we are calling w), the `size` argument is the number of trials (n), and the `prob` argument is π. We could have used `dbinom(1, 5, 0.6)` to obtain the same probability as long as the numerical values are in the same order as the arguments within the function (a full list of arguments and their order is available in the help for `dbinom()`). Generally, we will always specify the argument names in our code except with the most basic functions.

We find all of the probabilities for $w = 0, \ldots, 5$ by changing the `x` argument:

```
> dbinom(x = 0:5, size = 5, prob = 0.6)
[1] 0.01024 0.07680 0.23040 0.34560 0.25920 0.07776
```

where `0:5` means the integers 0 to 5 by 1. To display these probabilities in a more descriptive format, we save them into an object and print from a data frame using the `data.frame()` function:

```
> pmf <- dbinom(x = 0:5, size = 5, prob = 0.6)
> save <- data.frame(w = 0:5, prob = round(x = pmf, digits = 4))
> save
  w   prob
1 0 0.0102
2 1 0.0768
3 2 0.2304
4 3 0.3456
5 4 0.2592
6 5 0.0778
```

Note that we could have used different names than pmf and save for our objects if desired. The round() function rounds the values in the pmf object to 4 decimal places. We plot the PMF using the plot() and abline() functions:

```
> plot(x = save$w, y = save$prob, type = "h", xlab = "w", ylab =
    "P(W=w)", main = "Plot of a binomial PMF for n=5, pi=0.6",
    panel.first = grid(col = "gray", lty = "dotted"), lwd = 3)
> abline(h = 0)
```

Figure 1.1 gives the resulting plot. Appendix A.7.2 describes most of the arguments within plot(), so we provide only brief descriptions here. The x and y arguments specify the x- and y-axis values where we use the $ symbol to access parts of the save data frame. The type = "h" argument value specifies that vertical bars are to be plotted from 0 to the values given in the y argument. The main argument contains the plot title.[1] The abline() function plots a horizontal line at 0 to emphasize the bottom of each vertical line.

The simpler specification plot(x = save$w, y = save$prob, type = "h") produces a plot similar to Figure 1.1, but our extra arguments make the plot easier to interpret.

Assumptions

The binomial distribution is a reasonable model for the distribution of successes in a given number of trials as long as the process of observing repeated trials satisfies certain assumptions. Those assumptions are:

1. THERE ARE n IDENTICAL TRIALS. This refers to the process by which the trials are conducted. The action that results in the trial and the measurement taken must be the same in each trial. The trials cannot be a mixture of different types of actions or measurements.

2. THERE ARE TWO POSSIBLE OUTCOMES FOR EACH TRIAL. This is generally just a matter of knowing what is measured. However, there are cases where a response measurement has more than two levels, but interest lies only in whether or not one particular level occurs. In this case, the special level can be considered "success" and all remaining levels "failure."

---

[1] If it is desired to use a plot title with better notation, we could have used main = expression(paste("Plot of a binomial PMF for ", italic(n) == 5, " and ", italic(pi) == 0.6)) to obtain "Plot of the binomial PMF for $n = 6$ and $\pi = 0.6$." In this code, expression() allows us to include mathematical symbols and paste() combines these symbols with regular text. Please see Appendix A.7.4 for more information.

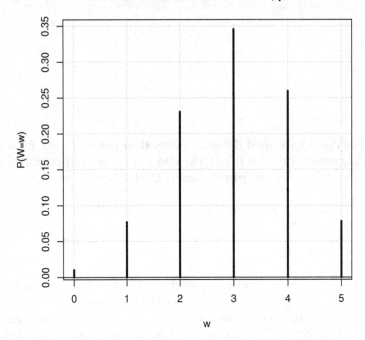

Figure 1.1: PMF for W.

3. THE TRIALS ARE INDEPENDENT OF EACH OTHER. In particular, there is nothing in the conduct of the trials that would cause any subset of trials to behave more similarly to one another. Counterexamples include (a) measuring trials in "clusters," where the units on which success or failure is measured are grouped together somehow before observation, and (b) trials measured in a time series, where trials measured close together in time might react more similarly than those measured far apart.

4. THE PROBABILITY OF SUCCESS REMAINS CONSTANT FOR EACH TRIAL. This means that all variables that can affect the probability of success need to be held constant from trial to trial. Because these variables are not always known in advance, this can be a very difficult condition to confirm. We often can only confirm that the "obvious" variables are not changing, and then merely assume that others are not as well.

5. THE RANDOM VARIABLE OF INTEREST W IS THE NUMBER OF SUCCESSES. Specifically, this implies that we are *not* interested in the order in which successes and failures occur, but rather only in their total counts.

The next two examples detail these assumptions with respect to applications, and they demonstrate how it can be difficult to assure that these assumptions are satisfied.

Example: Field goal kicking

In American and Canadian football, points can be scored by kicking a ball through a target area (goal) at each end of the field. Suppose an experiment is conducted where a placekicker successively attempts five field goal kicks during practice. A success occurs on one trial when the football is kicked over the crossbar and between the two uprights of the goal posts. A failure occurs when the football does not achieve this result (the football is kicked to the left or right of both uprights or falls short of the crossbar). We want to use these results to estimate the placekicker's true probability of success, so we record how many kicks are successful.

In order to use the binomial distribution here, the experiment needs to satisfy the following conditions:

1. THERE ARE n IDENTICAL TRIALS. In this case, $n = 5$ field goals are attempted. The action is always kicking a football in the same way, and the measurement of success is made the same way each time.

2. THERE ARE TWO POSSIBLE OUTCOMES FOR EACH TRIAL. Each field goal is a success or failure. Notice that we could further divide failures into other categories: too short, long enough but to the left of the uprights, long enough but to the right of the uprights, or some combinations of these. If our interest is only in whether the kick is successful, then differentiating among the modes of failure is not necessary.

3. THE TRIALS ARE INDEPENDENT OF EACH OTHER. Given that the field goals are attempted successively, this may be difficult to satisfy. For example, if one field goal attempt is missed to the left, the kicker may compensate by trying to kick farther to the right on the next attempt. On the other hand, the independence assumption may be approximately satisfied by a placekicker who tries to apply the exact same technique on each trial.

4. THE PROBABILITY OF SUCCESS REMAINS CONSTANT FOR EACH TRIAL. To make sure this is true, the field goals need to be attempted from the same distance under the same surrounding conditions. Weather conditions need to be constant. The same kicker, ball, and goalposts are used each time. We assume that fatigue does not affect the kicker for this small number of attempts. If the attempts occur close together in time, then it may be reasonable to assume that extraneous factors are reasonably constant as well, at least enough so that they do not have a substantial effect on the success of the field goals.

5. THE RANDOM VARIABLE OF INTEREST W IS THE NUMBER OF SUCCESSES. As we will see in Section 1.1.2, in order to estimate the probability of success, we need only to record W (=0, 1, 2, 3, 4, or 5) and not the entire sequence of trial results.

Example: Disease prevalence

The *prevalence* of a disease is the proportion of a population that is afflicted with that disease. This is equivalent to the probability that a randomly selected member of the population has the disease. Many public health studies are performed to understand disease prevalence, because knowing the prevalence is the first step toward solving societal problems caused by the disease. For example, suppose there is concern that a new infectious disease may be transmitted to individuals through blood donations. One way to examine the disease prevalence would be to take a sample a 1,000 blood donations and test each for the disease.

In order to use the binomial distribution here, this setting needs to satisfy the following conditions:

1. THERE ARE n IDENTICAL TRIALS. In this case, $n = 1000$ blood donations are examined. Each blood donation needs to be collected and tested the same way. In

particular, trials would *not* be identical if different diagnostic measures were used on different donations to determine presence of disease.

2. THERE ARE TWO POSSIBLE OUTCOMES FOR EACH TRIAL. Each blood donation is either positive or negative for the disease. Making this determination is not necessarily as straightforward as it may seem. Often, there is a continuous score reported from the results of an assay, such as the amount of an antigen in a specimen, and a threshold or cut-off point is used to make the positive or negative determination. In some instances, multiple thresholds may be used leading to responses such as positive, indeterminate, or negative.

3. THE TRIALS ARE INDEPENDENT OF EACH OTHER. This may be difficult to satisfy completely. For example, if married spouses are included in the sample, then presence of the disease in one spouse's donation may suggest a greater chance that the other spouse's donation will also test positive. Independence can be assured by random sampling from a large population of donations, but may always be in question when any non-random sampling method is used.

4. THE PROBABILITY OF SUCCESS REMAINS CONSTANT FOR EACH TRIAL. Each sampled donation needs to have the same probability of having the disease. This could be unreasonable if there are factors, such as risky behavior, that make donations for certain subpopulations more likely to have the disease than others, and if these subpopulations can be identified in advance. Similarly, if donations are collected over an extended period of time, prevalence may not be constant during the period.

5. THE RANDOM VARIABLE OF INTEREST W IS THE NUMBER OF SUCCESSES. There are $W = 0, \ldots, 1000$ possible positive blood donations. To estimate prevalence, we need to know how many positive donations there are, and not which ones are positive.

The previous examples show that it may be difficult to satisfy all of the assumptions for a binomial model. However, the binomial model may still be used as an approximation to the true model in a given problem, in which case the violated assumptions then would need to be identified in any stated results. Alternatively, if assumptions are not satisfied, there are other models and procedures that can be used to analyze binary responses. In particular, if the probability of success does not remain constant for each trial—for example, if disease probability is related to certain risky behaviors—we may be able to identify and measure the factors causing the variations and then use a regression model for the probability of success (to be discussed in Chapter 2).

Simulating a binomial sample

What does a sample from a binomial distribution look like? Of course, the observed values only can be $0, 1, \ldots, n$. The proportion of observed values that are $0, 1, \ldots, n$ are governed by the PMF and the parameter π within it. The mean and variance of these observed values are also controlled by the PMF and π. These properties are easily derived mathematically using basic definitions of mean and variance. In more complex problems, however, properties of statistics are much harder to derive mathematically.

We show in the next example how to *simulate* a sample using R so that we can check whether theory matches what actually happens. This example will also introduce Monte Carlo computer simulation as a valuable tool for evaluating a statistical procedure. All statistical procedures have assumptions underlying their mathematical framework. Monte

Carlo simulation is especially useful in assessing how well procedures perform when these assumptions are violated. For example, we may want to know if a confidence interval that is designed to work in large samples maintains its stated confidence level when the sample size is small.

A Monte Carlo simulation works by creating a computer version of the population we are studying, sampling from this virtual population in a prescribed way, performing the statistical analysis that we are studying, and measuring how it performs. The details of each step vary from one problem to the next. In all cases we draw a "large" number of samples from the virtual population. In so doing, the *law of large numbers* assures us that the average performance measured across the samples will be close to the true performance of the procedure in this context (a more mathematical definition of the law of large numbers is contained on p. 232 of Casella and Berger, 2002). We follow this prescription in the example below.

Example: Simulation with the binomial distribution in R (Binomial.R)

Below is the code that simulates 1,000 random observations of W from a binomial distribution with $\pi = 0.6$ and $n = 5$:

```
> set.seed(4848)
> bin5 <- rbinom(n = 1000, size = 5, prob = 0.6)
> bin5[1:10]
 [1] 3 2 4 1 3 1 3 3 3 4
```

The `set.seed()` function sets a seed number for the simulated observations. Without going into the details behind random number generation, the seed number specification allows us to obtain identical simulated observations each time the same code is executed.[2] The `rbinom()` function simulates the observations, where the `n` argument gives the number of observations (not n as in the number of trials). The `bin5` object contains 1,000 values, where the first 10 of which are printed.

The population mean and variance for W are

$$E(W) = n\pi = 5 \times 0.6 = 3$$

and

$$Var(W) = n\pi(1 - \pi) = 5 \times 0.6 \times 0.4 = 1.2.$$

We calculate the sample mean and variance of the simulated observations using the `mean()` and `var()` functions:

```
> mean(bin5)
[1] 2.991
> var(bin5)
[1] 1.236155
```

---

[2] It is best to not use the same seed number when doing other simulated data examples. A new seed number can be chosen from a random number table. Alternatively, a new seed number can be found by running `runif(n=1)` within R (this simulates one observation from a Uniform(0,1) distribution) and taking the first few significant digits.

The sample mean and variance are very similar to $E(W)$ and $Var(W)$, as expected. If a larger number of observations were simulated, say 10,000, we generally would expect these sample measures to be even closer to their population quantities due to the law of large numbers.

To examine how well the observed frequency distribution follows the PMF, we use `table()` to find the frequencies of each possible response and then use `hist()` to plot a histogram of the relative frequencies:

```
> table(x = bin5)
x
  0   1   2   3   4   5
 12  84 215 362 244  83
> hist(x = bin5, main = "Binomial with n=5, pi=0.6, 1000 bin.
    observations", probability = TRUE, breaks = c(-0.5:5.5), ylab
    = "Relative frequency")
> -0.5:5.5
[1] -0.5  0.5  1.5  2.5  3.5  4.5  5.5
```

For example, we see that $w = 3$ is observed with a relative frequency of $362/1000 = 0.362$. We had found earlier that $P(W = 3) = 0.3456$, which is very similar to the observed proportion. The histogram is in Figure 1.2, and its shape is very similar to the PMF plot in Figure 1.1. Note that the `probability = TRUE` argument gives the relative frequencies (`probability = FALSE` gives the frequencies, which is the default), and the `breaks` argument specifies the classes for the bars to be -0.5 to 5.5 by 1 (the bars will not be drawn correctly here without specifying `breaks`).

1.1.2 Inference for the probability of success

The purpose of this section is to estimate and make inferences about the probability of success parameter π from the Bernoulli distribution. We start by estimating the parameter using its maximum likelihood estimate, because it is relatively easy to compute and has properties that make it appealing in large samples. Next, confidence intervals for the true probability of success are presented and compared. Many different confidence intervals have been proposed in the statistics literature. We will present the simplest—but worst—procedure first and then offer several better alternatives. We conclude this section with hypothesis tests for π.

For those readers unfamiliar with estimation and inference procedures associated with the likelihood function, we encourage you to read Appendix B first for an introduction. We will reference specific parts of the appendix within this section.

Maximum likelihood estimation and inference

As described in Appendix B, a likelihood function is a function of one or more parameters conditional on the observed data. The likelihood function for π when $y_1, \ldots, y_n$ are observations from a Bernoulli distribution is

$$L(\pi|y_1,\ldots,y_n) = P(Y_1 = y_1) \times \cdots \times P(Y_n = y_n)$$
$$= \pi^w(1-\pi)^{n-w}. \tag{1.2}$$

Alternatively, when we only record the number of successes out of a number of trials, the likelihood function for π is simply $L(\pi|w) = P(W = w) = \binom{n}{w}\pi^w(1-\pi)^{n-w}$. The value of

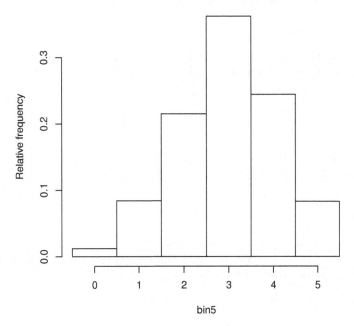

Figure 1.2: Histogram for the observed values of w.

π that maximizes the likelihood function is considered to be the most plausible value for the parameter, and it is called the maximum likelihood estimate (MLE). We show in Appendix B.3 that the MLE of π is $\hat{\pi} = w/n$, which is simply the observed proportion of successes. This is true for both $L(\pi|y_1, \ldots, y_n)$ or $L(\pi|w)$ because the $\binom{n}{w}$ contains no information about π.

Because $\hat{\pi}$ will vary from sample to sample, it is a statistic and has a corresponding probability distribution.[3] As with all MLEs, $\hat{\pi}$ has an approximate normal distribution in a large sample (see Appendix B.3.3). The mean of the normal distribution is π, and the variance is found from

$$\widehat{Var}(\hat{\pi}) = -E\left\{\frac{\partial^2 \log[L(\pi|W)]}{\partial \pi^2}\right\}^{-1}\bigg|_{\pi=\hat{\pi}}$$

$$= -E\left\{-\frac{W}{\pi^2} + \frac{n-W}{(1-\pi)^2}\right\}^{-1}\bigg|_{\pi=\hat{\pi}}$$

$$= \left[\frac{n}{\pi} - \frac{n}{1-\pi}\right]^{-1}\bigg|_{\pi=\hat{\pi}}$$

$$= \frac{\hat{\pi}(1-\hat{\pi})}{n}. \tag{1.3}$$

---

[3] In order for $\hat{\pi}$ to have a probability distribution, it needs to be a random variable. Thus, we are actually using $\hat{\pi} = W/n$ in this case. We could have defined W/n as $\hat{\Pi}$ instead, but this level of formality is unnecessary here. It will be apparent from a statistic's use whether it is a random or observed quantity.

where $\log(\cdot)$ is the natural log function (see Appendix B.3.4). We can write the distribution as $\hat{\pi} \dot\sim N(\pi, \widehat{Var}(\hat{\pi}))$ where $\dot\sim$ denotes "approximately distributed as." The approximation tends to be better as the sample size grows larger.

Wald confidence interval

Using this normal distribution, we can treat $(\hat{\pi} - \pi)/\widehat{Var}(\hat{\pi})^{1/2}$ as an approximate standard normal quantity (see Appendix B.5). Thus, for any $0 < \alpha < 1$, we have

$$P\left(Z_{\alpha/2} < \frac{\hat{\pi} - \pi}{\sqrt{\widehat{Var}(\hat{\pi})}} < Z_{1-\alpha/2}\right) \approx 1 - \alpha,$$

where Z_a is the ath quantile from a standard normal distribution (e.g., $Z_{0.975} = 1.96$). After rearranging terms and recognizing that $-Z_{\alpha/2} = Z_{1-\alpha/2}$, we obtain

$$P\left(\hat{\pi} - Z_{1-\alpha/2}\sqrt{\widehat{Var}(\hat{\pi})} < \pi < \hat{\pi} + Z_{1-\alpha/2}\sqrt{\widehat{Var}(\hat{\pi})}\right) \approx 1 - \alpha.$$

Now, we have an approximate probability that has the parameter π centered between two statistics. When we replace $\hat{\pi}$ and $\widehat{Var}(\hat{\pi})$ with observed values from the sample, we obtain the $(1-\alpha)100\%$ confidence interval for π

$$\hat{\pi} - Z_{1-\alpha/2}\sqrt{\hat{\pi}(1-\hat{\pi})/n} < \pi < \hat{\pi} + Z_{1-\alpha/2}\sqrt{\hat{\pi}(1-\hat{\pi})/n}.$$

This is the usual interval for a probability of success that is given in most introductory statistics textbooks. Confidence intervals based on the approximate normality of MLEs are called "Wald confidence intervals" because Wald (1943) was the first to show this property of MLEs in large samples.

When w is close to 0 or n, two problems occur with this interval:

1. Calculated limits may be less than 0 or greater than 1, which is outside the boundaries for a probability.

2. When $w = 0$ or 1, $\sqrt{\hat{\pi}(1-\hat{\pi})/n} = 0$ for $n > 0$. This leads to the lower and upper limits to be exactly the same (0 for $w = 0$ or 1 for $w = 1$).

We will discuss additional problems with the Wald interval shortly.

Example: Wald interval (CIpi.R)

Suppose $w = 4$ successes are observed out of $n = 10$ trials. The 95% Wald confidence interval for π is $0.4 \pm 1.96\sqrt{0.4(1-0.4)/10} = (0.0964, 0.7036)$, where we use the shorthand notation within parentheses to mean $0.0964 < \pi < 0.7036$. The R code below shows how these calculations are carried out:

```
> w <- 4
> n <- 10
> alpha <- 0.05
> pi.hat <- w/n
> var.wald <- pi.hat*(1-pi.hat)/n
> lower <- pi.hat - qnorm(p = 1-alpha/2) * sqrt(var.wald)
> upper <- pi.hat + qnorm(p = 1-alpha/2) * sqrt(var.wald)
> round(data.frame(lower, upper), 4)
  lower  upper
1 0.0964 0.7036
```

In the code, we use the `qnorm()` function to find the $1 - \alpha/2$ quantile from a standard normal distribution. We can calculate the interval quicker by taking advantage of how R performs vector calculations:

```
> round(pi.hat + qnorm(p = c(alpha/2, 1-alpha/2)) *
    sqrt(var.wald), 4)
[1] 0.0964 0.7036
```

See Appendix A.4 for a similar example.

The confidence interval is quite wide and may not be meaningful for some applications. However, it does give information on a range for π that may be useful in hypothesis testing situations. For example, a test of $H_0 : \pi = 0.5$ vs. $H_a : \pi \neq 0.5$ would not reject H_0 because 0.5 is within this range. If the test was instead $H_0 : \pi = 0.8$ vs. $H_a : \pi \neq 0.8$, there is evidence to reject the null hypothesis. Other ways to perform these tests with a test statistic and a p-value will be discussed shortly.

The inferences for π from the Wald confidence interval rely on the underlying normal distribution approximation for the maximum likelihood estimator. For this approximation to work well, we need a large sample, and, unfortunately, the sample size in the last example was quite small. Furthermore, notice that $\hat{\pi}$ can take on only 11 different possible values in the last example: $0/10, 1/10, ..., 10/10$, but a normal distribution is a continuous function. These problems lead the Wald conference interval to be *approximate*, in the sense that the probability that the interval covers the parameter (its *coverage* or *true confidence level*) is not necessarily equal to the stated level $1 - \alpha$. The quality of the approximation varies with n and π, and as we will see later, the Wald interval generally has coverage $< 1 - \alpha$. Such an interval is called a *liberal* interval. On the other hand, an interval with coverage in excess of the stated level is called *conservative*. While this latter property may seem to be a good quality, it can lead to intervals that are quite wide in comparison to others. We want confidence intervals that place the parameter within as narrow a range as possible, while maintaining at least the stated confidence level. If we wanted intervals that had greater coverage, we would have stated a higher confidence level!

There has been a lot of research on finding an interval like this for π, including Agresti and Caffo (2000), Agresti and Min (2001), Borkowf (2006), Brown et al. (2002), Henderson and Meyer (2001), Newcombe (2001), Suess et al. (2006), and Vos and Hudson (2005). Brown et al. (2001) present a thorough review of most competing intervals. We present their recommendations next along with our own thoughts on the best intervals.

Wilson confidence interval

When $n < 40$, Brown et al. (2001) recommend the Wilson interval or the Jeffreys interval because they maintain true confidence levels closer to the stated level than other intervals. The Wilson interval formula is found by examining the test statistic

$$Z_0 = \frac{\hat{\pi} - \pi_0}{\sqrt{\pi_0(1-\pi_0)/n}},$$

which is a *score test* statistic often used for a test of $H_0 : \pi = \pi_0$ vs. $H_a : \pi \neq \pi_0$, where $0 < \pi_0 < 1$ (see Appendix B.5). The variance in the denominator of Z_0 is computed assuming that the null hypothesis is true, rather than using the unrestricted estimate based on the data. This leads to the advantage that the denominator is not 0 whenever $w = 0$ or n. We can approximate the distribution of Z_0 with a standard normal to obtain $P(-Z_{1-\alpha/2} <$

$Z_0 < Z_{1-\alpha/2}) \approx 1 - \alpha$. Treating the approximation as an equality, the Wilson interval contains the set of all possible values of π_0 that satisfy the equation. Conversely, the set of all possible values for π_0 that lead to a rejection of the null hypothesis are outside of the confidence interval. The process of forming an interval from a hypothesis test procedure like this is often referred to as "inverting the test." See also Appendix B.5.2. Because the Wilson interval is based on a score test, it is often referred to as a *score interval* too.

The interval endpoints are found by setting Z_0 equal to $\pm Z_{1-\alpha/2}$, and applying the quadratic formula to solve for π_0. Thus, the $(1-\alpha)100\%$ Wilson interval is

$$\tilde{\pi} \pm \frac{Z_{1-\alpha/2}\sqrt{n}}{n + Z_{1-\alpha/2}^2} \sqrt{\hat{\pi}(1-\hat{\pi}) + \frac{Z_{1-\alpha/2}^2}{4n}}, \tag{1.4}$$

where

$$\tilde{\pi} = \frac{w + Z_{1-\alpha/2}^2/2}{n + Z_{1-\alpha/2}^2}$$

can be thought of as an adjusted estimate of π. This interval is named after Wilson (1927) who first proposed finding an interval for π in this manner. Note that the Wilson interval always has limits between 0 and 1.

The Wald and Wilson confidence intervals discussed so far are *frequentist* inference procedures. The "confidence" associated with these types of inference procedures comes about through repeating the process of taking a sample and calculating a confidence interval each time. This leads to the interpretation of

> We would expect $(1-\alpha)100\%$ of all similarly constructed intervals to contain the parameter π

for a $(1-\alpha)100\%$ confidence interval. Alternatively, a commonly used interpretation is

> With $(1-\alpha)100\%$ confidence, the parameter π is between <lower limit> and <upper limit>

where the appropriate lower and upper limits are inserted. Note that a single interval calculated from a sample either does or does not contain the parameter, so it is inappropriate to say it has a *probability* (other than 0 or 1) of containing the parameter. This is a confusing aspect to confidence intervals, causing them to be misinterpreted frequently in practice.

On the other hand, a Bayesian credible interval does have a $(1-\alpha)100\%$ probability of containing the parameter, because parameters are random variables in the Bayesian paradigm. A Jeffreys interval for π, also recommended by Brown et al. (2001), is one such Bayesian interval. We postpone its discussion until Section 6.6, when we describe Bayesian inference procedures in detail.

Agresti-Coull confidence interval

The Wilson interval is our preferred choice for a confidence interval for π. However, Brown et al. (2001) recommend the Agresti-Coull interval (Agresti and Coull 1998) for $n \geq 40$, primarily because it is a little easier to calculate by hand and it more closely resembles the popular Wald interval. The $(1-\alpha)100\%$ Agresti-Coull interval is

$$\tilde{\pi} - Z_{1-\alpha/2}\sqrt{\frac{\tilde{\pi}(1-\tilde{\pi})}{n + Z_{1-\alpha/2}^2}} < \pi < \tilde{\pi} + Z_{1-\alpha/2}\sqrt{\frac{\tilde{\pi}(1-\tilde{\pi})}{n + Z_{1-\alpha/2}^2}}.$$

The interval is essentially the Wald interval where $Z_{1-\alpha/2}^2/2$ successes and $Z_{1-\alpha/2}^2/2$ failures are added to the observed data. Specifically, for $\alpha = 0.05$, this means that about two successes and two failures are added because $Z_{1-0.05/2} = 1.96 \approx 2$. Similar to the Wald interval, this interval has the undesirable property that it may have limits less than 0 or greater than 1.

Example: Wilson and Agresti-Coull intervals (CIpi.R)

Suppose again $w = 4$ successes are observed out of $n = 10$ trials. For a 95% confidence interval, the adjusted estimate of π is

$$\tilde{\pi} = \frac{w + Z^2_{1-\alpha/2}/2}{n + Z^2_{1-\alpha/2}} = \frac{4 + 1.96^2/2}{10 + 1.96^2} = 0.4278.$$

The 95% Wilson interval limits are

$$\tilde{\pi} \pm \frac{Z_{1-\alpha/2}\sqrt{n}}{n + Z^2_{1-\alpha/2}}\sqrt{\hat{\pi}(1-\hat{\pi}) + \frac{Z^2_{1-\alpha/2}}{4n}}$$

$$= 0.4278 \pm \frac{1.96\sqrt{10}}{10 + 1.96^2}\sqrt{0.4(1-0.4) + \frac{1.96^2}{4 \times 10}}$$

leading to an interval of $0.1682 < \pi < 0.6873$. The 95% Agresti-Coull interval limits are

$$\tilde{\pi} \pm Z_{1-\alpha/2}\sqrt{\frac{\tilde{\pi}(1-\tilde{\pi})}{n + Z^2_{1-\alpha/2}}} = 0.4278 \pm 1.96\sqrt{\frac{0.4278(1-0.4278)}{10 + 1.96^2}}$$

leading to an interval of $0.1671 < \pi < 0.6884$. Both confidence intervals have limits that are quite similar in this case, but are rather different from the Wald interval limits of (0.0964, 0.7036) that we calculated earlier.

Continuing from the last example, below is how the calculations are performed in R:

```
> p.tilde <- (w + qnorm(p = 1-alpha/2)^2 / 2) / (n + qnorm(p =
    1-alpha/2)^2)
> p.tilde
[1] 0.4277533

> # Wilson C.I.
> round(p.tilde + qnorm(p = c(alpha/2, 1-alpha/2)) * sqrt(n) / (n
    + qnorm(p = 1-alpha/2)^2) * sqrt(pi.hat*(1-pi.hat) + qnorm(p =
    1-alpha/2)^2/(4*n)), 4)
[1] 0.1682 0.6873

> # Agresti-Coull C.I.
> var.ac <- p.tilde*(1-p.tilde) / (n + qnorm(p = 1-alpha/2)^2)
> round(p.tilde + qnorm(p = c(alpha/2, 1-alpha/2)) *
    sqrt(var.ac), 4)
[1] 0.1671 0.6884
```

After calculating $\tilde{\pi}$, we calculate the Wilson and Agresti-Coull intervals through one line of code for each. Note that executing part of a line of code can help highlight how it works. For example, one can execute `qnorm(p = c(alpha/2, 1-alpha/2))` to see that it calculates -1.96 and 1.96.

The `binom.confint()` function from the `binom` package can be used to simplify the calculations. Note that this package is not in the default installation of R, so it needs to be installed before its use (see Appendix A.5 for more information regarding package installation). Below is our use of the function:

```
> library(package = binom)
> binom.confint(x = w, n = n, conf.level = 1-alpha, methods =
    "all")
          method x  n      mean      lower     upper
1  agresti-coull 4 10 0.4000000 0.16711063 0.6883959
2     asymptotic 4 10 0.4000000 0.09636369 0.7036363
3          bayes 4 10 0.4090909 0.15306710 0.6963205
4        cloglog 4 10 0.4000000 0.12269317 0.6702046
5          exact 4 10 0.4000000 0.12155226 0.7376219
6          logit 4 10 0.4000000 0.15834201 0.7025951
7         probit 4 10 0.4000000 0.14933907 0.7028372
8        profile 4 10 0.4000000 0.14570633 0.6999845
9            lrt 4 10 0.4000000 0.14564246 0.7000216
10     prop.test 4 10 0.4000000 0.13693056 0.7263303
11        wilson 4 10 0.4000000 0.16818033 0.6873262
```

The function calculates 11 different intervals for π when the `methods = "all"` argument is used. The first, second, and eleventh intervals are the Agresti-Coull, Wald, and Wilson intervals, respectively. Please see the help for the function for more information on the other intervals. The end-of-chapter exercises discuss some of these in more detail.

Clopper-Pearson confidence interval

The Clopper-Pearson interval (Clopper and Pearson, 1934) is the last confidence interval for π that we will discuss in this section. While Brown et al. (2001) remark that the interval is "wastefully conservative and it is not a good choice for practical use," this interval does have a unique property that the other intervals do not: the true confidence level is always equal to or greater than the stated level. In order to achieve this true confidence level, the interval is usually wider than most other intervals for π.

The interval uses the relationship between the binomial distribution and the beta distribution to achieve its conservative confidence level (see #2.40 on p. 82 of Casella and Berger, 2002 for this distributional relationship). In fact, because the actual or *exact* distribution for W is used, the interval is called an *exact inference* procedure. There are many other exact inference procedures available for statistical problems, and some of these are discussed in Section 6.2.

To review the beta distribution, let V be a beta random variable. The probability density function (PDF) for V is

$$f(v; a, b) = \frac{\Gamma(a+b)}{\Gamma(a)\Gamma(b)} v^{a-1}(1-v)^{b-1}, \ 0 < v < 1 \tag{1.5}$$

where $a > 0$ and $b > 0$ are parameters and $\Gamma(\cdot)$ is the gamma function, $\Gamma(c) = \int_0^\infty x^{c-1} e^{-x} dx$ for $c > 0$. Note that $\Gamma(c) = (c-1)!$ for an integer c. The a and b parameters control the shape of the distribution. The distribution is right-skewed for $a > b$, and the distribution is left-skewed for $a < b$. When $a = b$, the distribution is symmetric about $v = 0.5$. Our program Beta.R gives a few example plots of the distribution. The α quantile of a beta distribution, denoted by v_α or beta$(\alpha; a, b)$, is found by solving

$$\alpha = \int_0^{v_\alpha} \frac{\Gamma(a+b)}{\Gamma(a)\Gamma(b)} v^{a-1}(1-v)^{b-1} dv$$

for v_α.

The $(1-\alpha)100\%$ Clopper-Pearson interval is simply quantiles from two beta distributions:

$$\text{beta}(\alpha/2;\ w,\ n-w+1) < \pi < \text{beta}(1-\alpha/2;\ w+1,\ n-w).$$

Due to the restriction $a > 0$, the lower endpoint cannot be computed if $w = 0$. In that case, the lower limit is taken to be 0. Similarly, the upper limit is taken to be 1 whenever $w = n$ due to $b > 0$. Because the remaining beta quantiles lie strictly between 0 and 1, the Clopper-Pearson interval respects the natural boundaries for probabilities.

Often, the Clopper-Pearson interval is given in terms of quantiles from an F-distribution rather than a beta distribution. This comes about through a relationship between the two distributions; see #9.21 on p. 454 of Casella and Berger (2002) for this relationship. Both formulas produce the same limits and beta quantiles are easy to compute, so we omit the F-based formula here.

There are a few variations on the Clopper-Pearson interval. The Blaker interval proposed in Blaker (2000, 2001) also guarantees the true confidence level is always equal to or greater than the stated level. An added benefit is that the interval is no wider than the Clopper-Pearson interval and is often narrower. A disadvantage is that the interval is more difficult to calculate and requires an iterative numerical procedure to find its limits. The CIpi.R program shows how to calculate the interval using the `binom.blaker.limits()` function of the `BlakerCI` package. Another variation on the Clopper-Pearson interval is the mid-p interval. This interval no longer guarantees the true confidence level to be greater than the stated level, but it will be shorter than the Clopper-Pearson interval while performing relatively well with respect to the stated confidence level (Brown et al., 2001). The CIpi.R program shows how to calculate this interval using the `midPci()` function of the `PropCIs` package.

Example: Clopper-Pearson interval (CIpi.R)

Suppose $w = 4$ successes are observed out of $n = 10$ trials again. The 95% Clopper-Pearson interval is beta$(0.025;\ 4,\ 7) < \pi <$ beta$(0.975;\ 5,\ 6)$. The `qbeta()` function in R calculates these quantiles resulting in an interval $0.1216 < \pi < 0.7376$. Notice that this is the widest of the intervals calculated so far.

Below is the R code used to calculate the interval:

```
> round(qbeta(p = c(alpha/2, 1-alpha/2), shape1 = c(w, w+1),
    shape2 = c(n-w+1, n-w)),4)
[1] 0.1216 0.7376

> binom.confint(x = w, n = n, conf.level = 1-alpha, methods =
    "exact")
    method x  n mean      lower     upper
1   exact  4 10  0.4 0.1215523 0.7376219
```

Within the `qbeta()` function call, the `shape1` argument is a and the `shape2` argument is b. We use vectors within `qbeta()` to find the quantiles. R matches each vector

Table 1.1: Confidence intervals for the hepatitis C prevalence.

Method	Interval	Length
Wald	(0.0157, 0.0291)	0.0134
Agresti-Coull	(0.0165, 0.0302)	0.0137
Wilson	(0.0166, 0.0301)	0.0135
Clopper-Pearson	(0.0162, 0.0302)	0.0140

value to produce the equivalent of separate `qbeta()` function runs for the lower and upper limits. The `binom.confint()` function is used as an alternative way to find the interval where `method = "exact"` gives the Clopper-Pearson interval. Note that this interval was also given earlier when we used `method = "all"`.

Example: Hepatitis C prevalence among blood donors (HepCPrev.R)

Blood donations are screened for diseases to prevent transmission from donor to recipient. To examine how prevalent hepatitis C is among blood donors, Liu et al. (1997) focused on 1,875 blood donations in Xuzhou City, China.[4] They observed that 42 donations tested positive for the antigen produced by the body when infected with the virus. The 95% Wilson interval is $0.0166 < \pi < 0.0301$, where we used the same type of R code as in the previous examples. Thus, with 95% confidence, the hepatitis C antigen prevalence in the Xuzhou City blood donor population is between 0.0166 and 0.0301.

In practice, we would only calculate one of the intervals discussed in this section. For demonstration purposes, Table 1.1 displays additional 95% confidence intervals. Due to the large sample size, we see that the intervals are similar with the Wald interval being the most different from the others. The lengths of the intervals are similar as well with the Clopper-Pearson interval being a little longer than the others.

Tests

When only one simple parameter is of interest, such as π here, we generally prefer confidence intervals over hypothesis tests, because the interval gives a range of possible parameter values. We can typically infer that a hypothesized value for a parameter can be rejected if it does not lie within the confidence interval for the parameter. However, there are situations where a fixed known value of π, say π_0, is of special interest, leading to a formal hypothesis test of $H_0 : \pi = \pi_0$ vs. $H_a : \pi \neq \pi_0$.

With regard to the Wilson interval, it was noted that the score test statistic

$$Z_0 = \frac{\hat{\pi} - \pi_0}{\sqrt{\pi_0(1 - \pi_0)/n}},$$

is often used in these situations. When the null hypothesis is true, Z_0 should have approximately a standard normal distribution, where the approximation is generally better for larger samples. The null hypothesis is rejected when an unusual value of Z_0 is observed

---

[4]The study's main purpose was to examine how well "group testing" would work to estimate overall disease prevalence. See Bilder (2009) for an introduction to group testing.

relative to this distribution, namely something less than $-Z_{1-\alpha/2}$ or greater than $Z_{1-\alpha/2}$. The p-value is a measure of how extreme the test statistic value is relative to what is expected when H_0 is true. This p-value is calculated as $2P(Z > |Z_0|)$ where Z has a standard normal distribution. Note that this test is equivalent to rejecting the null hypothesis when π_0 is outside of the Wilson interval. If desired, the `prop.test()` function can be used to calculate Z_0 (Z_0^2 is actually given) and a corresponding p-value; this is demonstrated in the CIpi.R program. This function also calculates the Wilson interval.

We recommend using the score test when performing a test for π. However, there are alternative testing procedures. In particular, the likelihood ratio test (LRT) is a general way to perform hypothesis tests, and it can be used here to test π (see Appendix B.5 for an introduction). Informally, the LRT statistic is

$$\Lambda = \frac{\text{Maximum of likelihood function under } H_0}{\text{Maximum of likelihood function under } H_0 \text{ or } H_a}.$$

For the specific test of $H_0 : \pi = \pi_0$ vs. $H_a : \pi \neq \pi_0$, the denominator is $\hat{\pi}^w(1-\hat{\pi})^{n-w}$ (using Equation 1.2), because the maximum possible value of the likelihood function occurs when it is evaluated at the MLE. The numerator is $\pi_0^w(1-\pi_0)^{n-w}$ because there is only one possible value of the likelihood function if the null hypothesis is true. The transformed statistic $-2\log(\Lambda)$ turns out to have an approximate χ_1^2 distribution in large samples if the null hypothesis is true. For this test, the transformed statistic can be re-expressed as

$$-2\log(\Lambda) = -2\left\{ w\log\left(\frac{\pi_0}{\hat{\pi}}\right) + (n-w)\log\left(\frac{1-\pi_0}{1-\hat{\pi}}\right) \right\}.$$

We reject the null hypothesis if $-2\log(\Lambda) > \chi^2_{1,1-\alpha/2}$, where $\chi^2_{1,1-\alpha/2}$ is the $1-\alpha/2$ quantile from a chi-square distribution with 1 degree of freedom (for example, $\chi^2_{1,0.95} = 3.84$ when $\alpha = 0.05$). The p-value is $P(A > -2\log(\Lambda))$ where A has a χ_1^2 distribution.

An alternative way to calculate a confidence interval for π is to invert the LRT in a similar manner as was done for the Wilson interval (see Exercise 13). This likelihood ratio (LR) interval is automatically calculated by the `binom.confint()` function of the `binom` package, where the `methods = "lrt"` argument value is used. We provide additional code in CIpi.R that shows how to find the interval without this function. The interval is generally harder to compute than the intervals that we recommend in this section and has no advantages over them. LR confidence intervals often are used in some more complicated contexts where better intervals are not available (this will be the case in Chapters 2 to 4). The interval is better than the Wald interval in most problems.

1.1.3 True confidence levels for confidence intervals

As discussed on p. 11, a confidence interval method may not actually achieve its stated confidence level. The reasons for this are explained shortly. Figure 1.3 provides a comparison of the true confidence levels for the Wald, Wilson, Agresti-Coull, and Clopper-Pearson intervals. For each plot, n is 40 and the stated confidence level is 0.95 ($\alpha = 0.05$). The true confidence level (coverage) for each interval method is plotted as a function of π. For example, the true confidence level at $\pi = 0.157$ is 0.8760 for the Wald, 0.9507 for the Wilson, 0.9507 for the Agresti-Coull, and 0.9740 for the Clopper-Pearson intervals, respectively. Obviously, none of these intervals achieve exactly the stated confidence level on a consistent basis. Below are some general conclusions from examining this plot:

- The Wald interval tends to be the farthest from 0.95 the most often. In fact, the true confidence level is often too low for it to be on the plot at extreme values of π.

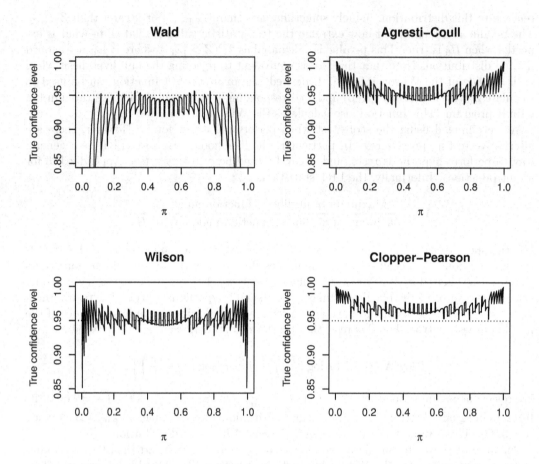

Figure 1.3: True confidence levels with $n = 40$ and $\alpha = 0.05$.

- The Agresti-Coull interval does a much better job than the Wald with its true confidence level usually between 0.93 and 0.98. For values of π close to 0 or 1, the interval can be very conservative.

- The Wilson interval performs a little better than the Agresti-Coull interval with its true confidence level generally between 0.93 and 0.97; however, for very extreme π, it can be very liberal. Note that this performance for extreme π can be improved by changing the lower interval limit to $-\log(1 - \alpha)/n$ when $w = 1$ and the upper interval limit to $1 + \log(1 - \alpha)/n$ when $w = n - 1$; see p. 112 of Brown et al. (2001) for justification. This small modification was used by the `binom.confint()` function in the past (version 1.0-5 of the package), but it is now no longer implemented as of version 1.1-1 of the package.

- The Clopper-Pearson interval has a true confidence level at or above the stated level, where it is generally oscillating between 0.95 and 0.98. For values of π close to 0 or 1, the interval can be very conservative.

Similar findings can be shown for other values of n and α. The R code used to construct Figure 1.3 is available in the ConfLevel4Intervals.R program, and it will be discussed shortly.

Why do these plots in Figure 1.3 have such strange patterns? It is all because of the discreteness of a binomial random variable. For a given n, there are only $n + 1$ possible intervals that can be formed, one for each value of $w = 0, 1, \ldots, n$. For a specific value of π, some of these intervals contain π and some do not. Thus, the true confidence level at π, say $C(\pi)$, is the sum of the binomial probabilities for all intervals that do contain π:

$$C(\pi) = \sum_{w=0}^{n} I(w) \binom{n}{w} \pi^w (1-\pi)^{n-w}, \qquad (1.6)$$

where $I(w) = 1$ if the interval formed with w contains π, and $I(w) = 0$ if not. Each of these binomial probabilities in Equation 1.6 changes slowly as π changes. As long as we do not move π past any interval limits, the true confidence level changes slowly too. However, as soon as π crosses over an interval limit, a mass of probability is suddenly either added to or subtracted from the true confidence level, resulting in the spikes that appear in all parts of Figure 1.3. We illustrate how to find the true confidence level and when these spikes occur in the next example.

Example: True confidence level for the Wald interval (ConfLevel.R)

We show in this example how to calculate a true confidence level for the Wald interval with $n = 40$, $\pi = 0.157$, and $\alpha = 0.05$. Below is a description of the process:

1. Find the probability of obtaining each possible value of w using the dbinom() function with $n = 40$ and $\pi = 0.157$,

2. Calculate the 95% Wald confidence interval for each possible value of w, and

3. Sum up the probabilities corresponding to those intervals that contain $\pi = 0.157$; this is the true confidence level.

Below is the R code:

```
> pi <- 0.157
> alpha <- 0.05
> n <- 40
> w <- 0:n
> pi.hat <- w/n
> pmf <- dbinom(x = w, size = n, prob = pi)
> var.wald <- pi.hat*(1-pi.hat)/n
> lower <- pi.hat - qnorm(p = 1-alpha/2) * sqrt(var.wald)
> upper <- pi.hat + qnorm(p = 1-alpha/2) * sqrt(var.wald)
> save <- ifelse(test = pi>lower, yes = ifelse(test = pi<upper,
    yes = 1, no = 0), no = 0)
> data.frame(w, pi.hat, round(data.frame(pmf, lower, upper),4),
    save)[1:13,]
   w pi.hat    pmf   lower  upper save
1  0  0.000 0.0011  0.0000 0.0000    0
2  1  0.025 0.0080 -0.0234 0.0734    0
3  2  0.050 0.0292 -0.0175 0.1175    0
4  3  0.075 0.0689 -0.0066 0.1566    0
5  4  0.100 0.1187  0.0070 0.1930    1
6  5  0.125 0.1591  0.0225 0.2275    1
7  6  0.150 0.1729  0.0393 0.2607    1
8  7  0.175 0.1564  0.0572 0.2928    1
9  8  0.200 0.1201  0.0760 0.3240    1
10 9  0.225 0.0795  0.0956 0.3544    1
```

```
11 10   0.250 0.0459   0.1158 0.3842         1
12 11   0.275 0.0233   0.1366 0.4134         1
13 12   0.300 0.0105   0.1580 0.4420         0

> sum(save*pmf)
[1] 0.875905
> sum(dbinom(x = 4:11, size = n, prob = pi))
[1] 0.875905
```

The code contains many of the same components that we have seen before. The main difference now is that we are calculating an interval for each possible value of w rather an interval for only one. One new part within the code is the `ifelse()` function. This function does a logical check for whether or not π is within each of the 41 intervals. For example, the second interval is $(-0.0234, 0.0734)$, which does not contain $\pi = 0.157$ so the `save` object has a value of 0 for its second element.

The data frame created at the end puts all of the calculated components together into a table. For example, we see that if $w = 3$, the corresponding interval does not contain π, but at $w = 4$ the corresponding interval does. By examining other parts of the data frame, we see that the intervals for $w = 4$ to 11 all contain $\pi = 0.157$. The probability that a binomial random variable is between 4 and 11 with $n = 40$ and $\pi = 0.157$ is 0.8759, which is the true confidence level. Obviously, the Wald confidence interval does not achieve its stated level of 95%.

Notice that the upper interval limit at $w = 3$ barely does not contain $\pi = 0.157$ and $P(W = 3) = 0.0689$. By using the same code with the change of `pi <- 0.156`, the upper limit at $w = 3$ now does contain $\pi = 0.156$, so that $P(W = 3) = 0.0706$ is included when summing probabilities for the true confidence level. Overall, $w = 3$ to 11 now have confidence intervals that contain $\pi = 0.156$ leading to a true confidence level of 0.9442! This demonstrates what we alluded to earlier as the cause for the spikes in Figure 1.3.

In simple problems like the present one, we can exactly determine the probabilities of each interval that contains a given π, so that the plots like in Figure 1.3 can be made exactly. In other cases, we may have to rely on Monte Carlo simulation. We explore the simulation approach next to enable us to compare an exact true confidence level to one estimated by simulation. This will be helpful later in the text when the simulation method is the only available method of assessment.

Example: Estimated true confidence level for the Wald interval (ConfLevel.R)

Suppose again that $n = 40$, $\pi = 0.157$, and $\alpha = 0.05$. Below is a description of the process to estimate the true confidence level through simulation:

1. Simulate 1,000 samples using the `rbinom()` function with $n = 40$ and $\pi = 0.157$,

2. Calculate the 95% Wald confidence interval for each sample, and

3. Calculate the proportion of intervals that contain $\pi = 0.157$; this is the estimated true confidence level.

Below is the R code:

```
> numb.bin.samples <- 1000   # Binomial samples of size n
> set.seed(4516)
> w <- rbinom(n = numb.bin.samples, size = n, prob = pi)
> pi.hat <- w/n
> var.wald <- pi.hat*(1-pi.hat)/n
> lower <- pi.hat - qnorm(p = 1-alpha/2) * sqrt(var.wald)
> upper <- pi.hat + qnorm(p = 1-alpha/2) * sqrt(var.wald)
> data.frame(lower, upper)[1:10,]
   w pi.hat      lower      upper
1  6  0.150  0.039344453  0.2606555
2  6  0.150  0.039344453  0.2606555
3  7  0.175  0.057249138  0.2927509
4  8  0.200  0.076040994  0.3239590
5  8  0.200  0.076040994  0.3239590
6  6  0.150  0.039344453  0.2606555
7  8  0.200  0.076040994  0.3239590
8  3  0.075 -0.006624323  0.1566243
9  5  0.125  0.022511030  0.2274890
10 4  0.100  0.007030745  0.1929693
> save <- ifelse(test = pi>lower, yes = ifelse(test = pi<upper,
    yes = 1, no = 0), no = 0)
> save[1:10]
 [1] 1 1 1 1 1 1 1 0 1 1
> mean(save)
[1] 0.878
```

Again, we are using much of the same code as in the past. The `ifelse()` function is used to check whether $\pi = 0.157$ is within each of the intervals. For example, we see that sample #8 results in $\hat{\pi} = 0.075$ and an interval of (-0.0066, 0.1566), which does not contain 0.157, so the corresponding value of `save` is 0. The mean of all the 0's and 1's in `save` is 0.878. This is our estimated true confidence level for the Wald interval at $n = 40$ and $\pi = 0.157$.

In this relatively simple simulation problem, we already know that the intervals for $w = 4, \ldots, 11$ contain $\pi = 0.157$ while the others do not. To see that the simulation is, indeed, estimating $P(4 \leq W \leq 11)$, the `table()` function is used to calculate the number of times each w occurs:

```
> counts <- table(w)
> counts
w
  1   2   3   4   5   6   7   8   9  10  11
  8  35  64 123 147 165 172 123  76  46  26
 12  13
 11   4
> sum(counts[4:11])/numb.bin.samples
[1] 0.878
```

For example, there were 64 out of the 1,000 observations that resulted in a $w = 3$. This is very similar to the $P(W = 3) = 0.0689$ that we obtained for the past example. Summing up the counts for $w = 4, \ldots, 11$ and dividing by 1000, we obtain the same estimate of 0.878 for the true confidence level.

The estimate of the true confidence level here is almost the same as the actual true confidence level found in the previous example. Due to using a large number of samples, the law of large numbers ensures that this will happen. We could even go as far as finding a confidence interval for the true confidence level! For this case, we have 878 "successes" out of 1,000 "trials." A 95% Wilson interval for the true confidence level itself is (0.8563, 0.8969), which happens to contain 0.8759, the known true confidence level.

Figure 1.3 provides the true confidence levels for $\pi = 0.001, ..., 0.999$ by 0.0005, where we use straight lines to fill in the missing confidence levels between plotting points at levels of π not used. In order to calculate all of these 1,997 different confidence levels for a particular interval, we repeat the same code as before, but now for each π by using a "for loop" within R. The next example illustrates this process.

Example: True confidence level plot (ConfLevel4Intervals.R, ConfLevelWald-Only.R)

With $n = 40$ and $\alpha = 0.05$, we calculate the true confidence levels for the Wald interval using the following code:

```
> alpha <- 0.05
> n <- 40
> w <- 0:n
> pi.hat <- w/n
> pi.seq <- seq(from = 0.001, to = 0.999, by = 0.0005)

> # Wald
> var.wald <- pi.hat*(1-pi.hat)/n
> lower.wald <- pi.hat - qnorm(p = 1-alpha/2) * sqrt(var.wald)
> upper.wald <- pi.hat + qnorm(p = 1-alpha/2) * sqrt(var.wald)

> # Save true confidence levels in a matrix
> save.true.conf <- matrix(data = NA, nrow = length(pi.seq), ncol
    = 2)

> # Create counter for the loop
> counter <- 1

> # Loop over each pi
> for(pi in pi.seq) {
    pmf <- dbinom(x = w, size = n, prob = pi)
    save.wald <- ifelse(test = pi>lower.wald, yes = ifelse(test =
        pi<upper.wald, yes = 1, no = 0), no = 0)
    wald <- sum(save.wald*pmf)
    save.true.conf[counter,] <- c(pi, wald)
    # print(save.true.conf[counter,])
    counter <- counter+1
}

> plot(x = save.true.conf[,1], y = save.true.conf[,2], main =
    "Wald", xlab = expression(pi), ylab = "True confidence level",
    type = "l", ylim = c(0.85,1))
> abline(h = 1-alpha, lty = "dotted")
```

We create a vector `pi.seq` which is a sequence of numbers from 0.001 to 0.999 by 0.0005. The `for(pi in pi.seq)` function code (often referred to as a "for loop") instructs R to take one π value out of `pi.seq` at a time. The code enclosed by braces then finds the true confidence level for this π. The `save.true.conf` object is a matrix that is created to have 1,997 rows and 2 columns. At first, all of its values are initialized to be "NA" within R. Its values are updated then one row at a time by inserting the value of π and the true confidence level. Finally, the `counter` object allows us to change the row number of `save.true.conf` within the loop.[5]

After the for loop, we use use the `plot()` function to plot the value of π on the x-axis and the true confidence level on the y-axis using the appropriate columns of `save.true.conf`. The `type = "l"` argument instructs R to construct a line plot where each π and true confidence level pair is connected by a line. The `abline()` function draws a horizontal dotted line at 0.95, which is the stated confidence level. Please see the upper left plot in Figure 1.3 for the final result. In order to construct all four plots in Figure 1.3, we insert the code for the other three intervals into the braces of the loop. We also add three columns to the `save.true.conf` matrix and use additional calls to the `plot()` function. Please see ConfLevel4Intervals.R for the code.

The `binom` package in R also can be used to calculate the true confidence levels. The `binom.coverage()` function calculates the true confidence level for one π at a time, and the `binom.plot()` function plots the true confidence levels over a set of different values of π. Examples of using these functions are in the programs for this example. Note that we purposely demonstrated the calculations without `binom.coverage()` first, because convenient functions like it are not available for other situations examined elsewhere in the textbook.

1.2 Two binary variables

We consider now the situation when the same Bernoulli trial is measured on units that can be classified into groups. The simplest such case is when a population consists of two groups, such as females and males, fresh- and salt-water fish, or American and foreign companies. Below are two examples with a binary response on trials that form two groups.

Example: Larry Bird's free throw shooting

A free throw is a shot in basketball where the shooter can shoot freely (unopposed by another player) from a specific location on the court. The shot is either made (a success) or missed (a failure). Most often during a National Basketball Association (NBA) game, free throws are shot in pairs. This means a free throw shooter has one attempt and then subsequently has a second attempt no matter what happens on the first.

The former NBA player Larry Bird was one of the most successful at making free throws during his career with a success rate of 88.6%. By comparison, the NBA

---

[5] If desired, the call to the `print()` function can be uncommented to see the progress during the loop. If this is done, it is best to turn off the "buffered output" in R: select Misc > Buffered output from the R main menu.

Table 1.2: Larry Bird's free throw outcomes; data source is Wardrop (1995).

		Second		
		Made	Missed	Total
First	Made	251	34	285
	Missed	48	5	53
	Total	299	39	338

Table 1.3: Salk vaccine clinical trial results; data source is Francis et al. (1955, p. 25).

	Polio	Polio free	Total
Vaccine	57	200,688	200,745
Placebo	142	201,087	201,229
Total	199	401,775	401,974

average during this time was about 75% (http://www.basketball-reference.com). Bird's outstanding success rate is among his many achievements, for which he has been recognized as one of the 50 greatest players in the history of the NBA (http://www.nba.com/history/players/50greatest.html). During the 1980-81 and 1981-82 NBA seasons, the outcomes from Bird's free throw attempts shot in pairs were recorded, and a summary is shown in Table 1.2. For example, Bird made both his first and second attempts 251 times. Also, Bird made the first attempt, but then subsequently missed the second attempt 34 times. Overall, he made the first attempt 285 times without regard to what happened on the second attempt. In total, Bird shot 338 pairs of free throw pairs during the season.

Basketball fans and commentators often speculate that the results of a second free throw might depend on the results of the first. For example, if a shooter misses the first attempt, will disappointment or determination lead to altering his/her approach for the second attempt? If so, then we should see that the probability of success on the second attempt is different depending on whether the first attempt was made or missed. Thus, the two groups in this problem are formed by the results of the first attempt, and the Bernoulli trials that we observe are the results of the second attempt. Given the data in Table 1.2, we will investigate if the second attempt outcome is dependent on what happens for the first attempt.

Example: Salk vaccine clinical trial

Clinical trials are performed to determine the safety and efficacy of new drugs. Frequently, the safety and efficacy responses are categorical in nature; for example, the efficacy response may be simply whether a drug cures or does not cure a patient of a disease. In order to ensure that a new drug is indeed better than doing nothing (patients sometimes get better without intervention), it is essential to have a control group in the trial. This is achieved in clinical trials by randomizing patients into two groups: new drug or control. The control group is often a placebo, which is administered just like the new drug but contains no medication.

One of the most famous and largest clinical trials ever performed was in 1954. Over 1.8 million children participated in the clinical trial to determine the effectiveness of the polio vaccine developed by Jonas Salk (Francis et al., 1955). While the actual design of the trial sparked debate (Brownlee, 1955; Dawson, 2004), we forgo this discussion and focus on the data obtained from the randomized, placebo-controlled portion of

Table 1.4: Probability and observed count structures for two independent binomial random variables.

		Response					Response		
		1	2				1	2	
Group	1	π_1	$1-\pi_1$	1	Group	1	w_1	$n_1 - w_1$	n_1
	2	π_2	$1-\pi_2$	1		2	w_2	$n_2 - w_2$	n_2
							w_+	$n_+ - w_+$	n_+

the trial. The data, given in Table 1.3, show that 57 out of the 200,745 children in the vaccine group developed polio during the study period, as opposed to 142 out of the 201,229 children in the placebo group. The question of interest for the clinical trial was "Does the vaccine help to prevent polio?" We will develop comparison measures in this section to answer this question.

1.2.1 Notation and model

The model and notation follow those used for a single binomial random variable in Section 1.1. We start by considering two separate Bernoulli random variables, Y_1 and Y_2, one for each group. The probabilities of success for the two groups are denoted by π_1 and π_2, respectively. We observe n_j trials of Y_j leading to w_j observed successes, $j = 1, 2$.[6] We replace a subscript with "+" to denote a sum across that subscript. Thus, $n_+ = n_1 + n_2$ is the total number of trials, and $w_+ = w_1 + w_2$ is the total number of observed successes. This notation is depicted in Table 1.4. The table on the right side of Table 1.4 is called a *two-way contingency table*, because it gives a listing of all possible cross-tabulations of two categorical variables. We cover more general forms of contingency tables in Chapters 3 and 4.

We denote the random variable representing the number of successes in group j by W_j and write its PMF as

$$P(W_j = w_j) = \binom{n_j}{w_j} \pi_j^{w_j} (1-\pi_j)^{n_j - w_j}, \quad w_j = 0, 1, \ldots, n_j, \quad j = 1, 2.$$

We assume that the two random variables Y_1 and Y_2 are independent, so that the outcome of one cannot affect the outcome of the other. In the Salk vaccine clinical trial, for example, this means that children assigned to receive vaccine cannot pass on immunity or disease to children in the placebo group and vice versa. This assumption is critical in what follows, and so this model is referred to as the *independent binomial model*. When independence is not satisfied, then other models need to be used that account for dependence between the random variables (e.g., see Section 1.2.6 for handling paired data).

When we want to simulate data from this model, we will use R code like what is shown below. This sampling procedure will be important shortly when we use these simulated counts to evaluate statistical inference procedures, like confidence intervals, to determine if they work as expected.

---

[6]More formally, we could define y_{ij} as the observed value for the ith independent trial in the jth group. This leads to $w_j = \sum_{i=1}^{n_j} y_{ij}$.

Example: Simulate counts in a contingency table (SimContingencyTable.R)

Consider the case of $\pi_1 = 0.2$, $\pi_2 = 0.4$, $n_1 = 10$, and $n_2 = 10$. The R code below shows how to simulate one set of counts for a contingency table.

```
> pi1 <- 0.2
> pi2 <- 0.4
> n1 <- 10
> n2 <- 10

> set.seed(8191)
> w1 <- rbinom(n = 1, size = n1, prob = pi1)
> w2 <- rbinom(n = 1, size = n2, prob = pi2)

> c.table <- array(data = c(w1, w2, n1-w1, n2-w2), dim = c(2,2),
    dimnames = list(Group = c(1,2), Response = c(1, 2)))
> c.table
     Response
Group 1 2
    1 1 9
    2 3 7
> c.table1[1,1]   # w1
[1] 1
> c.table1[1,2]   # n1-w1
[1] 9
> c.table1[1,]    # w1 and n1-w1
1 2
1 9
> sum(c.table1[1,])   # n1
[1] 10
```

Similar to Section 1.1, we use the `rbinom()` function to simulate values for w_1 and w_2. To form the contingency table (what we name `c.table`), we use the `array()` function. Its `data` argument contains the counts within the contingency table. These counts are concatenated together using the `c()` function. Notice that the data are entered by columns $(w_1, w_2, n_1 - w_1, n_2 - w_2)$. The `dim` argument specifies the contingency table's dimensions as (number of rows, number of columns), where the `c()` function is used again. Finally, the `dimnames` argument gives names for the row and column measurements. The names are given in a *list* format, which allows for a number of objects to be linked together (please see Appendix A.7.3 for more on lists if needed). In this case, the objects are `Group` and `Response` that contain the levels of the rows and columns, respectively.

For this particular sample, $w_1 = 1$, $n_1 - w_1 = 9$, $w_2 = 3$, and $n_2 - w_2 = 7$. We access these values from within the contingency table by specifying a row and column number with `c.table`. For example, `c.table[1,2]` is equal to $n_1 - w_1$. We omit a column or row number within [ ] to have a whole row or column, respectively, displayed. Summed counts are found by using the `sum()` function with the appropriate counts.

If we wanted to repeat this process, say, 1,000 times, the `n` argument of the `rbinom()` functions would be changed to 1,000. Each contingency table would need to be formed separately using the `array()` function. The corresponding program for this example provides the code.

Likelihood and estimates

The main interests in this problem are estimating the probabilities of success, π_1 and π_2, for each group and comparing these probabilities. Maximum likelihood estimation again provides a convenient and powerful solution. Because Y_1 and Y_2 are independent, so too are W_1 and W_2. The likelihood function formed by independent random variables is just the product of their respective likelihood functions. Hence, the likelihood function is $L(\pi_1, \pi_2 | w_1, w_2) = L(\pi_1 | w_1) \times L(\pi_2 | w_2)$. Maximizing this likelihood over π_1 and π_2 results in the "obvious" estimates $\hat{\pi}_1 = w_1/n_1$ and $\hat{\pi}_2 = w_2/n_2$, the sample proportions in the two groups. In other words, when the random variables are independent, each probability is estimated using only the data from its own group.

Example: Larry Bird's free throw shooting (Bird.R)

The purpose of this example is to estimate the probability of successes using a contingency table structure in R. If the Bernoulli trial results are already summarized into counts as in Table 1.2, then a contingency table is created in R using the `array()` function:

```
> c.table <- array(data = c(251, 48, 34, 5), dim = c(2,2),
    dimnames = list(First = c("made", "missed"), Second =
    c("made", "missed")))
> c.table
        Second
First    made missed
made     251    34
missed    48     5

> list(First = c("made", "missed"), Second = c("made", "missed"))
$First
[1] "made"   "missed"
$Second
[1] "made"   "missed"
```

Because the levels of row and column variables have names, we use these names within the `dimnames` argument.

The estimates of the probability of successes (or sample proportions) are found by taking advantage of how R performs calculations:

```
> rowSums(c.table)   # n1 and n2
  made missed
   285     53

> pi.hat.table <- c.table/rowSums(c.table)
> pi.hat.table
        Second
First         made     missed
  made   0.8807018 0.11929825
  missed 0.9056604 0.09433962
```

The `rowSums()` function find the sum of counts in each row to obtain n_1 and n_2. By taking the contingency table of counts divided by these row sums, we obtain $\hat{\pi}_1$ and $\hat{\pi}_2$ in the first column and $1 - \hat{\pi}_1$ and $1 - \hat{\pi}_2$ in the second column. Notice that R does this division correctly taking the counts in the first (second) row of c.table divided by the first (second) element of rowSums(c.table).

Data are often available as measurements on each trial, rather than as summarized counts. For example, the Larry Bird data would originally have consisted of 338 unaggregated pairs of first and second free throw results. Thus, the data might have appeared first as[7]

```
> head(all.data2)
   first  second
1   made    made
2 missed    made
3   made  missed
4 missed  missed
5   made    made
6 missed    made
```

In this example, first represents the group and second is the trial response. All 338 observations are stored in a data frame named all.data2 within the corresponding program for this example, and we print the first 6 observations using the head() function. We call this data format the *raw data* because it represents how the data looked before being processed into counts.

To form a contingency table from the raw data, we can use the table() or xtabs() functions:

```
> bird.table1 <- table(all.data2$first, all.data2$second)
> bird.table1
         made missed
  made    251     34
  missed   48      5
> bird.table1[1,1]    # w1
[1] 251

> bird.table2 <- xtabs(formula = ~ first + second, data =
    all.data2)
> bird.table2
        second
first    made missed
  made    251     34
  missed   48      5
> bird.table2[1,1]    # w1
[1] 251
```

In both cases, the functions produce a contingency table that is saved into an object to allow parts of it to be accessed as before. Note that the "xtabs" name comes about through an abbreviation of *crosstabulations*, which is a frequently used term to describe the joint summarization of multiple variables.

The estimated probability that Larry Bird makes his second free throw attempt is $\hat{\pi}_1 = 0.8807$ given that he makes the first and $\hat{\pi}_2 = 0.9057$ given he misses the first. In this sample, the probability of success on the second attempt is larger when the

---

[7]The actual order of Larry Bird's free throw results are not available. We present this as a hypothetical ordering to emulate what may have occurred.

first attempt is missed rather than made. This is somewhat counterintuitive to many basketball fans' perceptions that a missed first free throw should lower the probability of success on the second free throw. However, this is only for one sample. We would like to generalize to the population of all free throw attempts by Larry Bird. In order to make this generalization, we need to use statistical inference procedures. We discuss these next.

1.2.2 Confidence intervals for the difference of two probabilities

A relatively easy approach to comparing π_1 and π_2 can be developed by taking their difference $\pi_1 - \pi_2$. The corresponding estimate of $\pi_1 - \pi_2$ is $\hat{\pi}_1 - \hat{\pi}_2$. Each success probability estimate has a probability distribution that is approximated by a normal distribution in large samples: $\hat{\pi}_j \dot\sim N(\pi_j, \widehat{Var}(\hat{\pi}_j))$, where $\widehat{Var}(\hat{\pi}_j) = \hat{\pi}_j(1-\hat{\pi}_j)/n_j$, $j = 1, 2$. Because linear combinations of normal random variables are themselves normal random variables (Casella and Berger, 2002, p. 156), the probability distribution for $\hat{\pi}_1 - \hat{\pi}_2$ is approximated by $N(\pi_1 - \pi_2, \widehat{Var}(\hat{\pi}_1 - \hat{\pi}_2))$, where $\widehat{Var}(\hat{\pi}_1 - \hat{\pi}_2) = \hat{\pi}_1(1-\hat{\pi}_1)/n_1 + \hat{\pi}_2(1-\hat{\pi}_2)/n_2$.[8] This distribution forms the basis for a range of inference procedures.

The easiest confidence interval to form for $\pi_1 - \pi_2$ uses the normal approximation for $\hat{\pi}_1 - \hat{\pi}_2$ directly to create a Wald interval:

$$\hat{\pi}_1 - \hat{\pi}_2 \pm Z_{1-\alpha/2} \sqrt{\frac{\hat{\pi}_1(1-\hat{\pi}_1)}{n_1} + \frac{\hat{\pi}_2(1-\hat{\pi}_2)}{n_2}}.$$

Unfortunately, the Wald interval for $\pi_1 - \pi_2$ has similar problems with achieving the stated level of confidence as the Wald interval for π. We will investigate this shortly.

Due to these problems, a number of other confidence intervals for $\pi_1 - \pi_2$ have been proposed. Inspired by the good general performance of the Agresti-Coull interval for a single probability, Agresti and Caffo (2000) investigated various intervals constructed as Wald-type intervals on data with different numbers of added successes and failures. They found that adding one success and one failure for each group results in an interval that does a good job of achieving the stated confidence level. Specifically, let

$$\tilde{\pi}_1 = \frac{w_1 + 1}{n_1 + 2} \text{ and } \tilde{\pi}_1 = \frac{w_2 + 1}{n_2 + 2}$$

be the amended estimates of π_1 and π_2. Notice that unlike $\tilde{\pi}$ for the Agresti-Coull interval, the $\tilde{\pi}_1$ and $\tilde{\pi}_2$ estimates do not change when the confidence level changes. The $(1-\alpha)100\%$ Agresti-Caffo confidence interval for $\pi_1 - \pi_2$ is

$$\tilde{\pi}_1 - \tilde{\pi}_2 \pm Z_{1-\alpha/2} \sqrt{\frac{\tilde{\pi}_1(1-\tilde{\pi}_1)}{n_1 + 2} + \frac{\tilde{\pi}_2(1-\tilde{\pi}_2)}{n_2 + 2}}.$$

Overall, we recommend the Agresti-Caffo method for general use.

Other confidence interval methods though have been developed analogous to the single-parameter case discussed in Section 1.1.2. There is a score interval based on inverting the test statistic for a score test of $H_0 : \pi_1 - \pi_2 = d$ (i.e., determining for what values d

---

[8] This is an application of the following result: $Var(aU + bV) = a^2 Var(U) + b^2 Var(V) + 2ab Cov(U, V)$, where U and V are random variables and a and b are constants. If U and V are independent random variables, then $Cov(U, V) = 0$. See p. 171 of Casella and Berger (2002).

of $\pi_1 - \pi_2$ that the null hypothesis is not rejected). This turns out to be a considerably more difficult computational problem than in the single-parameter case, because H_0 does not actually specify the values of π_1 and π_2. Exercise 24 discusses how this interval is calculated. Similarly, there is a two-group Bayesian credible interval similar to Jeffreys method, but this involves distributions that are not as simple to compute as the standard normal. Details of this interval are available in Agresti and Min (2005a).

Example: Larry Bird's free throw shooting (Bird.R)

The purpose of this example is to calculate a confidence interval for the difference in the second free throw success probabilities given the first attempt outcomes. Continuing the code from earlier, we obtain the following:

```
> alpha <- 0.05
> pi.hat1 <- pi.hat.table[1,1]
> pi.hat2 <- pi.hat.table[2,1]

> # Wald
> var.wald <- pi.hat1*(1-pi.hat1) / sum(c.table[1,]) +
    pi.hat2*(1-pi.hat2) / sum(c.table[2,])
> pi.hat1 - pi.hat2 + qnorm(p = c(alpha/2, 1-alpha/2)) *
    sqrt(var.wald)
[1] -0.11218742  0.06227017

> # Agresti-Caffo
> pi.tilde1 <- (c.table[1,1] + 1) / (sum(c.table[1,]) + 2)
> pi.tilde2 <- (c.table[2,1] + 1) / (sum(c.table[2,]) + 2)
> var.AC <- pi.tilde1*(1-pi.tilde1) / (sum(c.table[1,]) + 2) +
    pi.tilde2*(1-pi.tilde2) / (sum(c.table[2,]) + 2)
> pi.tilde1 - pi.tilde2 + qnorm(p = c(alpha/2, 1-alpha/2)) *
    sqrt(var.AC)
[1] -0.10353254  0.07781192
```

The 95% Wald interval is $-0.1122 < \pi_1 - \pi_2 < 0.0623$, and the 95% Agresti-Caffo interval is $-0.1035 < \pi_1 - \pi_2 < 0.0778$. The intervals are somewhat similar with the Wald interval being shifted to the left of the Agresti-Caffo interval. Using the Agresti-Caffo interval, with 95% confidence, the difference in the second free throw success probabilities given the outcome of the first is between -0.1035 and 0.0778. Because this interval contains 0, we cannot detect a change in Bird's probability of a successful second free throw following made and missed first attempts. This means that either there is no difference, or there is a difference, but it was not detected in this sample. The latter situation could be caused by either bad luck (an unusual sample) or too small of a sample size.

The same confidence intervals can be obtained in other ways. First, we can use the following code when the data are not already within R via the array() function:

```
> w1 <- 251
> n1 <- 285
> w2 <- 48
> n2 <- 53
> alpha <- 0.05
> pi.hat1 <- w1/n1
> pi.hat2 <- w2/n2
```

```
> var.wald <- pi.hat1*(1-pi.hat1) / n1 +  pi.hat2*(1-pi.hat2) / n2
> pi.hat1 - pi.hat2 + qnorm(p = c(alpha/2, 1-alpha/2)) *
    sqrt(var.wald)
[1] -0.11218742  0.06227017
```

Second, the `prop.test()` function shown later gives the Wald confidence interval as part of its output (also used in Section 1.1.2). Finally, the `wald2ci()` function in the `PropCIs` package also calculates the Wald and Agresti-Caffo confidence intervals. Please see the corresponding program for example code.

To find a true confidence level for one of these confidence intervals, the joint probability distribution for all possible combinations of (W_1, W_2) is needed. This distribution is just the product of two binomial probabilities because these random variables are independent. For a given n_1 and n_2, there are $(n_1+1)(n_2+1)$ possible observed combinations of (w_1, w_2), and a confidence interval can be computed for each of these combinations. For set values of π_1 and π_2, some of these intervals contain $\pi_1 - \pi_2$ and some do not. Thus, the true confidence level at π_1 and π_2, $C(\pi_1, \pi_2)$, is the sum of the joint probabilities for all intervals that do contain $\pi_1 - \pi_2$:

$$C(\pi_1, \pi_2) = \sum_{w_2=0}^{n_2} \sum_{w_1=0}^{n_1} I(w_1, w_2) \binom{n_1}{w_1} \pi_1^{w_1}(1-\pi_1)^{n_1-w_1} \binom{n_2}{w_2} \pi_2^{w_2}(1-\pi_2)^{n_2-w_2}$$

where the indicator function $I(w_1, w_2)$ is 1 if the corresponding interval contains $\pi_1 - \pi_2$ and $I(w_1, w_2)$ is 0 otherwise. Calculation details are given in the next example.

Example: True confidence levels for the Wald and Agresti-Caffo intervals (ConfLevelTwoProb.R)

The true confidence level for the Wald interval can be found in a similar manner as discussed in Section 1.1.3. Consider the case of $\alpha = 0.05$, $\pi_1 = 0.2$, $\pi_2 = 0.4$, $n_1 = 10$, and $n_2 = 10$. To find all possible combinations of (w_1, w_2), we use the `expand.grid()` function, which finds all possible combinations of the arguments (separated by commas) within its parentheses. We repeat this same process to find all possible values of $(\hat{\pi}_1, \hat{\pi}_2)$ and $P(W_1 = w_1, W_2 = w_2)$. Below is the R code:

```
> alpha <- 0.05
> pi1 <- 0.2
> pi2 <- 0.4
> n1 <- 10
> n2 <- 10

> # All possible combinations of w1 and w2
> w.all <- expand.grid(w1 = 0:n1, w2 = 0:n2)

> # All possible combinations of pi^_1 and pi^_2
> pi.hat1 <- (0:n1)/n1
> pi.hat2 <- (0:n2)/n2
> pi.hat.all <- expand.grid(pi.hat1 = pi.hat1, pi.hat2 = pi.hat2)

> # Find joint probability for w1 and w2
> prob.w1 <- dbinom(x = 0:n1, size = n1, prob = pi1)
> prob.w2 <- dbinom(x = 0:n2, size = n2, prob = pi2)
```

```
> prob.all <- expand.grid(prob.w1 = prob.w1, prob.w2 = prob.w2)
> pmf <- prob.all$prob.w1*prob.all$prob.w2

> # P(W1 = w1, W2 = w2)
> head(data.frame(w.all, pmf = round(pmf,4)))
  w1 w2    pmf
1  0  0 0.0006
2  1  0 0.0016
3  2  0 0.0018
4  3  0 0.0012
5  4  0 0.0005
6  5  0 0.0002
```

For example, the probability of observing $P(W_1 = 1, W_2 = 0) = 0.0016$. Using these probabilities, we calculate the true confidence level for the interval:

```
> var.wald <- pi.hat.all[,1] * (1-pi.hat.all[,1]) / n1 +
    pi.hat.all[,2] * (1-pi.hat.all[,2]) / n2
> lower <- pi.hat.all[,1] - pi.hat.all[,2] - qnorm(p = 1-alpha/2)
    * sqrt(var.wald)
> upper <- pi.hat.all[,1] - pi.hat.all[,2] + qnorm(p = 1-alpha/2)
    * sqrt(var.wald)
> save <- ifelse(test = pi1-pi2 > lower, yes = ifelse(test =
    pi1-pi2 < upper, yes = 1, no = 0), no = 0)
> sum(save*pmf)
[1] 0.9281274
> data.frame(w.all, round(data.frame(pmf, lower, upper),4),
    save)[1:15,]
   w1 w2    pmf   lower  upper save
1   0  0 0.0006  0.0000 0.0000    0
2   1  0 0.0016 -0.0859 0.2859    0
3   2  0 0.0018 -0.0479 0.4479    0
4   3  0 0.0012  0.0160 0.5840    0
5   4  0 0.0005  0.0964 0.7036    0
6   5  0 0.0002  0.1901 0.8099    0
7   6  0 0.0000  0.2964 0.9036    0
8   7  0 0.0000  0.4160 0.9840    0
9   8  0 0.0000  0.5521 1.0479    0
10  9  0 0.0000  0.7141 1.0859    0
11 10  0 0.0000  1.0000 1.0000    0
12  0  1 0.0043 -0.2859 0.0859    1
13  1  1 0.0108 -0.2630 0.2630    1
14  2  1 0.0122 -0.2099 0.4099    1
15  3  1 0.0081 -0.1395 0.5395    0
```

All possible Wald intervals are calculated, and the `ifelse()` function is used to check if $\pi_1 - \pi_2 = 0.2 - 0.4 = -0.2$ is within each interval. The last data frame shows the first 15 intervals with the results from this check. The probabilities corresponding to the intervals that contain -0.2 are summed to produce the true confidence level of 0.9281, which is less than the stated level of 0.95.

We can also calculate the true confidence level while holding one of the probabilities constant and allowing the other to vary. Figure 1.4 gives a plot where $\pi_2 = 0.4$ and π_1 varies from 0.001 to 0.999 by 0.0005. We exclude the R code here because it is

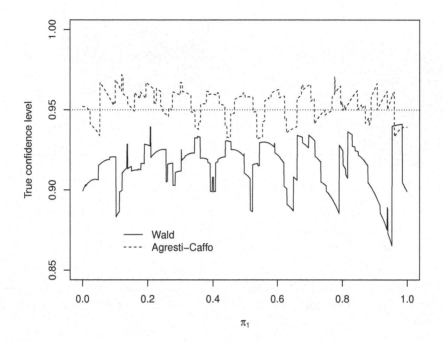

Figure 1.4: True confidence levels with $n_1 = 10$, $n_2 = 10$, $\pi_2 = 0.4$, and $\alpha = 0.05$.

quite similar to what was used in Section 1.1.3, where now we use the `for()` function to loop over the different values of π_1. Both the Agresti-Caffo and Wald lines are drawn on the plot simultaneously by using the `plot()` function first for the Wald true confidence levels and then using the `lines()` function for the Agresti-Caffo true confidence levels. The `legend()` function places the legend on the plot. Please see the program corresponding to this example for the code.

Figure 1.4 shows that the Wald interval never achieves the true confidence level! The Agresti-Caffo interval has a true confidence level between 0.93 and 0.97. We encourage readers to change the pre-set value for π_2 in the program to examine what happens in other situations. For example, the Wald interval is extremely liberal and the Agresti-Caffo interval is extremely conservative for small π_1 when $\pi_2 = 0.1$ with $n_1 = 10$, $n_2 = 10$, and $\alpha = 0.05$.

We can also allow π_2 to vary by the same increments as π_1 in order to produce a three-dimensional plot with the true confidence level on the third axis. The R code is in the corresponding program to this example. Two `for()` function calls—one loop for π_2 and one loop for π_1—are used within the code. Once all true confidence levels are found, the `persp3d()` function from the `rgl` package of R produces an interactive three-dimensional plot. Using the left and right mouse buttons inside the plot window, the plot can be rotated and zoomed in, respectively. Figure 1.5 gives separate plots for the Wald (left) and Agresti-Caffo (right) intervals. For both plots, a plane is drawn at the 0.95 stated confidence level. We can see that the Wald interval never achieves the stated confidence level, but the Agresti-Caffo interval does a much better job. We encourage the reader to construct these plots in order to see the surfaces better through rotating them.

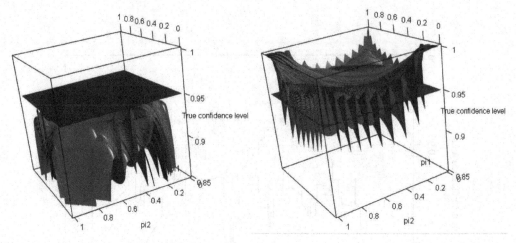

Figure 1.5: True confidence levels with $n_1 = 10$, $n_2 = 10$, and $\alpha = 0.05$. The left-side plot is for the Wald interval, and the right-side plot is for the Agresti-Caffo interval. Note that true confidence levels less than 0.85 are excluded from the Wald interval plot.

These true confidence level calculations can also be made through Monte Carlo simulation. We provide an example showing how this is done in our corresponding program.

1.2.3 Test for the difference of two probabilities

A formal test of the hypotheses $H_0: \pi_1 - \pi_2 = 0$ vs. $H_a: \pi_1 - \pi_2 \neq 0$ can be conducted, again in several ways. A Wald test uses a test statistic

$$Z_W = \frac{\hat{\pi}_1 - \hat{\pi}_2}{\sqrt{\frac{\hat{\pi}_1(1-\hat{\pi}_1)}{n_1} + \frac{\hat{\pi}_2(1-\hat{\pi}_2)}{n_2}}},$$

and compares this statistic against the standard normal distribution. Note that the denominator contains the estimated variance of $\hat{\pi}_1 - \hat{\pi}_2$ without regard to the null hypothesis.

Probability distributions for test statistics are generally computed assuming that null hypothesis is true. In the present context, that means that the two group probabilities are equal, and so a better estimated variance than what is in Z_W can be computed by assuming that $\pi_1 = \pi_2$. Notice that this condition implies that Y_1 and Y_2 have the same distribution. Thus, W_1 and W_2 are both counts of successes from the same Bernoulli random variable, and therefore w_1 and w_2 can be combined to represent w_+ successes in n_+ trials. Let $\bar{\pi} = w_+/n_+$ be the estimated probability of success when the null hypothesis is true. Then it can be shown that $\widehat{Var}(\hat{\pi}_1 - \hat{\pi}_2) = \bar{\pi}(1 - \bar{\pi})(1/n_1 + 1/n_2)$. This leads to a test based on comparing the statistic

$$Z_0 = \frac{\hat{\pi}_1 - \hat{\pi}_2}{\sqrt{\bar{\pi}(1-\bar{\pi})(1/n_1 + 1/n_2)}}$$

to a standard normal distribution. This is the score test.

A more general procedure that is used for comparing observed counts to estimated expected counts from any hypothesized model is the Pearson chi-square test. This particular test is often used to perform hypothesis tests in a contingency table setting, and we will spend much more time discussing it in Section 3.2. The model that is implied by our null hypothesis is a single binomial with n_+ trials and probability of success $\bar{\pi}$. The test statistic is formed by computing (observed count − estimated expected count)^2/(estimated expected count) over *all* observed counts, meaning here both successes and failures in the two groups. The estimated expected number of successes in group j under the null hypothesis model is $n_j\bar{\pi}$, and similarly the expected number of failures is $n_j(1-\bar{\pi})$. Thus, the Pearson chi-square test statistic is

$$X^2 = \sum_{j=1}^{2} \left(\frac{(w_j - n_j\bar{\pi})^2}{n_j\bar{\pi}} + \frac{(n_j - w_j - n_j(1-\bar{\pi}))^2}{n_j(1-\bar{\pi})} \right). \quad (1.7)$$

This can be simplified to

$$X^2 = \sum_{j=1}^{2} \frac{(w_j - n_j\bar{\pi})^2}{n_j\bar{\pi}(1-\bar{\pi})}.$$

The X^2 statistic has a distribution that is approximately χ_1^2 when n_1 and n_2 are large and when the null hypothesis is true. If the null hypothesis is false, the observed counts tend not to be close to what is expected when the null hypothesis is true; thus, large values of X^2 relative to the χ_1^2 distribution lead to a rejection of the null hypothesis. It can be shown that the Pearson chi-square and score test results are identical for this setting, because $X^2 = Z_0^2$ (see Exercise 25) and the χ_1^2 distribution is equivalent to the distribution of a squared standard normal random variable (e.g., $Z_{0.975}^2 = \chi_{1,0.95}^2 = 3.84$). For those readers unfamiliar with the latter result, please see p. 53 of Casella and Berger (2002) for a derivation if desired.

A LRT can also be conducted. The test statistic can be shown to be

$$-2\log(\Lambda) = -2\left[w_1\log\left(\frac{\bar{\pi}}{\hat{\pi}_1}\right) + (n_1-w_1)\log\left(\frac{1-\bar{\pi}}{1-\hat{\pi}_1}\right) + w_2\log\left(\frac{\bar{\pi}}{\hat{\pi}_2}\right) \right.$$
$$\left. +(n_2-w_2)\log\left(\frac{1-\bar{\pi}}{1-\hat{\pi}_2}\right)\right] \quad (1.8)$$

where we take $0 \times \log(\infty) = 0$ by convention. The null hypothesis is rejected if $-2\log(\Lambda) > \chi_{1,1-\alpha}^2$.

For all of these tests, the use of the standard normal or chi-squared distribution is based on large-sample approximations. The tests are asymptotically equivalent, meaning that they will give essentially the same results in very large samples. In small samples, however, the three test statistics can have distributions under the null hypothesis that are quite different from their approximations. Larntz (1978) compared the score, the LRT, and three other tests in various small-sample settings and found that the score test clearly maintains its size better than the others.[9] Thus, the score test is recommended here, as it was for testing the probability from a single group.

Example: Larry Bird's free throw shooting (Bird.R)

The purpose of this example is to show how to perform the score test, Pearson chi-square test, and LRT in R. We can use the `prop.test()` function to perform the score and Pearson chi-square tests:

---

[9]The size of a testing procedure is the probability that it rejects the null hypothesis when the null hypothesis is true. A test that holds the correct size is one that rejects at a rate equal to the type I error level of α.

```
> prop.test(x = c.table, conf.level = 0.95, correct = FALSE)

    2-sample test for equality of proportions without continuity
        correction

data:  c.table
X-squared = 0.2727, df = 1, p-value = 0.6015
alternative hypothesis: two.sided
95 percent confidence interval:
 -0.11218742  0.06227017
sample estimates:
   prop 1    prop 2
0.8807018 0.9056604
```

The argument value for x is the contingency table. Alternatively, we could have assigned x a vector with w_1 and w_2 within it and used a new argument n with a vector value of n_1 and n_2 (see corresponding program for an example). The correct = FALSE argument value guarantees that the test statistic is calculated as shown by Z_0; otherwise, a *continuity correction* is applied to help ensure that the test maintains its size at or below α.[10]

The output gives the test statistic value as $Z_0^2 = 0.2727$ and a p-value of $P(A > 0.2727) = 0.6015$, where A has a χ_1^2 distribution. The decision is to not reject the null hypothesis. The conclusion is that there is not a significant change in Bird's second free throw success percentage over the possible outcomes of the first attempt. Note that the chisq.test() function and the summary.table() method function also provide ways to perform the Pearson chi-square test. We will discuss these functions in Section 3.2.

The code for the LRT is shown below:

```
> pi.bar <- colSums(c.table)[1]/sum(c.table)
> log.Lambda <- c.table[1,1] * log(pi.bar / pi.hat.table[1,1]) +
    c.table[1,2] * log((1-pi.bar) / (1-pi.hat.table[1,1])) +
    c.table[2,1] * log(pi.bar / pi.hat.table[2,1]) + c.table[2,2]
    * log((1-pi.bar) /(1-pi.hat.table[2,1]))
> test.stat <- -2*log.Lambda
> crit.val <- qchisq(p = 0.95, df = 1)
> p.val <- 1-pchisq(q = test.stat, df = 1)
> round(data.frame(pi.bar, test.stat, crit.val, p.val, row.names
    = NULL), 4)
  pi.bar test.stat crit.val p.val
1 0.8846    0.2858   3.8415 0.593
```

Under the null hypothesis, the estimate of the probability of success parameter is found using the sum of the counts in the first column of c.table divided by the total sample size, and the result is put into pi.bar. The code for the log.Lambda object shows how to convert most of Equation 1.8 into the correct R syntax. The transformed test

---

[10] Note that the test statistic Z_0 is a discrete random variable. Modifications to Z_0 (or any other test statistic that is a discrete random variable) called *continuity corrections* are sometimes made, and they can be helpful when using a continuous distribution to approximate a discrete distribution. These corrections often lead to very conservative tests (i.e., reject the null hypothesis at a rate less than α when the null hypothesis is true), so they are not often used. Alternative procedures are discussed in Section 6.2.

statistic is $-2\log(\Lambda) = 0.2858$, and the p-value is $P(A > 0.2858) = 0.5930$. These are nicely printed using the `data.frame()` function, where the `row.names = NULL` argument value prevents the printing of an errant row name. The overall conclusion is the same as for the score test. Note that the test statistic and p-value could have been calculated a little more easily using the `assocstats()` function of the vcd package, and this function also gives the Pearson chi-square test statistic as well. We show how to use this function in the corresponding program to this example.

We conclude this example with a few additional notes:

- Notice that the success probability conditioning on the first free throw being missed was subtracted from the success probability conditioning on the first free throw being made. This is especially important to know if a one-side hypothesis test was performed or if 0 was outside of a confidence interval. For example, many basketball fans think that a missed first free throw has a negative impact on the second free throw outcome. If the lower limit of the interval had been positive (i.e., the whole interval is above 0), it would have confirmed this line of thinking with respect to Larry Bird.

- As with other applications of statistics, care needs to be taken when interpreting the results with respect to the population. For example, suppose we wanted to make some claims regarding all of Larry Bird's past, present, and future free throw pairs when the data was collected. Strictly speaking, a random sample would need to be taken from this entire population to formally make statistical inferences. Random samples are often not possible in a sports setting, as in our example where we have data from the 1980-1 and 1981-2 NBA seasons. Inference on a broader population of free throws may or may not be appropriate. For example, Larry Bird's free throw shooting success rate may have changed from year to year due to practice or injuries.

- In addition to the sampling problem, Larry Bird's career concluded in 1992. Therefore, the population data may be obtainable, and we could actually calculate population parameters such as π_1 and π_2. Statistical inference would not be necessary then.

1.2.4 Relative risks

The problem with basing inference on $\pi_1 - \pi_2$ is that it measures a quantity whose meaning changes depending on the sizes of π_1 and π_2. For example, consider two hypothetical scenarios where the probability of disease is listed for two groups of people, say for smokers (π_1) and for nonsmokers (π_2):

1. $\pi_1 = 0.51$ and $\pi_2 = 0.50$
2. $\pi_1 = 0.011$ and $\pi_2 = 0.001$.

In both cases $\pi_1 - \pi_2 = 0.01$. But in the first scenario, an increase of 0.01 due to smoking is rather small relative to the already sizable risk of disease in the nonsmoking population. On the other hand, scenario 2 has smokers with 11 times the chance of disease than nonsmokers. We need to be able to convey the relative magnitudes of these changes better than differences allow.

In this instance, a preferred way to compare two probabilities is through the *relative risk*, $RR = \pi_1/\pi_2$ (assuming $\pi_2 \neq 0$). For the example above, $RR = 0.011/0.001 = 11.0$ for the second scenario meaning that smokers are *11 times as likely* to have the disease than nonsmokers. Alternatively, we could say that smokers are 10 times *more* likely to have the disease than nonsmokers. On the other hand, for the first scenario, $RR = .51/.50 = 1.02$, indicating that smokers are just 2% more likely (or 1.02 times as likely) to have the disease. Notice that when $\pi_1 = \pi_2$, $RR = 1$.

These numerical values are based on population probabilities. To obtain an MLE for RR, we can make use of the invariance property of MLEs described in Appendix B.4 that allows us to substitute the observed proportions for the probabilities, $\widehat{RR} = \hat{\pi}_1/\hat{\pi}_2$, assuming $\hat{\pi}_2 \neq 0$. It is this estimate that is often given in news reports that state risks associated with certain factors such as smoking or obesity.

Because $\widehat{RR}$ is an MLE, inference can be carried out using the usual procedures. It turns out, however, that the normal approximation is rather poor for MLEs that are ratios, especially when the estimate in the denominator may have non-negligible variability as is the case here. Therefore, inference based on a normal approximation for $\widehat{RR}$ is not recommended. However, the normal approximation holds somewhat better for $\log(\widehat{RR}) = \log(\hat{\pi}_1) - \log(\hat{\pi}_2)$—the MLE for $\log(RR)$— so inference is generally carried out on the log scale. The variance estimate for $\log(\widehat{RR})$ can be derived by the delta method (Appendix B.4.2) as

$$\widehat{Var}(\log(\widehat{RR})) = \frac{1-\hat{\pi}_1}{n_1\hat{\pi}_1} + \frac{1-\hat{\pi}_2}{n_2\hat{\pi}_2} = \frac{1}{w_1} - \frac{1}{n_1} + \frac{1}{w_2} - \frac{1}{n_2}.$$

A $(1-\alpha)100\%$ Wald confidence interval for the population relative risk is found by first computing the confidence interval for $\log(\pi_1/\pi_2)$,

$$\log\left(\frac{\hat{\pi}_1}{\hat{\pi}_2}\right) \pm Z_{1-\alpha/2}\sqrt{\frac{1}{w_1} - \frac{1}{n_1} + \frac{1}{w_2} - \frac{1}{n_2}}.$$

The exponential transformation is then used to find the Wald interval for the relative risk itself:

$$\exp\left[\log\left(\frac{\hat{\pi}_1}{\hat{\pi}_2}\right) \pm Z_{1-\alpha/2}\sqrt{\frac{1}{w_1} - \frac{1}{n_1} + \frac{1}{w_2} - \frac{1}{n_2}}\right],$$

where $\exp(\cdot)$ is the inverse of the natural log function ($b = \exp(a)$ is equivalent to $a = \log(b)$). When w_1 and/or w_2 are equal to 0, the confidence interval cannot be calculated. One ad-hoc adjustment is to add a small constant, such as 0.5, to the 0 cell count and the corresponding row total. For example, if $w_1 = 0$, replace w_1 with 0.5 and n_1 with $n_1 + 0.5$. Exercise 32 will investigate how well the interval achieves its stated confidence level.

Example: Salk vaccine clinical trial (Salk.R)

The purpose of this example is to calculate the estimated relative risk and the confidence interval for the population relative risk in order to determine the effectiveness of the Salk vaccine in preventing polio. Below is the R code used to enter the data into an array and to perform the necessary calculations:

```
> c.table <- array(data = c(57, 142, 200688, 201087), dim =
    c(2,2), dimnames = list(Treatment = c("vaccine", "placebo"),
    Result = c("polio", "polio free")))
> c.table
         Result
Treatment polio polio free
  vaccine    57     200688
  placebo   142     201087
```

```
> pi.hat.table <- c.table/rowSums(c.table)
> pi.hat.table
          Result
Treatment         polio     polio free
  vaccine  0.0002839423     0.9997161
  placebo  0.0007056637     0.9992943

> pi.hat1 <- pi.hat.table[1,1]
> pi.hat2 <- pi.hat.table[2,1]

> round(pi.hat1/pi.hat2, 4)
[1] 0.4024
> round(1/(pi.hat1/pi.hat2), 4)    # inverted
[1] 2.4852

> alpha <- 0.05
> n1 <- sum(c.table[1,])
> n2 <- sum(c.table[2,])

> # Wald confidence interval
> var.log.rr <- (1-pi.hat1)/(n1*pi.hat1) +
    (1-pi.hat2)/(n2*pi.hat2)
> ci <- exp(log(pi.hat1/pi.hat2) + qnorm(p = c(alpha/2,
    1-alpha/2)) * sqrt(var.log.rr))
> round(ci, 4)
[1] 0.2959 0.5471
> rev(round(1/ci, 4))    # inverted
[1] 1.8278 3.3792
```

Defining index 1 to represent the vaccine group and 2 the placebo group, we find $\widehat{RR} = 0.40$. The estimated probability of contracting polio is only 0.4 times (or 40%) as large for the vaccine group than for placebo. Notice that we used the word *estimated* with this interpretation because we are using parameter estimates. The confidence interval is $0.30 < RR < 0.55$. Therefore, with 95% confidence, we can say that the vaccine reduces the *population* risk of polio by 45-70%. Also, notice the exponential function is calculated using exp() in R (for example, exp(1) is 2.718).

The ratio for the relative risk is often arranged so that $\widehat{RR} \geq 1$. This allows for an appealing interpretation that the group represented in the numerator has a risk that is, for example, "11 times as large" as the denominator group. This can be easier for a target audience to appreciate than the alternative, "0.091 times as large". However, the application may dictate which group should be the numerator regardless of the sample proportions. This was the case for the Salk vaccine example, where it is natural to think of the vaccine in terms of its risk reduction.

Also, relative risk is generally a more useful measure than the difference between probabilities when the probabilities are fairly small. It is of limited use otherwise. For example, if $\pi_2 = 0.8$, then the maximum possible relative risk is $1/0.8 = 1.25$. It is therefore useful to have an alternative statistic for comparing probabilities that is applicable regardless of the sizes of the probabilities. This is part of the motivation for the next measure.

1.2.5 Odds ratios

We have focused so far on using probabilities to measure the chance that an event will occur. *Odds* can be used as a similar measure. Odds are simply the probability of a success divided by the probability of a failure, $odds = \pi/(1-\pi)$. In some areas of application, such as betting, odds are used almost exclusively. For example, if a probability is $\pi = 0.1$, then the corresponding odds of success are $0.1/(1-0.1) = 1/9$. This will be referred to as "9-to-1 odds *against*," because the probability of failure is 9 times the probability of success. Notice that there is a 1-to-1 relationship between probability and odds: if you know one, you can find the other. Also notice that odds have no upper limit, unlike probabilities. Like relative risks, odds are estimated with MLEs by replacing the probabilities with their corresponding estimates.

When there are two groups, we can calculate the odds separately in each group: $odds_1 = \pi_1/(1-\pi_1)$ and $odds_2 = \pi_2/(1-\pi_2)$. Then a comparison of the odds is made using an *odds ratio*. Formally, it is defined as

$$OR = \frac{odds_1}{odds_2} = \frac{\pi_1/(1-\pi_1)}{\pi_2/(1-\pi_2)} = \frac{\pi_1(1-\pi_2)}{\pi_2(1-\pi_1)}.$$

The odds ratio can be estimated by substituting the parameters with their corresponding estimates to obtain the MLE:

$$\widehat{OR} = \frac{\widehat{odds_1}}{\widehat{odds_2}} = \frac{\hat{\pi}_1(1-\hat{\pi}_2)}{\hat{\pi}_2(1-\hat{\pi}_1)} = \frac{(w_1/n_1)[(n_2-w_2)/n_2]}{(w_2/n_2)[(n_1-w_1)/n_1]} = \frac{w_1(n_2-w_2)}{w_2(n_1-w_1)}.$$

When counts are written in the form of a contingency table, the estimated odds ratio is a product of the counts on the "diagonal" (top-left to bottom-right) of the table divided by a product of the counts off of the diagonal.

Interpretation

Determining whether or not an odds ratio is equal to 1, greater than 1, or less than 1 is often of interest. An odds ratio equal to 1 means that the odds in group 1 are the same as the odds in group 2. Thus, we can say that the odds are not dependent on the group; i.e., the odds of a success are *independent* of the group designation. An odds ratio greater than 1 means that the odds of a success are higher for group 1 than for group 2. The opposite is true for an odds ratio less than 1. Note that $OR = 1$ implies that $RR = 1$ and $\pi_1 = \pi_2$ and vice versa; see Exercise 28. Because of this equivalence, the hypothesis test procedures of Section 1.2.3 can equivalently be used to test $H_0 : OR = 1$ vs. $H_a : OR \neq 1$.

Because odds ratios are combinations of four probabilities, they are easy to misinterpret. The standard interpretation of an estimated odds ratio is as follows:

> The estimated odds of a success are $\widehat{OR}$ times as large as in group 1 than in group 2.

Obviously, "success," "group 1," and "group 2" would be replaced with meaningful terms in the context of the problem. As with relative risks, changing the order of the ratio, so that we have $\widehat{odds_2}/\widehat{odds_1}$, can sometimes make the interpretation easier. We refer to this as *inverting* the odds ratio. With this inverted odds ratio, we can now say that

> The estimated odds of a success are $1/\widehat{OR}$ times as large as in group 2 than in group 1.

The ratio of the *odds of a failure* for the two groups may be of interest too. Thus, $(1-\hat{\pi}_1)/\hat{\pi}_1$ is the odds of a failure for group 1, and $(1-\hat{\pi}_2)/\hat{\pi}_2$ is the odds of a failure for group 2. The ratio of group 1 to group 2 is

$$\frac{(1-\hat{\pi}_1)/\hat{\pi}_1}{(1-\hat{\pi}_2)/\hat{\pi}_2} = \frac{\hat{\pi}_2(1-\hat{\pi}_1)}{\hat{\pi}_1(1-\hat{\pi}_2)},$$

which is simply $1/\widehat{OR}$ from earlier. The interpretation then becomes

> The estimated odds of a failure are $1/\widehat{OR}$ times as large as in group 1 than in group 2.

Finally, we can also invert this ratio of two odds of a failure to obtain the interpretation of

> The estimated odds of a failure are $\widehat{OR}$ times as large as in group 2 than in group 1,

where the odds ratio is numerically the same as we had in the beginning.

This "symmetry" in the interpretation of odds ratios can be confusing at times, but it has a very distinct advantage. In *case-control studies*, measurements are taken retrospectively on fixed numbers of "cases" and "controls," which are equivalent to "successes" and "failures," respectively. That is, the proportion of cases and controls in the sample (e.g., patients who did or did not have a stroke) is chosen by the researcher and does not typically match the proportion of cases in the population. Whether or not they fall into group 1 or 2 (e.g., they did or did not take a particular medication in the past year) is the measurement of interest. In this context, the usual definitions of "group" and "response" are reversed: we want to know how taking medication might relate to the probability that someone had a stroke, but we measure how a having a stroke is related to the probability that they took the medication. If one computes an odds ratio in this instance, then it is easy to see that if the odds of medication use are x times as high for those who had a stroke than for those who did not have a stroke, then the odds of having a stroke are also x times as high for medication users as for non-medication users! Thus, we can get an answer to the question of interest without actually measuring the odds of having a stroke for either medication group. Exercise 29 explores this example further.

In fact, given only the odds ratio, we cannot estimate any of its constituent probabilities directly. It is incorrect to interpret an odds ratio as relating directly to the *probability* of success. We cannot say that "The estimated *probability* of success is $\widehat{OR}$ times as large as in group 1 than in group 2." This is one substantial drawback to using odds ratios instead of relative risks, because most people are more comfortable discussing probabilities than odds. However, it is easy to see that $OR \approx RR$ if the two groups' probabilities are both fairly low. In this case, both probabilities of failure are nearly 1 and cancel out in the odds ratio. For this reason, odds ratios can be used as a surrogate to relative risks when the successes being counted are relatively uncommon (e.g., both $\pi_j \leq 0.1$).

Confidence interval

Just as with relative risks, the probability distribution of $\log(\widehat{OR})$ is better approximated by a normal distribution than is the probability distribution of $\widehat{OR}$ itself. The variance of $\log(\widehat{OR})$ is again found using the delta method and has a particularly easy form:

$$\widehat{Var}(\log(\widehat{OR})) = \frac{1}{w_1} + \frac{1}{n_1 - w_1} + \frac{1}{w_2} + \frac{1}{n_2 - w_2}.$$

The $(1-\alpha)100\%$ Wald confidence interval for OR becomes

$$\exp\left[\log\left(\widehat{OR}\right) \pm Z_{1-\alpha/2}\sqrt{\frac{1}{w_1} + \frac{1}{n_1 - w_1} + \frac{1}{w_2} + \frac{1}{n_2 - w_2}}\right]. \quad (1.9)$$

Lui and Lin (2003) show the interval to be conservative with respect to its true confidence level, where the degree of its conservativeness is dependent on π_1, π_2, and n_+. The true confidence level is a little above the stated level most of the time. However, the interval is very conservative for small and large π_1 and π_2 and small n_+. These are cases that are likely to produce very small cell counts in the corresponding observed contingency table, and this leads to some instability in the estimates of OR and its corresponding variance.

Zero counts anywhere in the contingency table cause the estimated odds ratio to be 0 or undefined and the corresponding variance to be undefined. Small adjustments to the counts are often made in these situations. For example, we can use

$$\widetilde{OR} = \frac{(w_1 + 0.5)(n_2 - w_2 + 0.5)}{(w_2 + 0.5)(n_1 - w_1 + 0.5)}$$

as an estimate of the odds ratio. The addition of a 0.5 to each cell count leads the estimated odds ratio to be closer to 1 than without the adjustment. The estimated variance for $\log(\widehat{OR})$ can also be changed to have 0.5 added to each count within it. An adjusted Wald confidence interval can then be formed with $\widetilde{OR}$ and this new variance by making the appropriate substitutions in Equation 1.9. Note that some people advocate making this adjustment at all times even when there are no 0 counts. The main advantage of this approach is the confidence interval is shorter in length (due to the "additional" observations) than without this adjustment. Lui and Lin (2003) show this adjusted Wald interval to achieve a true confidence level very similar to the interval without the adjustment.

Example: Salk vaccine clinical trial (Salk.R)

The purpose of this example is to estimate the odds ratio describing the relationship between vaccine and polio. We have coded the data so that w_1 is the count of children contracting polio in the vaccine group and w_2 is the same count for placebo. Continuing the code from earlier, we obtain the following:

```
> OR.hat <- c.table[1,1] * c.table[2,2] / (c.table[2,1] *
    c.table[1,2])
> round(OR.hat, 4)
[1] 0.4024
> round(1/OR.hat, 4)
[1] 2.4863

> alpha <- 0.05
> var.log.or <- 1/c.table[1,1] + 1/c.table[1,2] + 1/c.table[2,1]
    + 1/c.table[2,2]
> OR.CI <- exp(log(OR.hat) + qnorm(p = c(alpha/2, 1-alpha/2)) *
    sqrt(var.log.or))
> round(OR.CI, 2)
[1] 0.30 0.55
> rev(round(1/OR.CI, 2))
[1] 1.83 3.38
```

We find $\widehat{OR} = 0.40$, so the estimated odds of contracting polio are 0.4 times (or 40%) as large as when the vaccine is given than when a placebo is given. The 95% confidence interval is $0.30 < OR < 0.55$. Because 1 is not within the interval, there is sufficient evidence to indicate the vaccine does decrease the true odds (and hence the probability) of contracting polio. Notice that we did not include the word *estimated* in the last sentence because we interpreted OR rather than $\widehat{OR}$.

Alternatively, we could interpret the inverted odds ratios in terms of *protection* due to vaccination: The estimated odds of being polio free are 2.49 times as large as when the vaccine is given than when the placebo is given. The 95% confidence interval for the odds ratio is (1.83, 3.38). Notice that the `rev()` function reverses the order of the elements within the object in parentheses. This is done so that the lower value of the confidence interval appears first in the resulting vector.

These calculated values are essentially the same as when we explored the same example using the relative risk. This is because (a) the probability of polio contraction is very low in both groups (about $1/1000$), and (b) the sample size is very large here. In smaller samples, the estimates $\widehat{RR}$ and $\widehat{OR}$ would still have been quite similar, but the confidence intervals would have differed. The difference in the confidence intervals stems from the difference in the variances: in very large samples, both $1/(n_j - w_j)$ and $-1/n_j$ are negligible relative to $1/w_j$, $j = 1, 2$.

Example: Larry Bird's free throw shooting (Bird.R)

The purpose of this example is to calculate an odds ratio comparing the odds of a successful second free throw attempt when the first is made to when the first is missed. The code for these calculations is available in the corresponding program. The estimated odds ratio is $\widehat{OR} = 0.77$, and the 95% confidence interval for OR is (0.29, 2.07). Because the estimated odds ratio is less than 1, we decided to invert it to help with its interpretation. The estimated odds of a successful second free throw attempt are 1.30 times as large as when the first free throw is missed than when the first free throw is made. Also, with 95% confidence, the odds of a successful free throw attempt are between 0.48 and 3.49 times as large as when the first free throw is missed than when the first free throw is made. Because 1 is within the interval, there is insufficient evidence to indicate an actual difference in the odds of success.

1.2.6 Matched pairs data

Section 1.2.2 examines a confidence interval for the difference of two probabilities. These intervals were developed using the fact that W_1 and W_2 are independent random variables. There are other situations where the two probabilities being compared are dependent random variables. This occurs with *matched pairs* data, where two binary response observations, say X and Y, are made on each sampled unit, and a desired comparison between success probabilities for X and Y involves two correlated statistics. Below is an example to illustrate this problem.

Example: Prostate cancer diagnosis procedures

Zhou and Qin (2005) discuss a study that was used to compare the diagnostic accuracy of magnetic resonance imaging (MRI) and ultrasound in patients who had

Table 1.5: Diagnoses established using MRI and ultrasound technology among individuals who truly have localized prostate cancer. Data source is Zhou and Qin (2005).

		Ultrasound		
		Localized	Advanced	
MRI	Localized	4	6	10
	Advanced	3	3	6
		7	9	16

been established as having localized prostate cancer by a gold standard test. The data from one study location is given in Table 1.5. We are interested in comparing the probability of a localized diagnosis by an MRI to the probability of a localized diagnosis by an ultrasound. Notice that because both procedures are performed on the same patients, these probability estimates—10/16 and 7/16, respectively—are based on the same counts, and hence their corresponding statistics are not independent. This violates the assumption of independent binomial counts used in the model from Section 1.2.2.

To develop the appropriate analysis methods, note that any combination of two measurements can occur for each subject. That is, we may observe $(X = 1, Y = 1), (X = 1, Y = 2), (X = 2, Y = 1)$, or $(X = 2, Y = 2)$, where the values of X and Y correspond to row and column numbers, respectively. Each of these events has probability $\pi_{ij} = P(X = i, Y = j)$, $i = 1, 2$, $j = 1, 2$. These are called *joint probabilities*. As the previous example highlights, we are interested in summaries of these joint probabilities, rather than in the joint probabilities themselves. That is, we are interested in the *marginal probabilities*, $\pi_{1+} = \pi_{11} + \pi_{12}$ and $\pi_{+1} = \pi_{11} + \pi_{21}$, so called because they relate to counts in the margins of the table. With respect to the prostate cancer diagnosis example, π_{1+} is the probability of a localized diagnosis by MRI and π_{+1} is the probability of a localized diagnosis by ultrasound.

Suppose n units are cross-classified according to X and Y. Then we observe counts n_{11}, n_{12}, n_{21}, and n_{22}, with n_{ij} being the number of joint occurrences of $X = i, Y = j$. This is just like the binomial setting, except with four possible outcomes rather than two. The model that describes the counts produced in this manner is called the *multinomial distribution*. We cover this model in greater detail in Chapter 3. For now, it suffices to know that MLEs and variances follow the same pattern as in the binomial model: $\hat{\pi}_{ij} = n_{ij}/n$ with $\widehat{Var}(\hat{\pi}_{ij}) = \hat{\pi}_{ij}(1-\hat{\pi}_{ij})/n$, $\hat{\pi}_{1+} = (n_{11} + n_{12})/n$ with $\widehat{Var}(\hat{\pi}_{1+}) = \hat{\pi}_{1+}(1-\hat{\pi}_{1+})/n$, and $\hat{\pi}_{+1} = (n_{11} + n_{12})/n$ with $\widehat{Var}(\hat{\pi}_{+1}) = \hat{\pi}_{+1}(1-\hat{\pi}_{+1})/n$, where $n = n_{11} + n_{12} + n_{21} + n_{22}$.

Comparing marginal probabilities in matched pairs is usually done through the difference $\pi_{1+} - \pi_{+1}$, because this comparison is mathematically much easier to work with than other possible comparisons.[11] If $\pi_{+1} - \pi_{1+} = 0$, the two marginal probabilities are equal. Alternatively, if $\pi_{+1} - \pi_{1+} > 0$, this means that success (first response category) is more likely to happen for Y than for X. The opposite is true if $\pi_{+1} - \pi_{1+} < 0$.

---

[11]In some situations, especially with very small marginal probabilities, the ratio of the two probabilities may be of more interest. See Bonett and Price (2006) for how to form intervals for ratios of marginal probabilities.

Confidence interval for the difference of marginal probabilities

To develop a confidence interval for $\pi_{+1} - \pi_{1+}$, we start with the MLE $\hat{\pi}_{+1} - \hat{\pi}_{1+}$ and find its probability distribution. Once again we make use of the facts that

1. Each estimate, $\hat{\pi}_{1+}$ and $\hat{\pi}_{+1}$, has a distribution that is approximated in large samples by a normal distribution with variance as given above, and

2. Differences of normal random variables are normal random variables.

However, this problem differs from the one laid out in Section 1.2.2 because the two random variables $\hat{\pi}_{1+}$ and $\hat{\pi}_{+1}$ are dependent. Therefore we have that[12]

$$\widehat{Var}(\hat{\pi}_{+1} - \hat{\pi}_{1+}) = \widehat{Var}(\hat{\pi}_{+1}) + \widehat{Var}(\hat{\pi}_{1+}) - 2\widehat{Cov}(\hat{\pi}_{+1}, \hat{\pi}_{1+}).$$

Using the underlying multinomial distribution for the cell counts, the covariance can be shown to be $\widehat{Cov}(\hat{\pi}_{+1}, \hat{\pi}_{1+}) = n^{-1}(\hat{\pi}_{11}\hat{\pi}_{22} - \hat{\pi}_{12}\hat{\pi}_{21})$.[13] Using the variances for the marginal probabilities given earlier, we obtain

$$\widehat{Var}(\hat{\pi}_{+1} - \hat{\pi}_{1+}) = \frac{\hat{\pi}_{+1}(1 - \hat{\pi}_{+1}) + \hat{\pi}_{1+}(1 - \hat{\pi}_{1+}) - 2(\hat{\pi}_{11}\hat{\pi}_{22} - \hat{\pi}_{12}\hat{\pi}_{21})}{n}$$
$$= \frac{\hat{\pi}_{21} + \hat{\pi}_{12} + (\hat{\pi}_{21} - \hat{\pi}_{12})^2}{n}.$$

This leads to the $(1 - \alpha)100\%$ Wald confidence interval for $\pi_{+1} - \pi_{1+}$:

$$\hat{\pi}_{+1} - \hat{\pi}_{1+} \pm Z_{1-\alpha/2}\sqrt{\widehat{Var}(\hat{\pi}_{+1} - \hat{\pi}_{1+})}.$$

Not surprisingly, the Wald interval has problems achieving the stated confidence level. There have been a number of other intervals proposed, and these are summarized and evaluated in articles such as Newcombe (1998), Tango (1998), and Zhou and Qin (2005). We focus on the interval proposed by Agresti and Min (2005b) due to its simplicity and performance. Motivated by the Agresti-Caffo interval, Agresti and Min (2005b) propose to add 0.5 to each count in the contingency table and compute a Wald confidence interval using these adjusted counts. Specifically, let $\tilde{\pi}_{ij} = (n_{ij} + 0.5)/(n+2)$, so that $\tilde{\pi}_{+1} = (n_{11} + n_{21} + 1)/(n+2)$ and $\tilde{\pi}_{1+} = (n_{11} + n_{12} + 1)/(n+2)$. The $(1-\alpha)100\%$ Agresti-Min interval is

$$\tilde{\pi}_{+1} - \tilde{\pi}_{1+} \pm Z_{1-\alpha/2}\sqrt{\frac{\tilde{\pi}_{+1}(1 - \tilde{\pi}_{+1}) + \tilde{\pi}_{1+}(1 - \tilde{\pi}_{1+}) - 2(\tilde{\pi}_{11}\tilde{\pi}_{22} - \tilde{\pi}_{12}\tilde{\pi}_{21})}{n+2}}.$$

Agresti and Min (2005b) examined adding other constants to each cell, and they concluded that 0.5 was best in terms of the true confidence level and expected length.[14]

---

[12] This is another application of the following result: $Var(aU + bV) = a^2 Var(U) + b^2 Var(V) + 2ab Cov(U, V)$, where U and V are random variables and a and b are constants. See p. 171 of Casella and Berger (2002).

[13] To show this result, note that $Cov(\hat{\pi}_{+1}, \hat{\pi}_{1+}) = Var(\hat{\pi}_{11}) + Cov(\hat{\pi}_{11}, \hat{\pi}_{12}) + Cov(\hat{\pi}_{11}, \hat{\pi}_{21}) + Cov(\hat{\pi}_{12}, \hat{\pi}_{21})$, and the covariance of terms like $Cov(\hat{\pi}_{11}, \hat{\pi}_{12})$ is $-n^{-1}\pi_{11}\pi_{12}$. Please see Appendix B.2.2 or Section 3.1 for more on the multinomial distribution, if needed.

[14] See Section 3 of their paper for further justification with respect to a score interval and a Bayesian credible interval. The main disadvantage of a score interval here is that it does not have a closed-form expression and it sometimes is longer than the Agresti-Min interval. The scoreci.mp() function in the PropCIs package calculates the score interval. We provide an example of its use in the program corresponding to the prostate cancer diagnosis example.

Example: Prostate cancer diagnosis procedures (DiagnosisCancer.R)

The purpose of this example is to compute confidence intervals for $\pi_{+1} - \pi_{1+}$. We begin by entering the data into R using the `array()` function:

```
> c.table <- array(data = c(4, 3, 6, 3), dim = c(2,2), dimnames =
    list(MRI = c("Localized", "Advanced"), Ultrasound =
    c("Localized", "Advanced")))
> c.table
            Ultrasound
MRI          Localized Advanced
  Localized          4        6
  Advanced           3        3

> n <- sum(c.table)
> pi.hat.plus1 <- sum(c.table[,1])/n
> pi.hat.1plus <- sum(c.table[1,])/n
> data.frame(pi.hat.plus1, pi.hat.1plus, diff = pi.hat.plus1 -
    pi.hat.1plus)
  pi.hat.plus1 pi.hat.1plus     diff
1       0.4375        0.625  -0.1875
```

The estimate of $\pi_{+1} - \pi_{1+}$ is -0.1875. To find the confidence interval for $\pi_{+1} - \pi_{1+}$, we use the `diffpropci.Wald.mp()` and `diffpropci.mp()` functions from the `PropCIs` package:

```
> library(package = PropCIs)
> diffpropci.Wald.mp(b = c.table[1,2], c = c.table[2,1], n =
    sum(c.table), conf.level = 0.95)

data:

95 percent confidence interval:
 -0.5433238  0.1683238
sample estimates:
[1] -0.1875

> diffpropci.mp(b = c.table[1,2], c = c.table[2,1], n =
    sum(c.table), conf.level = 0.95)

data:

95 percent confidence interval:
 -0.5022786  0.1689453
sample estimates:
[1] -0.1666667
```

The b and c arguments in the functions correspond to n_{12} and n_{21}, respectively, which leads to $\hat{\pi}_{+1} - \hat{\pi}_{1+} = \hat{\pi}_{21} - \hat{\pi}_{12}$ being calculated. The ordering of the estimates in this expression is important for determining that the interval is for $\pi_{+1} - \pi_{1+}$ rather than $\pi_{1+} - \pi_{+1}$. The 95% Wald and Agresti-Min intervals are $-0.5433 < \pi_{+1} - \pi_{1+} < 0.1683$ and $-0.5023 < \pi_{+1} - \pi_{1+} < 0.1689$, respectively. We show in the corresponding program for this example how to calculate these intervals by coding the formulas directly. Overall, we see that the interval is quite wide leaving much uncertainty about the actual difference—especially, note that 0 is within the interval.

Test for the difference between marginal probabilities

Especially for the prostate cancer diagnosis example, knowing whether $\pi_{+1} - \pi_{1+}$ is different from 0 is a very important special case of the matched pairs data problem. To formally test $H_0 : \pi_{+1} - \pi_{1+} = 0$ vs. $H_a : \pi_{+1} - \pi_{1+} \neq 0$, we could use the Wald test statistic of

$$Z_0 = \frac{\hat{\pi}_{+1} - \hat{\pi}_{1+} - 0}{\sqrt{\widehat{Var}(\hat{\pi}_{+1} - \hat{\pi}_{1+})}} = \frac{\hat{\pi}_{21} - \hat{\pi}_{12}}{\sqrt{n^{-1}[\hat{\pi}_{21} + \hat{\pi}_{12} + (\hat{\pi}_{21} - \hat{\pi}_{12})^2]}}.$$

Instead, a different test statistic is often used that incorporates simplifications which occur under the null hypothesis. Because $\pi_{+1} - \pi_{1+} = 0$ implies that $\pi_{21} = \pi_{12}$, the variance in the denominator can be simplified to $n^{-1}(\hat{\pi}_{21} + \hat{\pi}_{12})$. If we square the resulting statistic, we obtain what is known as McNemar's test statistic (McNemar, 1947):

$$M = \frac{(\hat{\pi}_{21} - \hat{\pi}_{12})^2}{n^{-1}(\hat{\pi}_{21} + \hat{\pi}_{12})} = \frac{(n_{21} - n_{12})^2}{n_{21} + n_{12}}.$$

Under the null hypothesis, the statistic has an approximate χ_1^2 distribution for large samples. When $\pi_{21} \neq \pi_{12}$, the null hypothesis is false, and we would expect this to be reflected by large values of $(\hat{\pi}_{21} - \hat{\pi}_{12})^2$ relative to the denominator. Thus, the null hypothesis is rejected when M has an unusually large observed value with respect to the χ_1^2 distribution; in other words, reject H_0 if $M > \chi_{1,1-\alpha}^2$.

Example: Prostate cancer diagnosis procedures (DiagnosisCancer.R)

We formally perform a test of $H_0 : \pi_{+1} - \pi_{1+} = 0$ vs. $H_a : \pi_{+1} - \pi_{1+} \neq 0$ using the mcnemar.test() function from the stats package in R:

```
> mcnemar.test(x = c.table, correct = FALSE)

        McNemar's Chi-squared test

data:   c.table
McNemar's chi-squared = 1, df = 1, p-value = 0.3173
```

The x argument is for the contingency table. The correct = FALSE argument value prevents the numerator from being modified to $(|n_{21} - n_{12}| - 1)^2$, which is sometimes used to help the χ_1^2 distribution approximation in small samples. We examine better alternatives to these modifications in Section 6.2 with exact inference procedures. Due to the large p-value, there is insufficient evidence to indicate that the marginal probabilities are different.

1.2.7 Larger contingency tables

While Section 1.2 focuses on two groups with binary responses, there are many instances where more than two groups exist (i.e., there are more than two rows within a contingency table). For example, in addition to comparing a newly developed drug to a placebo in a clinical trial, prior approved drugs could be included as well. The usual goal in this case is to show that the new drug performs better than the placebo and performs similarly or better than any other prior approved drugs. Once more rows are added to a contingency table, it is often easier to perform the analysis with a binary regression model as covered in Chapter 2. Typically, the same types of questions as in Section 1.2 can be answered with these regression models. Plus, more complex scenarios can be considered, such as including additional variables in the analysis.

There may also be more than two response categories. Combined with more than two rows, a contingency table with I rows and J columns can be formed. Independent binomial sampling can no longer be used as the underlying probability distribution framework. Rather, we most often use multinomial or Poisson distributions. We postpone discussion of these types of models until Chapters 3 and 4.

1.3 Exercises

1. There are many simple experiments that can be performed to estimate π for a particular problem, and we encourage you to perform your own. The parts below discuss some experiments that we or our students have performed in the past. For each part, discuss conditions that are needed in order to satisfy the five assumptions for using the binomial distribution outlined in Section 1.1.1.

    (a) One of the first solid foods that infants can feed to themselves is the cereal Cheerios. Early on, infants often lack the dexterity in their fingers to pick up an individual Cheerio and put it into their mouth. In order to estimate the probability of success for one infant, an experiment was designed where a Cheerio would be set on an infant's food tray for 20 successive trials at the dinner table. If the Cheerio made it into the infant's mouth, the response for the trial was considered a success; otherwise, the response was considered a failure. Out of these 20 trials, there were 9 Cheerios that made it into the infant's mouth.

    (b) In most billiards games, the person who breaks a rack of balls gets another turn if at least one ball is sunk, excluding the cue ball. This is advantageous for this person because there are a fixed number of balls that need to be sunk in order to win a game. In order to estimate one person's probability of success of sinking a ball on the break, there were 25 consecutive breaks performed in 8-ball billiards. Out of these 25, 15 breaks had a ball sunk.

    (c) The germination rate is the proportion of seeds that will sprout after being planted into soil. In order to estimate the germination rate for a particular type of sweet corn, 64 seeds were planted in a $3' \times 4'$ plot of land with fertile soil. The seed packet stated that sprouts should emerge within 7-14 days, and all planting and care guidelines given on the seed packet were followed as closely as possible. After waiting for three weeks, 48 seeds had sprouted out of the soil.

    (d) In order to estimate the proportion of residents in a town who drive alternative-fuel cars (e.g., electric, hybrid, or natural gas), a person records the number of cars passing through a certain intersection for a one-half hour time period. A total of 125 cars passed through the intersection at this time, where 14 were alternative-fuel vehicles.

2. Continuing Exercise 1, find the Wald, Agresti-Coull, Wilson, and Clopper-Pearson intervals using the given data. Interpret the intervals in the context of the experiments.

3. Richardson and Haller (2002) describe a classroom experiment where Milk Chocolate Hershey's Kisses are put into a cup and then poured onto a table. The purpose is to estimate the proportion of Kisses that land on their base. An example in the paper gives 39 out of 100 Kisses landing on their base (10 cups filled with 10 Kisses were used). Using this information, answer the following:

   (a) Describe the assumptions that would be needed in order to use one binomial distribution with the 100 Kisses.

   (b) Presuming the assumptions in part (a) are satisfied, compute the Wald, Agresti-Coull, Wilson, and Clopper-Pearson intervals to estimate the probability that a Kiss will land on its base. Interpret the intervals.

   (c) Which confidence interval method do you think is best for this type of setting? Why?

   (d) The experiment was repeated with the Milk Chocolate Kisses and also separately for Milk Chocolate with Almonds Kisses. In this experiment, 43 out of 100 Milk Chocolate Kisses landed on their base and 33 out of 100 Milk Chocolate with Almonds Kisses landed on their base. Compute the Wald and Agresti-Caffo intervals for the difference of two probabilities with these data. Is there sufficient evidence that a difference exists?

   (e) Again, which confidence interval method do you think is best? Why?

4. Rours et al. (2005) collected urine specimens from 750 asymptomatic pregnant women in Rotterdam, Netherlands, to estimate the prevalence of chlamydia among the corresponding population. Of the 750 specimens, 48 tested positive for the disease. Using this information, answer the following:

   (a) Describe the assumptions that would be needed in order to use one binomial distribution for the sample.

   (b) Presuming the assumptions in part (a) are satisfied, find a confidence interval to estimate the prevalence of chlamydia. Use the most appropriate confidence interval procedure for this problem, and interpret the results.

5. Continuing Exercise 4, in addition to estimating the overall prevalence of chlamydia, the authors evaluated the accuracy of two different diagnostic procedures: an automated COBAS AMPLICOR PCR system and the MagNA DNA Pure LC DNA Isolation kit. Table 1.6 summarizes their diagnoses in a contingency table. Use the Agresti-Min confidence interval procedure to estimate the difference in probabilities of positive test results. Also, use McNemar's test to perform a test of equality for these two probabilities. Interpret the results.[15]

6. Gen-Probe's Aptima Combo 2 Assay is used to simultaneously detect chlamydia and gonorrhea in individuals. Pages 19 and 27 of their product insert sheet (available at http://www.gen-probe.com/inserts) provide information on the accuracy of the

---

[15] Note that the authors combined the $27 + 17 + 4 = 48$ positive test results in order to estimate the prevalence. They make the assumption then that there were no false negative tests given jointly by the two diagnostic procedures.

Table 1.6: Chlamydia diagnoses from two testing procedures. Data source is Rours et al. (2005).

		MagNA		
		Positive	Negative	Total
COBAS	Positive	27	4	31
AMPLICOR	Negative	17	702	719
	Total	44	706	750

diagnostic test. We reproduce part of this information in the aptima.csv file available on the textbook's website. Each row of our data set contains information on the testing accuracy for specific combinations of disease (chlamydia, gonorrhea), gender (male, female), specimen type (swab, urine), and symptom status (symptomatic, asymptomatic). The first row of data along with some additional calculations are given below:

```
> aptima <- read.table(file = "C:\\data\\Aptima_combo.csv",
    header = TRUE, sep = ",")
> c.table <- array(data = c(aptima[1,6], aptima[1,7],
    aptima[1,9], aptima[1,8]), dim = c(2,2), dimnames =
    list(True = c("+", "-"), Assay = c("+", "-")))
> c.table
     Assay
True  +   -
  +  190   7
  -  15  464

> Se.hat <- c.table[1,1]/sum(c.table[1,]) # Sensitivity
> Sp.hat <- c.table[2,2]/sum(c.table[2,]) # Specificity
> data.frame(Se.hat, Sp.hat)
    Se.hat    Sp.hat
1 0.964467 0.9686848
```

Listed in the data set's first row, there were 676 symptomatic males that provided swab specimens for chlamydia testing. Of these 676 individuals, 190 tested positive and were judged truly positive. Note that the results from a combination of other testing procedures were used to make the "true" diagnosis. Similarly, 464 individuals were truly negative and tested negative. There were 22 errors with 15 positive tests for truly negative individuals and 7 negative tests for truly positive individuals.

The purpose of this problem is to calculate confidence intervals for *sensitivity* (S_e) and *specificity* (S_p) accuracy measures. The sensitivity is defined as the proportion of truly positive individuals that test positive, and the specificity is defined as the proportion of truly negative individuals that test negative. Ideally, it is desirable to have S_e and S_p as close to 1 as possible. Point estimates ($\hat{S}_e$ and $\hat{S}_p$) for these measures are given in the output. Using this background, complete the following:

(a) The product insert gives Clopper-Pearson intervals for all disease, gender, specimen, and symptom combinations. Suggest possible reasons why it gives only the Clopper-Pearson intervals rather than the other intervals examined in Section 1.1.2.

(b) Compute the Clopper-Pearson intervals for S_e and S_p among the symptomatic males that provided swab specimens for chlamydia testing. Interpret the intervals.

(c) Calculate the Clopper-Pearson intervals for the other disease/gender/specimen/symptom combinations and display them in an organized manner (e.g., they can be put into one data frame with the appropriate labeling). Note that the `for()` function can be used to simplify the calculation process.

For more on the sensitivity and specificity of diagnostic tests, please see Section 6.1.

7. In a study of insect physiology, eggs from a beneficial species of moth were put in boxes placed in chambers at different temperatures. There were 30 eggs placed into each box, and the number of eggs hatching after 30 days were counted. The first box at 10°C had 0 hatch, the first box at 15°C had 1 hatch, and the first box at 20°C had 25 hatch. The data is courtesy of Jim Nechols, Department of Entomology, Kansas State University.

   (a) Construct appropriate confidence intervals for the probability that an egg hatches at each temperature.

   (b) Assess informally whether the probabilities could be the same at each temperature. Explain your reasoning. Note that we will develop more formal ways of making these comparisons in Chapter 2.

8. Continuing Exercise 7, the researchers actually used 10 different boxes of 30 eggs at each temperature. The counts of hatched eggs for the 10 boxes at 15°C were 1, 2, 4, 1, 0, 0, 0, 12, 0, and 2.

   (a) Why do you suppose that the researchers used more than one box at each temperature?

   (b) Construct appropriate confidence intervals for the probability that an egg hatches in each box.

   (c) Based on the intervals, does it appear that the probability that an egg hatches is the same in all boxes? Explain your reasoning.

   (d) All 10 boxes were held in the same chamber at the same time, and the chamber was set to 15°C. How do you suppose it could happen that they give such different counts? Hint: See part (c).

   (e) Do you think that it would be appropriate to consider the data as $w = 22$ successes coming from a binomial distribution with $n = 300$ trials? Why or why not?

9. Suppose that 79 out of 80 people contacted in a survey of city residents oppose a new tax. Let π represent the probability that a randomly selected resident opposes the tax.

   (a) Compute a 95% Wald confidence interval for π. Interpret the results.

   (b) Compute a 95% Wilson confidence interval for π. Interpret the results.

   (c) Is it possible that $\pi = 1$? Explain.

   (d) Which interval do you prefer? Explain.

10. The Wilson interval is found through equating

$$Z_0 = \frac{\hat{\pi} - \pi_0}{\sqrt{\pi_0(1-\pi_0)/n}},$$

to $Z_{1-\alpha/2}$ and solving for π_0. Show that the solutions obtained are the Wilson interval limits given in Section 1.1.2. Note that the quadratic formula will be helpful to complete this problem.

11. Describe one advantage and one disadvantage of the Clopper-Pearson interval in comparison to the Wald, Agresti-Coull, and Wilson intervals.

12. Figure 1.3 compares true confidence levels at $n = 40$ and $\alpha = 0.05$ when constructing an interval for π. Reconstruct the plot for other values of n and α. Ideas for choosing possible values include:

    (a) Use levels of n and α found in other exercises in this section. Discuss how your confidence level findings affect the particular type of confidence interval method that you chose for that exercise.

    (b) Examine what happens to the true confidence levels as the sample size increases from $n = 40$. Conversely, examine what happens as the sample size decreases from $n = 40$.

    (c) Examine what happens when α is not 0.05. Do some intervals achieve the corresponding stated level better?

    Discuss your results with respect to the recommendations given in Brown et al. (2001).

13. There are many other proposed confidence intervals for π. One of these intervals mentioned in Section 1.1.2 was the LR interval. Using this interval, complete the following:

    (a) Verify the 95% LR confidence interval is $0.1456 < \pi < 0.7000$ when $n = 10$ and $w = 4$. Note that `binom.confint()` calculates this interval using the `methods = "lrt"` argument value. We also provide additional code to calculate the interval in CIpi.R.

    (b) Construct a plot of the true confidence levels similar to those in Figure 1.3. Use $n = 40$ and $\alpha = 0.05$, and vary π from 0.001 to 0.999 by 0.0005.

    (c) Compare the LR interval's true confidence level to those of the four other intervals discussed in Section 1.1.2. Which of these intervals is best? Explain.

14. Another interval for π is the logit confidence interval. This method finds a Wald confidence interval for $\log(\pi/(1-\pi))$, which is known as the *logit* transformation, and then transforms the endpoints back into an interval for π. The resulting interval is

$$\frac{\exp\left[\log\left(\frac{\hat{\pi}}{1-\hat{\pi}}\right) \pm Z_{1-\alpha/2}\sqrt{\frac{1}{n\hat{\pi}(1-\hat{\pi})}}\right]}{1+\exp\left[\log\left(\frac{\hat{\pi}}{1-\hat{\pi}}\right) \pm Z_{1-\alpha/2}\sqrt{\frac{1}{n\hat{\pi}(1-\hat{\pi})}}\right]}$$

    (a) What is the numerical range for $\log(\pi/(1-\pi))$? After transforming the endpoints back using the $\exp(\cdot)/[1+\exp(\cdot)]$ transformation, what is the new numerical range? Why is this new numerical range ideal?

(b) Verify the estimated variance of $\log(\hat{\pi}/(1-\hat{\pi}))$ is $[n\hat{\pi}(1-\hat{\pi})]^{-1}$ by using the delta method (see Appendix B.4.2).

(c) Verify the 95% logit confidence interval is $0.1583 < \pi < 0.7026$ when $n = 10$ and $w = 4$. Note that `binom.confint()` calculates this interval using the `methods = "logit"` argument value.

(d) Discuss the problems that occur when $\hat{\pi} = 0$ or 1. Propose a solution to solve the problems. How does `binom.confint()` take care of the problem?

(e) Using your solution from (d) or the solution given by `binom.confint()`, construct a plot of the true confidence levels similar to those in Figure 1.3. Use $n = 40$ and $\alpha = 0.05$, and vary π from 0.001 to 0.999 by 0.0005.

(f) Compare the logit interval's true confidence level to those of the four other intervals discussed in Section 1.1.2. Which of these intervals is best? Explain.

15. In addition to the true confidence level of a confidence interval, the expected length of an interval is important too. For example, the length of a Wald confidence interval when $w = 1$ and $n = 40$ is $0.0734 - (-0.0234) = 0.0968$ (see example on p. 19). The expected length is defined as

$$\sum_{w=0}^{n} L(\hat{\pi}) \binom{n}{w} \pi^w (1-\pi)^{n-w}$$

where $L(\hat{\pi})$ denotes the interval length using a particular $\hat{\pi}$. Because each length is multiplied by the corresponding probability that an interval is observed, we obtain the expected value of the length by summing these terms. Complete the following problems on the expected length.

(a) We would like a confidence interval to be as short as possible with respect to its expected length while having the correct confidence level. Why?

(b) For $n = 40$, $\pi = 0.16$, and $\alpha = 0.05$, find the expected length for the Wald interval and verify it is 0.2215.

(c) For $n = 40$ and $\alpha = 0.05$, construct a similar plot to Figure 1.3, but now using the expected length for the y-axis on the plots. Compare the expected lengths among the intervals. To help with the comparisons, you will want to fix the y-axis on each plot to be the same (use the `ylim = c(0, 0.35)` argument value in each `plot()` function call). Alternatively to having four plots in one R Graphics window, overlay the expected lengths for all four intervals on one plot. This can be done by using one `plot()` function for the first interval's expected length. Three calls to the `lines()` function can then be used to overlay the remaining intervals' expected lengths.

(d) Using Figure 1.3 and the plot from (c), which interval is best? Explain.

(e) Outline the steps that would be needed to find the estimated expected length using Monte Carlo simulation.

16. Figure 1.3 shows the actual true confidence level for the four intervals discussed in Section 1.1.3. Construct this plot now using Monte Carlo simulation with 1,000 samples taken for each π. When calculating the estimated true confidence level for a particular π, make sure to use the same samples for each confidence interval. This will help remove one source of simulation variability.

Table 1.7: Icing the kicker data. Data source is Berry and Wood (2004).

		Field goal		
		Success	Failure	Total
Strategy	No time-out	22	4	26
	Time-out	10	6	16
	Total	32	10	42

Table 1.8: Ethiopia HIV and condom use data. Data source is Aseffa et al. (1998).

		HIV		
		Positive	Negative	Total
Condom	Never	135	434	569
	Ever	15	9	24
	Total	150	443	539

17. Before a placekicker attempts a field goal in a pressure situation, an opposing team may call a time-out. The purpose of this time-out is to give the kicker more time to think about the attempt in the hopes that this extra time will cause him to become more nervous and lower his probability of success. This strategy is often referred to as "icing the kicker." Berry and Wood (2004) collected data from the 2002 and 2003 National Football League seasons to investigate whether or not the strategy actually lowers the probability of success when implemented. Table 1.7 contains the results from the 31-40 yard field goals during these seasons under pressure situations (attempt is in the last 3 minutes of a game and a successful field goal causes a lead change). Complete the following:

    (a) Calculate the Wald and Agresti-Caffo confidence intervals for the difference in probabilities of success conditioning on the strategy. Interpret the intervals.

    (b) Perform a score test, Pearson chi-square test, and LRT to test for the equality of the success probabilities.

    (c) Estimate the relative risk and calculate the corresponding confidence interval for it. Interpret the results.

    (d) Estimate the odds ratio and calculate the corresponding confidence interval for it. Interpret the results.

    (e) Is there sufficient evidence to conclude that icing the kicker is a good strategy to follow? Explain.

18. Aseffa et al. (1998) examine the prevalence of HIV among women visiting health care clinics in northwest Ethiopia. Along with testing individuals for HIV, additional information was collected on each woman such as condom use. Table 1.8 gives a contingency table summarizing the data.

    (a) Calculate the Wald and Agresti-Caffo confidence intervals for the difference in probabilities of being HIV positive based on condom use. Interpret the intervals.

    (b) Perform a score test, Pearson chi-square test, and LRT to test for the equality of the success probabilities.

    (c) Estimate the odds ratio and calculate the corresponding confidence interval for it. Interpret the estimate and the interval.

Table 1.9: HIV vaccine results from the modified intent-to-treat data. Data source is Rerks-Ngarm et al. (2009).

		Response		
		HIV	No HIV	Total
Treatment	Vaccine	51	8,146	8,197
	Placebo	74	8,124	8,198
	Total	125	16,270	16,395

(d) Generally, it is thought that condom use helps to prevent HIV transmission. Do the results here agree with this? If not, what factors may have led to these results? Note that Aseffa et al. (1998) give the estimated odds ratio and a corresponding Wald interval, but fail to interpret them.

(e) Table 3 of Aseffa et al. (1998) gives a large number of 2 × 2 contingency tables each examining the relationship between a current disease status (HIV, syphilis, and chlamydia) and explanatory variables (literacy, married, age of first sexual intercourse, past sexually transmitted disease history, condom use, and genital ulcer presence). For each table, the estimated odds ratio and corresponding Wald confidence interval is calculated. If you have access to the paper, replicate some of their calculations and develop conclusions about the associations between the variables.

19. On September 24, 2009, news reports hailed the findings from an HIV vaccine clinical trial as being the first time that a vaccine worked. These news reports often made front-page headlines in newspapers and lead stories on television newscasts (examples are available on the textbook's website). The vaccine actually was a combination of two previous tested vaccines, ALVAC-HIV and AIDSVAX B/E, that were not shown to prevent HIV infection in previous clinical trials by themselves. The purpose of this problem is for you to analyze the research data in order to draw your own conclusions. Table 1.9 summarizes the "modified intent-to-treat" data of Rerks-Ngarm et al. (2009) that was released to the media on September 24. Using the modified intent-to-treat data, complete the following:

(a) If you have access to Rerks-Ngarm et al. (2009), describe the population and sample used in the study. What is the population ultimately that the researchers would like to extend their conclusions to?

(b) Examine the benefits of using the vaccine over the placebo with the most appropriate measure(s) from Section 1.2. Use $\alpha = 0.05$.

(c) Suppose you were a reporter assigned to write a news report about the statistical results on September 24, 2009. Write a short news report that would be appropriate for your city's newspaper. Remember that most readers will not be familiar with the previously calculated statistical measures, so be sure to explain what they measure when writing the article.

20. Continuing Exercise 19, the Rerks-Ngarm et al. (2009) paper was published on October 20, 2009, on the *New England Journal of Medicine*'s website. This paper contained two other versions of the data named the "intent-to-treat" data and the "per-protocol" data, and these data sets are shown in Tables 1.10 and 1.11. The intent-to-treat data contains seven additional individuals that were not in the modified intent-to-treat data. These additional individuals were enrolled in the clinical trial and treated, but

Table 1.10: HIV vaccine results from the intent-to-treat data. Data source is Rerks-Ngarm et al. (2009).

		Response		
		HIV	No HIV	Total
Treatment	Vaccine	56	8,146	8,202
	Placebo	76	8,124	8,200
	Total	132	16,270	16,402

Table 1.11: HIV vaccine results from the per-protocol data. Data source is Rerks-Ngarm et al. (2009).

		Response		
		HIV	No HIV	Total
Treatment	Vaccine	36	6,140	6,176
	Placebo	50	6,316	6,366
	Total	86	12,456	12,542

later were found to have been HIV positive before the trial had started. The "per-protocol" data contains only those individuals that received all four treatments of ALVAC-HIV and both treatments of AIDSVAX B/E as specified in the treatment protocol for the clinical trial. Using each of these additional data sets, repeat (a) and (b).

21. Continuing Exercises 19 and 20, there were a number of news reports again regarding the HIV vaccine clinical trial when the Rerks-Ngarm et al. (2009) paper was published. Maugh (2009) said the following in a *Los Angeles Times* article:

    > A secondary analysis of data from the Thai AIDS vaccine trial—announced last month to much acclaim—suggests that the vaccine might provide some protection against the virus, but that the results are not statistically significant. In short, they could have come about merely by chance.

    Interestingly, news reports like this one were rarely publicized as much as the initial release of the modified intent-to-treat data results. Using this background, complete the following:

    (a) What might have been the reaction by the media if reports on all three data sets had been released at the same time?

    (b) What if $\alpha = 0.10$ or $\alpha = 0.01$ were used instead of $\alpha = 0.05$ for the computations in Exercises 19 and 20? Would the conclusions from the clinical trial change? Discuss.

    (c) Why do you think the intent-to-treat and per-protocol data results were less publicized? Should it have been more or less publicized given the reaction to the modified intent-to-treat analysis? Discuss.

    (d) Suppose again that you were a reporter assigned to write a news report about the statistical results. Write a short news report that would be appropriate for your city's newspaper that takes into account all of the available information. Remember that most readers will not be familiar with the previously calculated statistical measures, so be sure to explain what they measure when writing the article.

22. Do students whose native language is not English pick up on authors' efforts to use humor in illustrating points in textbooks? Students at one university were shown a passage from a textbook that was meant to be humorous. Of the 211 students in the study where English was not their native language, 118 thought the passage was humorous. Of the 206 students in the study where English was their native language, 155 thought the passage was humorous. Analyze these data to answer the research question. Note that this data set is from only one passage of many others used to examine humor and language for the study. The data set is courtesy of Melody Geddert, English Language Studies, Kwantlen Polytechnic University.

23. Matched pairs data often occurs when the same binary response is observed at two different time points, like what happens in the Larry Bird data. We used this data previously to compare the success probabilities for the second free throw given the outcome of the first free throw attempt. The purpose here instead is to compare the probability of success on the second attempt to the probability of success on the first attempt. This is of interest to determine on which free throw attempt Larry Bird is better. For example, it is reasonable to hypothesize a "warming up" effect that may lead to a larger probability of success on the second attempt.

    (a) Explain what $\pi_{+1} - \pi_{1+}$ means in the context of this problem.
    (b) Calculate the Wald and Agresti-Min intervals for $\pi_{+1} - \pi_{1+}$.
    (c) Perform a hypothesis test of $H_0 : \pi_{+1} - \pi_{1+} = 0$ vs. $H_a : \pi_{+1} - \pi_{1+} \neq 0$ using McNemar's test.
    (d) Interpret the intervals and the hypothesis test results in the context of the data.
    (e) Why would a hypothesis test of $H_0 : \pi_{+1} - \pi_{1+} \leq 0$ vs. $H_a : \pi_{+1} - \pi_{1+} > 0$ more closely correspond to the warming up effect hypothesis? Perform this test. Note that you will want to use $(n_{21} - n_{12})/\sqrt{n_{21} + n_{12}}$ as the test statistic and $Z_{1-\alpha/2}$ as the critical value.

24. Similar to the Wilson interval for π, an interval for $\pi_1 - \pi_2$ can be found using a test statistic from a hypothesis test for the parameters. Consider the test $H_0 : \pi_1 - \pi_2 = d$ vs. $H_a : \pi_1 - \pi_2 \neq d$ where d denotes the hypothesized value of the $\pi_1 - \pi_2$ difference. The interval is the set of all possible values of d such that

$$\frac{|\hat{\pi}_1 - \hat{\pi}_2 - d|}{\sqrt{\hat{\pi}_1^{(0)}(1 - \hat{\pi}_1^{(0)})/n_1 + \hat{\pi}_2^{(0)}(1 - \hat{\pi}_2^{(0)})/n_2}} < Z_{1-\alpha/2}$$

is satisfied. Note that $\hat{\pi}_1^{(0)}$ and $\hat{\pi}_2^{(0)}$ denote the MLEs of π_1 and π_2 under the constraint that $\pi_1 - \pi_2 = d$. This interval is called a *score* interval for the difference of two probabilities. Unfortunately, there is no closed form solution and iterative numerical procedures must be used to find the confidence interval limits. The `diffscoreci()` function from the `PropCIs` package performs these calculations. Complete the following using this package:

    (a) Calculate this score interval for the Larry Bird data. Compare the interval with the Wald and Agresti-Caffo intervals for it.
    (b) Score intervals can be calculated for other measures examined in this chapter. The following measures have functions in the `PropCIs` package that calculate these intervals: relative risk, `riskscoreci()`; odds ratio, `orscoreci()`; and difference of two probabilities in a matched pairs setting, `scoreci.mp()`. Calculate

these intervals for the Larry Bird data (see Exercise 23 for more on matched pairs data and the Larry Bird data).

(c) Examine how well these score intervals achieve the stated confidence level using similar methods as in Section 1.2.2.

25. With regard to the Pearson chi-square test statistic in Section 1.2.3, complete the following:

    (a) Show that

    $$\sum_{j=1}^{2} \frac{(w_j - n_j\bar{\pi})^2}{n_j\bar{\pi}} + \frac{(n_j - w_j - n_j(1 - \bar{\pi}))^2}{n_j(1 - \bar{\pi})} = \sum_{j=1}^{2} \frac{(w_j - n_j\bar{\pi})^2}{n_j\bar{\pi}(1 - \bar{\pi})},$$

    (b) Show that $X^2 = Z_0^2$.

26. What is the numerical range for the relative risk? What does it mean for the relative risk to be equal to 1?

27. What is the numerical range for an odds of a success? What does it mean for an odds to be equal to 1?

28. Show that $OR = 1$ when $\pi_1 = \pi_2$.

29. Section 1.2.5 discusses the symmetry properties of odds ratios and how these properties are useful for case-control studies. For the example given on p. 41 with respect to stroke and medication status, answer the following questions. Use probabilities expressed as words, such as "P(took medication and had no stroke)".

    (a) Write out the odds that a person took medication given that a person had a stroke. Similarly, write out the odds that a person took medication given that a person did not have a stroke.

    (b) With these odds expressions, write out the odds ratio comparing the odds of taking medication for the stroke group vs. the no stroke group. Show that it can be reorganized into the odds of stroke for the medication group vs. the no medication group.

30. Below are three results used to find McNemar's test statistic. Show that these results are true.

    (a) $\pi_{+1} - \pi_{1+} = \pi_{21} - \pi_{12}$.
    (b) $\pi_{+1}(1 - \pi_{+1}) + \pi_{1+}(1 - \pi_{1+}) - 2(\pi_{11}\pi_{22} - \pi_{12}\pi_{21}) = \pi_{21} + \pi_{12} + (\pi_{21} - \pi_{12})^2$
    (c) $\pi_{21} + \pi_{12} + (\pi_{21} - \pi_{12})^2 = \pi_{12} + \pi_{21}$ under the null hypothesis of $\pi_{+1} - \pi_{1+} = 0$

31. Figures 1.4 and 1.5 in Section 1.2.2 show the true confidence levels for the Wald and Agresti-Caffo intervals for the difference of two probabilities. Reconstruct the plots for other values of n_1, n_2, π_1, π_2, and α. Ideas for choosing possible values include:

    (a) Use levels of n_1 and n_2 found in other exercises in this section. Discuss how your confidence level findings affect the particular type of confidence interval method that you chose for that exercise.

(b) Examine what happens to the true confidence levels as the sample size increases from $n_1 = n_2 = 10$. Conversely, examine what happens as the sample size decreases from those same values.

(c) Examine what happens when α is not 0.05. Does either interval achieve the corresponding stated level better?

Discuss your results with respect to the recommendations given in Section 1.2.2.

32. The purpose of this problem is to examine the true confidence level for the Wald confidence interval for the relative risk. We recommend using ConfLevelTwoProb.R as a basis for your R program. Note that w_1 and/or w_2 counts that are equal to 0 require special attention when calculating the interval.

   (a) Find the true confidence level when $\pi_1 = 0.2$, $\pi_2 = 0.4$, $n_1 = 10$, $n_2 = 10$, and $\alpha = 0.05$.

   (b) Construct a plot of the true confidence levels similar to those in Figure 1.4. Use the same settings as in part (a), but allow π_1 to vary from 0.001 to 0.999 by 0.0005.

   (c) Construct a three-dimensional plot of the true confidence levels similar to those in Figure 1.5. Use the same settings as in part (a), but allow both π_1 and π_2 to vary from 0.001 to 0.999 by 0.005.

   (d) Discuss how well the confidence interval for the relative risk maintains the stated confidence level for appropriate values of π_1 and π_2 (remember that the usefulness of the relative risk decreases as π_1 and π_2 become large).

   (e) Repeat this problem using values of n_1 and n_2 that correspond to other exercises where the relative risk was used. Discuss the certainty that you have in your conclusions for those other exercises after seeing the results here.

33. Repeat Exercise 32, but now for the odds ratio and its corresponding Wald interval. Note that any counts within the contingency table structure which are equal to 0 require special attention when calculating the interval.

Chapter 2

Analyzing a binary response, part 2: regression models

Section 1.1 discusses how to estimate and make inferences about a single probability of success π. Section 1.2 generalizes this discussion to the situation of two probabilities of success that are now dependent on a level of a group. Chapter 2 completes the generalization to a situation where there are many different possible probabilities of success to estimate and perform inferences upon. Furthermore, we quantify how an explanatory variable with many possible levels (perhaps continuous rather than categorical) affects the probability of success. These generalizations are made through the use of binary regression models (also called binomial regression models).

2.1 Linear regression models

Before we introduce binary regression models, we need to review normal linear regression models. Let Y_i be the response variable for observations $i = 1, \ldots, n$. Also, consider the situation with p explanatory variables $x_{i1}, \ldots, x_{ip}$ that are each measured on the ith observation. We relate the explanatory variables to the response variable through a linear model:

$$Y_i = \beta_0 + \beta_1 x_{i1} + \cdots + \beta_p x_{ip} + \epsilon_i,$$

where ϵ_i for $i = 1, \ldots, n$ are independent and each has a normal probability distribution with mean 0 and variance σ^2. This leads to each Y_i for $i = 1, \ldots, n$ being independent with normal distributions. The $\beta_0, \ldots, \beta_p$ are *regression parameters* that quantify the relationships between the explanatory variables and Y_i. For example, if $\beta_1 = 0$, this says a linear relationship does not exist between the first explanatory variable and the response variable given the other variables in the model. Alternatively, if $\beta_1 > 0$, a positive relationship exists (an increase in the explanatory variable leads to an increase in the response variable), and if $\beta_1 < 0$, a negative relationship exists (an increase in the explanatory variable leads to a decrease in the response variable). Through taking the expectation of Y_i, we can also write out the model as

$$E(Y_i) = \beta_0 + \beta_1 x_{i1} + \cdots + \beta_p x_{ip},$$

where we assume the explanatory variables are conditioned upon or simply are constants. These model equations allow for different possible values of Y_i or $E(Y_i)$ to be functions of a set of explanatory variables. Thus, if we have $x_{i1}, \ldots, x_{ip}$ and somehow know all of the $\beta_0, \ldots, \beta_p$ parameter values, we could find what Y_i would be on average. Of course, we do not know $\beta_0, \ldots, \beta_p$ in actual applications, but we can estimate these parameters using least squares estimation or other more complicated procedures if desired.

In the context of binary responses, the quantity we want to estimate is the probability of success, π. Let Y_i be independent binary response variables for observations $i = 1, \ldots, n$, where a value of 1 denotes a success and a value of 0 denotes a failure. A normal distribution is obviously no longer an appropriate model for Y_i. Instead, similar to Section 1.1, a Bernoulli distribution describes Y_i very well but we now allow the probability of success parameter π_i to be different for each observation. Thus, $P(Y_i = y_i) = \pi_i^{y_i}(1-\pi_i)^{1-y_i}$ for $y_i = 0$ or 1 is the PMF for Y_i.

To find the MLEs for π_i, the likelihood function is

$$L(\pi_1,\ldots,\pi_n|y_1,\ldots,y_n) = P(Y_1 = y_1) \times \cdots \times P(Y_n = y_n)$$
$$= \prod_{i=1}^{n} P(Y_i = y_i)$$
$$= \prod_{i=1}^{n} \pi_i^{y_i}(1-\pi_i)^{1-y_i} \qquad (2.1)$$

Notice there are n different parameters for the n different observations. This means that the model is *saturated*: there are as many parameters as there are observations. The parameter estimates in this case turn out to be the same as the data ($\hat{\pi}_i = 0$ or 1), so there is nothing gained by working with this model.

Suppose instead that we have explanatory variables $x_{i1}, \ldots, x_{ip}$ as earlier, and we wish to relate the π_i's to these variables. We might propose a mathematical function, say $\pi_i = \pi(x_{i1}, \ldots, x_{ip})$, to describe the relationship. For example, in a simple setting, this function may designate two possible values of $\pi(x_{i1})$ depending on a lone binary value for x_{i1} (x_{i1} could denote the group number for observation i as we had in Section 1.2). We would then substitute $\pi(x_{i1}, \ldots, x_{ip})$ into Equation 2.1 to estimate its parameters. This would provide us with a model that can be used to estimate a probability of success as a function of any possible values for a set of explanatory variables. The next section shows that an appropriate choice of the function $\pi(x_{i1}, \ldots, x_{ip})$ leads to a full range of useful and convenient estimation and inference procedures.

2.2 Logistic regression models

The simplest type of function for $\pi(x_{i1}, \ldots, x_{ip})$ is the linear model $\pi_i = \beta_0 + \beta_1 x_{i1} + \cdots + \beta_p x_{ip}$. However, this model could lead to values of π_i less than 0 or greater than 1, depending on the values of the explanatory variables and regression parameters. Obviously, this is quite undesirable when estimating a probability. Fortunately, many non-linear expressions are available that force π_i to be between 0 and 1. The most commonly used expression is the *logistic regression model*:

$$\pi_i = \frac{\exp(\beta_0 + \beta_1 x_{i1} + \cdots + \beta_p x_{ip})}{1 + \exp(\beta_0 + \beta_1 x_{i1} + \cdots + \beta_p x_{ip})}. \qquad (2.2)$$

Notice that $\exp(\beta_0 + \beta_1 x_{i1} + \cdots + \beta_p x_{ip})$ is always positive and the numerator of Equation 2.2 is less than the denominator, which leads to $0 < \pi_i < 1$.

A logistic regression model can also be written as

$$\log\left(\frac{\pi_i}{1-\pi_i}\right) = \beta_0 + \beta_1 x_{i1} + \cdots + \beta_p x_{ip} \qquad (2.3)$$

through some algebraic manipulations of Equation 2.2. The left side of Equation 2.3 is the natural logarithm for the odds of a success, which we will be exploiting when interpreting the model in Section 2.2.3. This transformation of π_i is often referred to as the *logit transformation* (so named in analog to the "probit" model, discussed in Section 2.3), and it is simply denoted as logit(π_i) when writing out the left side of the model. The right side of Equation 2.3 is a linear combination of the regression parameters with the explanatory variables, and this is often referred to as the *linear predictor*. Equations 2.2 and 2.3 are equivalent statements of the logistic regression model.

We will often write the models without the i subscripts as

$$\pi = \frac{\exp(\beta_0 + \beta_1 x_1 + \cdots + \beta_p x_p)}{1 + \exp(\beta_0 + \beta_1 x_1 + \cdots + \beta_p x_p)} \text{ and } \text{logit}(\pi) = \beta_0 + \beta_1 x_1 + \cdots + \beta_p x_p$$

when we are not referring to particular observations. It will be clear based on the context when the probability of success is a function of explanatory variables (rather than how π was used in Section 1.1.1), even though the symbol π does not mention them explicitly.

Example: Plot of the logistic regression model (PiPlot.R)

The purpose of this example is to examine the shape of the logistic regression model when there is a single explanatory variable x_1. Consider the model $\pi = \exp(\beta_0 + \beta_1 x_1)/[1 + \exp(\beta_0 + \beta_1 x_1)]$, which is equivalently expressed as logit(π) = $\beta_0 + \beta_1 x_1$. Suppose that $\beta_0 = 1$ and $\beta_1 = 0.5$. Figure 2.1 shows this model plotted on the left. The plot on the right is the same, but with $\beta_1 = -0.5$. We can make the following generalizations from examining the model and these plots:

- $0 < \pi < 1$

- When $\beta_1 > 0$, there is a positive relationship between x_1 and π. When $\beta_1 < 0$, there is a negative relationship between x_1 and π.

- The shape of the curve is somewhat similar to the letter s (this shape is called "sigmoidal").

- The slope of the curve is dependent on the value of x_1. We can show this mathematically by taking the derivative of π with respect to x_1: $\partial \pi / \partial x_1 = \beta_1 \pi (1 - \pi)$.

- Above $\pi = 0.5$ is a mirror image of below $\pi = 0.5$.

We encourage readers to explore what happens when β_1 is increased or decreased by using the corresponding R code for this example. Specifically, with respect to the left plot, the curve becomes more steep in the middle as β_1 increases. Conversely, the curve becomes more flat in the middle as β_1 goes to toward 0.

Below is the R code used to create the plot on the left of Figure 2.1:

```
> par(mfrow = c(1,2))
> beta0 <- 1
> beta1 <- 0.5
> curve(expr = exp(beta0+beta1*x) / (1+exp(beta0+beta1*x)), xlim
    = c(-15, 15), col = "black", main = expression(pi ==
    frac(e^{1+0.5*x[1]}, 1+e^{1+0.5*x[1]})), xlab =
    expression(x[1]), ylab = expression(pi))
```

We use two new functions here. The `par()` function sets graphics parameters that control various plotting options. In our code, we use it to partition the graphics window into 1 row and 2 columns using the `mfrow` argument, which stands for "make

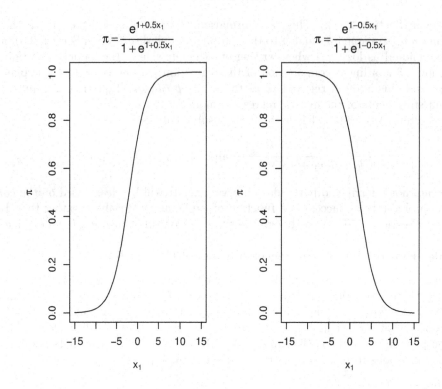

Figure 2.1: Logistic regression model for $\beta_0 = 1$ and $\beta_1 = 0.5$ and -0.5.

frame by row" (see Appendix A.7.3 for another example of its use). There are many other uses of the `par()` function, and we encourage readers to investigate these in the help for the function.

The second new function is the `curve()` function which is used to plot the model. This is a very useful function for plotting mathematical functions that vary over one variable. In our example, the `expr` argument contains the logistic regression model where the letter x must be used as the variable name for the variable plotted on the x-axis. By default, the mathematical function is evaluated at 101 equally spaced x-axis values within the range specified by `xlim`. These resulting 101 points are then joined by straight lines. Also within `curve()`, we use the `expression()` function with the title and axis labels in order to include Greek letters and fractions (see p. 491 in Appendix A for another example of `expression()`). We also include a simpler use of the `curve()` function in the corresponding program to this example.

2.2.1 Parameter estimation

Maximum likelihood estimation is used to estimate the regression parameters $\beta_0, \ldots, \beta_p$ of the logistic regression model. The initial estimation process is very similar to what is given in Appendix B.3.1 when there was a common value of π for all observations. Beginning from Equation 2.1, we substitute our model for π_i and take the natural logarithm to obtain the log-likelihood function:

$$\log\left[L(\beta_0, \ldots, \beta_p | y_1, \ldots, y_n)\right]$$

$$= \log\left(\prod_{i=1}^{n} \pi_i^{y_i}(1-\pi_i)^{1-y_i}\right) \qquad (2.4)$$

$$= \sum_{i=1}^{n} y_i \log(\pi_i) + (1-y_i)\log(1-\pi_i)$$

$$= \sum_{i=1}^{n} y_i \log\left(\frac{\exp(\beta_0 + \beta_1 x_{i1} + \cdots + \beta_p x_{ip})}{1 + \exp(\beta_0 + \beta_1 x_{i1} + \cdots + \beta_p x_{ip})}\right)$$

$$+ (1-y_i)\log\left(\frac{1}{1 + \exp(\beta_0 + \beta_1 x_{i1} + \cdots + \beta_p x_{ip})}\right)$$

$$= \sum_{i=1}^{n} y_i\left(\beta_0 + \beta_1 x_{i1} + \cdots + \beta_p x_{ip}\right) - \log\left(1 + \exp(\beta_0 + \beta_1 x_{i1} + \cdots + \beta_p x_{ip})\right). \qquad (2.5)$$

We take the derivatives of Equation 2.5 with respect to $\beta_0, \ldots, \beta_p$, set these equal to 0, and solve them simultaneously to obtain the parameter estimates $\hat{\beta}_0, \ldots, \hat{\beta}_p$. When the parameter estimates are included in the model, we can obtain the estimated probability of success as $\hat{\pi} = \exp(\hat{\beta}_0 + \hat{\beta}_1 x_1 + \cdots + \hat{\beta}_p x_p)/[1 + \exp(\hat{\beta}_0 + \hat{\beta}_1 x_1 + \cdots + \hat{\beta}_p x_p)]$.

Unfortunately, there are only a few simple cases where these parameter estimates have closed-form solutions; i.e., we cannot generally write out the parameter estimates in terms of the observed data like we could for the single probability estimate $\hat{\pi}$ in Section 1.1.2. Instead, we use iterative numerical procedures, as described in Appendix B.3.2, to successively find estimates of the regression parameters that increase the log-likelihood function. When the estimates change negligibly for successive iterations, this suggests that we have reached the peak of the log-likelihood function, and we say that they have *converged*. If the estimates continue to change noticeably up to a selected maximum number of iterations, the iterative numerical procedure has not converged, and those final parameter estimates should not be used. We will discuss convergence and non-convergence in more detail in Section 2.2.7.

Within R and most statistical software packages, *iteratively reweighted least squares* (IRLS) is the iterative numerical procedure used to find the parameter estimates. This procedure uses the weighted least squares criterion, which is commonly used for normal linear regression models when there is non-constant variance. The IRLS algorithm alternates between updating the weights and the parameter estimates in an iterative fashion until convergence is reached. The `glm()` function ("glm" stands for "generalized linear model") within R implements this parameter estimation procedure, and we show how to use this function in the next example. Later in this section, we show an alternative method to maximize the likelihood function using the `optim()` function.

Example: Placekicking (Placekick.R, Placekick.csv)

As discussed in Chapter 1, points can be scored in American football by a placekicker kicking a ball through a target area at an end of the field. A success occurs when the football is kicked over the crossbar and between the two uprights of the goal posts. The placekicker's team receives 1 or 3 points, where a point after touchdown (PAT) receives 1 point and a field goal receives 3 points. A placekick that is not successful receives 0 points.

Bilder and Loughin (1998) use logistic regression to estimate the probability of success for a placekick (PAT or field goal) in the National Football League (NFL). They examine a number of explanatory variables, including:

- `week`: Week of the season

- `distance`: Distance of the placekick in yards

- **change:** Binary variable denoting lead-change (1) vs. non-lead-change (0) placekicks; lead-changing placekicks are those that have the potential to change which team is winning the game (for example, if a field goal is attempted by a team that is losing by 3 points or less, they will no longer be losing if the kick is successful)

- **elap30:** Number of minutes remaining before the end of the half, with overtime placekicks receiving a value of 0

- **PAT:** Binary variable denoting the type of placekick, where a PAT attempt is a 1 and a field goal attempt is a 0

- **type:** Binary variable denoting outdoor (1) vs. dome (0) placekicks

- **field:** Binary variable denoting grass (1) vs. artificial turf (0) placekicks

- **wind:** Binary variable for placekicks attempted in windy conditions (1) vs. non-windy conditions (0); we define windy as a wind stronger than 15 miles per hour at kickoff in an outdoor stadium

- **good:** Binary variable denoting successful (1) vs. failed (0) placekicks; this is our response variable

There are 1,425 placekick observations collected during the 1995 NFL season that are available in the data set. Below is how the data are read into R (see Appendix A.7.1 for more information on reading data into R):

```
> placekick <- read.table(file = "C:\\data\\Placekick.csv",
    header = TRUE, sep = ",")
> head(placekick)
  week distance change  elap30 PAT type field wind good
1    1       21      1 24.7167   0    1     1    0    1
2    1       21      0 15.8500   0    1     1    0    1
3    1       20      0  0.4500   1    1     1    0    1
4    1       28      0 13.5500   0    1     1    0    1
5    1       20      0 21.8667   1    0     0    0    1
6    1       25      0 17.6833   0    0     0    0    1
```

The data are saved in a comma-delimited (.csv) format, so we use the **read.table()** function to read it in. The **file** argument specifies where the file is located on our computer. Note that double back slashes, not single back slashes, are required to separate folders and file names. The **header = TRUE** argument value instructs R to use the names in the first row of the file as the variable names. The **sep = ","** argument value states that the file is comma-delimited.

After reading the data set into R, it is saved into a data frame called **placekick**, and the first six observations are printed using the **head()** function. For example, the first placekick in the data set is from the first week of the season and was attempted at 21 yards with 24.7 minutes remaining in the half. Also, the first kick was a successful field goal attempted under lead-change conditions, on grass, in a domed stadium, and under non-windy conditions. Note that we do not consider the placekicker as a variable of interest (justification for this decision is given on p. 22 of Bilder and Loughin, 1998). We will examine the possibility of a placekicker effect in Exercise 1 from Section 6.5.

For now, we will use only the distance to estimate the probability of a successful placekick. Formally, Y is the response variable with a value of 1 for a success and

0 for a failure, and the explanatory variable x_1 denotes the distance in yards for the placekick. We will use the observed data to estimate the parameters in the model

$$\text{logit}(\pi) = \beta_0 + \beta_1 x_1,$$

which can also be written with "`distance`" replacing x_1 to help make the model statement more descriptive.

Below is how we use R to estimate the model with the `glm()` function:

```
> mod.fit <- glm(formula = good ~ distance, family =
    binomial(link = logit), data = placekick)
> mod.fit

Call:  glm(formula = good ~ distance, family = binomial(link =
    logit), data = placekick)

Coefficients:
(Intercept)       distance
     5.8121        -0.1150
Degrees of Freedom:1424 Total (i.e. Null); 1423 Residual
Null Deviance:       1013
Residual Deviance:775.7       AIC:  779.7
```

The results from `glm()` are saved into an object that we call `mod.fit`, which is a shortened version of "model fit" (we could choose a different name). The arguments within `glm()` are:

- `formula` — Specifies the model with a ~ separating the response and explanatory variables

- `family` — Gives the type of model to be fit where `binomial` states the response type (remember that Bernoulli is a special case of the binomial) and `logit` is the function on the left side of the model (we will discuss "link" functions in Section 2.3)

- `data` — Name of the data frame containing the variables

By printing the `mod.fit` object through executing `mod.fit` at the command prompt, we see that the estimated logistic regression model is $\text{logit}(\hat{\pi}) = 5.8121 - 0.1150\text{distance}$. Because there is a negative parameter estimate corresponding to `distance`, the estimated probability of success decreases as the distance increases.

Additional information is printed in the output, but there are actually many more items, called *components*, that are stored within the `mod.fit` object. Through the use of the `names()` function, we obtain the following list of these components:

```
> names(mod.fit)
 [1] "coefficients"        "residuals"            "fitted.values"

<OUTPUT EDITED>

[28] "method"              "contrasts"            "xlevels"

> mod.fit$coefficients
(Intercept)       distance
  5.8120790     -0.1150266
```

The `mod.fit` object is of a list format. As we have seen earlier, lists allow multiple objects to be linked together in one common place. To access components of a list, we use the format <object>$<component> without the < >. The example in the output shows `mod.fit$coefficients` contains $\hat{\beta}_0$ and $\hat{\beta}_1$. The results from many of the other components are obvious by their names. For example, `mod.fit$fitted.values` are the estimated probabilities for each observation ($\hat{\pi}_i$ for $i = 1, \ldots, n$) and `mod.fit$residuals` contains the residuals from the model's fit ($y_i - \hat{\pi}_i$ for $i = 1, \ldots, n$). Other components are not as obvious, and we will discuss many of them as we progress through this chapter. Note that descriptions of all components are available in the "Value" section of the help for `glm()`.

A summary of what is inside `mod.fit` is obtained from the `summary()` function:

```
> summary(mod.fit)
Call: glm(formula = good ~ distance, family = binomial(link =
    logit), data = placekick)

Deviance Residuals:
    Min        1Q    Median        3Q       Max
 -2.7441    0.2425    0.2425    0.3801    1.6092

Coefficients:
             Estimate Std. Error z value Pr(>|z|)
(Intercept)  5.812079   0.326158   17.82   <2e-16 ***
distance    -0.115027   0.008337  -13.80   <2e-16 ***
---
Signif. codes:  0 '***' 0.001 '**' 0.01 '*' 0.05 '.' 0.1 ' ' 1

(Dispersion parameter for binomial family taken to be 1)

    Null deviance: 1013.43  on 1424  degrees of freedom
Residual deviance:  775.75  on 1423  degrees of freedom
AIC: 779.75

Number of Fisher Scoring iterations: 5
```

The above output displays a lot of information about the model that we will describe throughout this chapter. For now, note that the values of $\hat{\beta}_0$ and $\hat{\beta}_1$ are displayed in the "Coefficients" table under the "Estimate" header. Also, it took 5 iterations to obtain these estimates as given by the last line in the output ("Fisher scoring" is equivalent to IRLS for logistic regression models).

Additional background into how R works is helpful for understanding the results from `summary()`. The *class* of the `mod.fit` object is `glm`. Associated with this class are a number of *method* functions that help to summarize the information within objects or that complete additional useful calculations. To see a list of these method functions, the `methods()` function is used:

```
> class(mod.fit)
[1] "glm" "lm"
> methods(class = glm)
 [1] add1.glm*              addterm.glm*            anova.glm

<OUTPUT EDITED>
```

```
[37] summary.glm            vcov.glm*              weights.glm*
    Non-visible functions are asterisked
```

We will use many of these method functions throughout this chapter. For now, notice the `summary.glm()` function. When the *generic* function `summary()` is run, R first finds the class of the object and then checks to see if there is a method function matching that class. In the case here, it looks for `summary.glm()`. Because this function exists, it is used to complete the main calculations. For more information on the class of an object along with generic and method functions, please see p. 488 in Appendix A.7.3.

If more than one explanatory variable is included in the model, the variable names can be separated by "+" symbols in the `formula` argument. For example, suppose we include the `change` variable in addition to `distance` in the model:

```
> mod.fit2 <- glm(formula = good ~ change + distance, family =
    binomial(link = logit), data = placekick)
> mod.fit2$coefficients
(Intercept)       change       distance
  5.8931801    -0.4477830     -0.1128887
```

The estimated logistic regression model is $\text{logit}(\hat{\pi}) = 5.8932 - 0.4478\text{change} - 0.1129\text{distance}$.

The estimated variance-covariance matrix (hereafter referred to as a "covariance matrix" for simplicity) for a vector of parameter estimates $\hat{\boldsymbol{\beta}} = (\hat{\beta}_0, \hat{\beta}_1, \ldots, \hat{\beta}_p)'$, $\widehat{Var}(\hat{\boldsymbol{\beta}})$, has the usual form

$$\begin{bmatrix} \widehat{Var}(\hat{\beta}_0) & \widehat{Cov}(\hat{\beta}_0, \hat{\beta}_1) & \cdots & \widehat{Cov}(\hat{\beta}_0, \hat{\beta}_p) \\ \widehat{Cov}(\hat{\beta}_0, \hat{\beta}_1) & \widehat{Var}(\hat{\beta}_1) & \cdots & \widehat{Cov}(\hat{\beta}_1, \hat{\beta}_p) \\ \vdots & \vdots & \ddots & \vdots \\ \widehat{Cov}(\hat{\beta}_0, \hat{\beta}_p) & \widehat{Cov}(\hat{\beta}_1, \hat{\beta}_p) & \cdots & \widehat{Var}(\hat{\beta}_p) \end{bmatrix}$$

and can be found in R using the `vcov()` function.[1] This matrix is found by inverting a matrix of the second partial derivatives (often referred to as a "Hessian matrix") of the log-likelihood function evaluated at the parameter estimates and multiplying this resulting matrix by -1; see Appendix B.3.4 for details and Section 1.1.2 on p. 9 for a more simple case. In the end, the resulting covariance matrix simplifies to $(\mathbf{X}'\mathbf{V}\mathbf{X})^{-1}$ where

$$\mathbf{X} = \begin{bmatrix} 1 & x_{11} & x_{12} & \cdots & x_{1p} \\ 1 & x_{21} & x_{22} & \cdots & x_{2p} \\ \vdots & \vdots & \vdots & \ddots & \vdots \\ 1 & x_{n1} & x_{n2} & \cdots & x_{np} \end{bmatrix}$$

and $\mathbf{V} = \text{Diag}\left(\hat{\pi}_i(1-\hat{\pi}_i)\right)$, an $n \times n$ matrix with $\hat{\pi}_i(1-\hat{\pi}_i)$ on the diagonal and 0's elsewhere. This covariance matrix has the same form as the one resulting from weighted least squares estimation in a linear model. The individual elements of the matrix do not have a simple form, so we do not present them here; see Exercise 30 for a discussion.

---

[1] Notice that `vcov.glm()` is listed as a method function in the `methods(class = glm)` output given in the last example. This is the function called by `vcov()` when it is applied to a `glm` class object.

Example: Placekicking (Placekick.R, Placekick.csv)

The purpose of this example is to show how to obtain the estimated covariance matrix for the regression parameter estimators. We use the `mod.fit` object from the model with `distance` as the only explanatory variable.

Below is the coefficients table again from the `summary(mod.fit)` output:

```
> round(summary(mod.fit)$coefficients,4)
            Estimate Std. Error z value Pr(>|z|)
(Intercept)   5.8121     0.3262 17.8198        0
distance     -0.1150     0.0083 -13.7975       0
```

We limit the displayed output by using the fact that `summary()` creates a list with `coefficients` as one component (if needed, please see the corresponding program for help). The `Std. Error` column gives the standard errors for the regression parameter estimators—$\widehat{Var}(\hat{\beta}_0)^{1/2}$ in the "(Intercept)" row and $\widehat{Var}(\hat{\beta}_1)^{1/2}$ in the "distance" row.

The `vcov()` function produces the estimated covariance matrix:

```
> vcov(mod.fit)
            (Intercept)      distance
(Intercept)  0.106378825 -2.604526e-03
distance    -0.002604526  6.950142e-05
> vcov(mod.fit)[2,2]   # Var-hat(beta-hat_1)
[1] 6.950142e-05
> summary(mod.fit)$coefficients[2,2]^2
[1] 6.950142e-05
```

We can extract the estimated variance for $\hat{\beta}_1$ by specifying the (2,2) element of the matrix. Thus, $\widehat{Var}(\hat{\beta}_1) = 0.0000695$, which is the square of 0.0083 given within the coefficients table.

For readers interested in the actual matrix calculations, we provide an example of how the covariance matrix is calculated by using $(\mathbf{X'VX})^{-1}$:

```
> pi.hat <- mod.fit$fitted.values
> V <- diag(pi.hat*(1-pi.hat))
> X <- cbind(1, placekick$distance)
> solve(t(X) %*% V %*% X)
              [,1]          [,2]
[1,]  0.106456753 -2.606250e-03
[2,] -0.002606250  6.953996e-05
```

We create the diagonal matrix $\mathbf{V} = \text{Diag}(\hat{\pi}_i(1-\hat{\pi}_i))$ by using the `diag()` function, where we first extract $\hat{\pi}_i$ from `mod.fit`. Next, we create the $\mathbf{X}$ matrix by using the `cbind()` function which puts a column of 1's before a column of the explanatory variable values.[2] Continuing, the `t()` function finds the transpose of the $\mathbf{X}$ matrix, the `%*%` syntax instructs R to perform matrix multiplication, and the `solve()` function finds the inverse of the matrix. The resulting matrix is practically the same as that produced by `vcov()`, where differences are due to rounding error.

---

[2] R *recycles* "1" n times so that the first column has the same number of elements in it as the second column.

As a reminder, we encourage readers to investigate individual parts of any set of code to help understand the coding process. For example, one could print part of V by executing V[1:3, 1:3] and see that the first three rows and columns of V are a diagonal matrix with values of $\hat{\pi}_1(1-\hat{\pi}_1)$, $\hat{\pi}_2(1-\hat{\pi}_2)$, and $\hat{\pi}_3(1-\hat{\pi}_3)$ as the (1,1), (2,2), and (3,3) elements, respectively. In addition, one can perform parts of the calculations within the solve() function to see the intermediate results.

While the glm() function will be used whenever we estimate a logistic regression model, it is instructive to be able to view the log-likelihood function while also fitting the model in a more general manner. In the next example, we illustrate how the log-likelihood function can be programmed directly into R and the general optimization function optim() can be used to find the parameter estimates. These programming techniques can be utilized for more complicated problems, such as those discussed in Chapter 6 where functions like glm() may not be available.

Example: Placekicking (Placekick.R, Placekick.csv)

We focus again on fitting the logistic regression model with only the distance explanatory variable. Below we create a function logL() to calculate the log-likelihood function for any given parameter values, explanatory variable values for x_1, and binary responses for Y:

```
> logL <- function(beta, x, Y) {
    pi <- exp(beta[1] + beta[2]*x)/(1 + exp(beta[1] + beta[2]*x))
    sum(Y*log(pi) + (1-Y)*log(1-pi))
}
> logL(beta = mod.fit$coefficients, x = placekick$distance, Y =
    placekick$good)
[1] -387.8725
> logLik(mod.fit)
'log Lik.' -387.8725 (df=2)
```

Inside the function, we find π_i through our model expression and then find

$$\log[L(\beta_0, \beta_1|y_1, \ldots, y_n)] = \sum_{i=1}^{n} y_i \log(\pi_i) + (1 - y_i)\log(1 - \pi_i)$$

using the sum() and log() functions, while taking advantage of how R performs addition and multiplication on vectors in an element-wise fashion. When we evaluate the function using the parameter estimates and the observed data, we obtain $\log[L(\beta_0, \beta_1|y_1, \ldots, y_n)] = -387.87$. As a check, we obtain the same value from the generic function logLik(), which extracts the maximum log-likelihood function value from mod.fit objects (the logLik.glm() method function is used here).

To maximize the log-likelihood function and find the corresponding MLEs, we use the optim() function. This is a very general optimization function that *minimizes* an R function with respect to a vector of parameters. Because we want to *maximize* the log-likelihood function instead, we use the control = list(fnscale = -1) argument value within optim(), which essentially instructs R to minimize the negative of logL(). Below is our code:

```
> # Find starting values for parameter estimates
> reg.mod <- lm(formula = good ~ distance, data = placekick)
> reg.mod$coefficients
(Intercept)    distance
 1.25202444  -0.01330212

> mod.fit.optim <- optim(par = reg.mod$coefficients, fn = logL,
    hessian = TRUE, x = placekick$distance, Y = placekick$good,
    control = list(fnscale = -1), method = "BFGS")
> names(mod.fit.optim)
[1] "par"         "value"        "counts"        "convergence"
[5] "message"     "hessian"
> mod.fit.optim$par
(Intercept)    distance
   5.8112544   -0.1150046
> mod.fit.optim$value
[1] -387.8725
> mod.fit.optim$convergence
[1] 0
> -solve(mod.fit.optim$hessian)
             (Intercept)      distance
(Intercept) -0.106482866   2.607258e-03
distance     0.002607258  -6.957463e-05
```

The optimization procedures within optim() need initial starting values for the regression parameters. We simply fit a linear regression model to the data using the lm() function and take the corresponding parameter estimates as these starting values (see Appendix A.7 for another example of using lm()). Within the call to optim(), we specify the initial parameter estimates using the par argument, and we specify the function to be maximized using the fn argument. Note that the first argument in the function named in fn must correspond to the initial parameter estimates; this is why beta was given as the first argument in logL(). The TRUE value for the hessian argument instructs R to obtain a numerical estimate of the Hessian matrix for the parameters. In other words, it estimates a matrix of second partial derivatives for the log-likelihood function, which can then be inverted to obtain the estimated covariance matrix for the parameter estimates (see Appendix B.3.4).[3] Finally, we specify the corresponding values of x and Y for the logL() function.

The object produced by optim() is a list that we save as mod.fit.optim. Components within the list include the parameter estimates ($par), the log-likelihood function's maximum value ($value), and the estimated covariance matrix ($hessian; we subsequently use solve() to find the estimated covariance matrix as the inverse of the Hessian matrix). All of these values are practically the same as those produced by glm(), where small differences are due to different convergence criteria for the iterative numerical procedures. The 0 value for mod.fit.optim$convergence means that convergence was achieved. For this implementation of optim(), we used the method = "BFGS" optimization procedure (similar to a Newton-Raphson procedure), but other

---

[3]The estimated covariance matrix obtained here is calculated as shown in Equation B.6, rather than as shown in Equation B.5 when an expectation is taken. For logistic regression, the resulting estimated covariance matrices are exactly the same no matter which equation is used. There are other models where the matrices will be different (e.g., see Exercise 28 and Section 6.1.2), but fortunately the two estimated covariance matrices are asymptotically equivalent (the same as $n \to \infty$).

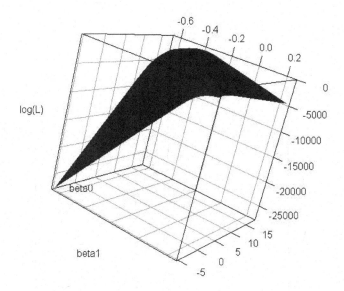

Figure 2.2: Three-dimensional plot of the log-likelihood function.

procedures are available within the function (see Gentle, 2009 or the help for `optim()` for details on various iterative numerical procedures).

The log-likelihood function can be plotted relatively easy for this example because there are only two parameters. The R code to create this plot is within the corresponding program for this example, and the code is very similar to the three-dimensional confidence level plots on p. 34 used to produce Figure 1.5. Two `for()` function calls—one for loop for the β_0 values and one for loop for the β_1 values—are used to evaluate the log-likelihood function.[4] We use the `persp3d()` function from the `rgl` package to produce the three-dimensional log-likelihood function surface in Figure 2.2. We also use the `contour()` function from the `graphics` package to plot contours of the log-likelihood function in Figure 2.3.[5] We can see from the plots (a little clearer from the contour plot) that the values which maximize the log-likelihood function are 5.81 for β_0 and -0.115 for β_1.

An alternative form of the response variable is available when there are multiple Bernoulli trials recorded at some or all combinations of explanatory variables. In particular, if there are $J < n$ unique combinations of explanatory variables, we can aggregate the individual trial results as in Section 1.2, so that we observe w_j successes in n_j trials each with $x_{j1}, \ldots, x_{jp}$ explanatory variable values for $j = 1, \ldots, J$. In other words, we now have J observations from a binomial distribution with probabilities of success π_j, $j = 1, \ldots, J$. Alternatively, the data may naturally arise in the form of w_j successes out of n_j trials. For example, in an experiment to determine a lethal dose level, a particular amount of pesticide (x_{j1}) could be applied to an enclosure with a known number of ants (n_j), where the number

---

[4]The `apply()` function can be used here instead of for loops. This code is available in the corresponding program for this example.

[5]For all possible values of β_0 and β_1, equal values of the log-likelihood function are joined with lines called *contours*. For example, the ellipse labeled -500 shows that $(\beta_0, \beta_1) = (2, -0.02)$, $(5.6, -0.07)$, and $(11.7, -0.27)$ all have values of $\log[L(\beta_0, \beta_1 | y_1, \ldots, y_n)]$ close to -500. Additional code in the corresponding program demonstrates this point.

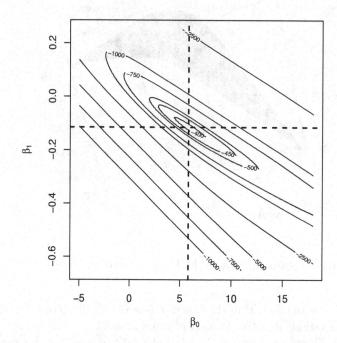

Figure 2.3: Contour plot of the log-likelihood function. Contours are drawn at -400, -450, -500, -750, -1000, -2500, -5000, -7500, and -10000.

of ants exterminated (w_j) would be recorded. By applying different amounts of pesticide to other enclosures, we have a binomial response setting for J treatment groups.

The same logistic regression model can be used to model the probability of success for the binomial response setting. The log-likelihood function is

$$\log[L(\beta_0,\ldots,\beta_p|w_1,\ldots,w_J)] = \sum_{j=1}^{J} \log\left[\binom{n_j}{w_j}\right] + w_j \log(\pi_j) + (n_j - w_j)\log(1 - \pi_j).$$

Because $\binom{n_j}{w_j}$ is a constant, this term does not change the estimated parameter values that result in the log-likelihood maximization. Therefore, the MLEs are the same as if the data instead consisted of $n = \sum_{j=1}^{J} n_j$ binary responses. We illustrate this in the next example.

Example: Placekicking (Placekick.R, Placekick.csv)

The purpose of this example is to show how to reform the placekicking data to a binomial response format and to show that the model parameter estimates are the same for this format. We focus only on the distance explanatory variable for this example.

The `aggregate()` function is used to find the number of successes and number of observations for each distance:

```
> w <- aggregate(formula = good ~ distance, data = placekick, FUN
    = sum)
```

```
> n <- aggregate(formula = good ~ distance, data = placekick, FUN
    = length)
> w.n <- data.frame(distance = w$distance, success = w$good,
    trials = n$good, proportion = round(w$good/n$good,4))
> head(w.n)
  distance success trials proportion
1       18       2      3     0.6667
2       19       7      7     1.0000
3       20     776    789     0.9835
4       21      19     20     0.9500
5       22      12     14     0.8571
6       23      26     27     0.9630
```

The formula argument within aggregate() is used the same way as within glm(). The FUN argument specifies how the response variable good will be summarized, where the sum() function adds the 0 and 1 responses for each distance and the length() function counts the number of observations for each distance. The final result is in the w.n data frame. For example, there are 2 successes out of 3 trials at a distance of 18 yards, which results in an observed proportion of successes of $2/3 \approx 0.6667$. Note that the reason for the large number of observations at 20 yards is because most PATs are attempted from this distance.

The logistic regression model is estimated by the glm() function with two changes to our previous code:

```
> mod.fit.bin <- glm(formula = success/trials ~ distance, weights
    = trials, family = binomial(link = logit), data = w.n)
> summary(mod.fit.bin)
Call: glm(formula = success/trials ~ distance, family =
    binomial(link = logit), data = w.n, weights = trials)

Deviance Residuals:
    Min      1Q   Median      3Q      Max
-2.0373 -0.6449  -0.1424  0.5004   2.2758

Coefficients:
             Estimate Std. Error z value Pr(>|z|)
(Intercept)  5.812080   0.326245   17.82   <2e-16 ***
distance    -0.115027   0.008338  -13.79   <2e-16 ***
---
Signif. codes:  0 '***' 0.001 '**' 0.01 '*' 0.05 '.' 0.1 ' ' 1

(Dispersion parameter for binomial family taken to be 1)

    Null deviance: 282.181  on 42  degrees of freedom
Residual deviance:  44.499  on 41  degrees of freedom
AIC: 148.46

Number of Fisher Scoring iterations: 4
```

First, we specify the response in the formula argument as success/trials. Notice this finds the observed proportion of success for each distance. Second, we use the weights argument to specify the number of observations per distance. The estimated logistic regression model is $\text{logit}(\hat{\pi}) = 5.8121 - 0.1150\text{distance}$, which is essentially the same as earlier (very small differences can occur in the parameter estimates due to the iterative numerical procedure).

2.2.2 Hypothesis tests for regression parameters

Hypothesis tests can be used to assess the importance of explanatory variable(s) in a model. For example, a test of $H_0 : \beta_r = 0$ vs. $H_a : \beta_r \neq 0$ evaluates the rth explanatory variable term in the model $\text{logit}(\pi) = \beta_0 + \beta_1 x_1 + \cdots + \beta_r x_r + \cdots + \beta_p x_p$. If $\beta_r = 0$, this means the corresponding term is excluded from the model. If $\beta_r \neq 0$, this means the corresponding term is included in the model. Equivalently, we can state the hypotheses as

$$H_0 : \text{logit}(\pi) = \beta_0 + \beta_1 x_1 + \cdots + \beta_{r-1} x_{r-1} + \beta_{r+1} x_{r+1} + \cdots + \beta_p x_p$$
$$H_a : \text{logit}(\pi) = \beta_0 + \beta_1 x_1 + \cdots + \beta_r x_r + \cdots + \beta_p x_p,$$

which can be helpful to emphasize the comparison of two different models. In general, the explanatory variables in the null-hypothesis model (also known as a *reduced model*) must all be in the alternative-hypothesis model (also known as a *full model*)

We will use a Wald test or a LRT to perform these tests. Appendix B.5.1 gives general details for how to perform them. We now focus on the specifics for the logistic regression setting.

Wald test

The Wald statistic

$$Z_0 = \frac{\hat{\beta}_r}{\sqrt{\widehat{Var}(\hat{\beta}_r)}}$$

is used to test $H_0 : \beta_r = 0$ vs. $H_a : \beta_r \neq 0$. If the null hypothesis is true, Z_0 has an approximate standard normal distribution for a large sample, and we reject the null hypothesis if Z_0 has an unusual observed value for this distribution. We define unusual by $|Z_0| > Z_{1-\alpha/2}$. The p-value is $2P(Z > |Z_0|)$ where Z has a standard normal distribution.

More than one parameter may be tested for equality to 0 in the null hypothesis. However, we forgo discussion of it because the Wald inference procedures here often encounter the same types of problems as those discussed in Chapter 1. In the context of hypothesis tests, this means that the stated type I error rate for a hypothesis test (i.e., α) is not the same as the actual type I error rate. The LRT typically performs better than the Wald test, so we focus on this procedure next.

Likelihood ratio test

The LRT statistic can be written informally as

$$\Lambda = \frac{\text{Maximum of likelihood function under } H_0}{\text{Maximum of likelihood function under } H_0 \text{ or } H_a}. \tag{2.6}$$

In the context of testing the equality of regression parameters to 0, the denominator in Equation 2.6 is the likelihood function evaluated at the MLEs for the model containing all $p+1$ regression parameters. The numerator in Equation 2.6 is the likelihood function evaluated at the MLEs for the model that excludes those variables whose regression parameters are set to 0 in the null hypothesis. For example, to test $H_0 : \beta_r = 0$ vs. $H_a : \beta_r \neq 0$, β_r would be held equal to 0, which means this corresponding explanatory variable would be excluded from the null-hypothesis model. Of course, the estimates for the remaining parameters need not be the same as with the alternative hypothesis model due to the differences between the two models.

If q regression parameters are set to 0 in the null hypothesis and if the null hypothesis is true, the $-2\log(\Lambda)$ statistic has an approximate χ_q^2 distribution for a large sample, and we reject the null hypothesis if $-2\log(\Lambda)$ has an unusually large observed value for this distribution. For example, if $\alpha = 0.05$ for the test of $H_0 : \beta_r = 0$ vs. $H_a : \beta_r \neq 0$, the rejection region is > 3.84, where $\chi^2_{1, 0.95} = 3.84$ is the 0.95 quantile from a chi-square distribution. The p-value is $P(A > -2\log(\Lambda))$ where A has a χ_1^2 distribution when $q = 1$.

We will most often use the generic functions anova() from the stats package and Anova() from the car package to perform these tests.[6] These functions test hypotheses of a similar structure to those typically seen in an analysis of variance (ANOVA), but can perform LRTs instead of the typical ANOVA F-tests. The two functions are based on model-comparison tests of different types. When used with one model fit object resulting from glm(), the anova() function computes *type 1* (sequential) tests, while Anova() computes *type 2* (partial) tests. The difference between types lies in the null-hypothesis models used for each term tested. With type 2 tests, each null-hypothesis model consists of all the other variables listed in the right side of the formula argument, ignoring any higher-order interactions that contain the term (see Section 2.2.5 for more on interactions). For type 1 tests, the null-hypothesis model contains only those variables listed in the formula argument *before* the tested term. Generally, type 2 tests are preferred unless there is some specific reason to consider a sequence of tests, such as in polynomial models. See Milliken and Johnson (2004) for further details.

The transformed LRT statistic $-2\log(\Lambda)$ has a simplified form. Suppose the estimated probability of successes under the null- and alternative-hypothesis models are denoted as $\hat{\pi}_i^{(0)}$ and $\hat{\pi}_i^{(a)}$, respectively. Similarly, we define a vector of the regression parameter estimates as $\hat{\boldsymbol{\beta}}^{(0)}$ and $\hat{\boldsymbol{\beta}}^{(a)}$. We have then

$$\begin{aligned}
-2\log(\Lambda) &= -2\log\left(\frac{L(\hat{\boldsymbol{\beta}}^{(0)}|y_1,\ldots,y_n)}{L(\hat{\boldsymbol{\beta}}^{(a)}|y_1,\ldots,y_n)}\right) \\
&= -2\left[\log\left(L(\hat{\boldsymbol{\beta}}^{(0)}|y_1,\ldots,y_n)\right) - \log\left(L(\hat{\boldsymbol{\beta}}^{(a)}|y_1,\ldots,y_n)\right)\right] \\
&= -2\sum_{i=1}^n y_i\log(\hat{\pi}_i^{(0)}) + (1-y_i)\log(1-\hat{\pi}_i^{(0)}) - y_i\log(\hat{\pi}_i^{(a)}) - (1-y_i)\log(1-\hat{\pi}_i^{(a)}) \\
&= -2\sum_{i=1}^n y_i\log\left(\frac{\hat{\pi}_i^{(0)}}{\hat{\pi}_i^{(a)}}\right) + (1-y_i)\log\left(\frac{1-\hat{\pi}_i^{(0)}}{1-\hat{\pi}_i^{(a)}}\right).
\end{aligned} \quad (2.7)$$

This formulation of the statistic can be easily coded into R. We show it and how to use the anova() and Anova() functions next.

Example: Placekicking (Placekick.R, Placekick.csv)

We begin this example by using the model that contains both change and distance,

$$\text{logit}(\pi) = \beta_0 + \beta_1\text{change} + \beta_2\text{distance},$$

---

[6] Functions commonly used by others include the drop1() function from the stats package and the lrtest() function from the lmtest package.

which was originally estimated in Section 2.2.1 and its results saved in the `mod.fit2` object. Below is our code and output for Wald tests:

```
> summary(mod.fit2)
Call: glm(formula = good ~ change + distance, family =
    binomial(link = logit), data = placekick)

Deviance Residuals:
    Min      1Q   Median      3Q     Max
 -2.7061  0.2282  0.2282  0.3750  1.5649

Coefficients:
            Estimate Std. Error z value Pr(>|z|)
(Intercept)  5.893181   0.333184  17.687  <2e-16 ***
change      -0.447783   0.193673  -2.312  0.0208 *
distance    -0.112889   0.008444 -13.370  <2e-16 ***
---
Signif. codes:  0 '***' 0.001 '**' 0.01 '*' 0.05 '.' 0.1 ' ' 1

(Dispersion parameter for binomial family taken to be 1)

    Null deviance: 1013.4  on 1424  degrees of freedom
Residual deviance:  770.5  on 1422  degrees of freedom
AIC: 776.5

Number of Fisher Scoring iterations: 6
```

Wald test statistics and p-values for each regression parameter are located in the `coefficients` table from `summary(mod.fit2)`. For example, to test the importance of the change explanatory variable, we use $H_0 : \beta_1 = 0$ vs. $H_a : \beta_1 \neq 0$ resulting in $Z_0 = -0.4478/0.1936 = -2.3124$ (`z value` column) and a p-value of $2P(Z > |-2.3124|) = 0.0208$ (`Pr(>|Z|)` column). Using $\alpha = 0.05$, we would reject the null hypothesis; however, we would not reject it using $\alpha = 0.01$. Thus, we say there is marginal evidence that `change` is important to include in the model given that `distance` is in the model. Note that it is important to include the "`distance` is in the model" part because the null and alternative hypotheses are equivalently stated as

$$H_0 : \text{logit}(\pi) = \beta_0 + \beta_1 \text{distance}$$
$$H_a : \text{logit}(\pi) = \beta_0 + \beta_1 \text{change} + \beta_2 \text{distance}.$$

With respect to testing the parameter for `distance` in the model containing `change`, the p-value is given as <2e-16 in the output, which means the p-value is less than 2×10^{-16}. Thus, there is strong evidence of the importance of `distance` given that `change` is in the model. In the corresponding program to this example, we show the R code needed to extract information from `mod.fit2` to perform the Wald tests without the help of `summary(mod.fit2)`.

Below is our code and output from `Anova()` for the LRTs:

```
> library(package = car)
> Anova(mod.fit2, test = "LR")
Analysis of Deviance Table (Type II tests)

Response: good
```

```
          LR Chisq Df Pr(>Chisq)
change       5.246  1    0.02200  *
distance   218.650  1    < 2e-16 ***
---
Signif. codes:  0 '***' 0.001 '**' 0.01 '*' 0.05 '.' 0.1 ' ' 1
```

The Anova() function produces LRTs using the test = "LR" argument value, which is the default. For the test of change with $H_0 : \beta_1 = 0$ vs. $H_a : \beta_1 \neq 0$, we obtain $-2\log(\Lambda) = 5.246$ with a p-value of $P(A > 5.246) = 0.0220$, and we reach the same conclusion as with the previous Wald test. The p-value for the distance test is given as < 2e-16 in the output, which again indicates there is strong evidence that distance is important (given that the model includes change).

The anova() function with the test = "Chisq" argument value also performs LRTs, but in a sequential manner based on the ordering of the explanatory variables in the model fit:

```
> anova(mod.fit2, test = "Chisq")
Analysis of Deviance Table

Model: binomial, link: logit

Response: good

Terms added sequentially (first to last)

         Df Deviance Resid. Df Resid. Dev P(>|Chi|)
NULL                     1424    1013.43
change    1   24.277   1423     989.15   8.343e-07 ***
distance  1  218.650   1422     770.50   < 2.2e-16 ***
---
Signif. codes:  0 '***' 0.001 '**' 0.01 '*' 0.05 '.' 0.1 ' ' 1
```

The change row of the function's output gives a test of

$$H_0 : \text{logit}(\pi) = \beta_0$$
$$H_a : \text{logit}(\pi) = \beta_0 + \beta_1 \text{change},$$

which results in a p-value of 8.34×10^{-7}. The distance row of the function's output gives a test of

$$H_0 : \text{logit}(\pi) = \beta_0 + \beta_1 \text{change}$$
$$H_a : \text{logit}(\pi) = \beta_0 + \beta_1 \text{change} + \beta_2 \text{distance},$$

which results in a p-value of less than 2×10^{-16}. The hypotheses and calculations for the test of distance are exactly the same as those using the Anova() function.

Another way to use the anova() function is to fit both the null- and alternative-hypothesis models and then use the corresponding model fit objects as argument values within the function. This will be especially useful later when we test null hypotheses that do not correspond to simple deletions of one term from the full model. To test the change variable given that distance is in the model, we can use the following code:

```
> mod.fit.Ho <- glm(formula = good ~ distance, family =
    binomial(link = logit), data = placekick)
> anova(mod.fit.Ho, mod.fit2, test = "Chisq")
Analysis of Deviance Table

Model 1: good ~ distance
Model 2: good ~ change + distance
  Resid. Df Resid. Dev Df Deviance P(>|Chi|)
1      1423     775.75
2      1422     770.50  1   5.2455   0.02200 *
---
Signif. codes:  0 '***' 0.001 '**' 0.01 '*' 0.05 '.' 0.1 ' ' 1
```

We obtain the same results as before with $-2\log(\Lambda) = 5.246$ and a p-value of $P(A > 5.246) = 0.0220$.

Equation 2.7 can be programmed directly into R to perform LRTs. While this generally is not needed for logistic regression applications, it is often helpful to go through the calculations in this manner to better understand what functions like anova() and Anova() are computing. Furthermore, there will be times later in this book where functions like these are not available. We demonstrate the LRT for the test of $H_0 : \text{logit}(\pi) = \beta_0$ vs. $H_a : \text{logit}(\pi) = \beta_0 + \beta_1 \textbf{change}$:

```
> mod.fit.Ho <- glm(formula = good ~ 1, family = binomial(link =
    logit), data = placekick)
> mod.fit.Ha <- glm(formula = good ~ change, family =
    binomial(link = logit), data = placekick)
> anova(mod.fit.Ho, mod.fit.Ha, test = "Chisq")
Analysis of Deviance Table

Model 1: good ~ 1
Model 2: good ~ change
  Resid. Df Resid. Dev Df Deviance P(>|Chi|)
1      1424    1013.43
2      1423     989.15  1   24.277 8.343e-07 ***
---
Signif. codes:  0 '***' 0.001 '**' 0.01 '*' 0.05 '.' 0.1 ' ' 1

> pi.hat.Ho <- mod.fit.Ho$fitted.values
> pi.hat.Ha <- mod.fit.Ha$fitted.values
> y <- placekick$good
> stat <- -2*sum(y*log(pi.hat.Ho/pi.hat.Ha) +
    (1-y)*log((1-pi.hat.Ho)/(1-pi.hat.Ha)))
> pvalue <- 1-pchisq(q = stat, df = 1)
> data.frame(stat, pvalue)
      stat       pvalue
1 24.27703 8.342812e-07
```

We fit both the null- and alternative-hypothesis models using glm(). Note that we use formula = good ~ 1 to estimate a model with only an intercept.[7] The anova() function gives a $-2\log(\Lambda) = 24.277$ value along with a p-value equal to 8.343×10^{-7}

---

[7]In contrast, if we wanted to exclude an intercept parameter from being estimated for a model, we would include a "-1" as part of a formula argument value along with any explanatory variable names.

as seen before. After creating objects for $\hat{\pi}_i^{(0)}$, $\hat{\pi}_i^{(a)}$, and y_i, we code Equation 2.7 using the fact that R performs mathematical operations on vectors element-wise.[8] To find the p-value, we use the `pchisq()` function and subtract it from 1 because we want the area to the right of the quantile but the function finds the area to its left. The resulting test statistic and p-value are the same as found earlier.

In the previous example's output, the word *deviance* and its abbreviation "dev" appeared a number of times. This word also appears in the `summary()` output as well. Deviance refers to the amount that a particular model deviates from another model as measured by the transformed LRT statistic $-2\log(\Lambda)$. For example, the $-2\log(\Lambda) = 5.246$ value used for testing `change` (given that `distance` is in the model) is a measure of how much the estimated probability of successes for the model excluding `change` deviate from those for the model including `change`.

The *residual deviance* is a particular form of the $-2\log(\Lambda)$ statistic that measures how much the probabilities estimated from a model of interest deviate from the observed proportions of successes (simply 0/1 or 1/1 for Bernoulli response data). These observed proportions are equivalent to estimating a logistic regression model where each observation is represented by one parameter, say $\text{logit}(\pi_i) = \gamma_i$ for $i = 1, \ldots, n$. This model is frequently referred to as the *saturated model*, because the number of parameters is equal to the number of observations, so that no additional parameters can be estimated. We will use the residual deviance as a measure of overall goodness of fit for a model in Chapter 5. The *null deviance* denotes how much the probabilities estimated from the model $\text{logit}(\pi_i) = \beta_0$ for $i = 1, \ldots, n$ deviate from the observed proportion of successes. Because $\text{logit}(\pi_i) = \beta_0$ contains only the intercept term (and thus just one parameter), every π_i is estimated to be the same value for this particular model (the MLE for π_i is the same as the MLE of a single probability success π that we estimated in Section 1.1.2).

Residual deviance statistics are often calculated as an intermediate step for performing a LRT to compare two models. For example, consider the hypotheses of

$$H_0 : \text{logit}(\pi^{(0)}) = \beta_0 + \beta_1 x_1$$
$$H_a : \text{logit}(\pi^{(a)}) = \beta_0 + \beta_1 x_1 + \beta_2 x_2 + \beta_3 x_3$$

where we use the superscript for π again to help differentiate between the two models, and suppose we are working with Bernoulli response data. The residual deviance for the model $\text{logit}(\pi^{(0)}) = \beta_0 + \beta_1 x_1$ tests $H_0 : \text{logit}(\pi^{(0)}) = \beta_0 + \beta_1 x_1$ vs. $H_a :$ Saturated model, using the quantity:

$$-2\log(\Lambda) = -2 \sum_{i=1}^{n} y_i \log\left(\frac{\hat{\pi}_i^{(0)}}{y_i}\right) + (1 - y_i)\log\left(\frac{1 - \hat{\pi}_i^{(0)}}{1 - y_i}\right). \tag{2.8}$$

Values of $y_i = 0$ or 1 in the fraction denominators do not cause problems due to y_i being a coefficient on the log function (i.e., 0 multiplied by another quantity is 0). The residual deviance for the model $\text{logit}(\pi^{(a)}) = \beta_0 + \beta_1 x_1 + \beta_2 x_2 + \beta_3 x_3$ tests $H_0 : \text{logit}(\pi^{(a)}) = \beta_0 + \beta_1 x_1 + \beta_2 x_2 + \beta_3 x_3$ vs. $H_a :$ Saturated model, using the quantity:

$$-2\log(\Lambda) = -2 \sum_{i=1}^{n} y_i \log\left(\frac{\hat{\pi}_i^{(a)}}{y_i}\right) + (1 - y_i)\log\left(\frac{1 - \hat{\pi}_i^{(a)}}{1 - y_i}\right). \tag{2.9}$$

---

[8]Remember that it is helpful to execute segments of code to determine what the code does. For example, one could first execute `y*log(pi.hat.Ho/pi.hat.Ha)` separately to see that a vector of zero and non-zero values are returned. When combined with `(1-y)*log((1-pi.hat.Ho)/(1-pi.hat.Ha))` and then through the use of `-2*sum()`, we obtain Equation 2.7.

When Equation 2.9 is subtracted from Equation 2.8, we obtain Equation 2.7, which is the correct transformed LRT statistic needed for the original set of hypotheses.

The residual deviance takes a slightly different form when we have binomial responses W_j for $j = 1, \ldots, J$. The statistic is

$$-2\log(\Lambda) = -2\sum_{j=1}^{J}\left[w_j\log\left(\frac{\hat{\pi}_j}{w_j/n_j}\right) + (n_j - w_j)\log\left(\frac{1-\hat{\pi}_j}{1-w_j/n_j}\right)\right], \quad (2.10)$$

where $\hat{\pi}_j$ is the estimate of the probability of success at each $j = 1, \ldots, J$. Note that the saturated model in this setting estimates the probability of success as w_j/n_j. Residual deviances for null- and alternative-hypothesis models can still be used in hypothesis tests in the same manner as in the previous Bernoulli response case.

Example: Placekicking (Placekick.R, Placekick.csv)

To perform the test of

$$H_0 : \text{logit}(\pi) = \beta_0 + \beta_1\text{distance}$$
$$H_a : \text{logit}(\pi) = \beta_0 + \beta_1\text{change} + \beta_2\text{distance},$$

we used the `anova()` function earlier to obtain $-2\log(\Lambda) = 5.246$. Alternatively, we can use the residual deviances from the two models:

```
> mod.fit.Ho <- glm(formula = good ~ distance, family =
    binomial(link = logit), data = placekick)
> df <- mod.fit.Ho$df.residual - mod.fit2$df.residual
> stat <- mod.fit.Ho$deviance - mod.fit2$deviance
> pvalue <- 1 - pchisq(q = stat, df = df)
> data.frame(Ho.resid.dev = mod.fit.Ho$deviance, Ha.resid.dev =
    mod.fit2$deviance, df = df, stat = round(stat,4), pvalue =
    round(pvalue,4))
  Ho.resid.dev Ha.resid.dev df    stat pvalue
1      775.745      770.4995  1  5.2455  0.022
```

We use the `deviance` and `df.residual` components of the model fit objects to obtain the residual deviances and corresponding degrees of freedom, respectively. The $-2\log(\Lambda)$ value and the p-value are both the same as earlier.

In this section, we illustrated a few different ways to perform a LRT. Generally, the `Anova()` and `anova()` functions are the easiest functions to use, and we recommend their use. We included other ways to perform the tests to help demonstrate more general methods and reinforce how the calculations are actually done. These more general methods will also be useful for other types of models where a `glm()`-like function does not exist for model fitting and no `Anova()` or `anova()` function exists for testing, which is the case for some sections in Chapters 3 and 6.

2.2.3 Odds ratios

In a linear model where $E(Y) = \beta_0 + \beta_1 x_1 + \cdots + \beta_p x_p$, the regression parameter β_r is interpreted as the change in the mean response for each 1-unit increase in x_r, holding the other variables in the model constant. In a logistic regression model the interpretation of regression parameters needs to account for the fact that they are related to the probability of success through $\text{logit}(\pi) = \beta_0 + \beta_1 x_1 + \cdots + \beta_p x_p$. Holding other variables constant, a 1-unit increase in x_r causes $\text{logit}(\pi)$ to change by β_r. This constant change in the log-odds of a success leads to a convenient way to use odds in the interpretation.

To see how this works, consider a logistic regression model with only one explanatory variable x. The odds of a success at a particular value of x are

$$Odds_x = \exp(\beta_0 + \beta_1 x).$$

If x is increased by $c > 0$ units, the odds of success become

$$Odds_{x+c} = \exp(\beta_0 + \beta_1(x+c)).$$

To determine how much the odds of success have changed by this c-unit increase, we find the odds ratio:
$$OR = \frac{Odds_{x+c}}{Odds_x} = \frac{\exp(\beta_0 + \beta_1(x+c))}{\exp(\beta_0 + \beta_1 x)} = \exp(c\beta_1).$$

Interestingly, the original value of the explanatory variable x is canceled out in the simplification; only the amount of increase c and the coefficient β_1 matter. The standard interpretation of this odds ratio is

The odds of a success change by $\exp(c\beta_1)$ times for every c-unit increase in x.

It is also common to say "increase" instead of "change" when $\exp(c\beta_1) > 1$, and "decrease" when $\exp(c\beta_1) < 1$. When x is binary with a coding of 0 and 1, c is always 1. The odds ratio interpretation then compares the odds of success at the level coded "$x = 1$" to the odds at "$x = 0$" in much the same way as in Section 1.2.5.

Model parameter estimates can be substituted for their corresponding parameters in OR to estimate the odds ratio. The estimated odds ratio becomes

$$\widehat{OR} = \exp(c\hat{\beta}_1),$$

and its interpretation is that the *estimated* odds of a success change by $\exp(c\hat{\beta}_1)$ times for every c-unit increase in x. Because the estimated odds ratio is a statistic, it will vary from sample to sample. Therefore, we need to find a confidence interval for OR in order to make inferences with a particular level of confidence. The likelihood-based procedures discussed in Appendix B.3.4 and B.5.2 provide the basis for the confidence intervals discussed next.

To find a Wald confidence interval for OR, we first find a confidence interval for $c\beta_1$ and then use the exponential function with the interval endpoints. To this end, we extract $\widehat{Var}(\hat{\beta}_1)$ from the estimated covariance matrix for the parameter estimators and form the $(1-\alpha)100\%$ Wald confidence interval for $c\beta_1$ as

$$c\hat{\beta}_1 \pm cZ_{1-\alpha/2}\sqrt{\widehat{Var}(\hat{\beta}_1)},$$

where we used $\widehat{Var}(c\hat{\beta}_1) = c^2\widehat{Var}(\hat{\beta}_1)$.[9] The $(1-\alpha)100\%$ Wald confidence interval for OR then becomes:

$$\exp\left(c\hat{\beta}_1 \pm cZ_{1-\alpha/2}\sqrt{\widehat{Var}(\hat{\beta}_1)}\right). \tag{2.11}$$

The standard interpretation of the confidence interval is

---

[9]In general, if a is a constant and Y is a random variable, we have $Var(aY) = a^2 Var(Y)$; see p. 60 of Casella and Berger (2002).

With $(1-\alpha)100\%$ confidence, the odds of a success change by an amount between <lower limit> to <upper limit> times for every c-unit increase in x,

where the appropriate numerical values are inserted within < >.

The Wald confidence interval generally has a true confidence level close to the stated confidence interval only when there are large samples. When the sample size is not large, profile LR confidence intervals generally perform better. For this setting, we find the set of β_1 values such that

$$-2\log\left(\frac{L(\tilde{\beta}_0, \beta_1|y_1,\ldots,y_n)}{L(\hat{\beta}_0, \hat{\beta}_1|y_1,\ldots,y_n)}\right) < \chi^2_{1,1-\alpha} \qquad (2.12)$$

is satisfied, where $\tilde{\beta}_0$ is the MLE of β_0 following specification of a value of β_1. In most settings, there are no closed-form solutions for the lower and upper limits, so iterative numerical procedures are needed to find them. Once the confidence interval limits for β_1 are found, say, "lower" and "upper," we use the exponential function and take into account a value of c to find the $(1-\alpha)100\%$ profile LR confidence interval for OR:

$$\exp(c \times \text{lower}) < OR < \exp(c \times \text{upper}).$$

A few additional notes are needed about odds ratios before proceeding to an example:

1. Inverting odds ratios less than 1 can be helpful for interpretation purposes.

2. A value of $c = 1$ is sometimes used as a default. However, this may not make sense, depending on the context of the data. For example, if $0.1 < x < 0.2$, there is little sense in reporting a change in odds using $c = 1$. Similarly, if $0 < x < 1000$, a value of $c = 1$ may be too small to be meaningful. Instead, an appropriate value of c should be chosen in the context of the explanatory variable. Absent any other guidance, taking c to be the standard deviation of x can be a reasonable choice.

3. When there is more than one explanatory variable, the odds ratio can be shown to be $\exp(c\beta_r)$ for x_r in the model. The same interpretation of the odds ratio generally applies, with the addition of "holding the other explanatory variables constant." If the model contains interaction, quadratic, or other transformations of the explanatory variables, the same interpretation may not apply. Furthermore, odds ratios for a categorical explanatory variable represented by multiple indicator variables do not have the same type of interpretation. We address how to approach transformations in Section 2.2.5 and categorical variables in Section 2.2.6.

4. In many problems, more than one odds ratio is computed. Creating $(1-\alpha)100\%$ confidence intervals for each odds ratio results in a probability less than $1-\alpha$ that *all* intervals cover their parameters. This is referred to as the problem of performing *simultaneous inference* or *multiple inference* on a *family* (i.e., a group of similar parameters). The problem becomes worse as the size of the family grows. The probability that all intervals cover their targets is the *familywise confidence level*. In order to maintain a familywise confidence level of $(1-\alpha)100\%$, the individual intervals need to be adjusted so that each has a somewhat higher confidence level. The *Bonferroni adjustment* is one simple way to achieve this: if the size of the family is g, then each confidence interval is found using a $1-\alpha/g$ confidence level. This method is easy to use but tends to be somewhat conservative: it produces intervals with a familywise confidence level greater than $(1-\alpha)$, and hence are wider than necessary. Better methods of control exist and are available in certain functions within R (examples to be discussed shortly).

Example: Placekicking (Placekick.R, Placekick.csv)

For this example, we use the `mod.fit` object from the logistic regression model that uses only `distance` as an explanatory variable. The estimated odds ratio for `distance` is calculated as

```
> exp(mod.fit$coefficients[2])
distance
0.8913424
> exp(-10*mod.fit$coefficients[2])
distance
3.159034
```

We see that $\exp(\hat{\beta}_1) = 0.8913$ with $c = 1$. Because a 1-yard increment is rather small (field-goal attempts range roughly from 20-60 yards), we instead focus on the change in the odds of success for a 10-yard increment. Also, because the estimated odds of success are lower for an increase in distance ($\exp(c\hat{\beta}_1) < 1$ for $c > 0$), we focus on $c = -10$ for our primary interpretation. Thus, we find that the estimated odds of success change by 3.16 times for every 10-yard decrease in the distance of the placekick. If a football coach were given a choice between attempting a 50-yard placekick or trying to gain 10 more yards and attempting a 40-yard placekick (or any other 10-yard decrease), the fact that the estimated odds of a success are 3.16 times larger for the shorter distance could factor into the coach's decision. Therefore, the preference would be for shorter placekicks. While football fans will not be surprised that shorter kicks have a higher chance of success, the amount by which the odds of success change is not widely known.

The `confint()` function provides a convenient way to calculate confidence intervals associated with regression parameters. This is a generic function that by default finds profile LR intervals for our setting. We use this function below to calculate the interval corresponding to the distance of the placekick:

```
> beta.ci <- confint(object = mod.fit, parm = "distance", level =
    0.95)
Waiting for profiling to be done...
> beta.ci
      2.5 %      97.5 %
-0.13181435 -0.09907104
> rev(exp(-10*beta.ci))    # OR C.I. for c = -10
   97.5 %     2.5 %
  2.693147  3.736478
> # Remove labels with as.numeric()
> as.numeric(rev(exp(-10*beta.ci)))
[1] 2.693147 3.736478
```

The 95% profile LR confidence interval for the `distance` parameter is $-0.1318 < \beta_1 < -0.0991$. Using $c = -10$, the 95% profile LR interval for the odds ratio is $2.69 < OR < 3.74$ where $OR = \exp(-10\beta_1)$. With 95% confidence, the odds of a success change by an amount between 2.69 to 3.74 times for every 10-yard decrease in the distance of the placekick. Because the interval is entirely above 1, there is sufficient evidence that a 10-yard decrease in distance increases the odds of a successful placekick. Note that we use `as.numeric()` in the previous code to prevent unnecessary labels from being printed.

In order to calculate a Wald interval, we need to use the specific method function `confint.default()`:

```
> beta.ci <- confint.default(object = mod.fit, parm = "distance",
    level = 0.95)
> beta.ci
              2.5 %       97.5 %
distance -0.1313664 -0.09868691
> rev(1/exp(beta.ci*10))   # Invert OR C.I. for c = 10
[1] 2.682822 3.719777
```

The 95% Wald confidence interval for the `distance` parameter is $-0.1314 < \beta_1 < -0.0987$. Using $c = -10$, we obtain the 95% Wald interval $2.68 < OR < 3.72$. This interval is similar to the profile LR interval due to the large sample size.

To see how the computations for the Wald interval are performed, the code below shows Equation 2.11 coded directly into R:

```
> beta.ci <- mod.fit$coefficients[2] + qnorm(p = c(0.025, 0.975))
    * sqrt(vcov(mod.fit)[2,2])
> beta.ci
[1] -0.13136638 -0.09868691
> rev(1/exp(beta.ci*10))
[1] 2.682822 3.719777
```

The `vcov()` function calculates the estimated covariance matrix for the parameter estimates using the information within `mod.fit`. By specifying `vcov(mod.fit)[2,2]`, we extract $\widehat{Var}(\hat{\beta}_1)$ from the matrix. The `mod.fit$coefficients[2]` syntax extracts $\hat{\beta}_1$ from the vector of parameter estimates. Putting these elements together, we calculate the confidence interval for β_1 and then the desired confidence interval for the odds ratio.

To see how the computations for the profile LR interval are performed, we include two sets of code in the corresponding program that compute the interval without the `confint()` function. Both rely on rewriting Equation 2.12 as

$$-2\left[\log\left(L(\tilde{\beta}_0, \beta_1|\mathbf{y})\right) - \log\left(L(\hat{\beta}_0, \hat{\beta}_1|\mathbf{y})\right)\right] - \chi^2_{1,0.95} < 0. \qquad (2.13)$$

The first method sequentially calculates the left-side of Equation 2.13 for a range of possible β_1 values. The smallest and largest values of β_1 that satisfy the inequality correspond to the lower and upper interval limits for β_1. The second method replaces the less than sign in Equation 2.13 with an equal sign and iteratively solves for β_1 values using the `uniroot()` function. After the lower and upper limits are found by either method, the `exp()` function is used to find the interval limits for the odds ratio.

2.2.4 Probability of success

Once a logistic regression model is estimated, it is often of interest to estimate the probability of success for a set of explanatory variable values. This can be done simply by substituting the parameter estimates into the model:

$$\hat{\pi} = \frac{\exp(\hat{\beta}_0 + \hat{\beta}_1 x_1 + \cdots + \hat{\beta}_p x_p)}{1 + \exp(\hat{\beta}_0 + \hat{\beta}_1 x_1 + \cdots + \hat{\beta}_p x_p)}.$$

Because $\hat{\pi}$ is a statistic, it will vary from sample to sample. Therefore, we need to find a confidence interval for π in order to make inferences with a particular level of confidence. Both Wald and profile LR intervals will be discussed in this section.

To help explain how the Wald interval for π is calculated, consider again a logistic regression model with only one explanatory variable. The estimated probability of success is then $\hat{\pi} = \exp(\hat{\beta}_0 + \hat{\beta}_1 x)/[1 + \exp(\hat{\beta}_0 + \hat{\beta}_1 x)]$. The normal distribution is a better approximation to the distribution of $\hat{\beta}_0$ and $\hat{\beta}_1$ than it is for $\hat{\pi}$ (see Exercise 31), so we proceed in a manner similar to finding the Wald interval for the odds ratio. We first find a confidence interval for the linear predictor, $\text{logit}(\pi) = \beta_0 + \beta_1 x$, and then transform this interval's endpoints into an interval for π using the $\exp(\cdot)/[1 + \exp(\cdot)]$ transformation. The $(1-\alpha)100\%$ Wald confidence interval for $\beta_0 + \beta_1 x$ is

$$\hat{\beta}_0 + \hat{\beta}_1 x \pm Z_{1-\alpha/2}\sqrt{\widehat{Var}(\hat{\beta}_0 + \hat{\beta}_1 x)}. \tag{2.14}$$

The estimated variance is found by

$$\widehat{Var}(\hat{\beta}_0 + \hat{\beta}_1 x) = \widehat{Var}(\hat{\beta}_0) + x^2 \widehat{Var}(\hat{\beta}_1) + 2x\widehat{Cov}(\hat{\beta}_0, \hat{\beta}_1),$$

where each variance and covariance term is available from the estimated covariance matrix of the parameter estimates.[10] Using the interval limits for $\beta_0 + \beta_1 x$, the $(1-\alpha)100\%$ Wald confidence interval for π is

$$\frac{\exp\left(\hat{\beta}_0 + \hat{\beta}_1 x \pm Z_{1-\alpha/2}\sqrt{\widehat{Var}(\hat{\beta}_0 + \hat{\beta}_1 x)}\right)}{1 + \exp\left(\hat{\beta}_0 + \hat{\beta}_1 x \pm Z_{1-\alpha/2}\sqrt{\widehat{Var}(\hat{\beta}_0 + \hat{\beta}_1 x)}\right)}. \tag{2.15}$$

Note that the lower (upper) limit for π uses the minus (plus) sign in the $\pm$ part of Equation 2.15.

When there are p explanatory variables in the model, the Wald interval for π is found in the same manner. The interval is

$$\frac{\exp\left(\hat{\beta}_0 + \hat{\beta}_1 x_1 + \cdots + \hat{\beta}_p x_p \pm Z_{1-\alpha/2}\sqrt{\widehat{Var}(\hat{\beta}_0 + \hat{\beta}_1 x_1 + \cdots + \hat{\beta}_p x_p)}\right)}{1 + \exp\left(\hat{\beta}_0 + \hat{\beta}_1 x_1 + \cdots + \hat{\beta}_p x_p \pm Z_{1-\alpha/2}\sqrt{\widehat{Var}(\hat{\beta}_0 + \hat{\beta}_1 x_1 + \cdots + \hat{\beta}_p x_p)}\right)},$$

where the variance expression is found in a similar manner as in the single variable case:

$$\widehat{Var}(\hat{\beta}_0 + \hat{\beta}_1 x_1 + \cdots + \hat{\beta}_p x_p) = \sum_{i=0}^{p} x_i^2 \widehat{Var}(\hat{\beta}_i) + 2 \sum_{i=0}^{p-1} \sum_{j=i+1}^{p} x_i x_j \widehat{Cov}(\hat{\beta}_i, \hat{\beta}_j),$$

with $x_0 = 1$. While this variance expression may be long, it is calculated automatically by the `predict()` function as illustrated in the next example.

Profile LR intervals are also calculated for $\text{logit}(\pi)$ first and then transformed into intervals for π. The same methodology as in Section 2.2.3 is used for calculating the first interval, but now the computation is more difficult because of the larger number of parameters involved. For example, in a model with one explanatory variable, $\text{logit}(\pi) = \beta_0 + \beta_1 x$ is a linear combination of β_0 and β_1. The numerator of $-2\log(\Lambda)$ involves maximizing the

---

[10]This is another application of the following result: $Var(aU + bV) = a^2 Var(U) + b^2 Var(V) + 2abCov(U, V)$, where U and V are random variables and a and b are constants. See p. 171 of Casella and Berger (2002).

likelihood function with a constraint for this linear combination, which is more difficult than constraining just one parameter. The problem becomes even more complicated when there are several explanatory variables, and in some cases the iterative numerical procedures may take excessively long to run or even fail to converge. Assuming that the computation is successful and an interval for $\beta_0 + \beta_1 x$ is found, we perform the $\exp(\cdot)/[1 + \exp(\cdot)]$ transformation to obtain an interval for π.

We use the `mcprofile` package (a user-contributed package that is not in the default R installation) to calculate profile likelihood intervals for π and for most other linear combinations of parameters that will be discussed in our book. We use this package because it is the most general package available to calculate these types of intervals. However, earlier versions of it produced questionable results at times. Version 0.1-7 of the package (the most current version at this writing) was used for all computations in our book, and these questionable results no longer occur. Still, we suggest the following course of action when computing profile LR confidence intervals with `mcprofile`:

1. Calculate a Wald interval.

2. Calculate a profile LR interval with the `mcprofile` package.

3. Use the profile LR interval as long as it is not outlandishly different than the Wald and there are no warning messages given by R when calculating the interval. Otherwise, use the Wald interval.

Our statements here about `mcprofile` are not meant to alarm readers about calculations performed by user-contributed packages in general. Rather, there are a vast number of R packages contributed by users, and many of them can perform calculations that no other software can; however, these contributed packages need to be used in a judicious manner. If possible, initial comparisons should be made between calculations performed by a contributed package and those computed in some other trustworthy way to make sure they agree. Also, because R is open source, the code for all functions is available for users to examine if needed. Line-by-line implementations of code within a function, while possibly arduous, can provide the needed assurances that the code works as desired. No other statistical software that is in wide use offers this opportunity for verification.

Example: Placekicking (Placekick.R, Placekick.csv)

For this example, we again use the `mod.fit` object from the logistic regression model that uses only `distance` as an explanatory variable. Below are three ways that the estimated probability of success can be calculated for a distance of 20 yards:

```
> linear.pred <- mod.fit$coefficients[1] +
    mod.fit$coefficients[2] * 20
> linear.pred
(Intercept)
   3.511546
> as.numeric(exp(linear.pred)/(1 + exp(linear.pred)))
0.9710145

> predict.data <- data.frame(distance = 20)
> predict(object = mod.fit, newdata = predict.data, type = "link")
1   3.511546
> predict(object = mod.fit, newdata = predict.data, type =
    "response")
1   0.9710145
```

```
> head(placekick$distance == 20)
[1] FALSE FALSE  TRUE FALSE  TRUE FALSE
> mod.fit$fitted.values[3]  # 3rd obs. has distance = 20
3  0.9710145
```

The first way directly calculates the linear predictor as $\hat{\beta}_0 + \hat{\beta}_1 x = 5.8121 - 0.1150 \times 20 = 3.5112$ resulting in $\hat{\pi} = \exp(3.5112)/[1 + \exp(3.5112)] = 0.9710$. Note that the "(Intercept)" heading is a label left over from mod.fit$coefficients[1], which can be removed by using as.numeric().

The second way to calculate $\hat{\pi}$ is to use the predict() function, and we recommend using this function most of the time for these types of calculations. The function is a generic function that actually uses the predict.glm() method function to perform calculations. To use predict(), a data frame must contain the explanatory variable values at which the estimates of π are desired. This data frame is included then in the newdata argument of predict(). Additionally, the object argument specifies where the model fit information from glm() is located, and the type = "response" argument value instructs R to estimate π. Alternatively, the type = "link" argument value instructs R to estimate $\beta_0 + \beta_1 x$.

If no data frame is given in the newdata argument of predict(), π is estimated for each observation in the data set. The result is exactly the same as given by mod.fit$fitted.values. Thus, a third way to estimate π is to look at values of mod.fit$fitted.values corresponding to the distance of 20 yards. Of course, estimates of π may be of interest for x values not in the data set, but for our current case the third observation corresponds to $x = 20$.

To calculate a Wald confidence interval for π, the easiest way is to use the predict() function to calculate $\hat{\beta}_0 + \hat{\beta}_1 x$ and $\widehat{Var}(\hat{\beta}_0 + \hat{\beta}_1 x)$ first. We then calculate the confidence interval for π using the appropriate code for Equations 2.14 and 2.15:

```
> alpha <- 0.05
> linear.pred <- predict(object = mod.fit, newdata =
    predict.data, type = "link", se = TRUE)
> linear.pred
$fit
1  3.511546
$se.fit
[1] 0.1732003
$residual.scale
[1] 1

> pi.hat <- exp(linear.pred$fit) / (1 + exp(linear.pred$fit))
> CI.lin.pred <- linear.pred$fit + qnorm(p = c(alpha/2,
    1-alpha/2)) * linear.pred$se
> CI.pi <- exp(CI.lin.pred)/(1+exp(CI.lin.pred))
> round(data.frame(predict.data, pi.hat, lower = CI.pi[1], upper
    = CI.pi[2])
  distance pi.hat  lower  upper
1       20  0.971 0.9598 0.9792
```

We use the se = TRUE argument value within predict() to find $\widehat{Var}(\hat{\beta}_0 + \hat{\beta}_1 x)^{1/2}$, the "standard error" for $\hat{\beta}_0 + \hat{\beta}_1 x$. Note that the use of this additional argument causes the resulting object to be a list of component vectors, whereas a single vector was produced when standard errors were not requested. The 95% Wald confidence interval for π is $0.9598 < \pi < 0.9792$; thus, the probability of success for the placekick is quite high at a distance of 20 yards. In the corresponding R program for this example, we also demonstrate how to use predict() with more than one distance and how to perform the calculations by coding Equation 2.15 directly into R.

To find a profile LR interval for π at a distance of 20 yards, we use the code below:

```
> library(package = mcprofile)
> K <- matrix(data = c(1, 20), nrow = 1, ncol = 2)
> K
     [,1] [,2]
[1,]    1   20

> # Calculate -2log(Lambda)
> linear.combo <- mcprofile(object = mod.fit, CM = K)
> # CI for beta_0 + beta_1 * x
> ci.logit.profile <- confint(object = linear.combo, level = 0.95)
> ci.logit.profile

   mcprofile - Confidence Intervals

level:          0.95
adjustment:     single-step

   Estimate lower upper
C1     3.51  3.19  3.87

> names(ci.logit.profile)
[1] "estimate"    "confint"    "CM"         "quant"
[5] "alternative" "level"      "adjust"
> exp(ci.logit.profile$confint)/(1 +
    exp(ci.logit.profile$confint))
      lower     upper
C1 0.9603164 0.9795039
```

After the initial call to the library() function for the mcprofile package, we create a matrix K that contains coefficients for the linear combination of interest. In this case, we want to first find a confidence interval for $1 \times \beta_0 + 20 \times \beta_1$. The mcprofile() function calculates $-2\log(\Lambda)$ for a large number of possible values of the linear combination, where the argument CM is short for "contrast matrix." The confint() function finds the 95% profile LR interval to be $3.19 < \beta_0 + 20\beta_1 < 3.87$. We use the $\exp(\cdot)/[1+\exp(\cdot)]$ transformation then to find the interval for π as $0.9603 < \pi < 0.9795$, which is very similar to the Wald interval due to the large sample size.

We include additional mcprofile examples in our corresponding program that do the following:

1. Calculate an interval for π at more than one distance,

2. Control the familywise confidence level at a specified level,

3. Construct an interval for π using a model that contains both distance and change, and

4. Use the `wald()` function in the `mcprofile` package to find the same Wald intervals as found earlier.

For a linear model with only one explanatory variable x, it is very common to examine a scatter plot of the data with the estimated linear regression model plotted upon it. The purpose of this type of plot is to obtain a visual assessment of trends in the data and to assess the fit of the model. Because the response variable in logistic regression is binary, constructing a plot like this is not very informative because all plotted points would be at $y = 0$ or 1 on the y-axis. However, we can plot the observed proportion of successes at each unique x instead to obtain a general understanding of how well the model fits the data (more formal assessments are discussed in Chapter 5). The estimated model should be expected to pass "near" these proportions as long as each is based on a large enough number of observations. A key to assessment with these types of plots then is knowing the number of observations that exist at each x, so that one can truly distinguish between a poor and good fit. In the extreme case of a continuous x, the number of observations would be 1 at each unique x and this plot generally would not be very useful.

Example: Placekicking (Placekick.R, Placekick.csv)

In an earlier example from Section 2.2.1, we found the number of successes out of the total number of trials for each distance. Using the resulting w.n data frame, we use the `plot()` and `curve()` functions here to plot the observed proportion of successes at each distance and then overlay the estimated logistic regression model:

```
> head(w.n)
  distance success trials proportion
1       18       2      3     0.6667
2       19       7      7     1.0000
3       20     776    789     0.9835
4       21      19     20     0.9500
5       22      12     14     0.8571
6       23      26     27     0.9630
> plot(x = w$distance, y = w$good/n$good, xlab = "Distance
    (yards)", ylab = "Estimated probability", panel.first =
    grid(col = "gray", lty = "dotted"))
> curve(expr = predict(object = mod.fit, newdata =
    data.frame(distance = x), type = "response"), col = "red", add
    = TRUE, xlim = c(18, 66))
```

The `curve()` function evaluates the `predict()` function at 101 distances within the x-axis limits by specifying `newdata = data.frame(distance = x)` for the `expr` argument. The resulting plot is given in Figure 2.4 (excluding the dot-dash lines).

We would also like to plot 95% Wald confidence intervals for π at each possible distance. These *bands* are added to the plot by using two more calls to `curve()`:

```
> ci.pi <- function(newdata, mod.fit.obj, alpha){
    linear.pred <- predict(object = mod.fit.obj, newdata =
        newdata, type = "link", se = TRUE)
    CI.lin.pred.lower <- linear.pred$fit - qnorm(p =
        1-alpha/2)*linear.pred$se
    CI.lin.pred.upper <- linear.pred$fit + qnorm(p =
        1-alpha/2)*linear.pred$se
```

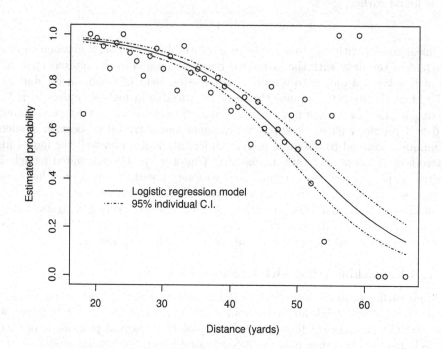

Figure 2.4: Estimated probability of success for a placekick.

```
    CI.pi.lower <- exp(CI.lin.pred.lower) / (1 +
        exp(CI.lin.pred.lower))
    CI.pi.upper <- exp(CI.lin.pred.upper) / (1 +
        exp(CI.lin.pred.upper))
    list(lower = CI.pi.lower, upper = CI.pi.upper)
  }

> # Test case
> ci.pi(newdata = data.frame(distance = 20), mod.fit.obj =
    mod.fit, alpha = 0.05)
$lower
      1
0.95977

$upper
        1
0.9791843

> # Plot C.I. bands
> curve(expr = ci.pi(newdata = data.frame(distance = x),
    mod.fit.obj = mod.fit, alpha = 0.05)$lower, col = "blue", lty
    = "dotdash", add = TRUE, xlim = c(18, 66))
> curve(expr = ci.pi(newdata = data.frame(distance = x),
    mod.fit.obj = mod.fit, alpha = 0.05)$upper, col = "blue", lty
    = "dotdash", add = TRUE, xlim = c(18, 66))
```

```
> # Legend
> legend(locator(1), legend = c("Logistic regression model", "95%
    individual C.I."), lty = c("solid", "dotdash"), col = c("red",
    "blue"), bty = "n")
```

Because the `expr` argument of `curve()` requires a function name as its value, we wrote a new function called `ci.pi()` that performs our confidence limit calculations. This new function essentially contains the same code as used earlier to calculate a Wald confidence interval for π, and it stores the lower and upper limits as separate components, so that each may be plotted with a separate call to `curve()`. The `ci.pi()` function can be used more generally to calculate Wald confidence intervals for π whenever they are needed. At the end of the code, we use the `legend()` function to clarify what each plotted line represents. The `locator(1)` argument within `legend()` instructs R to put the legend at a location that users click on with their mouse. Alternatively, x and y arguments can be given in `legend()` to specify the x- and y-coordinate values for the upper left corner of the legend box.

To informally assess how well our model fits the data, one may be tempted to compare how closely the plotted proportions in Figure 2.4 are to the estimated model. There are a number of proportions far from the estimated model; for example, examine the plotted proportion at a distance of 18 yards. However, there are only 3 observations at 18 yards, and it just so happened that 1 of the 3 is missed despite a high probability of success. With binary outcomes and a small number of observations for a particular value of x, this can happen and does not necessarily indicate a poor fit for the model.

Notice that the estimated model curve in Figure 2.4 generally goes through the middle of the plotted proportions. However, there are a number of proportions that are far from the estimated model, for example at 18, 56, and 59 yards. These are distances at which very few attempts were recorded—three at 18 and one each at 56 and 59—and so the sample proportions can vary considerably. The estimated model curve should not be expected to come as close to those points as it should to points based on large numbers of trials.

A better way to approach a plot like Figure 2.4 is to incorporate information about the number of trials n_j into every plotting point. This can be done with a bubble plot, where each plotting point is proportional in size to the number of trials. The `symbols()` function produces these plots in R:

```
> symbols(x = w$distance, y = w$good/n$good, circles =
    sqrt(n$good), inches = 0.5, xlab = "Distance (yards)", ylab =
    "Estimated probability", panel.first = grid(col = "gray", lty
    = "dotted"))
```

The x and y arguments within `symbols()` work the same way as in `plot()`. The `circles` argument controls the size of the plotting point where the maximum radius is specified in inches by the `inches` argument. For this plot, we use the number of observations at each unique distance to correspond to the plotting point size. The `sqrt()` function is used only to lessen the absolute differences between the values in `n$good`, and this does not need to be used in general.[11]

---

[11]There are 789 observations at a distance of 20 yards. The next largest number of observations is 30, which is at a distance of 32 yards. Due to the large absolute difference in the number of observations, this causes the plotting point at 20 yards to be very large and the other plotting points to be very small. To lessen this absolute difference, we use a transformation on `n$good`.

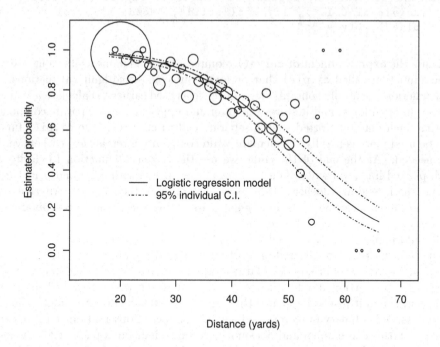

Figure 2.5: Bubble plot of the estimated probability of success for a placekick; the plotting point has a size proportional to the number of observations at a distance.

Figure 2.5 gives the final bubble plot with estimates and Wald confidence interval bands for π included. We can see that indeed the 18-yard placekicks should not cause concern due to the small size of the plotting point. There are similar points at the largest distances, where many of these include only one observation. As expected, the 20-yard placekicks have the largest plotting point, and it appears to be centered close to the estimated model. There are a few large plotting points, such as at 32 yards ($\hat{\pi} = 0.89$, observed proportion of $23/30 = 0.77$) and 51 yards ($\hat{\pi} = 0.49$, observed proportion of $11/15 = 0.73$), for which the model may not fit well. How to more formally assess these observations and others will be an important subject of Chapter 5 when we examine model diagnostic measures.

We provide additional code in the corresponding program which gives the same plot as shown in Figure 2.5 but with profile LR confidence interval bands rather than the Wald confidence interval bands. These two types of bands are indistinguishable on these plots due to the large sample size. One small advantage for using Wald bands here is that they are computed much more quickly than those from profile LR intervals.

2.2.5 Interactions and transformations for explanatory variables

Interactions and transformations extend the variety of shapes that can be proposed for relating π to the explanatory variables. Terms are created and added to the linear predictor in the logistic regression model in exactly the same way as they are used in linear regression (see Chapter 8 of Kutner et al., 2004 for a review). The most common and interpretable terms to add are pairwise (two-way) interactions and quadratic terms. While other types of terms can be added as well (e.g., three-way interactions and cubic transformations), we leave exploring these alternatives as exercises.

Interactions between explanatory variables are needed when the effect of one explanatory variable on the probability of success depends on the value for a second explanatory variable. There are a few ways to include these interactions in a formula argument for glm(). To describe them, suppose there are two explanatory variables named x1 and x2 in a data frame representing x_1 and x_2, and the goal is to estimate the model logit$(\pi) = \beta_0 + \beta_1 x_1 + \beta_2 x_2 + \beta_3 x_1 x_2$. Note that we adopt the convention that the explanatory variables themselves—their *main effects*—are always included in the model when the interaction is included. Below are equivalent ways to code the model:

1. formula = y ~ x1 + x2 + x1:x2

2. formula = y ~ x1*x2

3. formula = y ~ (x1 + x2)^2

In the first formula argument, the colon symbol denotes an interaction between x1 and x2. In the second formula argument the asterisk automatically creates all main effects and interactions among the connected variables. For the third formula argument, the ^2 creates all combinations of up to two variables from among the variables listed within the parentheses. If there is a third variable x3, then (x1 + x2 + x3)^2 is equivalent to x1*x2 + x1*x3 + x2*x3. Also, (x1 + x2 + x3)^3 is equivalent to x1*x2*x3.

Quadratic and higher order polynomials are needed when the relationship between an explanatory variable and logit(π) is not linear. To include a quadratic transformation in a formula argument, the identity function I() needs to be applied to the transformed explanatory variable.[12] For example, the model logit$(\pi) = \beta_0 + \beta_1 x_1 + \beta_2 x_1^2$ is coded as formula = y ~ x1 + I(x1^2). The I() function is needed because, the "^" symbol has a special meaning in formula arguments as we had just seen in the previous paragraph.

The inclusion of interaction terms and/or transformations causes odds ratios to depend on the numerical value of an explanatory variable. For example, consider the model logit$(\pi) = \beta_0 + \beta_1 x_1 + \beta_2 x_2 + \beta_3 x_1 x_2$. The odds ratio associated with changing x_2 by c units while holding x_1 constant is

$$OR = \frac{Odds_{x_2+c}}{Odds_{x_2}} = \frac{\exp(\beta_0 + \beta_1 x_1 + \beta_2(x_2+c) + \beta_3 x_1(x_2+c))}{\exp(\beta_0 + \beta_1 x_1 + \beta_2 x_2 + \beta_3 x_1 x_2)}$$
$$= \exp(c\beta_2 + c\beta_3 x_1) = \exp(c(\beta_2 + \beta_3 x_1)).$$

Thus, the increase or decrease in the odds of success for a c-unit change in x_2 depends on the level of x_1, which follows from the definition of an interaction. As another example, consider the model logit$(\pi) = \beta_0 + \beta_1 x_1 + \beta_2 x_1^2$. The odds ratio for a c-unit change in x_1 is

$$OR = \frac{Odds_{x_1+c}}{Odds_{x_1}} = \frac{\exp(\beta_0 + \beta_1(x_1+c) + \beta_2(x_1+c)^2)}{\exp(\beta_0 + \beta_1 x_1 + \beta_2 x_1^2)}$$
$$= \exp(c\beta_1 + 2cx_1\beta_2 + c^2\beta_2) = \exp(c\beta_1 + c\beta_2(2x_1+c)),$$

which depends on the value of x_1.

---

[12]I() is an "identity" function that instructs R to interpret its argument as it normally would. This function is used to provide argument values to other functions that may have a special interpretation for a symbol or object format.

Confidence intervals for these odds ratios can be more complicated to calculate, because they are generally based on linear combinations of regression parameters, and the `confint()` generic function cannot be used for the task. For the interaction example, where $OR = \exp(c(\beta_2 + \beta_3 x_1))$, the $(1-\alpha)100\%$ Wald confidence interval is

$$\exp\left(c(\hat{\beta}_2 + \hat{\beta}_3 x_1) \pm cZ_{1-\alpha/2}\sqrt{\widehat{Var}(\hat{\beta}_2 + \hat{\beta}_3 x_1)}\right) \quad (2.16)$$

with

$$\widehat{Var}(\hat{\beta}_2 + \hat{\beta}_3 x_1) = \widehat{Var}(\hat{\beta}_2) + x_1^2 \widehat{Var}(\hat{\beta}_3) + 2x_1 \widehat{Cov}(\hat{\beta}_2, \hat{\beta}_3). \quad (2.17)$$

Other Wald intervals can be calculated in a similar manner. Generally, the only challenging part is to find the variance for the linear combination of the parameter estimators, so we provide code to do this in the next example.

Profile LR intervals can be calculated for these odds ratios using the `mcprofile` package. Similar to the Wald interval, the most challenging part is to specify the correct linear combination of parameter estimators. The `wald()` function from this package also provides a somewhat automated way to compute Wald confidence intervals. We will discuss the calculation details in the next example.

Example: Placekicking (Placekick.R, Placekick.csv)

Generally, one would conjecture that the more time that a football is in the air, the more it is susceptible to the effect of wind. Also, longer placekicks generally have more time in the air than shorter kicks. Thus, it would be interesting to see whether there is an interaction between `distance` and `wind`. The `wind` explanatory variable in the data set is a binary variable for placekicks attempted in windy conditions (1) vs. non-windy conditions (0), where windy conditions are defined as a wind stronger than 15 miles per hour at kickoff in an outdoor stadium.[13]

We next estimate a model including this interaction term along with the corresponding main effects:

```
> mod.fit.Ho <- glm(formula = good ~ distance + wind, family =
    binomial(link = logit), data = placekick)
> mod.fit.Ha <- glm(formula = good ~ distance + wind +
    distance:wind, family = binomial(link = logit), data =
    placekick)
> summary(mod.fit.Ha)

Call: glm(formula = good ~ distance + wind + distance:wind,
    family = binomial(link = logit), data = placekick)

Deviance Residuals:
    Min      1Q   Median      3Q      Max
 -2.7291  0.2465  0.2465  0.3791  1.8647
```

---

[13] While wind direction information was available when we collected these data, the stadium orientation was not, making it impossible to account for wind direction relative to the placekicks. Even if orientation had been available, it might not have been useful due to the swirling wind effects that can occur within a stadium. Also, note that wind speed was dichotomized into windy vs. non-windy conditions due to the unknown wind speed inside domed stadiums (only one domed stadium recorded this information). Due to these stadiums' ventilation systems, it is not correct to assume the wind speed is 0.

```
Coefficients:
            Estimate Std. Error z value Pr(>|z|)
(Intercept)   5.684181   0.335962  16.919   <2e-16 ***
distance     -0.110253   0.008603 -12.816   <2e-16 ***
wind          2.469975   1.662144   1.486   0.1373
distance:wind -0.083735  0.043301  -1.934   0.0531 .
---
Signif. codes:  0 '***' 0.001 '**' 0.01 '*' 0.05 '.' 0.1 ' ' 1

(Dispersion parameter for binomial family taken to be 1)

Null deviance: 1013.43  on 1424  degrees of freedom
Residual deviance:  767.42  on 1421  degrees of freedom
AIC: 775.42

Number of Fisher Scoring iterations: 6

> # LRT - Anova(mod.fit.Ha) would also work here too.
> anova(mod.fit.Ho, mod.fit.Ha, test = "Chisq")
Analysis of Deviance Table

Model 1: good ~ distance + wind
Model 2: good ~ distance + wind + distance:wind
  Resid. Df Resid. Dev Df Deviance P(>|Chi|)
1      1422     772.53
2      1421     767.42  1   5.1097   0.02379 *
---
Signif. codes:  0 '***' 0.001 '**' 0.01 '*' 0.05 '.' 0.1 ' ' 1
```

The estimated logistic regression model including the main effects and the interaction is

$$\text{logit}(\hat{\pi}) = 5.6842 - 0.1103 \text{distance} + 2.4700 \text{wind}$$
$$-0.0837 \text{distance} \times \text{wind}.$$

For testing $H_0 : \beta_3 = 0$ vs. $H_a : \beta_3 \neq 0$, the Wald test gives a p-value of 0.0531 and the LRT gives a p-value of 0.0238. Both tests suggest there is marginal evidence to support a wind and distance interaction. The negative coefficient on the distance main effect indicates that when wind = 0, the probability of success decreases with increasing distance. The negative coefficient on the interaction indicates that this effect is exacerbated under windy conditions.

Figure 2.6 displays plots of the estimated logistic regression models for the wind = 0 and 1 cases. The left plot uses the estimated logistic regression model without the interaction (mod.fit.Ho contains the model estimates), and the right plot uses the estimated logistic regression model with the interaction (mod.fit.Ha contains the model estimates). The effect of the interaction is evident in the right plot by the reduction in the estimated probability of success for the longer placekicks when it is windy. To create these plots, we use the curve() function in a similar manner as we did in Section 2.2.4 for Figure 2.4. The code is included in the corresponding program for this example.

Because of the interaction in the model, we need to interpret the effect of wind at specific levels of distance. Similarly, we need to interpret the effect of distance at specific levels of wind. Below is the corresponding R code and output to obtain odds ratio estimates and 95% Wald confidence intervals:

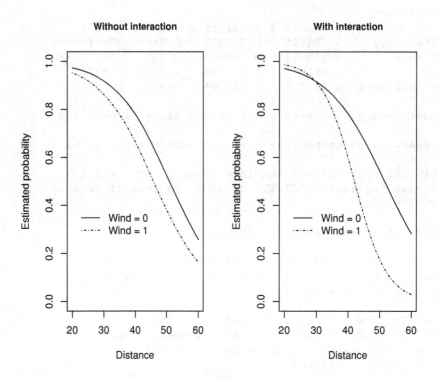

Figure 2.6: The estimated logistic regression model without (left) and with (right) a distance and wind interaction.

```
> beta.hat <- mod.fit.Ha$coefficients[2:4]
> c <- 1
> distance <- seq(from = 20, to = 60, by = 10)

> OR.wind <- exp(c*(beta.hat[2] + beta.hat[3]*distance))
> cov.mat <- vcov(mod.fit.Ha)[2:4,2:4]
> # Var(beta-hat_2 + distance*beta-hat_3)
> var.log.OR <- cov.mat[2,2] + distance^2*cov.mat[3,3] +
    2*distance*cov.mat[2,3]
> ci.log.OR.low <- c*(beta.hat[2] + beta.hat[3]*distance) -
    c*qnorm(p = 0.975)*sqrt(var.log.OR)
> ci.log.OR.up <- c*(beta.hat[2] + beta.hat[3]*distance) +
    c*qnorm(p = 0.975)*sqrt(var.log.OR)

> round(data.frame(distance = distance, OR.hat = 1/OR.wind,
    OR.low = 1/exp(ci.log.OR.up), OR.up = 1/exp(ci.log.OR.low)),2)
  distance OR.hat OR.low OR.up
1       20   0.45   0.09  2.34
2       30   1.04   0.40  2.71
3       40   2.41   1.14  5.08
4       50   5.57   1.54 20.06
5       60  12.86   1.67 99.13
```

We begin by extracting $\hat{\beta}_1, \hat{\beta}_2$, and $\hat{\beta}_3$ from mod.fit$coefficients. While this is not necessary, we do it here so that the resulting vector will have the same indices as the subscripts for the parameter estimates (i.e., `beta.hat[1]` is $\hat{\beta}_1$). The estimated odds ratios are within the `OR.wind` object, where we use $c = 1$ because `wind` is a binary variable and we examine distances of 20, 30, 40, 50, and 60 yards. We use Equations 2.16 and 2.17 to calculate $\widehat{Var}(\hat{\beta}_2 + \hat{\beta}_3 x_1)$ (or equivalently $\widehat{Var}(\log(\widehat{OR}))$, which is how we chose the name for this object) and the confidence interval for OR. For example, the 95% confidence interval for the `wind` odds ratio at a distance of 30 yards is $0.40 < OR < 2.71$. Because 1 is inside the interval, there is insufficient evidence to indicate that windy conditions have an effect on the success or failure of a placekick. On the other hand, the 95% confidence interval for the `wind` odds ratio at a distance of 50 yards is $1.54 < OR < 20.06$. Because 1 is outside the interval, there is sufficient evidence that `wind` has an effect on the success or failure of a placekick at this distance. As originally conjectured, the longer the distance, the more it is susceptible to the wind. Note that Exercise 20 discusses alternative code to calculate these odds ratios using the `multcomp` package.

Profile LR intervals for the odds ratios are obtained using `mcprofile()` and then `confint()`:

```
> K <- matrix(data = c(0, 0, 1, 20,
                       0, 0, 1, 30,
                       0, 0, 1, 40,
                       0, 0, 1, 50,
                       0, 0, 1, 60), nrow = 5, ncol = 4, byrow =
                       TRUE)
> # K <- cbind(0, 0, 1, distance)  # A quicker way to form K
> linear.combo <- mcprofile(object = mod.fit.Ha, CM = K)
> ci.log.OR <- confint(object = linear.combo, level = 0.95,
    adjust = "none")
> data.frame(distance, OR.hat = round(1/exp(ci.log.OR$estimate),
    2), OR.low = round(1/exp(ci.log.OR$confint$upper), 2), OR.up =
    round(1/exp(ci.log.OR$confint$lower), 2))
   distance Estimate OR.low  OR.up
C1       20     0.45   0.06   1.83
C2       30     1.04   0.34   2.42
C3       40     2.41   1.14   5.15
C4       50     5.57   1.68  22.79
C5       60    12.86   2.01 130.85
```

While the profile LR intervals are a little different than the Wald intervals, the same conclusions are reached. Note that the Wald confidence intervals can also be calculated with `wald()`:

```
> save.wald <- wald(object = linear.combo)
> ci.log.OR.wald <- confint(object = save.wald, level = 0.95,
    adjust = "none")
```

Exponentiating the intervals in `ci.log.OR.wald` (and then inverting the intervals), yields the same Wald intervals as calculated earlier.

The estimated odds ratios and corresponding confidence intervals are also found for `distance`, where `wind` is either 0 or 1. For 95% Wald intervals, the odds of a successful placekick change by an amount between 2.54 to 3.56 times for a 10-yard decrease in `distance` when it is not windy (`wind = 0`), and by an amount between 3.03 to 15.98 for a 10-yard decrease in `distance` when it is windy (`wind = 1`). For 95% profile LR

Table 2.1: Indicator variables for a 4-level categorical explanatory variable.

Level	x_1	x_2	x_3
A	0	0	0
B	1	0	0
C	0	1	0
D	0	0	1

intervals, the intervals are 2.55 to 3.58 (not windy) and 3.40 to 18.79 (windy) for a 10-yard decrease in `distance`. The corresponding R code is available in the program for this example.

2.2.6 Categorical explanatory variables

Categorical explanatory variables are represented in a logistic regression model the same way as they are in a linear regression model. If a variable has q categorical levels, then $q-1$ indicator variables can be used to represent it within a model. For example, we saw earlier that `change` in the placekicking data set was a categorical variable with two levels. It was represented in a logistic regression model with one indicator variable (0 for non-lead change placekicks and 1 for lead-change placekicks).

R treats categorical variables as a *factor* object type.[14] By default, R orders the levels using a numerical and then alphabetical ordering, where lowercase letters are ordered before uppercase letters. To see the ordering of any factor, the `levels()` function can be used. This ordering is important because most functions in R use it to construct indicator variables with the "set first level to 0" method of construction.[15] For example, Table 2.1 shows the default R coding for indicator variables of a categorical explanatory variable with the $q=4$ levels coded as "A," "B," "C," and "D." The first level "A" is the base level where all indicator variables are set to 0. The remaining levels each are assigned one indicator variable. In this case, the "B" level is represented by x_1 where $x_1 = 1$ for an observed level of "B" and $x_1 = 0$ otherwise. The "C" and "D" levels are defined in a similar manner for x_2 and x_3, respectively. The `relevel()` function in R can be used to define a new base level if desired, and this will be discussed further in the next example.

A categorical explanatory variable is represented in a logistic regression model using all of its indicator variables. For example, a logistic regression model with the 4-level categorical explanatory variable is written as

$$\text{logit}(\pi) = \beta_0 + \beta_1 x_1 + \beta_2 x_2 + \beta_3 x_3. \tag{2.18}$$

A somewhat less formal representation substitutes the level names for each of their corresponding indicator variables; i.e., $\text{logit}(\pi) = \beta_0 + \beta_1 B + \beta_2 C + \beta_3 D$. To test the importance of a categorical explanatory variable, all its corresponding regression parameters must be set equal to 0 in the null hypothesis. For example, we would test all three indicator variables in Equation 2.18 simultaneously with $H_0 : \beta_1 = \beta_2 = \beta_3 = 0$ vs. H_a : Not all equal to 0, where a LRT or other appropriate testing procedure is used.

---

[14] An exception can occur if the explanatory variable is coded numerically, like `change`. We will discuss how to handle these situations later in this section.

[15] The "set last level to 0" construction is used sometimes by other statistical software, like SAS. Other forms of construction exist; see Section 6.3.5 for an example.

Example: Control of the Tomato Spotted Wilt Virus (TomatoVirus.R, TomatoVirus.csv)

Plant viruses are often spread by insects. This occurs when insects feed on plants already infected with a virus and subsequently become carriers of the virus themselves. When these insects then feed on other plants, they may transmit this virus to these new plants.

To better understand one particular virus, the Tomato Spotted Wilt Virus, and how to control thrips that spread it, researchers at Kansas State University performed an experiment in a number of greenhouses.[16] One hundred uninfected tomato plants were put into each greenhouse. Within each greenhouse, one of two methods was used to introduce the virus to the clean plants:

1. Additional infected plants were placed among the clean ones, and then "uninfected" thrips were released to spread the virus (coded as `Infest = 1`)

2. Thrips that already carried the virus were released onto the clean plants (`Infest = 2`)

To examine ways of controlling the spread of the virus to plants, the researchers used one of three methods:

1. Biological control — Use predatory spider mites to attack the thrips (`Control = "B"`)

2. Chemical control — Use a pesticide to kill the thrips (`Control = "C"`)

3. No control (`Control = "N"`)

Among the plants that were originally clean, the number displaying symptoms of infection were recorded after 8 weeks for each greenhouse. Below is a portion of the data where each row of the `tomato` data frame represents a greenhouse:

```
> tomato <- read.csv(file = "C:\\data\\TomatoVirus.csv")
> head(tomato)
  Infest Control Plants Virus8
1      1       C    100     21
2      2       C    100     10
3      1       B    100     19
4      1       N    100     40
5      2       C    100     30
6      2       B    100     30
```

From row 1 of the data set, we see that 21 out of 100 originally uninfected plants in a greenhouse showed symptoms after 8 weeks, and these plants had infestation method #1 applied while trying to control the thrips with a chemical application.

Both the `Control` and `Infest` explanatory variables are categorical in nature. R automatically treats the `Control` variable as a factor (with a class name of `factor`) because of its character values:

---

[16] Data courtesy of Drs. James Nechols and David Margolies, Department of Entomology, Kansas State University.

```
> class(tomato$Control)
[1] "factor"
> levels(tomato$Control)
[1] "B" "C" "N"
> contrasts(tomato$Control)
  C N
B 0 0
C 1 0
N 0 1
```

The `contrasts()` function shows how R would represent `Control` with indicator variables in a model, similar to what was shown in Table 2.1. Thus, the B level of `Control` is the base level, and there are two indicator variables representing the levels C and N of `Control`. Notice that `Infest` is coded numerically. Because the variable has only two levels, it already is represented by two numerical levels (1 and 2 rather than 0 and 1), so there is no practical difference between using it as numerical or as a factor. However, if it had more than two levels, we would have needed to redefine its class to `factor` using the `factor()` function.[17] Below is the corresponding code:

```
> class(tomato$Infest)
[1] "numeric"
> levels(tomato$Infest)
NULL
> class(factor(tomato$Infest))
[1] "factor"
> levels(factor(tomato$Infest))
[1] "1" "2"
> contrasts(factor(tomato$Infest))
  2
1 0
2 1
> tomato$Infest <- factor(tomato$Infest)
> class(tomato$Infest)
[1] "factor"
```

The `factor()` function can be used throughout the analysis in a similar manner as above. For example, we can use it in the `formula` argument of `glm()` when specifying the variable. Alternatively, we simply replace the original version of `Infest` in the `tomato` data frame with a new factor version. Again, this would normally not be needed for a two-level explanatory variable, but we do it here for demonstration purposes.

Below is the output from estimating the logistic regression model with both the `Infest` and `Control` factors:

```
> mod.fit <- glm(formula = Virus8/Plants ~ Infest + Control,
    family = binomial(link = logit), data = tomato, weights =
    Plants)
> summary(mod.fit)
```

---

[17] An additional use of `factor()` is to change the level ordering, similar to `relevel()` but with more flexibility, and to change the level names. Examples are provided in the corresponding program to this example.

```
Call: glm(formula = Virus8/Plants ~ Infest + Control, family =
    binomial(link = logit), data = tomato, weights = Plants)

Deviance Residuals:
    Min      1Q  Median      3Q     Max
 -4.288  -2.425  -1.467   1.828   8.379

Coefficients:
             Estimate Std. Error z value Pr(>|z|)
(Intercept)   -0.6652     0.1018  -6.533 6.45e-11 ***
Infest2        0.2196     0.1091   2.013   0.0441 *
ControlC      -0.7933     0.1319  -6.014 1.81e-09 ***
ControlN       0.5152     0.1313   3.923 8.74e-05 ***
---
Signif. codes:  0 '***' 0.001 '**' 0.01 '*' 0.05 '.' 0.1 ' ' 1

(Dispersion parameter for binomial family taken to be 1)

    Null deviance: 278.69  on 15  degrees of freedom
Residual deviance: 183.27  on 12  degrees of freedom
AIC: 266.77

Number of Fisher Scoring iterations: 4
```

Because the response variable is given in a binomial form, we use the `weights` argument along with the success/trials formulation in the `formula` argument (see Section 2.2.1). The estimated logistic regression model is

$$\text{logit}(\hat{\pi}) = -0.6652 + 0.2196\text{Infest2} - 0.7933\text{ControlC} + 0.5152\text{ControlN},$$

where we use R's notation of `Infest2` for the indicator variable representing level 2 of `Infest` and "ControlC" and "ControlN" for the indicator variables representing levels "C" and "N" of `Control`, respectively.

Based on the positive estimated parameter for `Infest2`, the probability of showing symptoms is estimated to be larger in greenhouses where infestation method #2 is used. This is somewhat expected because the thrips are already virus carriers with this method. Also, based on the estimated parameters for `ControlC` and `ControlN`, the estimated probability of showing symptoms is lowest for the chemical control method and highest for when no control method is used. This model and interpretation assume that there is no interaction between the infestation and control methods. We next examine how to include these interactions and evaluate their importance.

Interactions

To represent an interaction between a categorical explanatory variable and a continuous explanatory variable in a model, all pairwise products of the variable terms are needed. For example, consider Equation 2.18 with an additional $\beta_4 v$ term added to it, where v generically denotes a continuous explanatory variable. The interaction between the 4-level categorical explanatory variable and v is represented by the inclusion of three pairwise products between v and x_1, x_2, and x_3 in the model with regression parameters as coefficients.

To represent a pairwise interaction between two categorical explanatory variables, all pairwise products between the two sets of indicator variables are included in the model with appropriate regression parameters as coefficients. For example, suppose that there is a 3-level factor X and a 3-level factor Z, represented by appropriate indicator variables x_1 and x_2 for X, and z_1, and z_2 for Z. The logistic regression model with the interaction between X and Z is

$$\text{logit}(\pi) = \beta_0 + \beta_1 x_1 + \beta_2 x_2 + \beta_3 z_1 + \beta_4 z_2 + \beta_5 x_1 z_1 + \beta_6 x_1 z_2 \quad (2.19)$$
$$+ \beta_7 x_2 z_1 + \beta_8 x_2 z_2.$$

To test the interaction, the null hypothesis is that all of the regression parameters for interactions terms are 0 ($H_0 : \beta_5 = \beta_6 = \beta_7 = \beta_8 = 0$ in our example) and the alternative is that at least one of these parameters is not 0. Examples are given next.

Example: Control of the Tomato Spotted Wilt Virus (TomatoVirus.R, TomatoVirus.csv)

In actual application, it is important to consider the possibility of an interaction between the infestation and control methods. The code below shows how to include this interaction and evaluate its importance:

```
> mod.fit.inter <- glm(formula = Virus8/Plants ~ Infest + Control
    + Infest:Control, family = binomial(link = logit), data =
    tomato, weights = Plants)
> summary(mod.fit.inter)

Call: glm(formula = Virus8/Plants ~ Infest + Control +
    Infest:Control, family = binomial(link = logit), data =
    tomato, weights = Plants)

Deviance Residuals:
    Min      1Q  Median      3Q     Max
  -3.466  -2.705  -1.267   2.811   6.791

Coefficients:
                 Estimate Std. Error z value Pr(>|z|)
(Intercept)       -1.0460     0.1316  -7.947 1.92e-15 ***
Infest2            0.9258     0.1752   5.283 1.27e-07 ***
ControlC          -0.1623     0.1901  -0.854    0.393
ControlN           1.1260     0.1933   5.826 5.68e-09 ***
Infest2:ControlC  -1.2114     0.2679  -4.521 6.15e-06 ***
Infest2:ControlN  -1.1662     0.2662  -4.381 1.18e-05 ***
---
Signif. codes:  0 '***' 0.001 '**' 0.01 '*' 0.05 '.' 0.1 ' ' 1

(Dispersion parameter for binomial family taken to be 1)

    Null deviance: 278.69  on 15  degrees of freedom
Residual deviance: 155.05  on 10  degrees of freedom
AIC: 242.55

Number of Fisher Scoring iterations: 4

> library(package = car)
> Anova(mod.fit.inter)
Analysis of Deviance Table (Type II tests)
```

```
Response: Virus8/Plants
               LR Chisq Df Pr(>Chisq)
Infest            4.060  1    0.0439 *
Control          91.584  2  < 2.2e-16 ***
Infest:Control   28.224  2  7.434e-07 ***
---
Signif. codes:  0 '***' 0.001 '**' 0.01 '*' 0.05 '.' 0.1 ' ' 1
```

The estimated logistic regression model is

$$\text{logit}(\hat{\pi}) = -1.0460 + 0.9258\text{Infest2} - 0.1623\text{ControlC} + 1.1260\text{ControlN}$$
$$- 1.2114\text{Infest2} \times \text{ControlC} - 1.1662\text{Infest2} \times \text{ControlN}.$$

A LRT to evaluate the importance of the interaction term tests the regression parameters corresponding to Infest2×ControlC and Infest2 × ControlN, say β_4 and β_5. The hypotheses are $H_0 : \beta_4 = \beta_5 = 0$ vs. $H_a : \beta_4 \neq 0$ and/or $\beta_5 \neq 0$. Equivalently, we could also write the hypotheses in terms of model comparisons:

$$H_0 : \text{logit}(\pi) = \beta_0 + \beta_1\text{Infest2} + \beta_2\text{ControlC} + \beta_3\text{ControlN}$$
$$H_a : \text{logit}(\pi) = \beta_0 + \beta_1\text{Infest2} + \beta_2\text{ControlC} + \beta_3\text{ControlN} +$$
$$\beta_4\text{Infest2} \times \text{ControlC} + \beta_5\text{Infest2} \times \text{ControlN}.$$

The test statistic is $-2\log(\Lambda) = 28.224$, and the p-value is 7.4×10^{-7} using a χ_2^2 approximation. Thus, there is strong evidence of an interaction between the infestation and control methods.

Odds ratios

Consider again the model given in Equation 2.18. The odds of a success at level B is $\exp(\beta_0 + \beta_1)$ because $x_1 = 1, x_2 = 0$, and $x_3 = 0$, and the odds of success at level A is $\exp(\beta_0)$ because $x_1 = 0, x_2 = 0$, and $x_3 = 0$. The resulting odds ratio comparing level B to A is

$$\frac{Odds_{x_1=1,x_2=0,x_3=0}}{Odds_{x_1=0,x_2=0,x_3=0}} = \frac{\exp(\beta_0 + \beta_1)}{\exp(\beta_0)} = \exp(\beta_1).$$

In a similar manner, one can show that the odds ratio comparing C to A is $\exp(\beta_2)$ and the odds ratio comparing D to A is $\exp(\beta_3)$. We can use the same technique to compare non-base levels. For example, the odds ratio comparing level B to level C is

$$\frac{Odds_{x_1=1,x_2=0,x_3=0}}{Odds_{x_1=0,x_2=1,x_3=0}} = \frac{\exp(\beta_0 + \beta_1)}{\exp(\beta_0 + \beta_2)} = \exp(\beta_1 - \beta_2).$$

Estimates and confidence intervals are formed in the same way as discussed in Sections 2.2.3 and 2.2.5. For example, the estimated odds ratio comparing level B to C is $\exp(\hat{\beta}_1 - \hat{\beta}_2)$, and the $(1 - \alpha)100\%$ Wald confidence interval for $\exp(\beta_1 - \beta_2)$ is

$$\exp\left(\hat{\beta}_1 - \hat{\beta}_2 \pm Z_{1-\alpha/2}\sqrt{\widehat{Var}(\hat{\beta}_1 - \hat{\beta}_2)}\right)$$

with

$$\widehat{Var}(\hat{\beta}_1 - \hat{\beta}_2) = \widehat{Var}(\hat{\beta}_1) + \widehat{Var}(\hat{\beta}_2) - 2\widehat{Cov}(\hat{\beta}_1, \hat{\beta}_2).$$

Table 2.2: Odds ratios comparing levels of X conditional on a level of Z.

Z	X	OR
A	B to A	$\exp(\beta_1)$
	C to A	$\exp(\beta_2)$
	B to C	$\exp(\beta_1 - \beta_2)$
B	B to A	$\exp(\beta_1 + \beta_5)$
	C to A	$\exp(\beta_2 + \beta_7)$
	B to C	$\exp(\beta_1 - \beta_2 + \beta_5 - \beta_7)$
C	B to A	$\exp(\beta_1 + \beta_6)$
	C to A	$\exp(\beta_2 + \beta_8)$
	B to C	$\exp(\beta_1 - \beta_2 + \beta_6 - \beta_8)$

When there are two categorical explanatory variables X and Z, comparing two levels for one of the variables is again performed by forming the ratios of the odds. When there is an interaction between the variables, these odds ratios depend on the level for the other explanatory variable. Consider again Equation 2.19, which models the log-odds of success against two 3-level categorical explanatory variables (say, with levels A, B, and C) and their interaction. A typical analysis focusing on X is to compute odds ratios to compare B to A, C to A, and B to C. Due to the interaction, each of these odds ratios needs to be computed conditional on a level of Z. Thus, there are 9 different odds ratios among levels of X. These odds ratios are displayed in Table 2.2. For example, the comparison of levels B and C of X at level B of Z is found by

$$\frac{Odds_{x_1=1,x_2=0,z_1=1,z_2=0}}{Odds_{x_1=0,x_2=1,z_1=1,z_2=0}} = \frac{\exp(\beta_0 + \beta_1 + \beta_3 + \beta_5)}{\exp(\beta_0 + \beta_2 + \beta_3 + \beta_7)} = \exp(\beta_1 - \beta_2 + \beta_5 - \beta_7).$$

Odds ratios for comparing levels of Z at each level of X can be found in the same way if needed.

Notice that if there is no interaction, then the parameters that define the interaction between X and Z—$\beta_5, \beta_6, \beta_7$, and β_8—are all 0. It is then apparent from Table 2.2 that the odds ratios comparing levels of X are the same at all levels of Z, and therefore need to be computed only once.

Example: Control of the Tomato Spotted Wilt Virus (TomatoVirus.R, TomatoVirus.csv)

Because the interaction between the infestation and control methods is significant, we would normally focus only on the model that includes the interaction. However, it is instructive first to see how calculations are performed using the model without the interaction. We will later perform the more complicated investigations with the model including the interaction.

To understand the effect of `Control` on plants displaying symptoms of infection, we compute estimated odds ratios for all possible comparisons among the levels:

```
> exp(mod.fit$coefficients[3:4])
ControlC ControlN
0.452342 1.674025

> # Control N vs. Control C
> exp(mod.fit$coefficients[4] - mod.fit$coefficients[3])
ControlN
3.700795
```

For example, the estimated odds ratio comparing no control to a biological control (the base level) is exp(0.5152) = 1.67. Thus, the estimated odds of plants showing symptoms are 1.67 times as large for using no control methods than using a biological control, where the infestation method is held constant. We invert this odds ratio because we prefer to make a statement about the effects of biological control relative to no control, rather than vice versa.[18] Thus, the estimated odds of plants showing symptoms are $1/1.67 = 0.5973$ times as large for using a biological control method than using no control methods, where the infestation method is held constant. The use of spider mites (biological control) is estimated to reduce the odds of a plant showing symptoms by approximately 40%.

Below is the code to find the corresponding confidence intervals:

```
> # Wald interval
> exp(confint.default(object = mod.fit, parm = c("ControlC",
    "ControlN"), level = 0.95))
              2.5 %     97.5 %
ControlC  0.3492898  0.5857982
ControlN  1.2941188  2.1654579

> # Profile likelihood ratio interval
> exp(confint(object = mod.fit, parm = c("ControlC", "ControlN"),
    level = 0.95))
Waiting for profiling to be done...
              2.5 %     97.5 %
ControlC  0.3486313  0.5848759
ControlN  1.2945744  2.1666574

> # Wald interval for Control N vs. Control C
> beta.hat <- mod.fit$coefficients[-1]  # Matches up beta indices
    with [i] to help avoid mistakes
> exp(beta.hat[3] - beta.hat[2])
ControlN
3.700795
> cov.mat <- vcov(mod.fit)[2:4,2:4]
> var.N.C <- cov.mat[3,3] + cov.mat[2,2] - 2*cov.mat[3,2]
> CI.betas <- beta.hat[3]- beta.hat[2] + qnorm(p = c(0.025,
    0.975))*sqrt(var.N.C)
> exp(CI.betas)
[1] 2.800422 4.890649
```

For example, the 95% Wald confidence interval comparing level "N" to level "B" is 1.29 to 2.17. Thus, with 95% confidence, the odds of plants showing symptoms are between 1.29 and 2.17 times as large when using no control methods rather than using a biological control (holding the infestation method constant). Alternatively, we could also say with 95% confidence, the odds of plants showing symptoms are between 0.46 and 0.77 times as large when using a biological control method rather than using no control methods (holding the infestation method constant). Thus, the use of spider mites (biological control) reduces the odds of a plant showing symptoms by approximately 23% to 54%.

---

[18] Alternatively, we could have set the base level to Control = "N" before estimating the model, which would lead to these types of comparisons more quickly.

In order to compare no control to chemical control, we needed to code $\exp(\hat{\beta}_3 - \hat{\beta}_2 \pm Z_{1-\alpha/2}\widehat{Var}(\hat{\beta}_2 - \hat{\beta}_3)^{1/2})$, where

$$\widehat{Var}(\hat{\beta}_2 - \hat{\beta}_3) = \widehat{Var}(\hat{\beta}_2) + \widehat{Var}(\hat{\beta}_3) - 2\widehat{Cov}(\hat{\beta}_2, \hat{\beta}_3),$$

into R. A simpler way to find the confidence interval is to use the `relevel()` function and change the base level to "C". After re-estimating the model, we can use the `confint.default()` and `confint()` functions to obtain the corresponding confidence interval. This process is demonstrated by code in the corresponding program to this example. We also show examples in the program of using the `mcprofile` package to find the confidence interval without `relevel()`.

Now consider the estimated logistic regression model that includes the interaction. To understand the effect of `Control` on the response, we will need to calculate odds ratios between pairs of `Control` levels, keeping the level of `Infest2` fixed at either 0 (`Infest = 1`) or 1 (`Infest = 2`). The odds ratio comparing level "N" to level "B" with `Infest2 = 0` is $\exp(\beta_3)$:

$$\frac{Odds_{\text{ControlC}=0,\text{ControlN}=1,\text{Infest2}=0}}{Odds_{\text{ControlC}=0,\text{ControlN}=0,\text{Infest2}=0}} = \frac{\exp(\beta_0 + \beta_3)}{\exp(\beta_0)} = \exp(\beta_3).$$

The odds ratio comparing level "N" to level "B" with `Infest2 = 1` is

$$\frac{Odds_{\text{ControlC}=0,\text{ControlN}=1,\text{Infest2}=1}}{Odds_{\text{ControlC}=0,\text{ControlN}=0,\text{Infest2}=1}} = \frac{\exp(\beta_0 + \beta_1 + \beta_3 + \beta_5)}{\exp(\beta_0 + \beta_1)} = \exp(\beta_3 + \beta_5).$$

Other odds ratios can be calculated in a similar manner. Below are all of the estimated odds ratios for `Control` holding `Infest2` constant:

```
> beta.hat <- mod.fit.inter$coefficients[-1]
> N.B.Infest2.0 <- exp(beta.hat[3])
> N.B.Infest2.1 <- exp(beta.hat[3] + beta.hat[5])
> C.B.Infest2.0 <- exp(beta.hat[2])
> C.B.Infest2.1 <- exp(beta.hat[2] + beta.hat[4])
> N.C.Infest2.0 <- exp(beta.hat[3] - beta.hat[2])
> N.C.Infest2.1 <- exp(beta.hat[3] - beta.hat[2] + beta.hat[5] -
    beta.hat[4])

> comparison <- c("N vs. B", "N vs. B", "C vs. B", "C vs. B",
        "N vs. C", "N vs. C")
> data.frame(Infest2 = c(0, 1, 0, 1, 0, 1), Control = comparison,
    OR.hat = round(c(N.B.Infest2.0, N.B.Infest2.1, C.B.Infest2.0,
    C.B.Infest2.1, N.C.Infest2.0, N.C.Infest2.1), 2))
  Infest2 Control OR.hat
1       0 N vs. B   3.08
2       1 N vs. B   0.96
3       0 C vs. B   0.85
4       1 C vs. B   0.25
5       0 N vs. C   3.63
6       1 N vs. C   3.79
```

For example, the estimated odds ratio comparing level "N" to level "B" with `Infest2 = 1` is $\exp(1.1260 - 1.1662) = 0.96$. Thus, the estimated odds of plants showing symptoms are 0.96 times as large for using no control than using a biological control when infected thrips are released into the greenhouse. We can also see why the interaction between Infest and Control was significant. The "N" vs. "B" and "C" vs. "B" odds ratios each differs by a large amount over the two levels of Infest.

We can use similar methods as earlier to find confidence intervals. Because there are many intervals to calculate involving many variances and covariances, it is easier to use the mcprofile package and its matrix formulation of coding:

```
> K <- matrix(data = c(0, 0,  0, 1,  0,  0,
                       0, 0,  0, 1,  0,  1,
                       0, 0,  1, 0,  0,  0,
                       0, 0,  1, 0,  1,  0,
                       0, 0, -1, 1,  0,  0,
                       0, 0, -1, 1, -1,  1), nrow = 6, ncol = 6,
                       byrow = TRUE)
> linear.combo <- mcprofile(object = mod.fit.inter, CM = K)
> ci.log.OR <- confint(object = linear.combo, level = 0.95,
    adjust = "none")
> data.frame(Infest2 = c(0, 1, 0, 1, 0, 1), comparison, OR =
    round(exp(ci.log.OR$estimate),2), OR.CI =
    round(exp(ci.log.OR$confint),2))
   Infest2 comparison Estimate OR.CI.lower OR.CI.upper
C1       0     N vs. B     3.08        2.12        4.52
C2       1     N vs. B     0.96        0.67        1.37
C3       0     C vs. B     0.85        0.58        1.23
C4       1     C vs. B     0.25        0.17        0.36
C5       0     N vs. C     3.63        2.47        5.36
C6       1     N vs. C     3.79        2.54        5.71

> save.wald <- wald(object = linear.combo)
> ci.logit.wald <- confint(object = save.wald, level = 0.95,
    adjust = "none")
> data.frame(Infest2 = c(0, 1, 0, 1, 0, 1), comparison, OR =
    round(exp(ci.log.OR$estimate),2), lower =
    round(exp(ci.logit.wald$confint[,1]),2), upper =
    round(exp(ci.logit.wald$confint[,2]),2))
   Infest2 comparison Estimate lower upper
C1       0     N vs. B     3.08  2.11  4.50
C2       1     N vs. B     0.96  0.67  1.38
C3       0     C vs. B     0.85  0.59  1.23
C4       1     C vs. B     0.25  0.17  0.37
C5       0     N vs. C     3.63  2.46  5.34
C6       1     N vs. C     3.79  2.53  5.68
```

The columns of K are ordered corresponding to the 6 regression parameters estimated by the model. Each row identifies the linear combination of parameters needed for a particular odds ratio. For example, row 2 corresponds to estimating the odds ratio comparing level "N" to level "B" with Infest2 = 1, which is $\exp(\beta_3 + \beta_5)$ (β_3 and β_5 are the fourth and sixth parameters in the model, respectively).

The 95% profile LR interval comparing level N to level B with Infest2 = 1 is (0.67, 1.37). Thus, with 95% confidence, the odds of plants showing symptoms are between 0.67 and 1.37 times as large for using no control than using a biological control when infected thrips are released in the greenhouse. Because 1 is within the interval, there is insufficient evidence to conclude a biological control is effective in this setting. Similar interpretations can be constructed for the other odds ratios. Notice the profile LR interval for comparing level N to level B with `Infest2` = 0 is (2.12, 4.52). Because the interval is above 1, there is sufficient evidence to conclude the biological control reduces the odds of plants showing symptoms when interspersing infected plants with uninfected thrips.

When there are many different odds ratios calculated, one should consider controlling the overall familywise confidence level as described on p. 84. For example, the Bonferroni adjustment could be applied to the $g = 6$ possible confidence intervals for `Control` by using an individual confidence level $1 - 0.05/6 = 0.9917$. This would guarantee that the overall familywise confidence level remains $\geq 95\%$. Using the argument value `adjust = "bonferroni"` and still using `level = 0.95` in `confint.mcprofile()` performs this adjustment automatically. A better form of error rate control is available through the argument value `adjust = "single-step"`, which is based on logic comparable to Tukey's studentized-range statistic for pairwise comparisons in ANOVA as described in Westfall and Young (1993). Results of both methods are given below for profile LR intervals:

```
> # Bonferroni
> ci.log.bon <- confint(object = linear.combo, level = 0.95,
    adjust = "bonferroni")
> # Single Step
> ci.log.ss <- confint(object = linear.combo, level = 0.95,
    adjust = "single-step")
> data.frame(Infest2 = c(0, 1, 0, 1, 0, 1), comparison, bon =
    round(exp(ci.log.bon$confint),2), ss =
    round(exp(ci.log.ss$confint),2))
   Infest2 comparison bon.lower bon.upper ss.lower ss.upper
C1       0     N vs. B      1.86      5.16     1.87     5.12
C2       1     N vs. B      0.59      1.56     0.60     1.55
C3       0     C vs. B      0.51      1.40     0.52     1.39
C4       1     C vs. B      0.15      0.41     0.15     0.41
C5       0     N vs. C      2.17      6.14     2.18     6.09
C6       1     N vs. C      2.22      6.59     2.24     6.54
```

The Bonferroni and single-step-adjusted intervals are very similar, although the single-step intervals are tighter on both ends than the Bonferroni. Both sets of intervals are wider than their unadjusted counterparts given in the previous output box, reflecting their larger individual confidence levels. The advantage of these adjusted intervals is that with either set we can claim that we are 95% confident that the process has covered *all* of the odds ratios, rather than merely having 95% confidence separately in each one.

While we have applied these confidence-level-control methods to the profile LR confidence intervals here, the same adjustment syntax can be applied to Wald intervals produced with this package, which is demonstrated in the program for this example.

2.2.7 Convergence of parameter estimation

The `glm()` function uses computational procedures that iterate until convergence is attained or until the maximum number of iterations is reached. The criterion used to determine convergence is the change in successive values of the residual deviance, rather than the change in successive estimates of the regression parameters. This creates a single, compact criterion that equitably balances the convergence of all parameter estimates simultaneously. If we let $G^{(k)}$ denote the residual deviance at iteration k, then convergence occurs when

$$\frac{|G^{(k)} - G^{(k-1)}|}{0.1 + |G^{(k)}|} < \epsilon,$$

where ϵ is some specified small number greater than 0. The numerator $|G^{(k)} - G^{(k-1)}|$ gives an overall measure of how much $\hat{\pi}_1, \ldots, \hat{\pi}_n$ change (thus, how much $\hat{\beta}_0, \ldots, \hat{\beta}_p$ change) from the previous iteration to the current iteration. The denominator $0.1 + G^{(k)}$ converts this approximately into a proportional change.

The glm() function provides a few ways to control how convergence is determined. First, its epsilon argument allows the user to state the value of ϵ. The default value is epsilon = 10^(-8). Second, the maxit argument states the maximum number of iterations allowed for the numerical procedure, where the default is maxit = 25. Third, the trace = TRUE argument value can be used to see the actual $G^{(k)}$ values for each iteration. The default is trace = FALSE.

Example: Placekicking (Placekick.R, Placekick.csv)

Although the default values work fine for the logistic regression model using distance as the only explanatory variable, we explore the changes that occur when we alter the argument values for epsilon, maxit, and trace:

```
> mod.fit <- glm(formula = good ~ distance, family =
    binomial(link = logit), data = placekick, trace = TRUE,
    epsilon = 0.0001, maxit = 50)
Deviance = 836.7715 Iterations - 1
Deviance = 781.1072 Iterations - 2
Deviance = 775.8357 Iterations - 3
Deviance = 775.7451 Iterations - 4
Deviance = 775.745 Iterations - 5

> mod.fit$control
$epsilon
[1] 1e-04

$maxit
[1] 50

$trace
[1] TRUE

> mod.fit$converged
[1] TRUE
```

The trace = TRUE argument value results in the residual deviances being printed along with the iteration number. The convergence criterion value for iteration $k = 5$ is

$$\frac{|G^{(k)} - G^{(k-1)}|}{0.1 + |G^{(k)}|} = \frac{|775.745 - 775.7451|}{0.1 + |775.745|} = 1.3 \times 10^{-7},$$

which is less than the stated $\epsilon = 0.0001$, so the iterative numerical procedure has stopped. For iteration $k = 4$, the convergence criterion value is 0.00012, which is greater than 0.0001, so this is why the procedure continues. The `control` and `converged` components of the `mod.fit` object are also printed in the output showing the convergence control values and that convergence was obtained.

We include another example in the corresponding program where `maxit = 3`. Because convergence is not achieved in so few iterations, R prints "Warning message: glm.fit: algorithm did not converge" after iteration $k = 3$. While convergence is not achieved, R still allows access to the parameter estimates obtained from the last iteration. As expected, these parameter estimates are different than those when convergence was obtained, and they should not be used.

When convergence does not occur, the first possible solution is to try a larger number of iterations. For example, if the default 25 iterations are not enough, try 50 iterations. If this does not work, there may be some fundamental problem with the data or the model that is interfering with the iterative numerical procedures. The most common problem occurs when an explanatory variable(s) perfectly separates the data between $y_i = 0$ and 1 values; this is often referred to as *complete separation*. In addition to a convergence warning message, `glm()` may also report "glm.fit: fitted probabilities numerically 0 or 1 occurred"; however, this statement alone is not always indicative of model fitting problems. We illustrate the complete separation problem in the next example.

Example: Complete separation (Non-convergence.R)

We begin by creating a simple data set with one explanatory variable x_1 such that $y = 0$ when $x_1 \leq 5$, and $y = 1$ when $x_1 \geq 6$. Thus, x_1 perfectly separates the two possible values of y. Below is the corresponding R code and output:

```
> set1 <- data.frame(x1 = c(1,2,3,4,5,6,7,8,9,10), y =
    c(0,0,0,0,0,1,1,1,1,1))
> set1
  x1 y
1  1 0
2  2 0

<OUTPUT EDITED>

10 10 1

> mod.fit1 <- glm(formula = y ~ x1, data = set1, family =
    binomial(link = logit), trace = TRUE)
Deviance = 4.270292 Iterations - 1
Deviance = 2.574098 Iterations - 2

<OUTPUT EDITED>

Deviance = 7.864767e-10 Iterations - 25
Warning messages:
1: glm.fit: algorithm did not converge
2: glm.fit: fitted probabilities numerically 0 or 1 occurred

> mod.fit1$coefficients
(Intercept)         x1
  -245.84732    44.69951
```

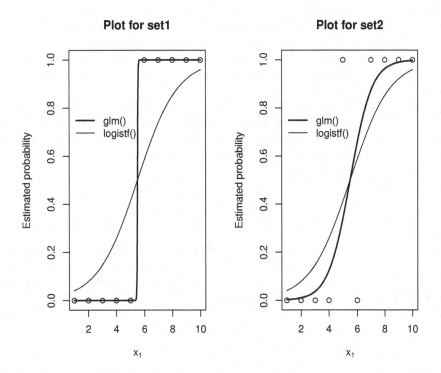

Figure 2.7: Plots showing when complete separation occurs (left) and when complete separation does not occur (right).

R indicates both that convergence did not occur and that the model is trying to create some estimates of π that are practically equal to 0 or 1. The left plot of Figure 2.7 shows the estimated model at the last iteration (code contained within the corresponding program). Because there is a separation between the $y = 0$ and 1 values, the estimated probability of success is appropriately equal to 0 up to $x_1 = 5$ and 1 for $x_1 = 6$ and beyond. The slope of the line between $x = 5$ and 6 will continue to get larger as the iterations are allowed to continue. If the `maxit` argument was set to a higher value than the default of 25, `glm()` will indicate convergence is attained at iteration 26 with $\hat{\beta}_0 = -256.85$ and $\hat{\beta}_1 = 46.70$. However, these estimates should not be used, because they still will be different for a larger number of iterations. For example, $\hat{\beta}_0 = -338.02$ and $\hat{\beta}_1 = 60.55$ for 34 iterations where a smaller value for `epsilon` allowed for a larger number of iterations. Estimates would continue to change for even more iterations, where software precision leads to an eventual maximum number of iterations that can be implemented (34 for this data set in R).

As an illustration, consider what happens if we reverse the y values at $x_1 = 5$ and 6 (see corresponding program for code; data contained in `set2`). Convergence occurs in 6 iterations, and Figure 2.7 (right side) shows a plot of the estimated model. The estimated slope is much smaller than before leading to most of the $\hat{\pi}_i$'s to not be 0 or 1.

Heinze and Schemper, 2002 outline a number of possible options for what to do when complete separation occurs. The most desirable options are to use either (1) exact logistic regression or (2) a modified likelihood function.[19] The first approach uses the exact distribution of each regression parameter estimator conditional on certain features of the data. We provide details of this procedure and other exact methods in Section 6.2. Second, because the likelihood function increases without bound during estimation, the function can be modified (sometimes referred to as being "penalized") to potentially prevent this problem from happening. We detail this approach next.

Firth (1993) proposes a modification to the likelihood function in order to lessen the bias of parameter estimates in fixed sample sizes.[20] Because complete separation leads to an extreme case of bias due to infinite parameter estimates, Firth's proposal is useful here. To help explain this procedure, we write the first derivative with respect to β_r of the log-likelihood function (Equation 2.5) as

$$\sum_{i=1}^{n}(y_i - \pi_i)x_{ir}$$

for $r = 0, \ldots, p$ and $x_{i0} = 1$. This equation is often referred to as the score function (see Appendix B). As discussed in Section 2.2.1, we normally would set the score function equal to 0 and solve for the regression parameters to produce their estimates. Firth suggests adding a penalty to the score function to produce

$$\sum_{i=1}^{n}(y_i - \pi_i + h_i(0.5 - \pi_i))x_{ir},$$

where h_i is the ith diagonal element from the *hat matrix* $\mathbf{H} = \mathbf{V}^{1/2}\mathbf{X}(\mathbf{X}'\mathbf{V}\mathbf{X})^{-1}\mathbf{X}'\mathbf{V}^{1/2}$ (see Section 2.2.1 for matrix definitions).[21] We again set this modified score function to 0 and solve for the regression parameters to produce their estimates. This penalty has the effect of changing y_i, $i = 1, \ldots, n$, from 0 or 1 to values slightly closer to 0.5. Therefore, it prevents any $\hat{\beta}_r$ from straying too far from 0, so that the iterative numerical procedures always converge to finite values (Heinze and Schemper, 2002). Wald and LR inference procedures are used in the same way as with ordinary maximum likelihood estimation. Heinze and Schemper (2002) show that convergence to finite parameter estimates always occurs using this modified score function.

Example: Complete separation (Non-convergence.R)

Below is our code used to estimate the same model as in the last example, but now using the modified score function with set1:

```
> library(package = logistf)
> mod.fit.firth1 <- logistf(formula = y ~ x1, data = set1)
> options(digits = 4)   # Controls printing in R Console window
> summary(mod.fit.firth1)
logistf(formula = y ~ x1, data = set1)
```

---

[19] Another potential solution involves Bayesian methods. We explore this approach in Exercise 7 under "Bayesian methods for categorical data" in Chapter 6.
[20] Bias measures the difference between the expected value of a statistic and the corresponding parameter it is estimating. Ideally, we would like the bias to be 0. Maximum likelihood estimators are approximately unbiased (bias is 0) for large samples, but there may be significant bias present with small samples; see Appendix B for a discussion on bias.
[21] Note that h_i helps to measure the influence of the ith observation on the parameter estimates. We discuss influence and the hat matrix in more detail in Section 5.2.3.

```
Model fitted by Penalized ML
Confidence intervals and p-values by Profile Likelihood

              coef se(coef) lower 0.95 upper 0.95 Chisq        p
(Intercept) -5.3386  3.3227   -30.7818    -0.8313 6.374 0.011579
x1           0.9706  0.5765     0.1962     5.5642 7.759 0.005345

Likelihood ratio test=7.759 on 1 df, p=0.005345, n=10
Wald test = 2.834 on 1 df, p = 0.09226

Covariance-Matrix:
       [,1]    [,2]
[1,] 11.040 -1.8282
[2,] -1.828  0.3324
> options(digits = 7)   # Default
```

The logistf() function from a package by the same name estimates the logistic regression model using Firth's method.[22] The parameter estimates converge, and the estimated model is $\text{logit}(\hat{\pi}) = -5.3386 + 0.9706 x_1$. Figure 2.7 plots the model (see corresponding program for code). Along with the parameter estimates, the summary() output gives information such as: $\widehat{Var}(\hat{\beta}_1)^{1/2} = 0.5765$, a 95% profile likelihood interval of $0.19 < \beta_1 < 5.56$, and a LRT for $H_0 : \beta_1 = 0$ vs. $H_a : \beta_1 \neq 0$ with a p-value = 0.0053, where all calculations use the modified likelihood in their formulations.

Objects resulting from logistf() have a class also named logistf. Most method functions used with glm class objects, like predict() and anova(), are no longer available for use. However, a number of useful items, like the estimates of π at the observed x values, are available within the resulting object (use names(mod.fit.firth1) to see these items). Also, the logistftest() function within the package performs LRTs in much the same manner as anova(), but now with the penalized likelihood function. Please see the examples in our corresponding program.

We conclude this section with three additional notes about complete separation:

1. Complete separation is not necessarily bad if you want to distinguish between the response levels of y. The problem is that the model estimated by maximum likelihood does not provide a good way to interpret the relationship between y and the explanatory variables.

2. It can be difficult to see complete separation graphically if there is more than one explanatory variable, because the separation may arise with respect to some linear combination of the explanatory variables rather than just one. There may even be times when the glm() function does not provide a warning. When parameter estimates have very large magnitudes with extremely large estimated standard deviations, this is a sign that complete separation may exist. Another symptom is that most or all estimated probabilities of success are essentially 0 or 1. Exercise 24 demonstrates this for a data set.

---

[22] In the summary() output, we limit the number of digits displayed by using the digits argument of the options() function. Otherwise, the output is quite wide. Note that we could have also used the width argument of options() to control the width of text within the R Console window.

3. A solution that is often used to address complete separation is to add a pseudo observation to the data in an attempt to force convergence of the parameter estimates. This is somewhat similar to what was done in Section 1.2.5, where we estimated an odds ratio for a contingency table with zero cell counts by adding a small constant to the cell counts. However, Firth's penalized likelihood already does this in a manner that is justified mathematically, so we recommend this latter approach or exact logistic regression instead.

2.2.8 Monte Carlo simulation

Through the use of Monte Carlo simulation, properties of the logistic regression model and associated inference procedures can be investigated. For example, in this section we will examine whether a normal distribution is a good approximation for the probability distribution of the regression parameter estimates. This will help us to determine whether Wald confidence intervals are appropriate.

Example: Simulating data (SamplingDist.R)

Before we can investigate the distribution of our estimators, we need to be able to simulate responses in a logistic regression setting. We arbitrarily consider the logistic regression model of $\text{logit}(\pi) = \beta_0 + \beta_1 x_1$, where we simulate x_1 from a uniform probability distribution that has a minimum and maximum value of 0 and 1, respectively. We choose values of β_0 and β_1 for our "true model" so that $\pi = 0.01$ when $x = 0$ and $\pi = 0.99$ when $x = 1$. Below is our code for finding the values of β_0 and β_1 that satisfy these requirements.

```
> # x=0: logit(pi) = beta_0 + beta_1*0 = beta_0
> pi.x0 <- 0.01
> beta0 <- log(pi.x0/(1-pi.x0))
> beta0
[1] -4.59512

> # x=1: logit(pi) = beta_0 + beta_1*1 = beta_0 + beta_1
> pi.x1 <- 0.99
> beta1 <- log(pi.x1/(1-pi.x1)) - beta0
> beta1
[1] 9.19024

> # Check
> exp(beta0 + beta1*c(0,1)) / (1 + exp(beta0 + beta1*c(0,1)))
[1] 0.01 0.99

> curve(expr = exp(beta0 + beta1*x) / (1 + exp(beta0 + beta1*x)),
    xlim = c(0,1), ylab = expression(pi), n = 1000, lwd = 3, xlab
    = expression(x[1]))
```

Figure 2.8 displays a plot of the true model.

To simulate data from the model, we utilize much of the same methodology as in Chapter 1, where the `rbinom()` function was used to simulate observations from a Bernoulli distribution. The main difference now is that we allow the probability of success parameter to vary as a function of an explanatory variable. We first sample $n = 500$ values from the uniform distribution using the `runif()` function in order to

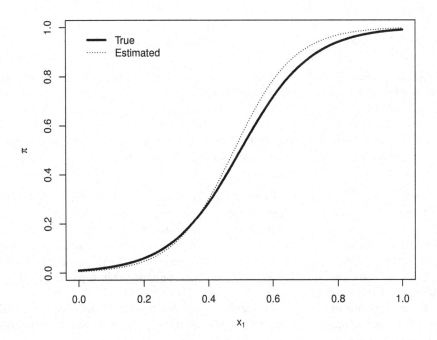

Figure 2.8: True and estimated model.

find values for x. These x values are inserted into our model to find π. We then use the rbinom() function to simulate one binary response y for each x. Below is our corresponding code:

```
> set.seed(8238)
> x <- runif(n = 500, min = 0, max = 1)
> pi <- exp(beta0 + beta1*x1) / (1 + exp(beta0 + beta1*x1))
> set.seed(1829)
> y <- rbinom(n = length(x1), size = 1, prob = pi)
> head(data.frame(y, x1, pi))
  y       x1         pi
1 1 0.4447197 0.37565323
2 1 0.9523950 0.98459621
3 1 0.5019858 0.50456242
4 0 0.1425656 0.03609259
5 0 0.2042915 0.06194091
6 0 0.2137520 0.06718933
> mod.fit <- glm(formula = y ~ x1, family = binomial(link =
    logit))
> mod.fit$coefficients
(Intercept)          x1
   -5.14340    10.71400
> beta.hat0 <- mod.fit$coefficients[1]
> beta.hat1 <- mod.fit$coefficients[2]
```

```
> curve(expr = exp(beta.hat0 + beta.hat1*x) / (1 + exp(beta.hat0
    + beta.hat1*x)), xlim = c(0,1), ylab = expression(pi), add =
    TRUE, col = "red", n = 1000)
> legend(x = 0, y = 1, legend = c("True", "Estimated"), lty =
    c(1,1), col = c("black", "red"), lwd = c(3, 1), bty = "n")
```

For example, the first simulated value for x_1 is 0.4447, which results in $\pi = 0.3757$. The simulated response for this x_1 is a success, $y = 1$. The logistic regression model is fit to the 500 (x,y) pairs, resulting in the estimate, $\text{logit}(\hat{\pi}) = -5.14 + 10.71x_1$. As expected, this is somewhat similar to the true model $\text{logit}(\pi) = -4.60 + 9.19x_1$. Note that we did not use a `data` argument in `glm()` because x and y are defined outside of a data frame. A plot of the estimated model is given in Figure 2.8.

The probability distributions of $\hat{\beta}_0$ and $\hat{\beta}_1$ can be found exactly by randomly generating infinitely many sets of data and then estimating the regression parameters separately on each set. Estimates can be summarized in a histogram or some other form that allows us to examine their distribution. As this is impractical, we can instead approximate the sampling distributions using a large number of data sets, where "large" typically depends on the computational time required to complete the task and the desired precision with which the sampling distribution is to be estimated.

We next carry out this process for our logistic regression model. Using the data simulation process from the last example, the next example discusses how to repeat it. This allows us to investigate the distribution of our estimators.

Example: Probability distribution of a regression parameter (SamplingDist.R)

We repeat the data set simulation process 10,000 times and estimate the model for each simulated data set of size $n = 500$. Generally, a smaller number of replicates for the simulation process would be sufficient; however, a larger value is chosen here in order to obtain good estimates of some extreme distributional quantiles. We do not include our code here due to space restrictions and similarities with the last example. Please see the corresponding program to this example. Note that most of our code is embedded within a function that we created, named `sim.all()`, so that we can repeat the simulation process for different sample sizes.

We would like to point out the structure of the simulation, because the same structure is used frequently for simulations in R. Data are stored in a matrix, where each column of the matrix is one sample data set of n observations. Our data matrix, `y.mat`, stores the binary responses in 500 rows and 10,000 columns. An analysis function is applied to each column of the matrix in one step using the `apply()` function. The syntax for `apply()` is: `apply(X = , MARGIN = , FUN = , ...)`. The X argument is used for the data matrix, the `MARGIN` argument value specifies whether the analysis function is to be applied to rows (=1) or columns (=2) of the matrix, and `FUN` denotes the analysis function to be applied. Alternatively, we could have used a `for()` function to loop over each data set, but `apply()` is more convenient.

Figure 2.9 plots the true model (as in Figure 2.8) with the estimated models from the first 100 data sets. The variability among the estimated models is evident from the plot, but they clearly show the same general trend as the true model from which they were sampled. Note that this picture is highly dependent on the sample size. For example, readers can re-run the simulation with a smaller sample size and see how the variability increases!

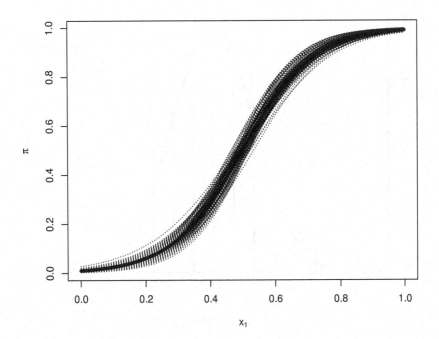

Figure 2.9: True model (black, thick line) and first 100 estimated models.

Table 2.3: Quantiles for the standard normal distribution and for the simulated distribution of z_0.

	0.005	0.025	0.05	0.90	0.95	0.995
Standard normal	-2.576	-1.960	-1.645	1.645	1.960	2.576
Simulated z_0	-2.691	-2.004	-1.666	1.615	1.897	2.382

This variability is further displayed in Figure 2.10. The figure gives a histogram of the 10,000 different $z_0 = (\hat{\beta}_1 - \beta_1)/\widehat{Var}(\hat{\beta}_1)^{1/2}$ values, where we have included a standard normal distribution overlay. Table 2.3 provides a quantile comparison for the simulated z_0 values and for a standard normal distribution. Overall, the standard normal distribution approximation is not too bad. There is some left skewness in the simulated distribution, and this leads to small differences in the extreme quantiles.

Knowing that the probability distribution of the parameter estimates is well-approximated by a normal distribution is important because we need normality for a Wald confidence interval for β_1. In this case, where the normal approximation appears reasonably good, the Wald interval's estimated true confidence level is 0.9520 when the stated level is 0.95 (calculation details are given in the corresponding program). A poor approximation would lead to an interval with a true confidence level quite different from the stated level. Calculation details are given in the corresponding program.

Figure 2.11 displays a histogram of the 10,000 different values of $-2\log(\Lambda) = -2\log[L(\tilde{\beta}_0, \beta_1|y_1, \ldots, y_{500})/L(\hat{\beta}_0, \hat{\beta}_1|y_1, \ldots, y_{500})]$, where we have included a χ_1^2 distribution overlay. Table 2.4 provides a comparison of the quantiles for the simulated $-2\log(\Lambda)$ values and for a χ_1^2 distribution. Similar to the results from z_0, we see that

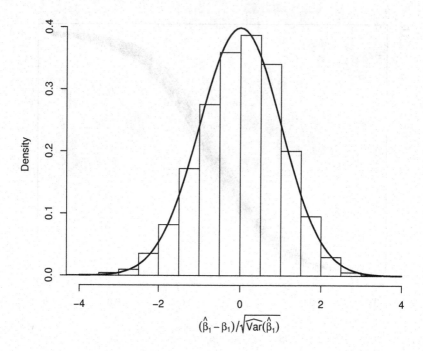

Figure 2.10: Histogram of z_0 with standard normal distribution overlay.

Table 2.4: Quantiles for the χ_1^2 distribution and for the simulated distribution of $-2\log(\Lambda)$.

	0.90	0.95	0.995
χ_1^2	2.706	3.841	7.879
Simulated $-2\log(\Lambda)$	2.750	3.896	8.408

the χ_1^2 distribution approximation is not too bad, but there is more right skewness in the simulated distribution than in a χ_1^2 distribution. This leads to small differences in the extreme quantiles. The estimated true confidence level for the profile LR interval is 0.9482 when the stated level is 0.95.

We have written the program so that readers can explore other sample sizes. For example, the normal and χ^2 distributions do not work as well for smaller sizes and work better for larger sample sizes.

There are many other investigations that can be performed for logistic regression models and associated procedures. For example, we can investigate the *consistency* of the parameter estimates. Estimators are consistent if they continue to get closer to their corresponding parameters as the sample size grows (Casella and Berger, 2002, Section 10.1). We can also investigate the performance of inference procedures for other parameters. For example, we can examine the Wald and profile LR intervals for π at a particular value of x. We encourage readers to perform these investigations on their own.

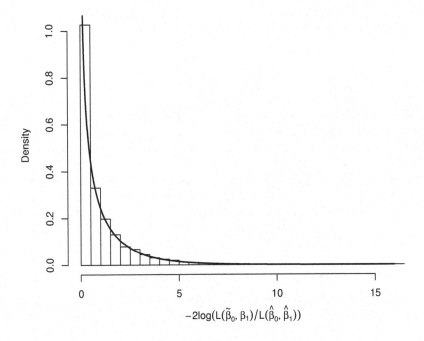

Figure 2.11: Histogram of $-2\log(\Lambda)$ with χ_1^2 distribution overlay.

2.3 Generalized linear models

Logistic regression models fall within a family of models called *generalized linear models* (GLMs). Each generalized linear model has three different parts:

1. RANDOM COMPONENT. This specifies the distribution for Y. For the logistic regression model, Y has a Bernoulli distribution, which is often generalized as a binomial.

2. SYSTEMATIC COMPONENT. This specifies a linear combination of the regression parameters with the explanatory variables, and it is often referred to as the *linear predictor*. For the logistic regression model, we have $\beta_0 + \beta_1 x_1 + \cdots + \beta_p x_p$.

3. LINK FUNCTION. This specifies how the expected value of the random component $E(Y)$ is linked to the systematic component. For the logistic regression model, we have $\text{logit}(\pi) = \beta_0 + \beta_1 x_1 + \cdots + \beta_p x_p$ where $E(Y) = \pi$ and the logit transformation is the link function.

Note that "linear" in generalized linear models comes from using a linear combination of the regression parameters with the explanatory variables in the systematic component. Alternative systematic components can involve more complex functional forms such as x^β. These would then be called generalized *nonlinear* models.

Link functions used with binary regression models

Other generalized linear models are used to model binary responses as well. These binary regression models have the same random and systematic components as the logistic regression model, but their link functions are different than the logit. The most important aspect of the link function in these cases is that its inverse must guarantee $E(Y)$ is between 0 and 1. For example, we saw in Section 2.2 that

$$\pi = \frac{\exp(\beta_0 + \beta_1 x_1 + \cdots + \beta_p x_p)}{1 + \exp(\beta_0 + \beta_1 x_1 + \cdots + \beta_p x_p)}$$

is always between 0 and 1. This guarantee is achieved by using a *cumulative distribution function* (CDF) as the inverse of the link function. In fact, the CDF of a logistic probability distribution is used for logistic regression, which results in the name for the model.

To review CDFs, suppose X is a continuous random variable with a *probability density function* (PDF) $f(x)$. The CDF $F(x)$ gives the area under the plotted PDF to the left of x. More formally, the CDF of X is $F(x) = P(X \leq x) = \int_{-\infty}^{x} f(u) du$.[23] Because all probabilities are between 0 and 1, $0 \leq F(x) \leq 1$ for $-\infty < x < +\infty$. We have used CDFs many times in R already through the use of functions such as pnorm() and pchisq(). For example, pnorm(q = 1.96) can be expressed as $F(1.96)$ where $F(\cdot)$ is the CDF of a standard normal distribution. Equivalently, $Z_{0.975} = 1.96$.

Example: Logistic probability distribution (Logistic.R)

A random variable X with a logistic probability distribution has a PDF of

$$f(x) = \frac{\sigma^{-1} e^{-(x-\mu)/\sigma}}{\left[1 + e^{-(x-\mu)/\sigma}\right]^2}$$

for $-\infty < x < \infty$ and parameters $-\infty < \mu < \infty$ and $\sigma > 0$. The CDF is

$$F(x) = \int_{-\infty}^{x} \frac{\sigma^{-1} e^{-(u-\mu)/\sigma}}{\left[1 + e^{-(u-\mu)/\sigma}\right]^2} du = \frac{1}{1 + e^{-(x-\mu)/\sigma}} = \frac{e^{(x-\mu)/\sigma}}{1 + e^{(x-\mu)/\sigma}}.$$

Figure 2.12 displays plots of $f(x)$ and $F(x)$ where $\mu = -2$ and $\sigma = 2$, and the code used to produce the plots is below:

```
> mu <- -2
> sigma <- 2

> # Examples for f(-2) and F(-2)
> dlogis(x = -2, location = mu, scale = sigma)
[1] 0.125
> plogis(q = -2, location = mu, scale = sigma)
[1] 0.5

> par(mfrow = c(1,2))
> curve(expr = dlogis(x = x, location = mu, scale = sigma), ylab
    = "f(x)", xlab = "x", xlim = c(-15, 15), main = "PDF", col =
    "black", n = 1000)
> abline(h = 0)

> curve(expr = plogis(q = x, location = mu, scale = sigma), ylab
    = "F(x)", xlab = "x", xlim = c(-15, 15), main = "CDF", col =
    "black", n = 1000)
```

---

[23] $f(u)$ rather than $f(x)$ is used as the integrand to avoid confusion between what is being integrated over and the limits of integration.

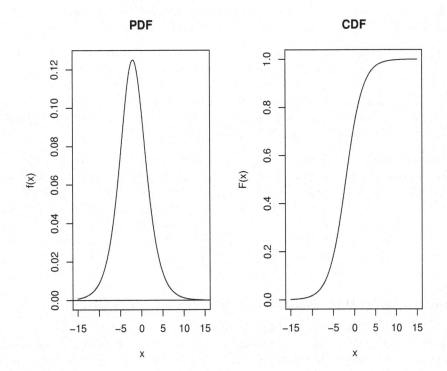

Figure 2.12: PDF and CDF for the logistic probability distribution with $\mu = -2$ and $\sigma = 2$.

The `dlogis()` and `plogis()` functions evaluate the PDF and CDF for a logistic distribution, respectively. We see that the distribution is symmetric and centered at μ. It can be shown that $E(X) = \mu$ and $Var(X) = \sigma^2 \pi^2/3$. In comparison to a normal distribution with the same mean and variance, the distributions are centered at the same location, but the logistic distribution is more spread out (fatter tails).

The plot of the CDF in Figure 2.12 should look very familiar because it is the same as given in the left plot of Figure 2.1. Notice that

$$F(x) = \frac{e^{(x-\mu)/\sigma}}{1 + e^{(x-\mu)/\sigma}} = \frac{e^{\beta_0 + \beta_1 x}}{1 + e^{\beta_0 + \beta_1 x}} = \pi$$

when $\beta_0 = -\mu \sigma^{-1}$ and $\beta_1 = \sigma^{-1}$. For example, with $\mu = -2$ and $\sigma = 2$, we obtain $\beta_0 = 1$ and $\beta_1 = 0.5$, which were the same values used for the left plot of Figure 2.1. Therefore, we see that logistic regression uses the CDF of a logistic distribution to relate the probability of success to the linear predictor. If we invert this relationship to solve for the linear predictor, we obtain

$$\beta_0 + \beta_1 x = \log\left(\frac{F(x)}{1 - F(x)}\right) = F^{-1}(\pi) \qquad (2.20)$$

We can denote the expression in Equation 2.20 by $F^{-1}(\pi)$, because it is the *inverse CDF* for a logistic distribution. Therefore, the link function of a logistic regression model is the inverse CDF of the logistic probability distribution.

Because the CDF always produces a value between 0 and 1, other inverse CDFs are used as link functions for binary regression models. While the logit link function is the one most prevalently used for binary regression, there are two others that are common:

1. Probit regression — The inverse CDF of a standard normal distribution is used as the link function. If we denote the CDF of a standard normal distribution as $\Phi(\cdot)$, the model can be written as $\pi = \Phi(\beta_0 + \beta_1 x_1 + \cdots + \beta_p x_p)$ or $\Phi^{-1}(\pi) = \beta_0 + \beta_1 x_1 + \cdots + \beta_p x_p$. Commonly, probit$(\pi)$ is used instead to denote $\Phi^{-1}(\pi)$ leading to the model written as probit$(\pi) = \beta_0 + \beta_1 x_1 + \cdots + \beta_p x_p$.[24] Note that probit$(\pi)$ is equivalent to Z_π, the standard normal quantile at probability π.

2. Complementary log-log regression — The inverse CDF from a Gumbel distribution (also known as the extreme value distribution) is used to form the link function. The CDF is $F(x) = \exp\{-\exp[-(x-\mu)/\sigma]\}$ for $-\infty < x < \infty$ and parameters $-\infty < \mu < \infty$ and $\sigma > 0$. Rather than setting π equal to the CDF as done for the logistic and probit regression models, the complementary log-log model sets π to be equal $1 - F(x)$, which is essentially the probability resulting from a complement of an event. We can write our model then as $\pi = 1 - \exp[-\exp(\beta_0 + \beta_1 x_1 + \cdots + \beta_p x_p)]$ or $\log[-\log(1-\pi)] = \beta_0 + \beta_1 x_1 + \cdots + \beta_p x_p$.[25]

Example: Compare three binary regression models (PiPlot.R)

Suppose the linear predictor has only one explanatory variable of the form $\beta_0 + \beta_1 x$. Figure 2.13 provides plots of the logistic, probit, and complementary log-log regression models. We choose different values of β_0 and β_1 for each model so that the mean is 2 and the variance 3 for random variables with the corresponding CDFs. Please see the program for the code. Both the logistic and probit regression models are symmetric about $\pi = 0.5$; i.e., above $\pi = 0.5$ is a mirror image of below $\pi = 0.5$. The complementary log-log model does not have this same symmetry, because a Gumbel distribution is a skewed distribution. The main difference between the logistic and probit models is that the logistic model rises a little more slowly toward $\pi = 1$ and falls a little more slowly to $\pi = 0$. This characteristic is due to a logistic distribution having more probability in its tails than a normal distribution.

Estimation and inference for binary regression models

The probit and complementary log-log models are estimated in the same way as the logistic regression model. The difference is that π_i is represented in the log-likelihood function of Equation 2.4 by the corresponding probit or complementary log-log model specification. Iterative numerical procedures are again needed to find the parameter estimates. Once the parameter estimates are found, the same inference procedures as used for the logistic regression model are available for the probit and complementary log-log models. These include the Wald tests and LRTs from Section 2.2.1 and the Wald and profile likelihood intervals from Sections 2.2.3 and 2.2.4. For example, we can use the anova() function for LRTs

---

[24]The term probit is a shortened version of "probability unit" (Hubert, 1992). Please see Chapter 8 of Salsburg (2001) for an interesting account of how this model was developed by Chester Bliss.
[25]By setting π equal to $1 - F(x)$, we are able to keep with standard conventions of what a positive and negative relationship mean with respect to the response variable. If π was set equal to $F(x)$ instead, the model would have a non-standard interpretation. Namely, increases in an explanatory variable, say x_r, where its corresponding β_r is positive, would lead to *decreases* in the probability of success. Conversely, decreases in x_r with β_r negative would lead to *increases* in the probability of success.

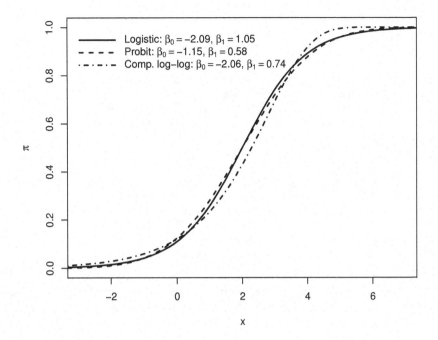

Figure 2.13: Logistic, probit, and complementary log-log models.

involving regression parameters and the `predict()` function for Wald confidence intervals for π.

One of the most important differences among the logistic, probit, and complementary log-log regression models arises when calculating odds ratios. In Section 2.2.3, we showed that the odds ratio for x in the model $\text{logit}(\pi) = \beta_0 + \beta_1 x$ is $OR = e^{c\beta_1}$ for a c-unit increase in x. A very important aspect of this odds ratio was that it is the same for any value of x. Unfortunately, this does not occur for probit and complementary log-log models. For example, consider the model $\text{probit}(\pi) = \beta_0 + \beta_1 x$. The odds of a success are then

$$Odds_x = \Phi(\beta_0 + \beta_1 x)/\left[1 - \Phi(\beta_0 + \beta_1 x)\right]$$

at a particular value of the explanatory variable x. If x is increased by $c > 0$ units, the odds of a success become

$$Odds_{x+c} = \Phi(\beta_0 + \beta_1 x + \beta_1 c)/\left[1 - \Phi(\beta_0 + \beta_1 x + \beta_1 c)\right].$$

When the ratio of these two odds is taken, the result does not have a closed form and depends on the value of x. Odds ratios from the complementary log-log model also depend on x (see Exercise 27). This is one of the main reasons why logistic regression models are used the most among binary regression models.

Example: Placekicking (Placekick.R, Placekick.csv)

The purpose of this example is to compare the estimated logistic, probit, and complementary log-log regression models with respect to the placekicking data set. We use `distance` as the only explanatory variable in the model. Our code to estimate these models is shown below:

```
> mod.fit.logit <- glm(formula = good ~ distance, family =
    binomial(link = logit), data = placekick)
> round(summary(mod.fit.logit)$coefficients, 4)
            Estimate Std. Error  z value Pr(>|z|)
(Intercept)   5.8121     0.3263  17.8133        0
distance     -0.1150     0.0083 -13.7937        0

> mod.fit.probit <- glm(formula = good ~ distance, family =
    binomial(link = probit), data = placekick)
> round(summary(mod.fit.probit)$coefficients, 4)
            Estimate Std. Error  z value Pr(>|z|)
(Intercept)   3.2071     0.1570  20.4219        0
distance     -0.0628     0.0043 -14.5421        0

> mod.fit.cloglog <- glm(formula = good ~ distance, family =
    binomial(link = cloglog), data = placekick)
> round(summary(mod.fit.cloglog)$coefficients, 4)
            Estimate Std. Error  z value Pr(>|z|)
(Intercept)   2.3802     0.1184  20.1011        0
distance     -0.0522     0.0037 -14.0763        0
```

To estimate the probit and complementary log-log models, we change the `link` argument value in `glm()` to its corresponding link function name. The estimated models are:

- logit($\hat{\pi}$) = 5.8121 − 0.1150distance

- probit($\hat{\pi}$) = 3.2071 − 0.0628distance

- log[−log(1 − $\hat{\pi}$)] = 2.3802 − 0.0522distance

Each model results in a negative estimate of the effect that distance has on the probability of success, and the Wald test of $H_0 : \beta_1 = 0$ vs. $H_a : \beta_1 \neq 0$ in each case results in very small p-values indicating strong evidence of the importance of distance.

Estimates of the probability of success can be found using a model's parameter estimates and the corresponding CDF. For example, using the probit model and a distance of 20 yards produces:

```
> lin.pred <- as.numeric(mod.fit.probit$coefficients[1] +
    mod.fit.probit$coefficients[2]*20)
> lin.pred
[1] 1.951195
> pnorm(q = lin.pred)
[1] 0.9744831
```

Thus, $\hat{\pi} = \Phi(3.2071 - 0.0628 \times 20) = \Phi(1.9512) = 0.9745$ when `distance` $= 20$.[26] These probability estimates can be found more easily using the `predict()` function:

---

[26]Using a different version of the notation, we have $Z_{0.9745} = 1.9512$.

```
> predict.data <- data.frame(distance = c(20, 35, 50))
> logistic.pi <- predict(object = mod.fit.logit, newdata =
    predict.data, type = "response")
> probit.pi <- predict(object = mod.fit.probit, newdata =
    predict.data, type = "response")
> cloglog.pi <- predict(object = mod.fit.cloglog, newdata =
    predict.data, type = "response")
> round(data.frame(predict.data, logistic.pi, probit.pi,
    cloglog.pi),4)
  distance logistic.pi probit.pi cloglog.pi
1       20      0.9710    0.9745     0.9777
2       35      0.8565    0.8436     0.8239
3       50      0.5152    0.5268     0.5477
```

Overall, we see that the differences in the probability of successes among the models are small—especially for the logistic and probit models. This may be surprising because the parameter estimates for the estimated models are quite different. For example, $\hat{\beta}_1$ for the complementary log-log model is less than half the $\hat{\beta}_1$ value for the logistic model. However, this is an inappropriate way to compare the models due to the differences in the way the regression parameters are used by their inverse link functions to form probabilities.

The similarities among the estimated models are seen further in Figure 2.14. In this figure, we provide a bubble plot with the plotting point proportional to the number of observations at each distance, which is the same as in Figure 2.5. Added to the plot are the three estimated models. Except for the very large distances, there is little difference among these models. Please see the corresponding program to this example for the code.

We end this example by examining estimated odds ratios for the three models:

```
> pi.hat <- data.frame(predict.data, logistic.pi, probit.pi,
    cloglog.pi)
> odds.x20 <- pi.hat[1, 2:4]/(1 - pi.hat[1, 2:4])
> odds.x35 <- pi.hat[2, 2:4]/(1 - pi.hat[2, 2:4])
> odds.x50 <- pi.hat[3, 2:4]/(1 - pi.hat[3, 2:4])

> OR.20.35 <- odds.x20/odds.x35
> OR.35.50 <- odds.x35/odds.x50

> data.frame(OR = c("20 vs. 35", "35 vs. 50"),
    round(rbind(OR.20.35, OR.35.50), 2))
         OR logistic.pi probit.pi cloglog.pi
1 20 vs. 35        5.61      7.08       9.36
2 35 vs. 50        5.61      4.84       3.86
```

The code calculates the estimated odds of a success for a distance of 20, 35, and 50 yards. By forming ratios of consecutive odds, we obtain the estimated odds ratios for 15-yard increments of distance. As expected, the odds ratios are exactly the same for the logistic regression model, but are different for the probit and complementary log-log models.

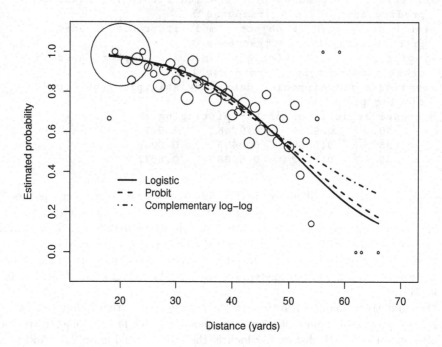

Figure 2.14: Bubble plot of the estimated probability of success for a placekick. The observed probability of success at each distance is plotted with the plotting point size being proportional to the number of observations at a distance. The estimated logistic, probit, and complementary log-log regression models are all given in the plot.

Random components used with generalized linear models

The Bernoulli distribution is the standard random component for regression models of binary responses. There are many other probability distributions that can be used, depending on the response type. For example, the standard linear regression model

$$E(Y) = \beta_0 + \beta_1 x_1 + \cdots + \beta_p x_p$$

uses the normal distribution for Y along with an identity link function (i.e., $E(Y)$ equals the linear predictor). This type of model is appropriate with a continuous response variable that has an approximate normal distribution conditional on the explanatory variables. Also, in Chapter 4, we will examine models for count responses that arise from mechanisms other than counting successes in repeated Bernoulli trials. In this setting, a Poisson random component is frequently used, where its mean is related to the explanatory variables through a log link,

$$\log[E(Y)] = \beta_0 + \beta_1 x_1 + \cdots + \beta_p x_p$$

This link function ensures that $E(Y) > 0$ for any combination of regression parameters and explanatory variable values.

2.4 Exercises

1. Beginning with the logistic regression model written as

$$\pi = \frac{\exp(\beta_0 + \beta_1 x_1 + \cdots + \beta_p x_p)}{1 + \exp(\beta_0 + \beta_1 x_1 + \cdots + \beta_p x_p)},$$

   show that it leads to the model written as $\text{logit}(\pi) = \beta_0 + \beta_1 x_1 + \cdots + \beta_p x_p$ and vice versa.

2. Show $\frac{d\pi}{dx_1} = \beta_1 \pi (1-\pi)$ for the model $\pi = \exp(\beta_0 + \beta_1 x_1)/[1+\exp(\beta_0+\beta_1 x_1)]$. Discuss how the slope changes as a function of x_1.

3. By modifying the code within PiPlot.R, describe what happens to the logistic regression model when β_1 is increased or decreased in the case of $\beta_1 < 0$.

4. The failure of an O-ring on the space shuttle Challenger's booster rockets led to its destruction in 1986. Using data on previous space shuttle launches, Dalal et al. (1989) examine the probability of an O-ring failure as a function of temperature at launch and combustion pressure. Data from their paper is included in the challenger.csv file. Below are the variables:

   - Flight: Flight number
   - Temp: Temperature (F) at launch
   - Pressure: Combustion pressure (psi)
   - O.ring: Number of primary field O-ring failures
   - Number: Total number of primary field O-rings (six total, three each for the two booster rockets)

   The response variable is O.ring, and the explanatory variables are Temp and Pressure. Complete the following:

   (a) The authors use logistic regression to estimate the probability an O-ring will fail. In order to use this model, the authors needed to assume that each O-ring is independent for each launch. Discuss why this assumption is necessary and the potential problems with it. Note that a subsequent analysis helped to alleviate the authors' concerns about independence.

   (b) Estimate the logistic regression model using the explanatory variables in a linear form.

   (c) Perform LRTs to judge the importance of the explanatory variables in the model.

   (d) The authors chose to remove Pressure from the model based on the LRTs. Based on your results, discuss why you think this was done. Are there any potential problems with removing this variable?

5. Continuing Exercise 4, consider the simplified model $\text{logit}(\pi) = \beta_0 + \beta_1 \text{Temp}$, where π is the probability of an O-ring failure. Complete the following:

   (a) Estimate the model.

(b) Construct two plots: (1) π vs. Temp and (2) Expected number of failures vs. Temp. Use a temperature range of 31° to 81° on the x-axis even though the minimum temperature in the data set was 53°.

(c) Include the 95% Wald confidence interval bands for π on the plot. Why are the bands much wider for lower temperatures than for higher temperatures?

(d) The temperature was 31° at launch for the Challenger in 1986. Estimate the probability of an O-ring failure using this temperature, and compute a corresponding confidence interval. Discuss what assumptions need to be made in order to apply the inference procedures.

(e) Rather than using Wald or profile LR intervals for the probability of failure, Dalal et al. (1989) use a parametric bootstrap to compute intervals. Their process was to (1) simulate a large number of data sets ($n = 23$ for each) from the estimated model of logit($\hat{\pi}$) = $\hat{\beta}_0 + \hat{\beta}_1$Temp; (2) estimate new models for each data set, say logit($\hat{\pi}^\star$) = $\hat{\beta}_0^\star + \hat{\beta}_1^\star$Temp; and (3) compute $\hat{\pi}^\star$ at a specific temperature of interest. The authors used the 0.05 and 0.95 observed quantiles from the $\hat{\pi}^\star$ simulated distribution as their 90% confidence interval limits. Using the parametric bootstrap, compute 90% confidence intervals separately at temperatures of 31° and 72°.[27]

(f) Determine if a quadratic term is needed in the model for the temperature.

6. Continuing Exercise 5, investigate if the estimated probability of failure would have significantly changed if a probit or complementary log-log regression model was used instead of a logistic regression model.

7. Exercise 17 of Chapter 1 examined data from Berry and Wood (2004) to determine if if an "icing the kicker" strategy implemented by the opposing team would reduce the probability of success for a field goal. Additional data collected for this investigation are included in the placekick.BW.csv file. Below are descriptions of the variables available in this file:

- GameNum: Identifies the year and game
- Kicker: Last name of kicker
- Good: Response variable ("Y" = success, "N" = failure)
- Distance: Length in yards of the field goal
- Weather: Levels of "Clouds", "Inside", "SnowRain", and "Sun"
- Wind15: 1 if wind speed is ≥ 15 miles per hour and the placekick is outdoors, 0 otherwise.
- Temperature: Levels of "Nice" (40°F < temperature < 80° or inside a dome), "Cold" (temperature $\leq$ 40° and outdoors), and "Hot" (temperature $\geq$ 80° and outdoors)
- Grass: 1 if kicking on a grass field, 0 otherwise
- Pressure: "Y" if attempt is in the last 3 minutes of a game and a successful field goal causes a lead change, "N" otherwise

---

[27] The bootstrap methods used here correspond to what is known as a "percentile" interval. There are better ways to form confidence intervals using the bootstrap, namely the BC$_a$ and studentized intervals. Please see Davison and Hinkley (1997) for a discussion and the boot() and boot.ci() functions in the boot package for implementation.

- Ice: 1 if Pressure = 1 and a time-out is called prior to the attempt, 0 otherwise

Notice that these variables are similar but not all are exactly the same as given for the placekicking data described in Section 2.2.1 (e.g., information was collected on field goals only, so there is no PAT variable). Using this new data set, complete the following:

(a) When using a `formula` argument value of `Good ~ Distance` in `glm()`, how do you know if R is modeling the probability of success or failure? Explain.

(b) Estimate the model from part (a), and plot it using the `curve()` function.

(c) Add to the plot in part (b) the logit$(\hat{\pi}) = 5.8121 - 0.1150$distance model estimated in Section 2.2.1. Notice that the models are quite similar. Why is this desirable?

8. Continuing Exercise 7, use the `Distance`, `Weather`, `Wind15`, `Temperature`, `Grass`, `Pressure`, and `Ice` explanatory variables as linear terms in a new logistic regression model and complete the following:

(a) Estimate the model and properly define the indicator variables used within it.

(b) The authors use "Sun" as the base level category for `Weather`, which is not the default level that R uses. Describe how "Sun" can be specified as the base level in R.

(c) Perform LRTs for all explanatory variables to evaluate their importance within the model. Discuss the results.

(d) Estimate an appropriate odds ratio for distance, and compute the corresponding confidence interval. Interpret the odds ratio.

(e) Estimate all six possible odds ratios for `Weather`, and compute the corresponding confidence intervals. Interpret the odds ratios. Is a value of 1 found to be within any of the intervals? Relate this result to the LRT performed earlier for `Weather`.

(f) Continuing part (e), discuss how the overall familywise error rate could be controlled for all of the confidence intervals. Carry out the necessary computations.

(g) The purpose of this part is for you to estimate the probability of success for a field goal attempted by the Dallas Cowboys on December 12, 2011, in their game against the Arizona Cardinals. With the score tied and seven seconds remaining, the Dallas placekicker, Dan Bailey, attempted a 49-yard field goal that was successful. Unfortunately for Bailey, his coach, Jason Garrett, called a time out right before the attempt, so Bailey had to re-kick. Bailey was not successful the second time, so Garrett essentially "iced" his own kicker, and the Cardinals went on to win the game in overtime.[28]

    i. Discuss the appropriateness of using the model here to estimate the probability success.

    ii. The explanatory variable values for this placekick are: `Distance` = 49 yards, `Wind15` = 0 (5 miles per hour), `Grass` = 1, `Pressure` = "Y", and `Ice` = 1. The game was played in Arizona's retractable roof domed stadium, and

---

[28] A play-by-play description of the game is available at http://www.nfl.com/liveupdate/gamecenter/55351/ARI_Gamebook.pdf. Interestingly, if Dallas had won one additional game during the regular season, the New York Giants would not have made the playoffs and subsequently would not have won the Super Bowl.

information about whether or not the roof was open is unavailable. For this part, assume then Weather = "Sun" and Temperature = "Nice". Find the estimated probability of success for this field goal.

iii. Compute the corresponding confidence interval for the probability of success.

(h) Is there evidence to conclude that icing the kicker is a good strategy to follow?

9. Continuing Exercise 7, consider the model of logit$(\pi) = \beta_0 + \beta_1$Distance$+ \beta_2$Wind15$+ \beta_3$Distance $\times$ Wind15, where π is the probability of success. Complete the following:

    (a) Estimate the models with and without the Distance $\times$ Wind15 interaction.

    (b) Construct a plot similar to Figure 2.4 in order to see the effect of the interaction. Does the interaction appear to be important?

    (c) Using the model with the interaction term, compute and interpret odds ratios to better understand the effect of Wind15 on the success or failure of a placekick.

    (d) Perform a LRT to examine the importance of the interaction term. Compare the results here to what was found earlier in Section 2.2.5. Suggest possible reasons for the similarities and/or differences.

10. Exercise 6 of Chapter 1 examined the effectiveness of Gen-Probe's Aptima Combo 2 Assay used to detect chlamydia and gonorrhea in individuals. For the chlamydia testing results with respect to sensitivity, complete the following:

    (a) Use a logistic regression model with Gender, Specimen, and Symptoms_Status as explanatory variables to estimate the sensitivity of the assay. Use only linear terms in the model. Describe the effect that each explanatory variable has on the sensitivity of the assay.

    (b) The BD Probetec Assay is another assay used to detect chlamydia and gonorrhea in individuals. Unfortunately, the assay is known to have a much lower sensitivity for females when a urine specimen is tested than when a swab specimen is tested. The same effect does not occur for males. Describe why this effect corresponds to an interaction between gender and specimen type.

    (c) Construct a plot of the observed sensitivity vs. gender by conditioning on specimen type and symptomatic status. Below is code that will get you started on constructing the plot (our original data are in a data frame called CT2)

```
CT2$gender.numb <- ifelse(test = CT2$Gender == "Female",
    yes = 0, no = 1)
CT2$sensitivity <- CT2$True_positive/CT2$Se.trials
plot(x = CT2$gender.numb, y = CT2$sensitivity, xlab =
    "Gender", ylab = "Sensitivity", type = "n", xaxt = "n")
axis(side = 1, at = c(0,1), labels = c("Female", "Male"))
Swab.Symptomatic <- CT2$Specimen == "Swab" &
    CT2$Symptoms_Status == "Symptomatic"
lines(x = CT2$gender.numb[Swab.Symptomatic],
    CT2$sensitivity[Swab.Symptomatic], type = "o", lty =
    "solid", col = "black")
# Use three more calls to lines() for the other specimen
    type and symptomatic status combinations. Also, include
    an appropriate legend.
```

What does this plot suggest about the possibility of a `Gender` and `Specimen` type interaction?

(d) Perform a hypothesis test to determine if the same type of interaction that occurs with the BD Probetec Assay also occurs with the Gen-Probe's Aptima Combo 2 Assay.

(e) Does the plot in (c) suggest other possible interactions may exist? If so, perform hypothesis tests for these interactions.

11. Continuing Exercise 10 using the model with all three explanatory variables as linear terms and a `Gender` and `Specimen` interaction, complete the following:

    (a) Construct confidence intervals for sensitivity at all possible combinations of `Gender`, `Specimen`, and `Symptoms_Status` levels. Control the familywise error level at an appropriate level.

    (b) The methodology discussed in Section 1.1.2 could also be used to calculate intervals for each combination of `Gender`, `Specimen`, and `Symptoms_Status` without the use of a logistic regression model. Discuss the differences between these methods and those used for part (a) here.

    (c) Discuss the differences in methodology between the intervals calculated for part (c) of Exercise 6 in Chapter 1 and those intervals constructed for part (a) here.

12. Continuing Exercise 10, complete the following:

    (a) Perform the same type of analysis as in parts (a), (c), and (e) for the specificity when testing chlamydia.

    (b) Perform the same type of analysis as in parts (a), (c), and (e) for the sensitivity and specificity when testing for gonorrhea.

13. The Larry Bird free throw shooting data discussed in Section 1.2 can be examined in a logistic regression context. The purpose here is to estimate the probability of success for the second free throw attempt given what happened on the first free throw attempt. Complete the following:

    (a) The `c.table` object created in Bird.R contains the contingency table format of the data. To create the data frame format needed for `glm()`, we can re-enter the data in a new data frame with

    ```
    > bird <- data.frame(First = c("made", "missed"), success =
        c(251, 48), trials = c(285, 53))
    > bird
      First success trials
    1   made     251    285
    2 missed      48     53
    ```

    or transform it directly with

    ```
    > bird1 <- as.data.frame(as.table(c.table))
    > trials <- aggregate(formula = Freq ~ First, data = bird1,
        FUN = sum)
    > success <- bird1[bird1$Second == "made", ]
    > bird2 <- data.frame(First = success$First, success =
        success$Freq, trials = trials$Freq)
    ```

```
> bird2
   First success trials
1   made     251     285
2 missed      48      53
```

The second method is more general and can work with larger contingency tables like those examined in Chapters 3 and 4. Implement each line of code in the second method above and describe what occurs.

(b) Estimate a logistic regression model for the probability of success on the second attempt, where First is the explanatory variable.

(c) Estimate the odds ratio comparing the two levels of First. Calculate both Wald and profile LR intervals for the odds ratio. Compare these calculated values with those obtained in the Larry Bird example of Section 1.2.5. If you use confint() to compute the intervals, make sure to use the correct value for the parm argument (same name as the indicator variable given by summary(mod.fit)).

(d) Perform a hypothesis test of $H_0 : \beta_1 = 0$ vs. $H_a : \beta_1 \neq 0$ using Wald and LR statistics. Compare these results to those found for the Larry Bird example in Section 1.2.3.

(e) Discuss why similarities and/or differences occur between the calculations here using logistic regression and the corresponding calculations in Chapter 1.

14. Similar to Exercise 13, the Salk vaccine clinical trial data discussed in Section 1.2 can be examined in a logistic regression context. Complete the following:

(a) Starting from the contingency table format of the data given as

```
c.table <- array(data = c(57, 142, 200688, 201087), dim =
    c(2,2), dimnames = list(Treatment = c("vaccine",
    "placebo"), Result = c("polio", "polio free")))
```

create a data frame in a similar format as bird2 in Exercise 13. Name this new data frame polio.

(b) Use the levels() function to examine ordering of the levels in polio$treatment. Note that these are not in alphabetical order due to the ordering given originally in the array() function.

(c) Estimate a logistic regression model for the probability of being polio free (second column of the contingency table) where the type of treatment is the explanatory variable. Specifically state the indicator variable coding for the explanatory variable.

(d) Estimate an odds ratio comparing the two treatments and calculate corresponding confidence intervals. Compare your results to those found in Section 1.2.5.

(e) Determine if there is a difference in the probability of being polio free based on the treatment.

(f) Suppose additional explanatory variables, such as geographical location and socioeconomic status, are available for each individual, and these are included in the model as well. Discuss why using a logistic regression model would be an easier way to assess the effectiveness of the treatment rather than a contingency table(s) format.

15. The relative risk and the difference between two probabilities of success can be calculated using a logistic regression model. For this problem, consider the logistic regression model of $\text{logit}(\pi) = \beta_0 + \beta_1 x_1$, where x_1 is 0 or 1, and complete the following:

   (a) Find the difference of the two probabilities of success in terms of the model.
   (b) Find the relative risk in terms of the model.
   (c) Through the delta method described in Appendix B, the variance of the estimated statistics in (a) and (b) can be found. Find these estimated variances in terms of $\widehat{Var}(\hat{\beta}_0)$, $\widehat{Var}(\hat{\beta}_1)$, and $\widehat{Cov}(\hat{\beta}_0, \hat{\beta}_1)$.

16. The `deltaMethod()` function of the `car` package provides a convenient way to apply the delta method to find variances of mathematical functions. The syntax of the function is

    ```
    deltaMethod(object = <model fit object>, g = <mathematical
        function>, parameterNames = <Vector of parameter names>).
    ```

    The `parameterNames` argument uses the same parameter names as given in the `g` argument. These names do not need to match those given by `glm()`, but they need to be in the same order as in the model fit object. Apply `deltaMethod()` to the Larry Bird and Salk vaccine data of Exercises 13 and 14, respectively, as follows:

   (a) Let $\pi_{\text{First=missed}}$ and $\pi_{\text{First=made}}$ denote the probability of success on the second attempt given what happened on the first attempt for Larry Bird. The following code estimates $\pi_{\text{First=missed}} - \pi_{\text{First=made}}$ and $Var(\hat{\pi}_{\text{First=missed}} - \hat{\pi}_{\text{First=made}})^{1/2}$:

    ```
    library(package = car)
    deltaMethod(object = mod.fit, g = "exp(beta0 + beta1*1)/(1
        + exp(beta0 + beta1*1)) - exp(beta0)/(1+exp(beta0))",
        parameterNames = c("beta0", "beta1"))
    ```

    where `mod.fit` contains the results from fitting the model with `glm()`. Execute this code and use it to find the corresponding Wald confidence interval for the difference of the two probabilities. Compare your estimate and confidence interval to those for the Larry Bird example in Section 1.2.2.

   (b) Let π_{vaccine} and π_{placebo} denote the probability of being polio free given the type of treatment received. Use the `deltaMethod()` function to estimate the relative risk given by $(1-\pi_{\text{placebo}})/(1-\pi_{\text{vaccine}})$ and its corresponding variance. Calculate a Wald confidence interval for the relative risk with these results. Compare your estimate and confidence interval to those for the Salk vaccine example in Section 1.2.4.

   (c) Repeat part (b), but now using $\log[(1-\pi_{\text{placebo}})/(1-\pi_{\text{vaccine}})]$.

17. Consider the model $\text{logit}(\pi) = \beta_0 + \beta_1 \text{distance}$ for the placekicking data set introduced in Section 2.2.1. Assuming the model estimation results from `glm()` are stored in `mod.fit`, complete the following:

   (a) Show that $\widehat{Var}(\hat{\pi})^{1/2}$ is computed as the same value using both `predict()` and `deltaMethod()` as shown below:

```
predict(object = mod.fit, newdata = data.frame(distance =
    x), type = "response", se.fit = TRUE)

library(package = car)
deltaMethod(object = mod.fit, g = "exp(beta0 + beta1*x)/(1
    + exp(beta0 + beta1*x))", parameterNames = c("beta0",
    "beta1"))
```

where x is replaced by a particular value of `distance`.

(b) Through the help of the `deltaMethod()` function, estimate $\pi_{x+c} - \pi_x$, where π_x is the probability of success at a distance of x yards and c is a constant increment in the distance. Do the estimates change as a function of x for a constant c? Compute the corresponding Wald confidence intervals for $\pi_{x+c} - \pi_x$.

(c) Through the help of the `deltaMethod()` function, estimate π_{x+c}/π_x for a few different values of x and compute corresponding Wald confidence intervals. Do the estimates change as a function of x for a constant c?

18. Consider the model $\text{logit}(\pi) = \beta_0 + \beta_1\text{distance} + \beta_2\text{distance}^2$ for the placekicking data set introduced in Section 2.2.1. Use odds ratios to interpret the effect that distance has on the success or failure of a placekick.

19. Thorburn et al. (2001) examine hepatitis C prevalence among healthcare workers in the Glasgow area of Scotland. These healthcare workers were categorized into the following occupational groups: (1) exposure prone (e.g., surgeons, dentists, surgical nurses), (2) fluid contact (e.g., non-surgical nurses), (3) lab staff, (4) patient contact (e.g., pharmacists, social workers), and (5) no patient contact (e.g., clerical). The collected data are available in the healthcare_worker.csv file. Is there evidence of an occupational group effect on hepatitis status? If there is sufficient evidence of an effect, use the appropriate odds ratios to make comparisons among the groups. If there is not sufficient evidence of an effect, discuss why this may be a preferred result.

20. An example in Section 2.2.5 examined using $\text{logit}(\pi) = \beta_0 + \beta_1\text{distance} + \beta_2\text{wind} + \beta_3\text{distance} \times \text{wind}$ for the placekicking data set, and Wald confidence intervals were calculated to interpret the odds ratio for wind. Because of the interaction, confidence intervals were found at a few specific distances using R code that directly programmed the interval equations. A different way to find these confidence intervals in R is through the `multcomp` package (Bretz et al., 2011).

(a) Implement the following continuation of code from the example. Verify the data frame contains the estimated wind odds ratio and the 95% Wald confidence interval when the `distance` is 20 yards.

```
library(package = multcomp)
K <- matrix(data = c(0, 0, 1, 1*20), nrow = 1, ncol = 4,
    byrow = TRUE)
linear.combo <- glht(model = mod.fit.Ha, linfct = K)
ci.log.OR <- confint(object = linear.combo, calpha =
    qnorm(0.975))
data.frame(estimate = exp(-ci.log.OR$confint[1]), lower =
    exp(-ci.log.OR$confint[3]), upper =
    exp(-ci.log.OR$confint[2]))
```

(b) Describe what each line of the code in (a) calculates given the argument values specified. You may want to examine the functions' help along with the online vignettes for the package at http://cran.r-project.org/web/packages/multcomp/index.html. Note that the elements of the c() function are put in an order corresponding to $\hat{\beta}_0$, $\hat{\beta}_1$, $\hat{\beta}_2$, and $\hat{\beta}_3$.

(c) Modify the code to complete the same calculations for distances of 30, 40, 50, and 60 yards.

(d) Continue to implement the code below:

```
K <- matrix(data = c(1, 0, 0, 0,  0, 1, 0, 0,  0, 0, 1, 0,
    0, 0, 0, 1), nrow = 4, ncol = 4, byrow = TRUE)
linear.combo <- glht(model = mod.fit.inter, linfct = K)
summary(object = linear.combo, test = adjusted(type =
    "none"))
```

and compare it to what summary(mod.fit.Ha) gives.

(e) The multcomp package contains numerous built-in adjustments for simultaneous inference including those used with mcprofile described on p. 110. The default for both tests and confidence intervals is to use single-step adjustment. Thus, the syntax to compute confidence intervals with a fixed familywise error rate for these distances is just confint(object = linear.combo). Compute these intervals, and explain what statement can be made about the familywise confidence level for covering the true odds ratios. Compare the intervals to the unadjusted intervals and comment on differences.

21. Rather than finding the probability of success at an explanatory variable value, it is often of interest to find the value of an explanatory variable given a desired probability of success. This is referred to as *inverse prediction*. One application of inverse prediction involves finding the amount of a pesticide or herbicide needed to have a desired kill rate when applied to pests or plants. The *lethal dose level* x_π (commonly called "LDz", where $z = 100\pi$) is defined as

$$x_\pi = [\text{logit}(\pi) - \beta_0]/\beta_1$$

for the logistic regression model $\text{logit}(\pi) = \beta_0 + \beta_1 x$. The estimated value of x_π is $\hat{x}_\pi = [\text{logit}(\pi) - \hat{\beta}_0]/\hat{\beta}_1$. Complete the following:

(a) Show how x_π is derived by solving for x in the logistic regression model.

(b) A $(1-\alpha)100\%$ confidence interval for x_π is the set of all possible values of x such that

$$\frac{\left|\hat{\beta}_0 + \hat{\beta}_1 x - \text{logit}(\pi)\right|}{\sqrt{\widehat{Var}(\hat{\beta}_0) + x^2 \widehat{Var}(\hat{\beta}_1) + 2x\widehat{Cov}(\hat{\beta}_0, \hat{\beta}_1)}} < Z_{1-\alpha/2} \quad (2.21)$$

is satisfied. Describe how this confidence interval for x_π is derived. Note that there is generally no closed-form solution for the confidence interval limits, which leads to the use of iterative numerical procedures.

(c) The dose.p() function of the MASS package and the deltaMethod() function of the car package can calculate $\hat{x}_\pi$ and $\widehat{Var}(\hat{x}_\pi)^{1/2}$, where the estimated variance is found using the delta method. Using this statistic and its corresponding variance, show how a $(1-\alpha)100\%$ Wald confidence interval for x_π can be formed.

(d) Derive the form of x_π using probit and complementary log-log models.

22. Turner et al. (1992) uses logistic regression to estimate the rate at which picloram, a herbicide, kills tall larkspur, a weed. Their data was collected by applying four different levels of picloram to separate plots, and the number of weeds killed out of the number of weeds within the plot was recorded. The data are in the file picloram.csv.[29] Complete the following:

   (a) Estimate a logistic regression model using the picloram amount as the explanatory variable and the number of weeds killed as the response variable.

   (b) Plot the observed proportion of killed weeds and the estimated model. Describe how well the model fits the data.

   (c) Estimate the 0.9 kill rate level ("LD90") for picloram. Add lines to the plot in (b) to illustrate how it is found (the `segments()` function can be useful for this purpose).

   (d) To calculate a confidence interval for the dosage that yields a 0.9 kill rate, below is a set of R code:

```
root.func <- function(x, mod.fit.obj, pi0, alpha) {
  beta.hat <- mod.fit.obj$coefficients
  cov.mat <- vcov(mod.fit.obj)
  var.den <- cov.mat[1,1] + x^2*cov.mat[2,2] +
    2*x*cov.mat[1,2]
  abs(beta.hat[1] + beta.hat[2]*x - log(pi0/(1-pi0))) /
    sqrt(var.den) - qnorm(1-alpha/2) }

lower <- uniroot(f = root.func, interval =
  c(min(set1$picloram), LD.x), mod.fit.obj = mod.fit, pi0
  = 0.9, alpha = 0.95)
lower  # lower$root contains the lower bound

upper <- uniroot(f = root.func, interval = c(LD.x,
  max(set1$picloram)), mod.fit.obj = mod.fit, pi0 = 0.9,
  alpha = 0.95)
upper  # upper$root contains the upper bound
```

that uses Equation 2.21 in its calculation. Describe what each line of the R code does and run it to find a 95% confidence interval for $x_{0.9}$.

   (e) What amount of picloram should be used in order to have a 0.9 kill rate?

   (f) The data for this problem consist of only four different dosage levels of picloram. What assumptions are needed in order for the model to provide a good estimate of x_π?

23. Repeat the analysis in Exercise 22 using the probit and complementary log-log models. Do the results change in comparison to using logistic regression?

---

[29] The experiment was performed in three separate runs, which are denoted by the `rep` variable in the data file. The authors estimate a logistic regression model without accounting for the `rep` variable, and we do the same here. We include `rep` in a re-analysis of the data in Exercise 2 of Chapter 6.

24. Potter (2005) and Heinze (2006) examine a urinary incontinence study that examined three explanatory variables and their relation to incontinence after a drug was administered. Unfortunately, the author does not provide an exact description of what the response variable coding represents (most likely, the response $y = 1$ denotes continent and $y = 0$ denotes incontinent) and does not give detailed descriptions of the explanatory variables (they only say that x_1, x_2, and x_3 are measurements on the lower urinary tract). However, the main reason to examine these data here is to see an example of where complete separation occurs when these explanatory variables are included in a logistic regression model. Using the data in the file incontinence.csv, complete the following.

   (a) Estimate the logistic regression models $\text{logit}(\pi) = \beta_0 + \beta_1 x_r$ for $r = 1, 2, 3$ and $\pi = P(Y = 1)$ using maximum likelihood estimation. Plot the estimated models along with the observed values. Are there any problems with complete separation?

   (b) Estimate the logistic regression model $\text{logit}(\pi) = \beta_0 + \beta_1 x_1 + \beta_2 x_2 + \beta_3 x_3$ using maximum likelihood estimation. Using the parameter estimates from the last iteration of glm(), estimate the probability of success for each observation. Discuss the problems that occur.

   (c) Estimate the same model as in (b), but now using the modified likelihood function as discussed in Section 2.2.7. Do any problems occur with the estimation? Compare the results here to those found in (b).

   (d) Using the model fit from (c), interpret the relationship between the explanatory variables and the response.

25. Section 2.2.7 discusses the criteria for how R determines if and when the convergence of parameter estimates takes place. Discuss why the residual deviance is used for this purpose. Suggest other ways to determine convergence.

26. What does the na.action argument of the glm() function control and what is its default? Discuss possible problems that may occur with this default value.

27. Consider the complementary log-log regression model of $\log[-\log(1-\pi)] = \beta_0 + \beta_1 x_1$. Show that the odds ratio comparing different levels of x_1 cannot be written without the explanatory variable being presented.

28. Section 2.2.1 gives an example of how to estimate a logistic regression model by creating an R function to calculate the log-likelihood function and then maximize it using optim(). Following this example, write an R function that calculates the log-likelihood function for the probit regression model. Use this function with optim() to estimate $\text{probit}(\pi) = \beta_0 + \beta_1 \text{distance}$ with the placekicking data set. Compare the estimated standard errors resulting from the use of optim() to those obtained from glm(). Repeat this process with the corresponding complementary log-log regression model.

29. Show that when Equation 2.9 is subtracted from Equation 2.8, Equation 2.7 is the result.

30. For a simple model of $\text{logit}(\pi_i) = \beta_0 + \beta_1 x_{i1}$, one can show:

   - $\widehat{Var}(\hat{\beta}_0) = m^{-1} \sum_{i=1}^{n} x_{i1}^2 \hat{\pi}_i (1 - \hat{\pi}_i)$
   - $\widehat{Var}(\hat{\beta}_1) = m^{-1} \sum_{i=1}^{n} \hat{\pi}_i (1 - \hat{\pi}_i)$

- $\widehat{Cov}(\hat{\beta}_0, \hat{\beta}_1) = -m^{-1} \sum_{i=1}^{n} x_{i1} \hat{\pi}_i (1 - \hat{\pi}_i)$

where $m = \left[\sum_{i=1}^{n} x_{i1}^2 \hat{\pi}_i (1 - \hat{\pi}_i)\right] \left[\sum_{i=1}^{n} \hat{\pi}_i (1 - \hat{\pi}_i)\right] - \left[\sum_{i=1}^{n} x_{i1} \hat{\pi}_i (1 - \hat{\pi}_i)\right]^2$. Complete the following:

(a) Show how these formulas are found using $(\mathbf{X}'\mathbf{V}\mathbf{X})^{-1}$. Note that the inverse of a 2×2 matrix $\begin{bmatrix} a & b \\ c & d \end{bmatrix}$ is $m^{-1} \begin{bmatrix} d & -b \\ -c & a \end{bmatrix}$ where $m = ad - bc$.

(b) Using these formulas, calculate the estimated covariance matrix for the $\text{logit}(\hat{\pi}) = 5.8121 - 0.1150\texttt{distance}$ model used with the placekicking example. Verify this matrix is the same as if vcov() was used to calculate it.

31. Rather than using Equation 2.15 to form a confidence interval for π, a normal approximation for $\hat{\pi}$ can be made directly resulting in a Wald interval of $\hat{\pi} \pm Z_{1-\alpha/2} \widehat{Var}(\hat{\pi})^{1/2}$, where $\widehat{Var}(\hat{\pi})$ is found using the delta method. For large samples and π not close to 0 or 1, this interval will be very similar to Equation 2.15. However, when this does not happen, this interval can have problems. Discuss what these problems are and why this interval is not desirable to use in general practice.

32. Suppose a logistic regression model with three binary explanatory variables is used to estimate the probability of success. This model includes all three linear terms, the three two-way interactions, and the one three-way interaction. Derive the odds ratio to examine the odds of a success across two levels of one of the explanatory variables and provide its interpretation. Give the variance of the odds ratio estimator.

Chapter 3

Analyzing a multicategory response

Chapters 1 and 2 investigated a response variable that had two possible category choices. Chapter 3 generalizes this to a setting where the response variable value is chosen from a fixed set of more than two category choices. For example, response options could be of the form:

1. Five-level Likert scale — Strongly disagree, disagree, neutral, agree, or strongly agree,

2. Chemical compounds in drug discovery experiments — Positive, blocker, or neither,

3. Cereal shelf-placement in a grocery store — Bottom, middle, or top,

4. Canadian political party affiliation — Conservative, New Democratic, Liberal, Bloc Quebecois, or Green, and

5. Beef grades — Prime, choice, select, standard, utility, and commercial.

For these examples, some responses are ordinal (e.g., Likert scale) and some are not (e.g., chemical compounds). We will investigate both ordinal and nominal (unordered) multicategory responses within this chapter.

In each of the above examples, an observed unit fits into exactly one category. For example, a chemical compound cannot be both a positive and a blocker. There are other situations where a unit may fit simultaneously into more than one category, such as in "choose all that apply" survey questions. We will investigate these "multiple response" problems separately in Section 6.4.

3.1 Multinomial probability distribution

The multinomial probability distribution is the extension of the binomial distribution to situations where there are more than two categories for a response. Let Y denote the categorical response random variable with levels $j = 1, \ldots, J$, where each category has probability $\pi_j = P(Y = j)$ such that $\sum_{j=1}^{J} \pi_j = 1$. If there are n identical trials with responses $Y_1, \ldots, Y_n$, then we can define random variables N_j, $j = 1, \ldots, J$, such that N_j counts the number of trials responding with category j. That is, $N_j = \sum_{i=1}^{n} I(Y_i = j)$, where $I(\cdot) = 1$ when the condition in parentheses is true, and $= 0$ otherwise. Let $n_1, \ldots, n_J$ denote the observed response count for category j with $\sum_{j=1}^{J} n_j = n$. The probability mass function (PMF) for observing a particular set of counts $n_1, \ldots, n_J$ is

$$P(N_1 = n_1, \ldots, N_J = n_J) = \frac{n!}{\prod_{j=1}^{J} n_j!} \prod_{j=1}^{J} \pi_j^{n_j}, \qquad (3.1)$$

which is known as the *multinomial probability distribution*. Notice that when $J = 2$, the distribution simplifies to the binomial distribution as described in Section 1.1.1, where $n_1 = w$, $n_2 = n - w$, $\pi_1 = \pi$, and $\pi_2 = 1 - \pi$ in the notation of that section.

We use maximum likelihood estimation to obtain estimates of $\pi_1, \ldots, \pi_J$. The likelihood function is simply Equation 3.1, and the MLE for each π_j is $\hat{\pi}_j = n_j/n$ (i.e., the observed proportion for each category).

Example: Multinomial simulated sample (Multinomial.R)

To picture what data from a multinomial distribution might look like, we simulate a sample of $n = 1000$ trials from a multinomial distribution with $\pi_1 = 0.25$, $\pi_2 = 0.35$, $\pi_3 = 0.2$, $\pi_4 = 0.1$, and $\pi_5 = 0.1$. We use the `rmultinom()` function to simulate one set (n = 1) of 1000 trials (size = 1000) from a multinomial distribution with probabilities listed in the `prob` argument:

```
> pi.j <- c(0.25, 0.35, 0.2, 0.1, 0.1)
> set.seed(2195)   # Set a seed to be able to reproduce the sample
> n.j <- rmultinom(n = 1, size = 1000, prob = pi.j)
> data.frame(n.j, pihat.j = n.j/1000, pi.j)
  n.j pihat.j pi.j
1 256   0.256 0.25
2 365   0.365 0.35
3 195   0.195 0.20
4  82   0.082 0.10
5 102   0.102 0.10
```

The simulated counts are saved into an object named `n.j`. After dividing these counts by n, we see that the observed proportions are very similar to the actual parameter values.

If we wanted to simulate m sets of 1000 trials from the same multinomial distribution, we could change the `n` argument value to m. In this situation, there would be m columns of observations produced by `rmultinom()`. For example, the result for $m = 5$ is

```
> set.seed(9182)
> n.j <- rmultinom(n = 5, size = 1000, prob = pi.j)
> n.j
     [,1] [,2] [,3] [,4] [,5]
[1,]  259  259  237  264  247
[2,]  341  346  374  339  341
[3,]  200  188  198  191  210
[4,]   92  106   89  108  107
[5,]  108  101  102   98   95
```

As we can see, there is variability in the counts from one sample set to the next. This same type of variability should be expected in actual samples with multinomial responses.

The corresponding program for this example also shows how to use the `dmultinom()` function to evaluate the PMF for a given set of counts $n_1, \ldots, n_J$.

Table 3.1: $I \times J$ contingency table.

		Y				
		1	2	$\cdots$	J	Total
	1	n_{11}	n_{12}	$\cdots$	n_{1J}	n_{1+}
X	2	n_{21}	n_{22}	$\cdots$	n_{2J}	n_{2+}
	$\vdots$	$\vdots$	$\vdots$	$\ddots$	$\vdots$	$\vdots$
	I	n_{I1}	n_{I2}	$\cdots$	n_{IJ}	n_{I+}
	Total	n_{+1}	n_{+2}	$\cdots$	n_{+J}	n

3.2 $I \times J$ contingency tables and inference procedures

Section 1.2 discusses how counts from two binomial distributions can be analyzed in the form of a 2×2 contingency table. We expand this discussion now to allow for more than two rows and/or columns. We assume that two categorical variables, X and Y, are measured on each unit with levels of $i = 1, \ldots, I$ for X and $j = 1, \ldots, J$ for Y. Our sample consists of n units that are cross-classified according to their levels of X and Y. The counts of units for which the combination $(X = i, Y = j)$ is observed are denoted by n_{ij}, and these represent observations of the random variables N_{ij}, $i = 1, \ldots, I$, $j = 1, \ldots, J$. Our goal is to use the observed counts to make inferences about the distribution of the N_{ij} and to identify any relationships between X and Y. To this end, we can create a contingency table summarizing these counts by letting X be the row variable and Y be the column variable. Table 3.1 provides the general contingency table. Note that we use a $+$ subscript to indicate a sum; for example, $n_{+j} = \sum_{i=1}^{I} n_{ij}$ is the total for column j. We let n_{++} be denoted simply by n.

In the discussion that follows, we examine two of the models that are typically used to represent counts in a contingency table.

3.2.1 One multinomial distribution

First, consider the situation where a fixed sample size of n units are sampled from a large population. We define $\pi_{ij} = P(X = i, Y = j)$. This is the same multinomial setting as in Section 3.1 with n trials and IJ possible responses, except now each possible response is a combination of two variables instead of just one. We assume that each sampled unit has one and only one X and Y category combination, so that $\sum_{i=1}^{I} \sum_{j=1}^{J} \pi_{ij} = 1$.

Maximum likelihood estimation for these probabilities proceeds analogously to what was done in Section 3.1, with minor modifications to the notation to account for the second variable. The PMF for $N_{11}, \ldots, N_{IJ}$ becomes

$$P(N_{11} = n_{11}, \ldots, N_{IJ} = n_{IJ}) = \frac{n!}{\prod_{i=1}^{I} \prod_{j=1}^{J} n_{ij}!} \prod_{i=1}^{I} \prod_{j=1}^{J} \pi_{ij}^{n_{ij}}, \quad (3.2)$$

which is also the likelihood function for a sample of size n. The MLE of π_{ij} is the estimated proportion $\hat{\pi}_{ij} = n_{ij}/n$.

Marginal distributions for X and for Y may be found as well. The marginal probability for level i of X is $\pi_{i+} = P(X = i)$ for $i = 1, \ldots, I$. The marginal distribution of X is therefore multinomial with n trials and probabilities $\pi_{1+}, \ldots, \pi_{I+}$ and results in the marginal counts $n_{1+}, \ldots, n_{I+}$. Similarly, the marginal probability for category j of Y is $\pi_{+j} = \sum_{i=1}^{I} \pi_{ij}$ for

$j = 1, \ldots, J$. The marginal distribution of Y is multinomial with n trials and probabilities $\pi_{+1}, \ldots, \pi_{+J}$, and results in the marginal counts $n_{+1}, \ldots, n_{+J}$. Note that $\sum_{i=1}^{I} \pi_{i+} = 1$ and $\sum_{j=1}^{J} \pi_{+j} = 1$. The MLEs of π_{i+} and π_{+j} are the corresponding row and column proportions, $\hat{\pi}_{i+} = n_{i+}/n$ and $\hat{\pi}_{+j} = n_{+j}/n$, respectively.

When the outcome of X does not have an effect on the probabilities for the outcomes of Y, we say that Y is *independent* of X. As a result of independence, the probability of any joint outcome $(X = i, Y = j)$ factors into the marginal probabilities for $X = i$ and $Y = j$: $\pi_{ij} = \pi_{i+}\pi_{+j}$. Independence simplifies the probability structure within a contingency table by reducing the number of unknown parameters to the $(I - 1) + (J - 1) = I + J - 2$ marginal probabilities. Note that the "-1" parts occur due to the $\sum_{i=1}^{I} \pi_{i+} = 1$ and $\sum_{j=1}^{J} \pi_{+j} = 1$ constraints on the probabilities. Without independence, there are $IJ - 1$ unknown probability parameters, where the "-1" part occurs due to the $\sum_{i=1}^{I} \sum_{j=1}^{J} \pi_{ij} = 1$ constraint on the probabilities. Thus, there is a reduction of $(IJ - 1) - (I + J - 2) = (I - 1)(J - 1)$ parameters compared to the same table without independence. We will examine how to perform a hypothesis test for independence shortly.

3.2.2 I multinomial distributions

An alternative model is needed when samples of sizes n_{i+}, $i = 1, \ldots, I$ are deliberately taken from each of I different groups. In this case, the marginal counts n_{i+} are fixed by design, so we have a separate J-category multinomial distribution in each of the I groups, where $n = \sum_{i=1}^{I} n_{i+}$. Each of these distributions has its own set of probability parameters. Define $P(Y = j|X = i) = \pi_{j|i}$ as the conditional probability of observing response category j given that a unit is from group i. Note that $\sum_{j=1}^{J} \pi_{j|i} = 1$ for each $i = 1, \ldots, I$. The conditional joint distribution of $N_{i1}, \ldots, N_{iJ}$ has PMF

$$P(N_{i1} = n_{i1}, \ldots, N_{iJ} = n_{iJ}|N_{i+} = n_{i+}) = \frac{n_{i+}!}{\prod_{j=1}^{J} n_{ij}!} \prod_{j=1}^{J} \pi_{j|i}^{n_{ij}}$$

for each $i = 1, \ldots, I$. Assuming that I different samples are taken independently of one another, the likelihood for the parameters is just the product of I multinomial distributions,

$$\prod_{i=1}^{I} \frac{n_{i+}!}{\prod_{j=1}^{J} n_{ij}!} \prod_{j=1}^{J} \pi_{j|i}^{n_{ij}}. \tag{3.3}$$

As a result, this model is often referred to as the *product multinomial* model. The MLE of $\pi_{j|i}$ is $\hat{\pi}_{j|i} = n_{ij}/n_{i+}$. Notice that the same estimates result from applying the definition of conditional probability, $P(Y = j|X = i) = P(X = i, Y = j)/P(X = i)$, to the estimates from the one multinomial model: $\hat{\pi}_{j|i} = \hat{\pi}_{ij}/\hat{\pi}_{i+} = (n_{ij}/n)/(n_{i+}/n) = n_{ij}/n_{i+}$. In fact, one can start with Equation 3.2, condition on $n_{1+}, \ldots, n_{I+}$, and obtain the same PMF as given in Equation 3.3. See Exercise 3 for more details.

Independence of X and Y in the context of a product multinomial model means that the conditional probabilities for each Y are equal across the rows of the table. That is, for each j, $\pi_{j|1} = \cdots = \pi_{j|I} = \pi_{+j}$. Note that this condition is mathematically equivalent to independence as defined for the one multinomial model. To show this equivalence, we again need to use the definition of conditional probability. Because X is not random in the product multinomial model, we define π_{i+} to be the fixed proportion of the total sample that is taken from group i. Then $\pi_{j|i} = \pi_{ij}/\pi_{i+}$ and $\pi_{j|i} = \pi_{+j}$ together imply that $\pi_{ij} = \pi_{i+}\pi_{+j}$.

Example: Multinomial simulated sample (Multinomial.R)

We simulate samples here for both the one and product multinomial models to help readers understand the relationships between them. The data are simulated using rmultinom() as in Section 3.1. The main difference is in how we define probabilities to correspond to cells of an $I \times J$ contingency table.

Below is how we simulate a sample of size $n = 1000$ for a 2×3 contingency table corresponding to the one multinomial model where $\pi_{11} = 0.2, \pi_{21} = 0.3, \pi_{12} = 0.2, \pi_{22} = 0.1, \pi_{13} = 0.1$, and $\pi_{23} = 0.1$:

```
> # Probabilities entered by column for array()
> pi.ij <- c(0.2, 0.3, 0.2, 0.1, 0.1, 0.1)
> pi.table <- array(data = pi.ij, dim = c(2,3), dimnames = list(X
    = 1:2, Y = 1:3))
> pi.table   # pi_ij
   Y
X    1   2   3
  1 0.2 0.2 0.1
  2 0.3 0.1 0.1

> set.seed(9812)
> save <- rmultinom(n = 1, size = 1000, prob = pi.ij)
> c.table1 <- array(data = save, dim = c(2,3), dimnames = list(X
    = 1:2, Y = 1:3))
> c.table1
   Y
X    1   2   3
  1 191 206  94
  2 311  95 103
> c.table1/sum(c.table1)
   Y
X       1     2     3
  1 0.191 0.206 0.094
  2 0.311 0.095 0.103
```

For example, $n_{11} = 191$ and $n_{23} = 103$. We see that each $\hat{\pi}_{ij}$ is quite similar to its corresponding π_{ij} as would be expected for a sample this large. Also, notice that the row marginal totals vary—$n_{1+} = 491$ and $n_{2+} = 509$—and are very close to what would be expected considering that $\pi_{1+} = \pi_{2+} = 0.5$.

For the I multinomial setting, we again simulate a sample for a 2×3 contingency table (thus, $I = 2$). With this model, we need to draw samples of fixed size separately for each row. We also need to re-express the cell probabilities as conditional probabilities that sum to 1 in each row, which results in $\pi_{1|1} = 0.4, \pi_{2|1} = 0.4, \pi_{3|1} = 0.2, \pi_{1|2} = 0.6, \pi_{2|2} = 0.2$, and $\pi_{3|2} = 0.2$. Unlike the one multinomial model, the row totals are not specified by the probabilities on the table, so we can choose them to suit the needs of the problem. We arbitrarily select $n_{1+} = 400$ and $n_{2+} = 600$. Two separate calls to the `rmultinom()` function are used to generate samples independently from the two separate probability distributions:

```
> pi.cond <- pi.table/rowSums(pi.table)
> pi.cond   # pi_j|i
   Y
X     1   2   3
  1 0.4 0.4 0.2
  2 0.6 0.2 0.2

> set.seed(8111)
> save1 <- rmultinom(n = 1, size = 400, prob = pi.cond[1,])
> save2 <- rmultinom(n = 1, size = 600, prob = pi.cond[2,])

> c.table2 <- array(data = c(save1[1], save2[1], save1[2],
    save2[2], save1[3], save2[3]), dim = c(2,3), dimnames = list(X
    = 1:2, Y = 1:3))
> c.table
   Y
X     1   2   3
  1 162 159  79
  2 351 126 123

> rowSums(c.table2)
  1   2
400 600
> c.table2/rowSums(c.table2)   # Estimate of pi_j|i
   Y
X       1      2      3
  1 0.405 0.3975 0.1975
  2 0.585 0.2100 0.2050

> round(c.table1/rowSums(c.table1),4)   # From 1 multinomial
   Y
X       1      2      3
  1 0.389 0.4196 0.1914
  2 0.611 0.1866 0.2024
```

We again see that each $\hat{\pi}_{j|i}$ is quite similar to its respective $\pi_{j|i}$ as would be expected.

If we wanted to simulate the data under independence, we could find π_{ij}'s that satisfy $\pi_{ij} = \pi_{i+}\pi_{+j}$ for each i, j pair and use the `rmultinom()` function to simulate a sample from one 6-category multinomial distribution. Equivalently, we could find π_{+j} for $j = 1, 2$, and 3, and simulate data from two three-category multinomial distributions using two calls to the `rmultinom()` function. We demonstrate this process in the corresponding program for this example.

3.2.3 Test for independence

A test for independence,

$$H_0 : \pi_{ij} = \pi_{i+}\pi_{+j} \text{ for each } i,j$$
$$H_a : \pi_{ij} \neq \pi_{i+}\pi_{+j} \text{ for some } i,j$$

can be performed using a Pearson chi-square test and a LRT. These tests were already shown in Section 1.2.3 as tests for the equality of two binomial success probabilities, which is equivalent to a test for independence in a product multinomial model with $I = J = 2$. The Pearson chi-square test statistic is again formed by summing (observed count − estimated expected count)^2/(estimated expected count) across all cells of the contingency table. The observed count is n_{ij}. The expected count under independence is $n\hat{\pi}_{i+}\hat{\pi}_{+j} = n_{i+}n_{+j}/n$ for the one multinomial model, or equivalently $n_{i+}\hat{\pi}_{+j} = n_{i+}n_{+j}/n$ under the product multinomial model. This leads to the test statistic

$$X^2 = \sum_{i=1}^{I}\sum_{j=1}^{J} \frac{(n_{ij} - n_{i+}n_{+j}/n)^2}{n_{i+}n_{+j}/n}.$$

Note that X^2 is equivalent to Equation 1.7 when $I = 2$ and $J = 2$. The X^2 statistic has an approximate $\chi^2_{(I-1)(J-1)}$ distribution in large samples when the null hypothesis is true. When the null hypothesis is false, we expect large deviations between the observed and expected counts (relative to the size of $n_{i+}n_{+j}/n$), which lead to large values of the X^2 statistic. Therefore, we reject the null hypothesis of independence between X and Y when $X^2 > \chi^2_{(I-1)(J-1),1-\alpha}$.

The likelihood ratio is formed in the usual way as

$$\Lambda = \frac{\text{Maximum of likelihood function under } H_0}{\text{Maximum of likelihood function under } H_0 \text{ or } H_a}.$$

The computation of Λ is based on Equation 3.2, using $\hat{\pi}_{ij} = \hat{\pi}_{i+}\hat{\pi}_{+j} = n_{i+}n_{+j}/n$ in the numerator and $\hat{\pi}_{ij} = n_{ij}/n$ in the denominator to estimate each π_{ij}. Applying the usual transformation with Λ, we have

$$-2\log(\Lambda) = 2\sum_{i=1}^{I}\sum_{j=1}^{J} n_{ij} \log\left(\frac{n_{ij}}{n_{i+}n_{+j}/n}\right),$$

where we take $0 \times \log 0 = 0$ by convention. Like X^2, the transformed LRT statistic has a large-sample $\chi^2_{(I-1)(J-1)}$ distribution when H_0 is true and uses the same rejection rule.

The degrees of freedom used for the test of independence are found by calculating

(Number of parameters under H_a) − (Number of parameters under H_0).

This is a general way to find degrees of freedom for any model-comparison test. As shown in Section 3.2.1, we need to estimate $I - J - 2$ parameters when the null hypothesis of independence holds and $IJ - 1$ when it does not. Thus, the degrees of freedom for the test of independence are $(IJ - 1) - (I + J - 2) = (I - 1)(J - 1)$.

Both the LRT and the Pearson chi-square test generally give similar results in large samples. However, at times, their values may differ considerably in smaller samples, leading to ambiguity if their values lie in opposite sides of the rejection region. There have been a number of recommendations given for what constitutes a "large enough" sample to obtain a good χ^2 approximation. The most common criteria are $n_{i+}n_{+j}/n > 1$ or > 5 for all cells of the contingency table. These criteria may not be satisfied when there are very small cell

Table 3.2: Bloating severity after eating a fiber-enriched cracker; data source is DASL (http://lib.stat.cmu.edu/DASL/Stories/HighFiberDietPlan.html).

		Bloating severity			
		None	Low	Medium	High
	None	6	4	2	0
Fiber source	Bran	7	4	1	0
	Gum	2	2	3	5
	Both	2	5	3	2

counts in many cells of the table. For example, a row for which the marginal count n_{i+} is not much larger than the number of columns cannot possibly have "large" expected cell counts in all columns. In these instances, a Monte Carlo simulation or the exact testing procedures described in Section 6.2 can provide a visual assessment to determine if the $\chi^2_{(I-1)(J-1)}$ distributional approximation is appropriate. This is our preferred approach whenever there is any doubt regarding the χ^2 approximation, and we provide an example of its implementation shortly.

Example: Fiber-enriched crackers (Fiber.R, Fiber.csv)

Dietary fiber is a healthful compound that is found in many vegetables and grains. Heavily processed foods are low in fiber, so it is sometimes added to such foods to make them more nutritious. The Data and Story Library (DASL) describes the results of a study where individuals were given a new type of fiber-enriched cracker. The participants ate the crackers and then a meal. Shortly afterward, the participants were instructed to describe any bloating that they experienced. Table 3.2 gives the resulting data in a 4 × 4 contingency table. There were four different types of crackers based on their fiber source, and there were four levels of bloating severity reported. While not specified in the data description at DASL, we assume that each participant was assigned to only one fiber source group. Also, the data description does not mention whether the fact that there are 12 participants for each fiber source was by design or by chance, so it is not clear whether a one-multinomial or product-multinomial model might be more appropriate. Fortunately, as noted at the end of Section 3.2.2, this distinction has no bearing on the analysis results. Finally, while the data given at DASL was in a 4 × 4 contingency table, we can see that fiber source could actually be split into two separate explanatory variables, one indicating whether bran is present and one indicating whether gum is present. We explore alternative analysis methods that account for this "factorial" structure in Section 3.3.

We begin by reading the data into R and summarizing the data in a contingency table form using the `xtabs()` function:

```
> diet <- read.csv(file = "C:\\data\\Fiber.csv")
> head(diet)
  fiber  bloat count
1  bran   high     0
2   gum   high     5
3  both   high     2
4  none   high     0
5  bran medium     1
6   gum medium     3
```

```
> diet$fiber <- factor(x = diet$fiber, levels = c("none", "bran",
   "gum", "both"))
> diet$bloat <- factor(x = diet$bloat, levels = c("none", "low",
   "medium", "high"))
> diet.table <- xtabs(formula = count ~ fiber + bloat, data =
   diet)
> diet.table
       bloat
fiber   none low medium high
  none     6   4      2    0
  bran     7   4      1    0
  gum      2   2      3    5
  both     2   5      3    2
```

We use the `factor()` function to change the order of the levels for both variables to match that given in Table 3.2. We could instead have entered the contingency table directly into R using the `array()` function and arranged the array in the desired order. We provide an example of this in our corresponding program.

We would like to determine if bloating severity is related to the type of fiber. If there is a relation, this would indicate to the cracker manufacturer that some types of fiber may cause more severe bloating and should be avoided. We use a test for independence to make this determination. The hypotheses for the test are $H_0 : \pi_{ij} = \pi_{i+}\pi_{+j}$ for each i,j vs. $H_a : \pi_{ij} \neq \pi_{i+}\pi_{+j}$ for some i,j, where π_{ij} is the probability that a randomly selected person is assigned to fiber source i and experiences bloating level j. R provides a number of functions to test these hypotheses, including: (1) `chisq.test()`, (2) `assocstats()` from the vcd package, and (3) the generic function `summary()` to summarize a table object. Normally, we would only use one of these functions, but we show all three for illustrative purposes:

```
> ind.test <- chisq.test(x = diet.table, correct = FALSE)
Warning message:
In chisq.test(diet.table, correct = FALSE) :
  Chi-squared approximation may be incorrect
> ind.test

        Pearson's Chi-squared test

data:  diet.table
X-squared = 16.9427, df = 9, p-value = 0.04962

> library(package = vcd)
> assocstats(x = diet.table)
                     X^2 df  P(> X^2)
Likelihood Ratio 18.880   9  0.026230
Pearson          16.943   9  0.049621

Phi-Coefficient   : 0.594
Contingency Coeff.: 0.511
Cramer's V        : 0.343

> class(diet.table)
[1] "xtabs" "table"
> summary(diet.table)
```

```
Call: xtabs(formula = count ~ fiber + bloat, data = diet)
Number of cases in table: 48
Number of factors: 2
Test for independence of all factors:
        Chisq = 16.943, df = 9, p-value = 0.04962
        Chi-squared approximation may be incorrect

> qchisq(p = 0.95, df = 9)
[1] 16.91898
```

Note that we include the `correct = FALSE` argument value in `chisq.test()` to prevent a continuity correction from being applied. Please see Section 1.2.3 for why we avoid these corrections. The Pearson chi-square test statistic is $X^2 = 16.94$, the critical value at the $\alpha = 0.05$ level is $\chi^2_{0.95,9} = 16.92$, and the p-value is $P(A > 16.94) = 0.0496$ where $A \sim \chi^2_9$. Because the p-value is small, but not extremely so, we would say there is moderate evidence against independence (thus, moderate evidence of dependence). The LRT leads to a similar conclusion with $-2\log(\Lambda) = 18.88$ and a p-value of 0.0262.

There are a number of very small cell counts in the contingency table, so there may be some concern about the appropriateness of the χ^2_9 approximation. The expected cell counts under independence for this table are:

```
> ind.test$expected
      bloat
fiber  none  low  medium  high
  none 4.25 3.75   2.25  1.75
  bran 4.25 3.75   2.25  1.75
  gum  4.25 3.75   2.25  1.75
  both 4.25 3.75   2.25  1.75
```

which satisfies the $n_{i+}n_{+j}/n > 1$ criterion given earlier, but not $n_{i+}n_{+j}/n > 5$. The `chisq.test()` and `summary()` functions even print a warning message about the appropriateness of the χ^2_9 approximation. This warning is given by the functions whenever $n_{i+}n_{+j}/n < 5$ for any cell.

One way to examine the distributional approximation is through a Monte Carlo simulation. This involves simulating a large number of contingency tables of size n whose probabilities satisfy the null hypothesis of independence. Ideally, one should follow the same type of sampling method as the assumed model. For a one-multinomial model, we can use one call to `rmultinom()`, where we set each π_{ij} (`prob` argument) to $n_{i+}n_{+j}/n^2$ (the estimated expected count under the null hypothesis divided by the sample size). For a product-multinomial model, we can use I different calls to `rmultinom()` by setting $\pi_{j|i}$ equal to $\hat{\pi}_{+j}$. We chose to use the one multinomial model here due to its simplicity and to the unknown nature of the actual sampling method.

The Pearson chi-square test statistic, say X^{2*}, is calculated for each contingency table of simulated counts. These X^{2*} values can be summarized through a plot in a number of ways, such as a histogram, an empirical CDF plot[1], and a QQ-plot[2],

---

[1] For a sample of observations $w_1, w_2, \ldots, w_m$, the empirical CDF at w is the proportion of observations at or below w: $\hat{F}(w) = \#\{w_j \leq w\}/m$, where "#" means "number of."

[2] Ordered observed values from a statistic are plotted against the corresponding quantiles from a particular probability distribution of interest. For example, the observed median (provided the number of observed values is odd) would be plotted against the 0.5 quantile from the probability distribution. If the plotted points follow close to a straight line of $y = x$, where y (x) indicates what is plotted on the y-axis (x-axis), the statistic approximately follows the probability distribution.

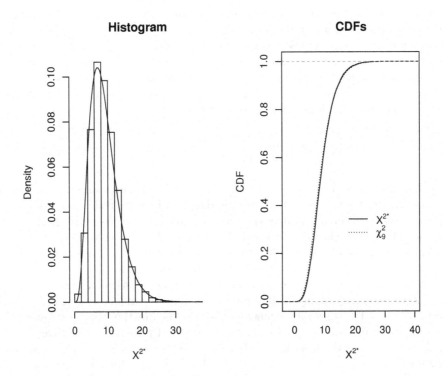

Figure 3.1: Plots of X^{2*} and a χ_9^2 distribution.

in order to view an estimated distribution for the statistic. Using a χ_9^2 overlay (or examining the deviation from a straight line in the QQ-plot), one can judge if the distribution approximation is appropriate. The same procedure can also be applied to the LRT statistic.

The Monte Carlo simulation using 10,000 simulated data sets is implemented with code in the corresponding program for this example. Figure 3.1 shows the histogram and the empirical CDF plot with the χ_9^2 distribution overlay. We see general agreement in both plots between the distribution of X^{2*} and a χ_9^2 distribution. The quantiles at $\alpha = 0.90, 0.95$, and 0.99 are 14.57, 16.61, and 21.04 for X^{2*} and 14.68, 16.92, and 21.67 for χ_9^2, respectively. Thus, there is general agreement between the two distributions.

It is interesting to note that the Monte Carlo simulation performed here is actually a form of the bootstrap. The bootstrap is applied here as a nonparametric procedure that provides estimates of a CDF for a statistic. For this example, we can obtain a bootstrap estimate of the p-value for the test by calculating the proportion of simulated X^{2*} values greater than or equal to $X^2 = 16.94$. The result is 0.0446, which is in general agreement with the p-value from the Pearson chi-square test. For a more in-depth discussion of the bootstrap, please see Davison and Hinkley (1997).

When independence is rejected, there are a number of ways to examine the association between X and Y. In Section 1.2 with 2×2 contingency tables, we could simply calculate an odds ratio to summarize the degree of association; i.e., the amount by which the odds ratio is > 1 or < 1. Now, with $I \times J$ contingency tables, one odds ratio does not summarize the association. Instead, one approach is to calculate odds ratios for 2×2 parts of a contingency table that are of special interest. Unless there is firm guidance from the context of the problem, there is potential for creating many different odds ratios, and hence potential for inflated error rates for multiple tests or confidence intervals.

As an alternative to odds ratios, various forms of residuals may be calculated on each cell to measure the how far the observed cell count deviates from what would be expected under independence. The basic residual in cell (i, j) has the form $n_{ij} - n_{i+}n_{+j}/n$. A different version of residual, the *standardized Pearson residual*, is

$$\frac{n_{ij} - n_{i+}n_{+j}/n}{\sqrt{(1 - n_{i+}/n)(1 - n_{+j}/n)n_{i+}n_{+j}/n}},$$

where the denominator is $\widehat{Var}(N_{ij} - N_{i+}N_{+j}/n)^{1/2}$. The stdres component of an object resulting from chisq.test() provides these standardized Pearson residuals. The advantage of a standardized Pearson residual is that its distribution can be approximated by the standard normal.[3] This means that values beyond, say, ± 2 can be considered "unusual" and should occur only in about 5% of cells. Cells where large standardized Pearson residuals occur indicate potential sources of the association between X and Y.

Residuals and odds ratios are somewhat ineffective in tables where I and/or J are much larger than 2, because they do not lead easily to a systematic understanding of the relationship between X and Y. However, displaying the residuals in the same rows and columns as the contingency table can sometimes reveal patterns of positive or negative values that hint at a larger cause for the association. For example, if residuals in a given row show an increasing (or decreasing) trend across increasing levels of an ordinal Y, then this indicates that this row's responses tend toward larger (or smaller) levels of Y than in other rows where this does not occur. Alternatively, if one cell has a residual that is large and of the opposite sign from others in the same row and column, then this one cell appears to violate independence. Unfortunately, patterns of residuals are not always apparent. Regression modeling is a more systematic approach to assess the association between variables. Sections 3.3 and 3.4 cover a variety of model structures that allow us to better understand the association between X and Y. Furthermore, ordinal properties of multinomial response categories can be taken into account more easily with these models and point to sources of association that may not be noticeable otherwise. Odds ratios and residuals can still be examined within these regression models.

3.3 Nominal response regression models

We now move to the problem of modeling the probabilities of a categorical response variable Y with response categories $j = 1, \ldots, J$ using explanatory variables $x_1, \ldots, x_p$. In Section 1.2.5, we defined odds for a binary response as $P(\text{success})/P(\text{failure})$. More generally, for a multinomial response, we can define odds to be a comparison of *any* pair of response categories. For example, $\pi_j/\pi_{j'}$ is the odds of category j relative to j'. A popular regression model for multinomial responses is developed then by selecting one response category as the base level and forming the odds of the remaining $J - 1$ categories against this level. For example, let $j = 1$ represent the base level category, and form the odds

---

[3]A *Pearson residual* has the same numerator, but the denominator is $\widehat{Var}(N_{ij})^{1/2} = \sqrt{(1 - n_{i+}/n)(1 - n_{+j}/n)}$, which is not as well approximated by a standard normal distribution.

π_j/π_1 for $j = 2, \ldots, J$. These odds are then modeled as a function of explanatory variables using a generalized form of logistic regression. Notice that if $J = 2$, we have $\log(\pi_2/\pi_1) = \log(\pi_2/(1 - \pi_2))$, which is equivalent to $\log(\pi/(1 - \pi))$ as in logistic regression (response category 2 becomes the "success" category).

Specifically, a *multinomial regression model* (also known as a *baseline-category logit model*) relates a set of explanatory variables to each log-odds according to

$$\log(\pi_j/\pi_1) = \beta_{j0} + \beta_{j1}x_1 + \cdots + \beta_{jp}x_p, \tag{3.4}$$

for $j = 2, \ldots, J$. Notice that the first subscript on the β parameters corresponds to the response category, which allows each response's log-odds to relate to the explanatory variables in a different way. Also, notice that by subtracting the appropriate log-odds we can rewrite Equation 3.4 to compare any pair of response categories. For example, to find $\log(\pi_j/\pi_{j'})$ where $j' \neq 1$ and $j' \neq j$, we compute

$$\begin{aligned}\log(\pi_j/\pi_{j'}) &= \log(\pi_j) - \log(\pi_{j'}) \\ &= \log(\pi_j) - \log(\pi_{j'}) + \log(\pi_1) - \log(\pi_1) \\ &= \log(\pi_j/\pi_1) - \log(\pi_{j'}/\pi_1) \\ &= (\beta_{j0} - \beta_{j'0}) + (\beta_{j1} - \beta_{j'1})x_1 + \cdots + (\beta_{jp} - \beta_{j'p})x_p.\end{aligned} \tag{3.5}$$

Thus, the choice of base level is not important and can be made based on convenience or interpretation.

The probabilities for each individual category also can be found in terms of the model. We can re-write Equation 3.4 as

$$\pi_j = \pi_1 \exp(\beta_{j0} + \beta_{j1}x_1 + \cdots + \beta_{jp}x_p)$$

using properties of logarithms. Noting that $\pi_1 + \pi_2 + \cdots + \pi_J = 1$, we have

$$\pi_1 + \pi_1 \exp(\beta_{20} + \beta_{21}x_1 + \cdots + \beta_{2p}x_p) + \cdots + \pi_1 \exp(\beta_{J0} + \beta_{J1}x_1 + \cdots + \beta_{Jp}x_p) = 1.$$

By factoring out the common π_1 in each term, we obtain an expression for π_1:

$$\pi_1 = \frac{1}{1 + \sum_{j=2}^{J} \exp(\beta_{j0} + \beta_{j1}x_1 + \cdots + \beta_{jp}x_p)}. \tag{3.6}$$

This leads to a general expression for π_j:

$$\pi_j = \frac{\exp(\beta_{j0} + \beta_{j1}x_1 + \cdots + \beta_{jp}x_p)}{1 + \sum_{j=2}^{J} \exp(\beta_{j0} + \beta_{j1}x_1 + \cdots + \beta_{jp}x_p)} \tag{3.7}$$

for $j = 2, \ldots, J$.

Parameters for the model are estimated using maximum likelihood. For a sample of observations, y_i that denote the category response and corresponding explanatory variables $x_{i1}, \ldots, x_{ip}$ for $i = 1, \ldots, m$, the likelihood function is simply the product of m multinomial distributions with probability parameters as given by Equations 3.6 and 3.7. Iterative numerical procedures are used then to find the parameter estimates. The `multinom()` function from the `nnet` package performs the necessary computations. As mentioned in the help for this function, re-scaling the explanatory variables to be within 0 and 1 can help at times with convergence of the parameter estimates. We illustrate how to make this transformation in Exercise 12.

The estimated covariance matrix for the parameter estimates is found using standard likelihood procedures as outlined in Appendix B.3.4. Wald and LR-based inference methods are performed in the same ways as for likelihood procedures in earlier chapters. However, profile likelihood ratio intervals are difficult to obtain in R, because neither a `confint()` method function nor the `mcprofile` package have been extended to calculate these types of intervals for objects resulting from `multinom()`. For this reason, we focus on using Wald intervals in this section.

Example: Wheat kernels (Wheat.R, Wheat.csv)

The presence of sprouted or diseased kernels in wheat can reduce the value of a wheat producer's entire crop. It is important to identify these kernels after being harvested but prior to sale. To facilitate this identification process, automated systems have been developed to separate healthy kernels from the rest. Improving these systems requires better understanding of the measurable ways in which healthy kernels differ from kernels that have sprouted prematurely or are infected with a fungus ("Scab"). To this end, Martin et al. (1998) conducted a study examining numerous physical properties of kernels—density, hardness, size, weight, and moisture content—measured on a sample of wheat kernels from two different classes of wheat, hard red winter (hrw) and soft red winter (srw). Each kernel's condition was also classified as "Healthy," "Sprout," or "Scab" by human visual inspection. In the data provided by the authors of this paper, we have measurements from 275 wheat kernels. Below is a portion of the data:

```
> wheat <- read.csv(file = "C:\\data\\Wheat.csv")

> head(wheat, n = 3)
  class  density  hardness   size   weight  moisture    type
1   hrw 1.349253  60.32952 2.30274 24.6480 12.01538  Healthy
2   hrw 1.287440  56.08972 2.72573 33.2985 12.17396  Healthy
3   hrw 1.233985  43.98743 2.51246 31.7580 11.87949  Healthy

> tail(wheat, n = 3)
    class   density  hardness    size  weight moisture type
273   srw 0.8491887  34.06615 1.40665 12.0870 11.92744 Scab
274   srw 1.1770230  60.97838 1.05690  9.4800 12.24046 Scab
275   srw 1.0305543  -9.57063 2.05691 23.8185 12.64962 Scab
```

We begin with useful plots to initially examine the data (the code is given in our program). Figure 3.2 shows a parallel coordinates plot constructed using the `parcoord()` function of the `MASS` package. The horizontal axis gives each variable name, and plotted above these names are the corresponding values for each kernel. These values are scaled so that the minimum and maximum values for each variable appear at the bottom and top, respectively, of the vertical axis. A line is drawn for each kernel indicating its position across the variables. For example, the kernel that has the smallest weight (lowest line for `weight`) also has the second smallest size (second lowest line for `size`).[4] We have used different line types to represent the three types of kernel, so that we can understand some of the features that differentiate the healthy, sprout, and scab conditions. We see that

1. Scab kernels generally have smaller density, size, and weight values,

---

[4]Because it is difficult to identify particular kernel numbers, we provide code in the corresponding program that shows how to highlight a kernel. For example, one can see that kernel #269 is the kernel with the smallest weight.

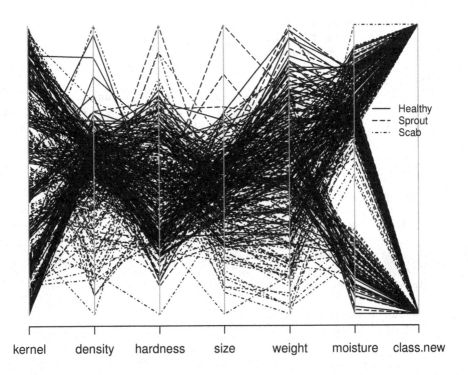

Figure 3.2: Parallel coordinate plot of the wheat data. The kernel variable corresponds to the observation number in the data set. The `class.new` variable is a 1 for soft red winter wheat and 0 for hard red winter wheat. We recommend viewing this plot in color (see program for code) if the version in your book is only in black and white.

2. Healthy kernels may have higher densities,

3. There is much overlap for healthy and sprout kernels, and

4. The moisture content appears to be related to hard or soft red winter wheat class.

To examine these data more formally, we estimate a multinomial regression model using the wheat class and the five measurements on each kernel as explanatory variables:

```
> levels(wheat$type)   # The 3 response categories
[1] "Healthy" "Scab"    "Sprout"

> library(package = nnet)
> mod.fit <- multinom(formula = type ~ class + density + hardness
    + size + weight + moisture, data = wheat)
# weights:  24 (14 variable)
initial    value 302.118379
```

```
iter  10 value 234.991271
iter  20 value 192.127549
final  value 192.112352  converged

> summary(mod.fit)
Call: multinom(formula = type ~ class + density + hardness + size
   + weight + moisture, data = wheat)

Coefficients:
        (Intercept)   classsrw    density      hardness       size
Scab       30.54650  -0.6481277  -21.59715  -0.01590741   1.0691139
Sprout     19.16857  -0.2247384  -15.11667  -0.02102047   0.8756135
            weight     moisture
Scab     -0.2896482   0.10956505
Sprout   -0.0473169  -0.04299695

Std. Errors:
        (Intercept)   classsrw    density      hardness       size
Scab      4.289865   0.6630948   3.116174   0.010274587   0.7722862
Sprout    3.767214   0.5009199   2.764306   0.008105748   0.5409317
            weight     moisture
Scab     0.06170252  0.1548407
Sprout   0.03697493  0.1127188

Residual Deviance: 384.2247
AIC: 412.2247
```

Output from levels(wheat$type) shows us that "Healthy" is stored as the first level of type, so multinom() uses it as the base level. Also, note that an indicator variable is created for class, where a 1 is for "srw" and a 0 for "hrw". These choices by R follow the usual convention of making the first level of a factor the base level. From the coefficients table in the output, we obtain the estimated model as

$$\log(\hat{\pi}_{\text{Scab}}/\hat{\pi}_{\text{Healthy}}) = 30.55 - 0.65\text{srw} - 21.60\text{density} - 0.016\text{hardness}$$
$$+ 1.07\text{size} - 0.29\text{weight} + 0.11\text{moisture}$$

and

$$\log(\hat{\pi}_{\text{Sprout}}/\hat{\pi}_{\text{Healthy}}) = 19.17 - 0.22\text{srw} - 15.12\text{density} - 0.021\text{hardness}$$
$$+ 0.88\text{size} - 0.047\text{weight} - 0.043\text{moisture}.$$

Hypothesis tests of the form $H_0 : \beta_{jr} = 0$ vs. $H_a : \beta_{jr} \neq 0$ can be performed by Wald tests or LRTs. For example, a Wald test of $H_0 : \beta_{21} = 0$ vs. $H_a : \beta_{21} \neq 0$, which corresponds to the class explanatory variable in the log-odds for scab vs. healthy, results in

$$Z_0 = \frac{\hat{\beta}_{21}}{\sqrt{\widehat{Var}(\beta_{21})}} = \frac{-0.6481}{0.6631} = -0.98$$

and a p-value of $2P(Z > |-0.98|) = 0.33$. Thus, there is not sufficient evidence that hard and soft red winter wheat have different effects on the scab or healthy status of the kernels given the other explanatory variables are in the model.

Hypotheses that are more interesting explore the effects of a given explanatory variable on *all* response categories. For example, does class of wheat have *any* effect on the probabilities of healthy, scab, or sprout kernels? To investigate this, we need to test whether *all* parameters corresponding to the explanatory variable of interest, say x_r, are zero: $H_0 : \beta_{jr} = 0, j = 2, \ldots, J$. In the case of the class explanatory variable, we use $H_0 : \beta_{21} = \beta_{31} = 0$ vs. $H_a : \beta_{21} \neq 0$ and/or $\beta_{31} \neq 0$. This test is easily carried out with LR methods using the Anova() function:

```
> library(package = car)
> Anova(mod.fit)
Analysis of Deviance Table (Type II tests)

Response: type
          LR Chisq Df Pr(>Chisq)
class       0.964   2    0.61751
density    90.555   2  < 2.2e-16 ***
hardness    7.074   2    0.02910 *
size        3.211   2    0.20083
weight     28.230   2  7.411e-07 ***
moisture    1.193   2    0.55064
---
Signif. codes:  0 '***' 0.001 '**' 0.01 '*' 0.05 '.' 0.1 ' ' 1
```

The transformed test statistic is $-2\log(\Lambda) = 0.964$, and the corresponding p-value is 0.6175 using a χ_2^2 distributional approximation. Because of the large p-value, there is not sufficient evidence to indicate that the class of wheat is important given that the other variables are in the model. Separate tests for density, hardness, and weight in the output all indicate at least marginal evidence of importance for these explanatory variables. Note that the anova() function may also be used to perform these tests. We demonstrate its use in the corresponding program to this example.

We can use names(mod.fit) to obtain a listing of the objects within the mod.fit list. These include deviance (residual deviance) and convergence (convergence indicator). The fitted.values component contains the estimated probabilities of healthy, scab, or sprout for each kernel. Alternatively, the predict() function can compute these estimated probabilities using the type = "probs" argument value:

```
> pi.hat <- predict(object = mod.fit, newdata = wheat, type =
    "probs")
> head(pi.hat)
    Healthy         Scab       Sprout
1 0.8552110 0.046396827 0.09839221
2 0.7492553 0.021572158 0.22917255
3 0.5172800 0.068979903 0.41374011
4 0.8982064 0.006740716 0.09505287
5 0.5103245 0.176260796 0.31341473
6 0.7924907 0.015304122 0.19220522
```

For example, the estimated probability of being healthy for the first kernel is

$$\hat{\pi}_{\text{Healthy}} = [1 + \exp(30.55 - 0.65 \times 0 + \cdots + 0.11 \times 12.02) + \exp(19.17 - 0.22 \times 0 + \cdots - 0.043 \times 12.02)]^{-1}$$
$$= 0.8552.$$

We include code in our corresponding program to demonstrate how these probabilities are computed manually using Equations 3.6 and 3.7 instead of predict(). The predict() function can also be used to find the most probable response category for each kernel using the type = "class" argument value, which is the default.

Computing confidence intervals for π_{Healthy}, π_{Scab}, and π_{Sprout} is not as easy as one might expect. The predict() function does not provide the estimated variances needed to compute a Wald confidence interval for π_j. The reason is that these probabilities form a probability distribution themselves. A joint confidence region would be needed in $J - 1$ dimensions (the "-1" is due to $\sum_{j=1}^{J} \pi_j = 1$) to properly obtain an "interval" for these probabilities.[5] Because a joint confidence region can be difficult to calculate and interpret, an alternative is one-at-a-time confidence intervals for each π_j and use these as an approximation. This allows one to still take into account the estimator variability in some manner. These one-at-a-time Wald confidence intervals can be calculated as $\hat{\pi}_j \pm Z_{1-\alpha/2} \widehat{Var}(\hat{\pi}_j)^{1/2}$ through the help of the deltaMethod() function first described in Exercise 16 of Chapter 2. Below are the calculations for the first observation of the data set and π_{Healthy}:

```
> # Obtain observation values, ";" ends a command
> x1 <- 0;          x2 <- wheat[1,2]; x3 <- wheat[1,3]
> x4 <- wheat[1,4]; x5 <- wheat[1,5]; x6 <- wheat[1,6]

> g.healthy <- "1/(1 + exp(b20 + b21*x1 + b22*x2 + b23*x3 +
    b24*x4 + b25*x5 + b26*x6) + exp(b30 + b31*x1 + b32*x2 + b33*x3
    + b34*x4 + b35*x5 + b36*x6))"
> calc.healthy <- deltaMethod(object =  mod.fit, g = g.healthy,
    parameterNames = c("b20", "b21", "b22", "b23", "b24", "b25",
    "b26", "b30", "b31", "b32", "b33", "b34", "b35", "b36"))
> calc.healthy$Estimate   # pi-hat_Healthy
[1] 0.855211
> calc.healthy$SE   # sqrt(Var-hat(pi-hat_Healthy))
[1] 0.05999678
> alpha <- 0.05
> calc.healthy$Estimate + qnorm(p = c(alpha/2, 1-alpha/2)) *
    calc.healthy$SE
[1] 0.7376194 0.9728025
```

The g.healthy object contains a character string for the formula used to calculate π_{Healthy}, where we use bjr to denote β_{jr} and xr to denote x_r for $j = 2, 3$; $r = 0, \ldots, 6$. There can be more efficient ways to enter character strings like these in R, and we provide an example within the corresponding program. The deltaMethod() function calculates $\hat{\pi}_{\text{Healthy}}$ and uses the delta method (see Appendix B.4.2) to calculate $\widehat{Var}(\hat{\pi}_j)^{1/2}$. The 95% confidence interval for π_{Healthy} is (0.7376, 0.9728). The 95% confidence intervals for π_{Scab} and π_{Sprout} are (-0.0067, 0.0995) and (0.0143, 0.1825), respectively (code is given in the program). Notice that these types of intervals can be below 0 or above 1. Also, notice that combinations of values for $\pi_{\text{Healthy}}, \pi_{\text{Scab}}$, and π_{Sprout} that all lie inside their respective intervals could result in $\sum_{j=1}^{3} \pi_j \neq 1$, because these intervals do not form a joint confidence region.

The estimated model can be plotted by including the estimated probability for each wheat type on the y-axis and one of the explanatory variables on the x-axis, while holding the other variables within the model constant. Because the plot is conditional

---

[5]See http://thread.gmane.org/gmane.comp.lang.r.general/16832/focus=16834 for a discussion of this issue.

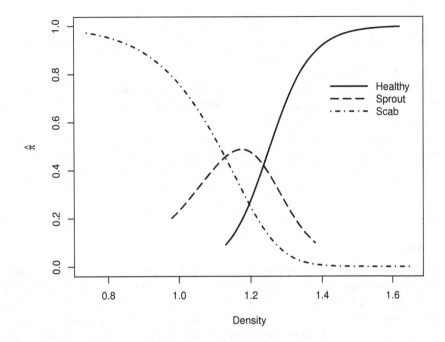

Figure 3.3: Estimated multinomial regression model for the wheat data where density is the only explanatory variable included in the model.

on these variable values held constant, more than one plot should usually be made at different constant values. We discuss plots like this in Exercise 19. Instead for this example, we re-estimate a model with `density` as the only explanatory variable:

$$\log(\hat{\pi}_{\text{Scab}}/\hat{\pi}_{\text{Healthy}}) = 29.38 - 24.56 \text{density}$$

and

$$\log(\hat{\pi}_{\text{Sprout}}/\hat{\pi}_{\text{Healthy}}) = 19.12 - 15.48 \text{density}.$$

The model is plotted in Figure 3.3 (see the program for the code), where the estimated probabilities for each kernel condition are drawn between the smallest and the largest observed density values for that condition. From the plot, we see that the estimated scab probability is the largest for the smaller density kernels. The estimated healthy probability is the largest for the high density kernels. For density levels in the middle, sprout has the largest estimated probability. The parallel coordinates plot in Figure 3.2 on p. 155 displays similar findings where the density levels tend to follow the scab < sprout < healthy ordering.

Transformations and interaction terms can be included within `multinom()` in a similar manner as with `glm()`. We provide an example in the corresponding program that shows how to estimate parameters in $\log(\pi_j/\pi_{\text{Healthy}}) = \beta_{j0} + \beta_{j1}\text{SRW} + \beta_{j2}\text{density} + \beta_{j3}\text{density}^2 + \beta_{j4}\text{SRW} \times \text{density}$. After 100 iterations, convergence is not attained. However, when the number of iterations is increased to 500 (using the `maxit` argument), convergence is reached.

Using the default coding of "Healthy" as the base level of the response was ideal in this example so that comparisons could be made more easily with each of the two non-healthy conditions. However, there may be other situations where it makes sense to change the default ordering. The `relevel()` function can be used as it was in Section 2.2.6 for this purpose. For example, we can make "Sprout" the base level with `wheat$type <- relevel(x = wheat$type, ref = "Sprout")` and then re-estimate the model using `multinom()`.

Estimation of multinomial regression models can also be done using the general model fitting function `vglm()` from the VGAM package (Yee, 2010). We provide code showing how to use this function in the corresponding program for this example. An advantage to using `vglm()` is that it can sometimes obtain faster convergence of parameter estimates (see Exercise 12 for an example). Also, the function can be used to fit a wide variety of model types for categorical data. We focus on the `multinom()` function here because its package, nnet, is part of the default installation of R.

3.3.1 Odds ratios

Because the log-odds are modeled directly in a multinomial regression model, odds ratios are useful for interpreting an explanatory variable's relationship with the response. As described in Section 2.2.3 for logistic regression models, odds ratios for numerical explanatory variables represent the change in odds corresponding to a c-unit increase in a particular explanatory variable. The only difference with multinomial models is that the odds are now formed as a comparison between two of the J response categories.

For example, suppose the terms in Equation 3.4 are distinct explanatory variables (not transformations or interactions). The odds of a category j response vs. a category 1 response are $\exp(\beta_{j0} + \beta_{j1}x_1 + \cdots + \beta_{jp}x_p)$. These odds change by $\exp(c\beta_{jr})$ times for every c-unit increase in x_r, holding the other variables constant. Similarly, from Equation 3.5, the odds of a category j vs. a category j' response ($j \neq j'$, $j > 1$, and $j' > 1$) change by $\exp(c(\beta_{jr} - \beta_{j'r}))$ times for every c-unit increase in x_r while holding the other variables in the model constant.

Maximum likelihood estimates of odds ratios are obtained by replacing the regression parameters with their corresponding estimates. Wald and LR-based inference methods for odds ratios are used in the same ways as discussed in earlier chapters.

Example: Wheat kernels (Wheat.R, Wheat.csv)

The odds in this problem are constructed as $P(Y = j)/P(Y = 1)$, $j = 2, 3$, where category 1 = healthy, 2 = scab, and 3 = sprout. The estimated odds ratios for scab vs. healthy or sprout vs. healthy are calculated for each explanatory variable as $\widehat{OR} = \exp(c\hat{\beta}_{jr})$ for $j = 2, 3$ and $r = 1, \ldots, 6$. To choose appropriate values of c, we set c to be equal to one standard deviation for each continuous explanatory variable. Each of these standard deviations is found using the `apply()` function, which applies the `sd()` function to specific columns (MARGIN = 2) of the wheat data frame. For class, we choose $c = 1$ because it is a binary variable for soft or hard red winter wheat. Below are the estimated standard deviations, followed by the odds ratios:

```
> sd.wheat <- apply(X = wheat[,-c(1,7,8)], MARGIN = 2, FUN = sd)
> c.value <- c(1, sd.wheat)    # class = 1 is first value
> round(c.value,2)
        density  hardness      size    weight  moisture
```

```
      1.00         0.13        27.36         0.49         7.92         2.03

> # beta.hat_jr for r = 1, ..., 6 and j = 2, 3, where beta.hat_jr
    = coefficients(mod.fit)[j-1,r+1]
> beta.hat2 <- coefficients(mod.fit)[1,2:7]
> beta.hat3 <- coefficients(mod.fit)[2,2:7]

> # Odds ratios for j = 2 vs. j = 1 (scab vs. healthy)
> round(exp(c.value*beta.hat2), 2)
  density  hardness      size    weight  moisture
     0.52      0.06      0.65      1.69      0.10      1.25
> round(1/exp(c.value*beta.hat2), 2)
  density  hardness      size    weight  moisture
     1.91     17.04      1.55      0.59      9.90      0.80

> # Odds ratios for j = 3 vs. j = 1 (sprout vs. healthy)
> round(exp(c.value*beta.hat3), 2)
  density  hardness      size    weight  moisture
     0.80      0.14      0.56      1.54      0.69      0.92
> round(1/exp(c.value*beta.hat3), 2)
  density  hardness      size    weight  moisture
     1.25      7.28      1.78      0.65      1.45      1.09
```

We use the `coefficients()` function to obtain the estimated regression parameter values because there is no `coefficients` component in objects produced by `multinom()`. Interpretations of the estimated odds ratios include:

- The estimated odds of a scab vs. a healthy kernel change by 0.06 times for a 0.13 increase in the density holding the other variables constant. Equivalently, we can say that the estimated odds of a scab vs. a healthy kernel change by 17.04 times for a 0.13 *decrease* in the density holding the other variables constant. The estimated odds of a sprout vs. a healthy kernel change by 7.28 times for a 0.13 decrease in the density holding the other variables constant.

- The estimated odds of a scab vs. healthy kernel change by 9.90 times for a 7.92 decrease in the weight holding the other variables constant. The estimated odds of a sprout vs. healthy kernel change by 1.45 times for a 7.92 decrease in the weight holding the other variables constant.

We see that the larger the density and weight, the more likely a kernel is healthy. We can relate these results back to Figure 3.2 on p. 155. The plot shows that kernels with low density and weight are often of the scab condition rather than healthy condition. We see a similar separation between sprout and healthy for the density measurements, but not as much so for weight. Interpretations of the other variables are part of Exercise 11.

As with the `glm()` function, there are a number of method functions associated with objects resulting from `multinom()`:

```
> class(mod.fit)
[1] "multinom" "nnet"
> methods(class = multinom)
[1] add1.multinom*           anova.multinom*
[3] coef.multinom*           confint.multinom*
[5] drop1.multinom*          extractAIC.multinom*
```

```
 [7] logLik.multinom*           model.frame.multinom*
 [9] predict.multinom*          print.multinom*
[11] summary.multinom*          vcov.multinom*

   Non-visible functions are asterisked

> sqrt(vcov(mod.fit)[2,2])   # sqrt(Var-hat(beta-hat_21))
[1] 0.6630948
```

Note that we previously used coefficients(), the equivalent to coef.multinom() to extract the estimated regression parameters.

Confidence intervals can be constructed for the regression parameters using confint(). Unfortunately, the confint.multinom() method function that is called by the generic confint() does not produce profile LR intervals. The Wald intervals for the odds ratios are given below:

```
> conf.beta <- confint(object = mod.fit, level = 0.95)
> conf.beta   # Results are in a 3D array
, , Scab
              2.5 %  97.5 %
(Intercept)   22.14   38.95
classsrw      -1.95    0.65
density      -27.70  -15.49
hardness      -0.04    0.00
size          -0.44    2.58
weight        -0.41   -0.17
moisture      -0.19    0.41

, , Sprout
              2.5 %  97.5 %
(Intercept)   11.78   26.55

<OUTPUT EDITED>

> ci.OR2 <- exp(c.value*conf.beta[2:7,1:2,1])
> ci.OR3 <- exp(c.value*conf.beta[2:7,1:2,2])

> round(data.frame(low = ci.OR2[,1], up = ci.OR2[,2]), 2)
          low   up
classsrw  0.14  1.92
density   0.03  0.13
hardness  0.37  1.12
size      0.80  3.55
weight    0.04  0.26
moisture  0.67  2.32

> round(data.frame(low = 1/ci.OR2[,2], up = 1/ci.OR2[,1]),
    2)[c(2,5),]
          low    up
density   7.64  38.00
weight    3.80  25.79

> round(data.frame(low = ci.OR3[,1], up = ci.OR3[,2]), 2)
          low   up
```

```
classsrw  0.30 2.13
density   0.07 0.28
hardness  0.36 0.87
size      0.91 2.59
weight    0.39 1.22
moisture  0.58 1.44

> round(data.frame(low = 1/ci.OR3[,2], up = 1/ci.OR3[,1]),
    2)[c(2,3),]
          low    up
density  3.57 14.82
hardness 1.15  2.74
```

The `confint()` function stores the confidence limits in a three-dimensional array using a [row, column, table] format. Row $r+1$ of each table contains the lower and upper limits for explanatory variable r in columns 1 and 2, respectively. For example, to access the scab vs. healthy intervals for `density` ($r = 2$), we need elements [3, 1:2, 1]. In order to find the intervals for the odds ratios, we multiply each interval endpoint by the appropriate constant and then use the `exp()` function.[6] The density odds ratios can be interpreted as follows:

> With 95% confidence, the odds of a scab instead of a healthy kernel change by 7.64 to 38.00 times when density is decreased by 0.13 holding the other variables constant. Also, with 95% confidence, the odds of a sprout instead of a healthy kernel change by 3.57 and 14.82 times when density is decreased by 0.13 holding the other variables constant.

For the scab vs. healthy comparison, only the density and weight odds ratio confidence intervals do not include 1. For the sprout vs. healthy comparison, only the density and hardness odds ratio confidence intervals do not include 1. Odds ratios comparisons can be made for scab vs. sprout too. This is the subject of Exercise 11.

3.3.2 Contingency tables

The multinomial regression model provides a convenient way to perform the LRT for independence described in Section 3.2.3. We can treat the row variable X as a categorical explanatory variable (see Section 2.2.6) by constructing $I-1$ indicator variables $x_2, \ldots, x_I$ representing levels $2, \ldots, I$ of X. Note that we exclude the label x_1 for convenience so that the indicator variable indexes match the levels of X. Using Y as the response variable with category probabilities $\pi_1, \ldots, \pi_J$, the multinomial regression model for the table is

$$\log(\pi_j/\pi_1) = \beta_{j0} + \beta_{j2}x_2 + \cdots + \beta_{jI}x_I, \tag{3.8}$$

for $j = 2, \ldots, J$. This model implies that $\log(\pi_j/\pi_1) = \beta_{j0}$ when $X = 1$ and $\log(\pi_j/\pi_1) = \beta_{j0} + \beta_{ji}$ when $X = i$ for $i = 2, \ldots, I$. Therefore, the log-odds between two columns depends on the row of the table. In addition, note that there are $I(J-1)$ regression parameters in

---

[6]Note that the `c.value*conf.beta[2:7,1:2,1]` segment of code multiplies each row of `conf.beta[2:7,1:2,1]` by its corresponding element in `c.value`. See the code in the corresponding program for an example.

total for the model. Because $\sum_{j=1}^{J} \pi_j = 1$ for each X (i.e., only $J-1$ response probabilities need to be estimated in each row), the model is saturated.

Independence between X and Y removes this row dependence, so that the log-odds between two columns is constant across rows. This implies the simplified model

$$\log(\pi_j/\pi_1) = \beta_{j0}.$$

Thus, testing for independence is equivalent to testing $H_0 : \beta_{j2} = \cdots = \beta_{jI} = 0$ for each $j = 2, \ldots, J$, vs. $H_a :$ At least one $\beta_{ji} \neq 0$ for some j and i. This test is easily performed using a LRT.

When the null hypothesis is rejected, the next step is to investigate the sources of dependence. As in Section 3.2.3, calculating odds ratios for selected 2×2 sections of the contingency table can sometimes help to explain the nature of the association. These odds ratios are easily obtained from the regression model parameters as described in the previous section. For example, the estimated odds for comparing response categories $Y = j$ to $Y = 1$ are $\exp(\hat\beta_{ji})$ times as large for $X = i$ than $X = 1$. Odds ratios involving rows and columns other than the first are found as exponentiated linear combinations of regression parameters. For example, the estimated odds for comparing response categories 2 and 3 are $\exp[(\hat\beta_{24} - \hat\beta_{34}) - (\hat\beta_{25} - \hat\beta_{35})]$ times as large in row 4 as they are in row 5. Confidence intervals can be formed for these odds ratios as described in Section 3.3.1.

Example: Fiber-enriched crackers (Fiber.R, Fiber.csv)

We use multinomial regression to repeat the calculations for the LRT for independence that were presented in the example of Section 3.2.3. We start by fitting the model using multinom() with bloating severity as the response variable and fiber source as the explanatory variable:

```
> library(package = nnet)
> mod.fit.nom <- multinom(formula = bloat ~ fiber, weights =
    count, data = diet)
# weights:  20 (12 variable)
initial  value 66.542129
iter  10 value 54.519963
iter  20 value 54.197000
final  value 54.195737
converged

> summary(mod.fit.nom)
Call: multinom(formula = bloat ~ fiber, data = diet, weights =
    count)

Coefficients:
       (Intercept)   fiberbran    fibergum  fiberboth
low     -0.4057626  -0.1538545   0.4055575   1.322135
medium  -1.0980713  -0.8481379   1.5032639   1.503764
high   -12.4401085  -4.1103893  13.3561038  12.440403

Std. Errors:
       (Intercept)   fiberbran    fibergum  fiberboth
low      0.6455526   0.8997698   1.190217   1.056797
medium   0.8163281   1.3451836   1.224593   1.224649
high   205.2385583 1497.8087307 205.240263 205.240994
```

Residual Deviance: 108.3915
AIC: 132.3915

The function recognizes that `fiber` is a factor with levels ordered as "none", "bran", "gum", and "both". Therefore, it fits the model

$$\log(\pi_j/\pi_{\text{none}}) = \beta_{j0} + \beta_{j1}\text{bran} + \beta_{j2}\text{gum} + \beta_{j3}\text{both},$$

where `bran`, `gum`, and `both` represent indicator variables for the corresponding fiber source categories, and $j = 2, 3, 4$ represents response categories `low`, `medium`, and `high`, respectively. The `weights = count` argument is used within `multinom()` because each row of `diet` represents contingency table counts (stored in the variable "count") rather than observations from individual trials.

The `summary()` generic function reports the parameter estimates and corresponding standard errors, $\widehat{Var}(\hat{\beta}_{ji})^{1/2}$. Note that the residual deviance is given as 108.3915 rather than 0 even though a saturated model is being fit. This is a side effect of how `multinom()` uses the `weights` argument to create a likelihood function. For this contingency table setting, the likelihood function should be a product of 4 multinomial PMFs, one for each fiber source, where the number of trials is 12 for each PMF. Then the response category probabilities estimated by the model will be identical to the sample proportions in the table, and this causes a residual deviance of zero. The `multinom()` function instead reformulates the counts into individual multinomial trials (i.e., the "raw data" format as examined in Section 1.2.1). This leads to 48 rows in the data frame, where each row represents one individual's fiber and bloating values. The likelihood function becomes the product of 48 multinomial PMFs with one trial for each PMF. Each individual's sample proportions for the bloating response categories are either 1, if the individual experienced that level of bloating, or 0 otherwise. The estimated probabilities from the model do not match these proportions, so the residual deviance differs from 0. However, despite this different formulation of the likelihood function by `multinom()`, we will not have any problems in the analysis that follows. For these data and in general, parameter estimates, standard errors, and LRT results for explanatory variables are the same no matter which likelihood function and data format is used. We will discuss these two data formats further in Chapter 5.

The test for independence of fiber source and bloating severity compares $H_0 : \beta_{j2} = \beta_{j3} = \beta_{j4} = 0$ for each $j = 2, 3, 4$ vs. H_a : Not all $\beta_{ji} = 0$ for some j and i. The LRT is performed by `Anova()`:

```
> library(package = car)
> Anova(mod.fit.nom)
# weights:  8 (3 variable)
initial  value 66.542129
final  value 63.635876
converged

Analysis of Deviance Table (Type II tests)
Response: bloat
      LR Chisq Df Pr(>Chisq)
fiber    18.9   9     0.026 *
---
Signif. codes:  0 '***' 0.001 '**' 0.01 '*' 0.05 '.' 0.1 ' ' 1
```

The LRT results in $-2\log(\Lambda) = 18.9$ with a p-value of 0.026. Thus, there is moderate evidence of dependence between type of fiber and severity of bloating. Notice that these values match what was found earlier using the `assocstats()` function.

To examine the sources of association, we can start by examining the estimated parameters of the model. For example, the estimated log-odds comparing the low bloating severity category to no bloating is

$$\log(\hat{\pi}_{\text{low}}/\hat{\pi}_{\text{none}}) = -0.41 - 0.15\text{bran} + 0.41\text{gum} + 1.32\text{both}.$$

The parameter estimates suggest that using gum, with or without bran, leads to a larger odds of low bloating relative to no bloating than does using no fiber. On the other hand, there appears to be a somewhat smaller effect in the opposite direction due to using bran instead of no fiber. We compute the estimated odds ratios as follows:

```
> round(exp(coefficients(mod.fit.nom)[,-1]), 2)
       fiberbran fibergum fiberboth
low         0.86      1.5      3.75
medium      0.43      4.5      4.50
high        0.02 631658.3 252812.49
```

For example, the estimated odds of having low bloating rather than none is $\exp(-0.15) = 0.86$ times as large for using bran as a fiber source than using no fiber at all. Of course, final conclusions about the effects of fiber need to account for the variability of the parameter estimators, which we will do shortly.

Notice that the high bloating row in the table of odds ratios contains two very large estimates. Also, notice the table of standard errors from `summary(mod.fit.nom)` contains very large values in the high bloating row. In fact, if one were to decrease the convergence criterion value given by the `reltol` argument (default is `reltol = 10^(-8)`) for `multinom()`, one could see the parameter estimates are increasing toward infinity for the `fibergum` and `fiberboth` columns of the high bloating row despite the `converged` message printed in the output. These problems occur due to the 0 counts present in the corresponding cells of Table 3.2. We saw in Section 1.2.5 that 0 counts can cause difficulties with estimating odds ratios, and these problems occur here as well. An ad-hoc solution similar to what was used in Section 1.2.5 is to add 0.5 to the 0 cell counts. Alternatively, a small constant, like 0.5, could be added to all cells of the contingency table. In either case, the regression parameter estimates are biased toward 0, and hence the odds ratios are biased toward 1. However, the reduction in the variance of the estimates is substantial, and there are not necessarily better solutions within the framework of using maximum likelihood estimation for the multinomial regression model.

The code below shows how 0.5 is added to the 0 cells using the `ifelse()` function. We then re-estimate the model:

```
> diet$count2 <- ifelse(test = diet$count == 0, yes = diet$count
    + 0.5, no = diet$count)
> mod.fit.nom2 <- multinom(formula = bloat ~ fiber, weights =
    count2, data = diet)
# weights:  20 (12 variable)
initial  value 67.928424
iter  10 value 58.549878
final  value 58.394315
converged
```

```
> sum.fit <- summary(mod.fit.nom2)
> round(sum.fit$coefficients, 4)
       (Intercept) fiberbran fibergum fiberboth
low        -0.4055   -0.1541   0.4055    1.3218
medium     -1.0986   -0.8473   1.5041    1.5040
high       -2.4849   -0.1542   3.4012    2.4849
> round(sum.fit$standard.errors, 4)
       (Intercept) fiberbran fibergum fiberboth
low         0.6455    0.8997   1.1902    1.0567
medium      0.8165    1.3452   1.2247    1.2247
high        1.4719    2.0759   1.6931    1.7795
```

We see that the estimated parameters and corresponding standard errors are very similar to before except that the high bloating values are much closer to 0 due to the added constant. Below are the estimated odds ratios and some of the corresponding confidence intervals:

```
> round(exp(coefficients(mod.fit.nom2)[,-1]), 2)
       fiberbran fibergum fiberboth
low         0.86      1.5      3.75
medium      0.43      4.5      4.50
high        0.86     30.0     12.00
> conf.beta <- confint(object = mod.fit.nom2, level = 0.95)
> round(exp(conf.beta[2:4,,3]),2)   # compare high to no bloating
          2.5 %  97.5 %
fiberbran  0.01   50.13
fibergum   1.09  828.46
fiberboth  0.37  392.52
```

These odds ratios are the same as what we found earlier, except for high bloating. The confidence intervals for that response category are rather wide, calling their usefulness into question. This happens because the odds ratios and confidence intervals are based on such small counts that there is considerable uncertainty regarding the population values. Additional calculations for odds ratios comparing other levels of bloating are given in the corresponding program.

It is interesting to note that there is an apparent trend among the estimated odds ratios for gum vs. none and both vs. none. We see that as the bloating severity increases, so do the estimated odds ratios. This may not be a coincidence. Notice that the bloating severity response is actually an ordinal variable. In Section 3.4 we will show how to adapt our modeling to ordinal response variables that allow a potentially more powerful, yet parsimonious, analysis to be conducted. An additional advantage to modeling the ordinal structure is that the impact of individual 0 counts is substantially lessened.

Interestingly, the estimated probabilities, odds, and odds ratios produced from Equation 3.8 are exactly the same as those obtained directly from the contingency table (as long as there are no 0 counts). For instance, in the last example, the estimated odds ratio of having low bloating rather than none is $n_{11}n_{22}/(n_{12}n_{21}) = 6 \times 4/(4 \times 7) = 0.86$ times as large for using bran as a fiber source than using no fiber at all. This estimated odds ratio value is exactly the same as $\exp(-0.15) = 0.86$ computed using the estimated multinomial regression model. In this sense, the multinomial model seems like a complicated way to perform the analysis. This is true for the relatively simple problem of relating a multinomial response to a single categorical explanatory variable. The strength of the modeling procedure is it can be used in more complex problems where the simple table-based analysis cannot. In particular, the multinomial regression model can be used to assess effects of arbitrarily many variables on the response (subject to adequate sample size). The variables can be numerical as shown earlier in this section, categorical as shown in this sub-section, or any combination of these.

When additional categorical explanatory variables are available, we can examine the data in higher dimensional contingency tables. For example, consider a setting with a three-dimensional contingency table summarizing the counts for categorical variables X, Y, and Z, with levels $i = 1, \ldots, I$, $j = 1, \ldots, J$, and $k = 1, \ldots, K$, respectively. An analysis using the simpler tools of Section 3.2 requires that separate contingency tables summarizing any pair of variables (e.g., X and Y) be formed for each level of the third variable (e.g., Z). In some cases there is an "obvious" way to do this, such as when X represents a treatment variable, Y a response variable, and Z a blocking or stratification variable. Then modeling the relationship between X and Y controlling for the level of Z is the natural approach to the analysis. A Pearson or LR test for independence can be conducted in each table, and subsequently odds ratios can be estimated and other follow-up analyses performed. However, this approach is limited in its scope. It allows assessments only of the association between X and Y conditional on the level of Z. It makes no attempt to compare these associations across Z-levels, nor to combine them into a single association that applies to all Z-levels. Although table-based methods do exist to address the latter issue (see Section 6.3.5 of Agresti, 2002), they still do not allow consideration of the association between Z and Y, which is of interest when Z is another explanatory variable rather than a stratification variable.

The multinomial regression framework (and the Poisson regression model structure introduced in Section 4.2.4) allows these more general questions to be answered. For example, a general model allowing for association of both X and Z with Y is

$$\log(\pi_j/\pi_1) = \beta_{j0} + \beta_{j2}^X x_2 + \cdots + \beta_{jI}^X x_I + \beta_{j2}^Z z_2 + \cdots + \beta_{jK}^Z z_K, \, j = 2, \ldots, J$$

where we use $x_2, \ldots, x_I$ and $z_2, \ldots, z_K$ as indicator variables for the levels given in their indexes. The superscript X or Z is added to the β parameters to make clear which variable's association with Y is described by the parameter. Excluding either the X and/or Z indicator variables from the model corresponds to independence between that explanatory variable and Y. An interaction between X and Z could be included in the model as well to allow each variable's association with Y to vary among the levels of the other explanatory variable. Adding additional categorical explanatory variables is a straightforward extension of this model, as is adding numerical explanatory variables.

Example: Fiber-enriched crackers (Fiber.R, Fiber.csv)

As mentioned in Section 3.2, the fiber explanatory variable is actually constructed from combinations of two separate variables: bran ("yes" or "no") and gum ("yes" or "no"). This structure is often referred to as a 2×2 factorial.[7] It is typical in these structures to ask questions regarding separate effects of each constituent variable on

---

[7] In general an $a \times b$ factorial consists of all combinations of one variable with a levels and another with b levels. The variables making up the combinations are called "factors" (see, for example, Milliken and Johnson, 2004 for details).

the responses, as well as to look for interactions between the variables. To pursue this analysis, we first transform the explanatory variable `fiber` into `bran` and `gum` using the `ifelse()` function:

```
> diet$bran <- factor(ifelse(test = diet$fiber == "bran" |
    diet$fiber == "both", yes = "yes", no = "no"))
> diet$gum <- factor(ifelse(test = diet$fiber == "gum" |
    diet$fiber == "both", yes = "yes", no = "no"))
> head(diet, n = 4)
   fiber  bloat count bran gum
1   bran   high     0  yes  no
2    gum   high     5   no yes
3   both   high     2  yes yes
4   none   high     0   no  no
```

The single vertical line | joining conditions within the `test` argument means "or", so that the first line assigns `bran` a "yes" value if `fiber` = "bran" or `fiber` = "both"; otherwise, `bran` is set to "no". Also, we enclose `ifelse()` within `factor()` so that the result is a `factor` class type rather than `character`. If this is not done, warning messages will be produced by `multinom()` as it performs this conversion internally.

Next, we fit the multinomial model that includes effects for `bran`, `gum`, and their interaction:

```
> mod.fit.nom.inter <- multinom(formula = bloat ~ bran + gum +
    bran:gum, weights = count, data = diet)
# weights:  20 (12 variable)
initial  value 66.542129
iter  10 value 54.406806
iter  20 value 54.196639
final  value 54.195746
converged

> summary(mod.fit.nom.inter)
Call:
multinom(formula = bloat ~ bran + gum + bran:gum, data = diet,
    weights = count)

Coefficients:
        (Intercept)      branyes       gumyes  branyes:gumyes
low      -0.4063036   -0.1532825    0.4076874       1.0674733
medium   -1.0973831   -0.8503013    1.5033271       0.8503872
high    -12.4349471   -2.0291752   13.3518350       1.1121073

Std. Errors:
        (Intercept)     branyes       gumyes  branyes:gumyes
low       0.6456363    0.899801     1.190275         1.58416
medium    0.8160908    1.345541     1.224613         1.86471
high    204.7028906  561.368770   204.704601       561.37028

Residual Deviance: 108.3915
AIC: 132.3915

> Anova(mod.fit.nom.inter)
```

```
Analysis of Deviance Table (Type II tests)

Response: bloat
         LR Chisq Df Pr(>Chisq)
bran       2.5857  3   0.460010
gum       16.2897  3   0.000989 ***
bran:gum   0.4880  3   0.921514
--- Signif. codes:  0 '***' 0.001 '**' 0.01 '*' 0.05 '.' 0.1 ' ' 1

> logLik(mod.fit.nom.inter)
'log Lik.' -54.19575 (df=12)
```

The LRT for the interaction gives a p-value of 0.9215 indicating there is not sufficient evidence of an interaction. Furthermore, we see that `bran` has a large p-value, but `gum` has a small p-value. Thus, there is sufficient evidence to indicate that gum has an effect on the bloating severity.

Because the saturated model is being fit here, the problems with zero counts in the table that were noted in the previous example remain a problem in this model. In the end, these difficulties are not of concern for the LRT, because the $-2\log(\Lambda)$ statistic will change very little for a larger number of iterations.[8]

As an alternative to the multinomial structure of this chapter, analysis of contingency tables is frequently performed assuming the counts have a Poisson distribution (or variants of it). We will discuss these Poisson regression models in Chapter 4.

3.4 Ordinal response regression models

Many categorical response variables have a natural ordering to their levels. For example, a response variable may be measured using a Likert scale with categories "strongly disagree," "disagree," "neutral," "agree," or "strongly agree." If response levels can be arranged so that category 1 < category 2 < $\cdots$ < category J in some conceptual scale of measurement (e.g., amount of agreement), then regression models can incorporate this ordering through a variety of logit transformations of the response probabilities. In this section, we focus on modeling cumulative probabilities based on the category ordering.

The cumulative probability for category j of Y is $P(Y \leq j) = \pi_1 + \cdots + \pi_j$ for $j = 1, \ldots, J$. Note that $P(Y \leq J) = 1$. Regression models for ordinal multinomial responses can examine the effects of explanatory variables $x_1, \ldots, x_p$ on the log-odds of cumulative probabilities, also called *cumulative logits*,

$$\text{logit}(P(Y \leq j)) = \log\left(\frac{P(Y \leq j)}{1 - P(Y \leq j)}\right) = \log\left(\frac{\pi_1 + \cdots + \pi_j}{\pi_{j+1} + \cdots + \pi_J}\right).$$

In particular, the *proportional odds model* is a special model that assumes that the logit of these cumulative probabilities changes linearly as the explanatory variables change, and

---

[8]For the two 0 count occurrences, the multinomial regression model estimate of π_j is already very close to 0. The estimate can only get a little closer to 0 if the convergence criterion was made more strict than the default used here.

also that the slope of this relationship is the same regardless of the category j. Formally, the model is stated as

$$\text{logit}(P(Y \leq j)) = \beta_{j0} + \beta_1 x_1 + \cdots + \beta_p x_p, \; j = 1, \ldots, J-1. \tag{3.9}$$

Notice that there are no j subscripts on the parameters $\beta_1, \ldots, \beta_p$. The model assumes that the effects of the explanatory variables are the same regardless of which cumulative probabilities are used to form the log odds. Thus, the name "proportional odds" derives from each odds being a multiple of $\exp(\beta_{j0})$.

For a fixed j, increasing x_r by c units changes every log-odds in Equation 3.9 by $c\beta_r$ when holding other explanatory variables constant. On the other hand, the difference in the log-odds between response categories j and j' is constant, $\beta_{j0} - \beta_{j'0}$, and does not depend on the values of $x_1, \ldots, x_p$ when they are held fixed. These results relate directly to odds ratios, as detailed in Section 3.4.1. Also, notice that the odds must increase as j increases, because we put progressively more probability in the numerator, $P(Y \leq j)$. This implies that $\beta_{10} < \cdots < \beta_{J-1,0}$.

Probabilities for observing a particular response category j are found by noting that

$$\begin{aligned} \pi_j &= P(Y = j) \\ &= P(Y \leq j) - P(Y \leq j-1) \end{aligned} \tag{3.10}$$

where $P(Y \leq 0) = 0$, $P(Y \leq J) = 1$, and

$$P(Y \leq j) = \frac{\exp(\beta_{j0} + \beta_1 x_1 + \cdots + \beta_p x_p)}{1 + \exp(\beta_{j0} + \beta_1 x_1 + \cdots + \beta_p x_p)}.$$

For example, the probability for category 1 is

$$\pi_1 = P(Y \leq 1) - P(Y \leq 0) = \frac{\exp(\beta_{10} + \beta_1 x_1 + \cdots + \beta_p x_p)}{1 + \exp(\beta_{10} + \beta_1 x_1 + \cdots + \beta_p x_p)},$$

and the probability for category J is

$$\pi_J = P(Y \leq J) - P(Y \leq J-1) = 1 - \frac{\exp(\beta_{J-1,0} + \beta_1 x_1 + \cdots + \beta_p x_p)}{1 + \exp(\beta_{J-1,0} + \beta_1 x_1 + \cdots + \beta_p x_p)}.$$

Example: Proportional odds model plots (CumulativeLogitModelPlot.R)

The purpose of this example is to examine features of the proportional odds model. Consider the model $\text{logit}(P(Y \leq j)) = \beta_{j0} + \beta_1 x_1$, where $J = 4, \beta_{10} = 0, \beta_{20} = 2, \beta_{30} = 4$, and $\beta_1 = 2$. Figure 3.4 plots the cumulative probabilities and individual category probabilities for this model. Notice that the cumulative probability curves in the left plot have exactly the same shape, because β_1 is the same for each response category j. The horizontal shift is due to the different values for β_{j0}. For example, $P(Y \leq j) = 0.5$ when $\beta_{j0} + \beta_1 x_1 = 0$, i.e., when $x_1 = -\beta_{j0}/\beta_1$.

Probabilities of response for individual categories, $\pi_j, j = 1, 2, 3, 4$, are found at each x_1 from the vertical distances between consecutive curves in the left plot, along with $P(Y \leq 0) = 0$ and $P(Y \leq 4) = 1$. These are shown in the plot on the right in Figure 3.4, which also illustrates the orderings among the categories for Y with respect to x_1. For example, the smallest values of x_1 lead to $Y = 4$ having the largest probability. As x_1 grows in size, the probability of observing $Y = 3$ becomes larger until eventually it has the largest probability for a particular range of x_1. Similar outcomes occur for $Y = 2$ and $Y = 1$ in this order.

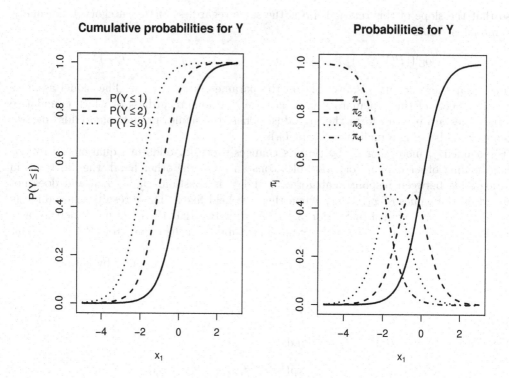

Figure 3.4: Proportional odds model for $\beta_{10} = 0, \beta_{20} = 2, \beta_{30} = 4,$ and $\beta_1 = 2$.

The code for these plots is in the corresponding program to this example. We encourage readers to change elements of the code to see what happens as values of $\beta_{10}, \beta_{20}, \beta_{30}$, and β_1 change. For example, decreasing the absolute distance between β_{10} and β_{20} can lead to a case where π_2 is never the largest probability for any value of x_1 (see Exercise 6).

Parameters of the proportional odds model are estimated using maximum likelihood. Similar to Section 3.3, the likelihood function for a sample of size m is simply the product of m multinomial distributions with probability parameters expressed as functions of the explanatory variables as given by Equations 3.9 and 3.10. Iterative numerical procedures are used then to fit the model. The polr() function from the MASS package performs the necessary computations. It is important to ensure that the levels of the categorical response are ordered in the desired way when using polr(); otherwise, the ordering of the levels of Y will not be correctly taken into account. We will demonstrate how to check and, if necessary, change the ordering in the next example. The covariance matrix for the regression parameter estimates is found using standard likelihood procedures as outlined in Appendix B.3.4.

Wald and LR-based inference procedures are performed in the usual ways as well. The hypotheses associated with tests of individual regression parameters are $H_0 : \beta_r = 0$ vs. $H_a : \beta_r \neq 0$. If the null hypothesis is true, this says that the $J-1$ log-odds comparing $P(Y \leq j)$ to $P(Y > j)$ do not depend on x_r, holding all other explanatory variables constant. If the alternative hypothesis is true, then the log-odds for each cumulative probability grow larger or smaller with x_r depending on the sign of β_r. In turn, this imposes an ordering on the individual category probabilities such as that seen in the right plot of Figure 3.4.

Contrast the previous test involving β_r to a corresponding test involving the multinomial regression model with a nominal response from Section 3.3, $H_0: \beta_{2r} = \cdots = \beta_{Jr} = 0$ vs. H_a: At least one $\beta_{jr} \neq 0$. The alternative hypothesis model places fewer constraints on how x_r relates to individual category probabilities through its use of more parameters than the proportional odds model. Hence, when the proportional odds assumptions are applicable, the alternative hypothesis model for multinomial regression describes the relationship between the response and the explanatory variables less efficiently.

Example: Wheat kernels (Wheat.R, Wheat.csv)

Healthy kernels are the most desirable type of wheat kernel. One could also reason that presence of disease makes scab kernels less desirable than sprouted kernels. Thus, we use the ordering of scab $(Y = 1) <$ sprout $(Y = 2) <$ healthy $(Y = 3)$ and fit a proportional odds model to these data.

We begin by creating the required ordering of the levels for the type variable. There are several ways to accomplish this, and we show one approach below that is fairly straightforward:

```
> levels(wheat$type)
[1] "Healthy" "Scab"    "Sprout"
> wheat$type.order <- factor(wheat$type, levels = c("Scab",
    "Sprout", "Healthy"))
> levels(wheat$type.order)
[1] "Scab"    "Sprout"  "Healthy"
```

The new variable of wheat named type.order contains the factor with the proper ordering of its levels. The proportional odds model is then estimated using polr():

```
> library(package = MASS)
> mod.fit.ord <- polr(formula = type.order ~ class + density +
    hardness + size + weight + moisture, data = wheat, method =
    "logistic")
> summary(mod.fit.ord)

Re-fitting to get Hessian

Call: polr(formula = type.order ~ class + density + hardness +
    size + weight + moisture, data = wheat, method = "logistic")

Coefficients:
            Value Std. Error t value
classsrw   0.17370   0.391764  0.4434
density   13.50534   1.713009  7.8840
hardness   0.01039   0.005932  1.7522
size      -0.29253   0.413095 -0.7081
weight     0.12721   0.029996  4.2411
moisture  -0.03902   0.088396 -0.4414

Intercepts:
               Value   Std. Error t value
Scab|Sprout    17.5724 2.2460     7.8237
Sprout|Healthy 20.0444 2.3395     8.5677
```

```
Residual Deviance: 422.4178
AIC: 438.4178

> library(package = car)
> Anova(mod.fit.ord)
Analysis of Deviance Table (Type II tests)

Response: type.order
         LR Chisq Df Pr(>Chisq)
class       0.197  1    0.65749
density    98.437  1  < 2.2e-16 ***
hardness    3.084  1    0.07908 .
size        0.499  1    0.47982
weight     18.965  1  1.332e-05 ***
moisture    0.195  1    0.65872
---
Signif. codes:  0 '***' 0.001 '**' 0.01 '*' 0.05 '.' 0.1 ' ' 1
```

The method = "logistic" argument value instructs R to use the logit transformation on the cumulative probabilities. The polr() function estimates the model

$$\text{logit}(P(Y \leq j)) = \beta_{j0} - \eta_1 x_1 - \cdots - \eta_p x_p, \tag{3.11}$$

where $-\eta_r$ is β_r in Equation 3.9. Due to these notational differences, we therefore need to reverse the sign of all the estimated η_r values in the output in order to state the estimated model. This results in the model

$$\text{logit}(\widehat{P}(Y \leq j)) = \hat{\beta}_{j0} - 0.17\text{srw} - 13.51\text{density} - 0.010\text{hardness}$$
$$+ 0.29\text{size} - 0.13\text{weight} + 0.039\text{moisture},$$

where $\hat{\beta}_{10} = 17.57$ and $\hat{\beta}_{20} = 20.04$. The t value column in the Coefficients table provides the Wald statistics for testing $H_0 : \beta_r = 0$ vs. $H_a : \beta_r \neq 0$ for $r = 1, \ldots, 6$. The Anova() function provides the corresponding LRTs. Because of the large test statistic values for density and weight, there is sufficient evidence that these are important explanatory variables. There is marginal evidence that hardness is important too. As with other tests like this in previous chapters, each of these tests is conditional on the other variables being in the model.

The predict() function finds the estimated probabilities or classes for each response category:

```
> pi.hat.ord <- predict(object = mod.fit.ord, type = "probs")
> head(pi.hat.ord)
        Scab     Sprout    Healthy
1 0.03661601 0.2738502 0.6895338
2 0.03351672 0.2576769 0.7088064
3 0.08379891 0.4362428 0.4799583
4 0.01694278 0.1526100 0.8304472
5 0.11408176 0.4899557 0.3959626
6 0.02874814 0.2308637 0.7403882

> head(predict(object = mod.fit.ord, type = "class"))
[1] Healthy Healthy Healthy Healthy Sprout Healthy
Levels: Scab Sprout Healthy
```

For example, the estimated probability of being healthy for the first observation is

$$\hat{\pi}_{\text{Healthy}} = 1 - \frac{\exp(20.04 - 0.17 \times 0 + \cdots + 0.04 \times 12.02)}{1 + \exp(20.04 - 0.17 \times 0 + \cdots + 0.04 \times 12.02)}$$
$$= 0.6895.$$

This probability and others are found by using the `type = "probs"` argument in `predict()`.[9] A `type = "class"` argument (the default) gives the predicted classification based on its largest estimated probability. For the first kernel, `Healthy` has the largest estimated probability, so the kernel is predicted to be of this condition (which is a correct prediction).

As with model fit objects from `multinom()`, the `predict()` function does not provide the estimated variances needed to form Wald confidence intervals for π_{Scab}, π_{Sprout}, or π_{Healthy}. In addition, the `deltaMethod.polr()` method function does not allow for any mathematical functions of the intercept terms, and further problems arise because `polr()` estimates a model as defined in Equation 3.11 rather than Equation 3.9. For these reasons, we have constructed a new function `deltaMethod.polr2()` that calculates $\hat{\pi}_j$ and $\widehat{Var}(\hat{\pi}_j)$:

```
> deltaMethod.polr2 <- function(object, g)  {
    # All beta-hat's where the slope estimates are adjusted
    beta.hat <- c(-object$coefficients, object$zeta)

    # Count the number of slopes and intercepts
    numb.slope <- length(object$coefficients)
    numb.int <- length(object$zeta)

    # Name the corresponding parameters
    names(beta.hat) <- c(paste("b", 1:numb.slope, sep=""),
       paste("b", 1:numb.int, "0", sep=""))

    # Fix covariance matrix - All covariances between slopes and
       intercepts need to be multiplied by -1
    cov.mat <- vcov(object)
    # Above diagonal
    cov.mat[1:numb.slope, (numb.slope + 1):(numb.slope +
       numb.int)] <-
       -cov.mat[1:numb.slope, (numb.slope + 1):(numb.slope +
          numb.int)]
    # Below diagonal
    cov.mat[(numb.slope + 1):(numb.slope + numb.int),
       1:numb.slope] <-
       -cov.mat[(numb.slope + 1):(numb.slope + numb.int),
          1:numb.slope]

    # deltaMethod.default() method function completes calculations
    deltaMethod(object = beta.hat, g = g, vcov. = cov.mat)
 }
```

---

[9]We include code in our corresponding program showing how these probabilities are computed without using `predict()`. This is done to illustrate the coding of Equation 3.10 into R's syntax.

The function is used in a very similar manner as `deltaMethod()` in Section 3.3. The model fit object from `polr()` is given in the `object` argument and the character string for $\hat{\pi}_j$ is given in the `g` argument. However, there is no `parameterNames` argument because the parameter names are automatically created within the function. Only names of the form `bj0` and `br` can be used for β_{j0} and β_r, respectively. We calculate an interval for π_{Healthy} as follows:

```
> x1 <- 0;              x2 <- wheat[1,2]; x3 <- wheat[1,3]
> x4 <- wheat[1,4]; x5 <- wheat[1,5]; x6 <- wheat[1,6]

> g.healthy <- "1 - exp(b20 + b1*x1 + b2*x2 + b3*x3 + b4*x4 +
    b5*x5 + b6*x6) / (1 + exp(b20 + b1*x1 + b2*x2 + b3*x3 + b4*x4
    + b5*x5 + b6*x6))"
> calc.healthy <- deltaMethod.polr2(object = mod.fit.ord, g =
    g.healthy)
Re-fitting to get Hessian
> calc.healthy$Estimate   # pi-hat_Healthy
[1] 0.6895338
> calc.healthy$SE    # sqrt(Var-hat(pi-hat_Healthy))
[1] 0.08199814
> alpha <- 0.05
> calc.healthy$Estimate + qnorm(p = c(alpha/2, 1-alpha/2)) *
    calc.healthy$SE
[1] 0.5288204 0.8502472
```

The 95% confidence interval for π_{Healthy} is (0.5288, 0.8502). The 95% confidence intervals for π_{Scab} and π_{Sprout} are (0.0061, 0.0671) and (0.1380, 0.4097), respectively (see program for code).

A plot of the estimated probabilities for each category vs. an individual explanatory variable (holding the other variables in the model constant) can be used to interpret the model. Exercise 19 examines plots of this form. We simplify the problem here by again using `density` as the only explanatory variable in fitting a new proportional odds model:

$$\text{logit}(\widehat{P}(Y \leq j)) = \hat{\beta}_{j0} - 15.64\text{density}$$

with $\hat{\beta}_{10} = 17.41$ and $\hat{\beta}_{20} = 19.63$ (see program for code). Figure 3.5 gives a plot of this model along with the multinomial regression model estimated earlier that treats the response as nominal. These models produce somewhat similar probability curves with the same scab < sprout < healthy ordering for density with both models. This provides some reassurance that the ordering used for the proportional odds model is appropriate and that the assumption of proportional odds among cumulative probabilities is reasonable.

Below are additional notes about using proportional odds models in R:

1. Transformations and interaction terms are included into `polr()` using the same syntax as in `glm()` and `multinom()`.

2. There are a number of other functions that can estimate a proportional odds model. We show how to use the `lrm()` function of the `rms` package in the corresponding program to this example. We also show how to use the `vglm()` function of the `VGAM` function in a later example within this section.

3. The `mcprofile` package cannot be used for LR-based inference methods with objects resulting from `polr()`.

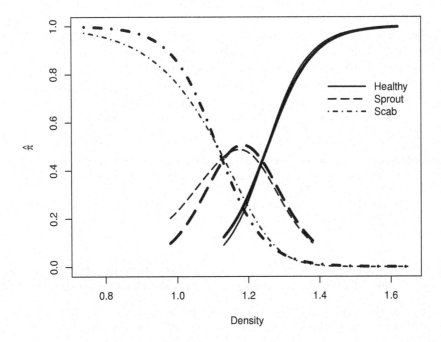

Figure 3.5: Estimated proportional odds (thicker line) and multinomial (thinner line) regression models for the wheat data where density is the only explanatory variable included in the model.

3.4.1 Odds ratios

Odds ratios based on the cumulative probabilities are easily formed because the proportional odds model equates log-odds to the linear predictor. For example, the odds ratio involving one explanatory variable, say x_1, is

$$\frac{Odds_{x_1+c, x_2, \ldots, x_p}(Y \leq j)}{Odds_{x_1, \ldots, x_p}(Y \leq j)} = \frac{e^{\beta_{j0} + \beta_1(x_1+c) + \cdots + \beta_p x_p}}{e^{\beta_{j0} + \beta_1 x_1 + \cdots + \beta_p x_p}} = e^{c\beta_1},$$

where we let $Odds_{x_1, \ldots, x_p}(Y \leq j)$ denote the odds of observing category j or smaller for Y. Notice that this odds ratio formulation assumes $x_2, \ldots, x_p$ are unchanged. The formal interpretation of the odds ratio is

> The odds of $Y \leq j$ vs. $Y > j$ change by $e^{c\beta_1}$ times for a c-unit increase in x_1 while holding the other explanatory variables in the model constant.

Interestingly, the odds ratio stays the same no matter what response category is used for j. This key feature of the proportional odds model occurs due to the absence of a j subscript on the slope parameters $\beta_1, \ldots, \beta_p$.

Example: Wheat kernels (Wheat.R, Wheat.csv)

The estimated odds ratios for each explanatory variable are calculated as $\widehat{OR} = \exp(c\hat{\beta}_r)$ for $r = 1, \ldots, 6$. Similarly to Section 3.3, we set c to be equal to one standard deviation for each continuous explanatory variable and $c = 1$ for the srw indicator variable. Recall that we also need to multiply the estimated regression parameters produced by polr() by -1. Below are the calculations:

```
> round(c.value, 2)    # class = 1 is first value
   density  hardness      size    weight  moisture
      1.00      0.13     27.36      0.49      7.92      2.03

> round(exp(c.value * (-mod.fit.ord$coefficients)), 2)
   density  hardness      size    weight  moisture
      0.84      0.17      0.75      1.15      0.37      1.08
> round(1/exp(c.value * (-mod.fit.ord$coefficients)), 2)
   density  hardness      size    weight  moisture
      1.19      5.89      1.33      0.87      2.74      0.92
```

In the last line of code, we re-express the odds ratios with respect to decreases in their respective explanatory variables. Interpretations of the estimated odds ratios include:

- The estimated odds of a scab ($Y \leq 1$) vs. sprout or healthy ($Y > 1$) response are 0.84 times as large for soft rather than hard red winter wheat, holding the other variables constant.

- The estimated odds of a scab vs. sprout or healthy response change by 5.89 times for a 0.13 decrease in the density, holding the other variables constant.

- The estimated odds of a scab vs. sprout or healthy response change by 2.74 times for a 7.92 decrease in the weight, holding the other variables constant.

Because of the proportional odds assumption, each of these statements also applies to the odds of a scab or sprout vs. healthy response. For this reason, it is common to interpret odds ratios, such as for density, by saying:

> The estimated odds of kernel quality being below a particular level change by 5.89 times for a 0.13 decrease in the density, holding the other variables constant.

Overall, we see that the larger the density and weight, the more likely a kernel is from a healthier category. We can again relate these results back to Figure 3.2 on p. 155. The plot shows small density, size, and weight kernels often have a scab condition. There is much more overlap among kernel classifications for the other explanatory variables in the plot, which corresponds to estimated odds ratios close to 1. Further interpretations of these other variables are part of Exercise 11.

The polr() function creates objects that have a "polr" class type. The methods() function with this class type shows what method functions are available:

```
> class(mod.fit.ord)
[1] "polr"
> methods(class = polr)
[1] anova.polr*            Anova.polr*
[3] confint.polr*          deltaMethod.polr*
```

```
 [5] extractAIC.polr*        linearHypothesis.polr*
 [7] logLik.polr*            model.frame.polr*
 [9] nobs.polr*              predict.polr*
[11] print.polr*             profile.polr*
[13] simulate.polr*          summary.polr*
[15] vcov.polr*
```

We have already used the `summary.polr()` and `Anova.polr()` functions. The corresponding method function for `confint()` can be used to construct profile LR confidence intervals for the parameters $\beta_1, \ldots, \beta_p$. Results from this function are used then to find profile LR intervals for the odds ratios:

```
> conf.beta <- confint(object = mod.fit.ord, level = 0.95)
Waiting for profiling to be done...

Re-fitting to get Hessian

> ci <- exp(c.value*(-conf.beta))
> round(data.frame(low = ci[,2], up = ci[,1]), 2)
          low   up
classsrw 0.39 1.81
density  0.11 0.26
hardness 0.55 1.03
size     0.77 1.72
weight   0.23 0.58
moisture 0.76 1.54

> round(data.frame(low = 1/ci[,1], up = 1/ci[,2]), 2)
          low   up
classsrw 0.55 2.57
density  3.87 9.36
hardness 0.97 1.83
size     0.58 1.29
weight   1.73 4.40
moisture 0.65 1.31
```

The density odds ratio interpretation is:

> With 95% confidence, the odds of kernel quality being below a particular level change by 3.87 to 9.36 times when density is decreased by 0.13, holding the other variables constant.

The confidence intervals for density and weight are the only intervals that do not contain 1. Thus, there is sufficient evidence that both of these explanatory variables are important.

Although we generally prefer LR intervals, Wald confidence intervals may be calculated as well. However, the `confint.default()` function cannot be applied. We present code to calculate the Wald interval directly in the corresponding program for this example.

3.4.2 Contingency tables

Section 3.2 discusses how to test for independence in a contingency table using a multinomial regression model for a nominal response. A similar test for independence can also be performed using the proportional odds model, but the alternative hypothesis specifies the type of dependence that may be present. This is because the proportional odds model assumes a specific structure to the association between a categorical explanatory variable X and a response variable Y, and hence uses fewer parameters than the nominal model to summarize this association.

In particular, for an $I \times J$ contingency table with ordinal categories of Y, the proportional odds model is

$$\text{logit}(P(Y \leq j)) = \beta_{j0} + \beta_2 x_2 + \ldots + \beta_I x_I, \quad j = 1, \ldots, J-1 \qquad (3.12)$$

where $x_2, \ldots, x_I$ are indicator variables for rows $2, \ldots, I$, respectively. Any $\beta_i \neq 0$ for $i = 2, \ldots, I$ means that the odds involving cumulative probabilities for Y are not the same in rows 1 and i. Lower categories of Y are more likely to be observed in row i than in row 1 if $\beta_i > 0$, and less likely if $\beta_i < 0$. Thus, independence is tested as $H_0 : \beta_2 = \beta_3 = \ldots = \beta_I = 0$ vs. H_a : any $\beta_i \neq 0$, $i = 2, \ldots, I$. If the null hypothesis of independence is rejected, the association is summarized with the help of the signs and values of $\hat{\beta}_2, \ldots, \hat{\beta}_I$, along with confidence intervals for the corresponding parameters.

Notice that Equation 3.12 contains only $(I-1)+(J-1)$ parameters, compared to $I(J-1)$ in the corresponding multinomial regression model. The reduction in parameters comes from the proportional odds assumption, which specifies that the difference in the logits of cumulative probabilities for any two rows is controlled by just one parameter, β_i, regardless of j. Thus, the test for independence using the proportional odds model is more powerful than a test for independence using the multinomial regression model when the association does actually follow this structure. If it does not, then the test using the proportional odds model may fail to detect any association, even when it is very strong.

Example: Fiber-enriched crackers (Fiber.R, Fiber.csv)

When examining a 4×4 contingency table for the fiber-enriched cracker data, we found previously that there was moderate evidence of an association between bloating severity and fiber source. Because bloating severity is measured in an ordinal manner (none < low < medium < high), a proportional odds model allows us to perform a potentially more powerful analysis. Using fiber as the explanatory variable with appropriately constructed indicator variables (see Section 3.3 example), our model is

$$\text{logit}(P(Y \leq j)) = \beta_{j0} + \beta_2 \text{bran} + \beta_3 \text{gum} + \beta_4 \text{both}.$$

where j corresponds to levels 1 (none), 2 (low), and 3 (medium) of bloating severity. Notice that this model involves a smaller number of parameters than the multinomial regression model of Section 3.3. We estimate the model using the `polr()` function:

```
> library(package = MASS)
> levels(diet$bloat)
[1] "none"   "low"    "medium" "high"
> mod.fit.ord <- polr(formula = bloat ~ fiber, weights = count,
    data = diet, method = "logistic")
> summary(mod.fit.ord)

Re-fitting to get Hessian

Call: polr(formula = bloat ~ fiber, data = diet, weights = count,
    method = "logistic")
```

```
Coefficients:
             Value  Std. Error  t value
fiberbran  -0.3859      0.7813   -0.494
fibergum    2.4426      0.8433    2.896
fiberboth   1.4235      0.7687    1.852

Intercepts:
              Value   Std. Error  t value
none|low     0.0218       0.5522   0.0395
low|medium   1.6573       0.6138   2.7002
medium|high  3.0113       0.7249   4.1539

Residual Deviance: 112.2242
AIC: 124.2242
```

The estimated model is

$$\text{logit}(\widehat{P}(Y \leq j)) = \hat{\beta}_{j0} + 0.3859\text{bran} - 2.4426\text{gum} - 1.4235\text{both},$$

where $\hat{\beta}_{10} = 0.0218$, $\hat{\beta}_{20} = 1.6573$, and $\hat{\beta}_{30} = 3.0113$.

A LRT for independence tests $H_0 : \beta_2 = \beta_3 = \beta_4 = 0$ vs. H_a : any $\beta_i \neq 0$, $i = 2, 3, 4$. This produces:

```
> library(package = car)
> Anova(mod.fit.ord)
Analysis of Deviance Table (Type II tests)

Response: bloat
      LR Chisq Df Pr(>Chisq)
fiber   15.048  3   0.001776 **
```

Because $-2\log(\Lambda) = 15.048$ is large relative to a χ_3^2 distribution (p-value = 0.0018), there is strong evidence that association exists in the form of a trend among the log-odds for the cumulative probabilities. Remember that with the multinomial regression model of Section 3.3 and the tests for independence given in Section 3.2, there was only moderate evidence of association. By narrowing the type of association considered by our test, we now have more power to reject the null hypothesis.

The estimated odds ratios for comparing each fiber source to using no fiber are $\widehat{OR} = \exp(\hat{\beta}_r)$ for $r = 2, 3$, and 4. Below are these odds ratios and the corresponding profile LR intervals:

```
> round(exp(-coefficients(mod.fit.ord)), 2)
fiberbran  fibergum fiberboth
     1.47      0.09      0.24

> conf.beta <- confint(object = mod.fit.ord, level = 0.95)
Waiting for profiling to be done...

Re-fitting to get Hessian

> ci <- exp(-conf.beta)
> round(data.frame(low = ci[,2], up = ci[,1]), 2)
          low    up
```

```
fiberbran 0.32 7.06
fibergum  0.02 0.43
fiberboth 0.05 1.05
> round(data.frame(low = 1/ci[,1], up = 1/ci[,2]), 2)
           low    up
fiberbran  0.14   3.15
fibergum   2.32  65.01
fiberboth  0.95  19.82
```

Note that inverting the confidence intervals in the last set of code expresses the effect of each fiber source in terms of the odds of *more severe* bloating. Because the profile LR interval for gum does not contain 1, there is sufficient evidence to indicate that using gum as a fiber source increases bloating severity; however, there is not sufficient evidence that bran increases bloating severity because its confidence interval contains 1. When combining both sources of fiber, there is marginal evidence of an effect, because the lower bound of the interval is close to 1. Formally, we can write an interpretation for the gum vs. no fiber source confidence intervals as:

> With 95% confidence, the odds of bloating severity being above a particular level are between 2.32 and 65.01 times as large when using gum as a fiber source than using no fiber.

To compare gum directly to bran, the simplest way is to use the `relevel()` function to change the base level of the fiber source variable. Similar code to that presented above can then be used to re-estimate the model and calculate the profile LR intervals (code is within the program). We find that

> With 95% confidence, the odds of bloating severity being below a particular level are between 0.01 and 0.30 times as large when using gum as a fiber source than when using bran.

Thus, this indicates that gum leads to more bloating than bran.

Overall, if given a choice between the two fiber sources and their combination, bran would be preferred if the only criterion was bloating severity.

3.4.3 Non-proportional odds model

A proportional odds model is one of the preferred ways to account for an ordered multinomial response, because slope regression parameters are constant over the response categories. While this can greatly simplify the model, it imposes the assumption that association affects the log-odds of cumulative probabilities the same way for all $j = 1, \ldots, J-1$. This may not be true in all situations. An alternative model that relaxes this assumption is

$$\text{logit}(P(Y \leq j)) = \beta_{j0} + \beta_{j1}x_1 + \cdots + \beta_{jp}x_p, \quad (3.13)$$

where $j = 1, \ldots J - 1$. Notice that all the regression parameters are now allowed to vary across the levels of Y. Equation 3.13 is referred to as the *non-proportional odds model*.

Because the proportional odds model is a special case of Equation 3.13, we can test the proportional odds assumption through the hypotheses $H_0 : \beta_{1r} = \cdots = \beta_{J-1,r}$ for $r = 1, \ldots, p$ vs. H_a : Not all equal. The test is conducted as a LRT where the degrees of freedom for the reference χ^2 distribution are found from the difference in the number of parameters in the two models, $(p+1)(J-1) - (p+J-1) = p(J-2)$.

Rejecting the proportional odds assumption suggests that the non-proportional odds model may be preferred. Estimated probabilities and odds ratios have a different form due to the extra parameters, and Exercise 7 discusses these formulations. However, the proportional odds model may still be preferred due to its smaller number of parameters. Each parameter estimated adds variability to the subsequent estimation of probabilities and odds ratios, so one may well get odds ratio estimates that are closer to the truth by using a smaller model with a minor defect than by using a much larger model without the defect. Also, a very large sample size could result in a rejection of the null hypothesis even though the data deviate only slightly from the proportional odds assumption.

Failing to reject the proportional odds hypothesis is not proof that it is true. However, it does offer some assurance that a proportional odds model provides a reasonable approximation to true relationships between Y and the explanatory variables. We will examine additional ways to evaluate a model's fit in Chapter 5.

Example: Fiber-enriched crackers (Fiber.R, Fiber.csv)

Unfortunately, the polr() function does not provide a way to test the proportional odds assumption. We instead use the vglm() of the VGAM package to help perform the test. First, we estimate the proportional odds model and examine a summary of the results:

```
> library(package = VGAM)
> mod.fit.po <- vglm(formula = bloat ~ fiber, family =
    cumulative(parallel = TRUE), weights = count, data =
    diet[diet$count != 0,])
> summary(mod.fit.po)

Call:
vglm(formula = bloat ~ fiber, family = cumulative(parallel =
    TRUE), data = diet[diet$count != 0, ], weights = count)

Pearson Residuals:
                 Min       1Q  Median      3Q     Max
logit(P[Y<=1]) -2.9213 -0.81492 -0.46031 1.47861 4.6688
logit(P[Y<=2]) -3.7464 -1.41839  0.55098 1.09119 2.3751
logit(P[Y<=3]) -3.0383  0.28798  0.33008 0.77698 1.8433

Coefficients:
              Value   Std. Error  z value
(Intercept):1  0.021823  0.55654   0.039212
(Intercept):2  1.657339  0.62533   2.650334
(Intercept):3  3.011276  0.72766   4.138292
fiberbran      0.385935  0.79412   0.485994
fibergum      -2.442585  0.82591  -2.957463
fiberboth     -1.423481  0.78090  -1.822868

Number of linear predictors:  3

Names of linear predictors: logit(P[Y<=1]), logit(P[Y<=2]),
    logit(P[Y<=3])

Dispersion Parameter for cumulative family:   1

Residual Deviance: 112.2242 on 36 degrees of freedom
```

```
Log-likelihood: -56.11209 on 36 degrees of freedom

Number of Iterations: 6
```

The estimated model is practically the same as found earlier with `polr()`. Within the call to `vglm()`, the `family = cumulative` argument value specifies a cumulative logit model, where the `parallel = TRUE` argument value denotes the proportional odds version. Also, the `weights` argument is used to denote the number of individuals that each row of the data frame represents. If the data set had alternatively been formulated so that each row represented an individual observation (i.e., a person testing a cracker), the `weights` argument would not be needed. Note that a quirk of `vglm()` requires the removal of all 0 counts from the data frame before using the function.

The VGAM package uses a different class structure than we have seen so far in this book or in Appendix A. This class structure, often referred to as *S4*, produces objects with components in *slots*.[10] For example, to access a list of all slots within the `mod.fit.po` object, we use the `slotNames()` function:

```
> slotNames(mod.fit.po)    # Like names() in S3
[1] "extra"                 "family"                "iter"

<OUTPUT EDITED>

[37] "y"

> mod.fit.po@coefficients  # Like <object name>$coefficients in S3
(Intercept):1 (Intercept):2 (Intercept):3     fiberbran
   0.02182272    1.65733908    3.01127560    0.38593487
      fibergum     fiberboth
   -2.44258450   -1.42348056
```

To access specific slots, the `@` symbol is used in the form `<object name>@<slot name>`. Thus, `mod.fit.po@coefficients` contains the estimated parameter values for the model.

Next we fit the non-proportional odds model

$$\text{logit}(P(Y \leq j)) = \beta_{j0} + \beta_{j2}\text{bran} + \beta_{j3}\text{gum} + \beta_{j4}\text{both}$$

using `vglm()` with the `parallel = FALSE` argument value:

```
> mod.fit.npo <- vglm(formula = bloat ~ fiber, family =
    cumulative(parallel = FALSE), weights = count, data =
    diet[diet$count != 0,])
> # summary(mod.fit.npo)    # Excluded to save space
> round(mod.fit.npo@coefficients, 2)
(Intercept):1 (Intercept):2 (Intercept):3   fiberbran:1
         0.00          1.61         18.02          0.34
```

---

[10] R and its corresponding packages are written in a form very similar to the S programming language. This language was first developed in the 1970s at Bell Laboratories. Versions 3 and 4 of S are emulated by R. Version 3 (S3) is predominant, and this is what is used primarily in our book. Version 4 is used by the VGAM package. For more information on versions 3 and 4 of S, please see Chapter 9 of Chambers (2010).

```
       fiberbran:2    fiberbran:3      fibergum:1      fibergum:2
              0.79           0.00           -1.61           -2.30
        fibergum:3    fiberboth:1     fiberboth:2     fiberboth:3
            -17.68          -1.61           -1.27          -16.41
```

Notice that there are now $p(J-2) = 6$ additional parameters compared to the proportional odds model. The hypotheses for the test of the proportional odds assumption are $H_0 : \beta_{1i} = \beta_{2i} = \beta_{3i}$ for $i = 2, 3, 4$ vs. H_a : Not all equal. Neither Anova() nor anova() has been implemented in VGAM, so we calculate $-2\log(\Lambda)$ by finding the difference between the model's two residual deviances (see Section 2.2.2 for an example involving logistic regression):

```
> tran.LR <- deviance(mod.fit.po) - deviance(mod.fit.npo)
> df <- mod.fit.po@df.residual - mod.fit.npo@df.residual
> p.value <- 1 - pchisq(q = tran.LR, df = df)
> data.frame(tran.LR, df, p.value)
   tran.LR df   p.value
1 3.832754  6 0.6992972
```

We use the deviance() function to extract the residual deviances, because there is no @deviance slot available. The LRT results in a statistic of $-2\log(\Lambda) = 3.83$ and a p-value of 0.6993. Thus, there is not sufficient evidence to indicate the proportional odds assumption has been violated.

A problem with using non-proportional odds models for general use is that the model does not adequately constrain the parameters to prevent $P(Y \leq j) < P(Y \leq j')$ for $j > j'$. Thus, the cumulative probabilities can decrease at some point causing an individual category probability to be less than 0. This violation of probability rules occurs because the effect an explanatory variable has on $\text{logit}(P(Y \leq j))$ can change for each j; i.e., the $\beta_{j1}, \ldots, \beta_{jp}$ parameters can vary freely over the levels of Y. For this reason, caution needs to be used with these models to make sure that nonsensical probabilities do not occur. The next example illustrates a case where a non-proportional odds model is inappropriate.

Example: Non-proportional odds model plots (CumulativeLogitModelPlot.R)

Consider the model $\text{logit}(P(Y \leq j)) = \beta_{j0} + \beta_{j1}x_1$, where $\beta_{10} = 0, \beta_{20} = 2, \beta_{30} = 4, \beta_{11} = 2, \beta_{21} = 3, \beta_{31} = 6$, and $J = 4$. Compared to the example from p. 172, the $j = 1$ case is the same here, but the $j = 2$ and 3 cases have different "slope" parameters ($\beta_{21} \neq \beta_{11}$ and $\beta_{31} \neq \beta_{11}$). Figure 3.6 shows plots of the cumulative and individual category probabilities. Notice that the cumulative probability curve for $P(Y \leq 3)$ crosses the other two curves. Thus, the model allows for $P(Y \leq 3) < P(Y \leq 2)$ and $P(Y \leq 3) < P(Y \leq 1)$ for some values of x_1. As a result, π_3 is less than 0 for some values of x_1. Please see the corresponding program for the code to create these plots.

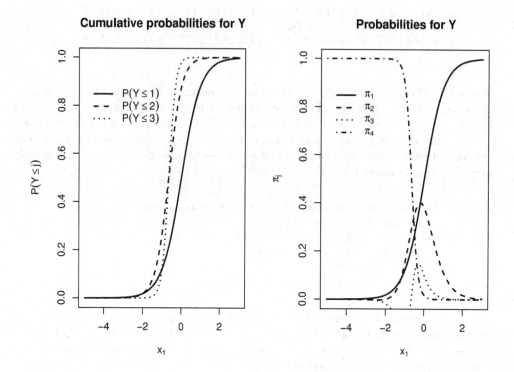

Figure 3.6: Non-proportional odds model for $\beta_{10} = 0$, $\beta_{20} = 2$, $\beta_{30} = 4$, $\beta_{11} = 2$, $\beta_{21} = 3$, and $\beta_{31} = 6$.

3.5 Additional regression models

Sections 3.3 and 3.4 give the two most common ways to model multinomial responses. There are many other approaches. For example, we are not limited to only using a logit transformation of an odds. The probit and complementary log-log transformations shown in Section 2.3 can be used instead. The `polr()` function estimates these models by using the appropriate `method` argument value within the function call.

Another model for multinomial responses is the *adjacent-categories model*:

$$\log(\pi_{j+1}/\pi_j) = \beta_{j0} + \beta_{j1}x_1 + \cdots + \beta_{jp}x_p,$$

where $j = 1, \ldots, J - 1$. This model is similar to the multinomial regression model stated in Equation 3.5, but it also allows one to take advantage of an ordinal response due to the successive comparisons of the $j + 1$ and j categories. To further take advantage of ordinal properties in a response, we can use a simplified version of the model:

$$\log(\pi_{j+1}/\pi_j) = \beta_{j0} + \beta_1 x_1 + \cdots + \beta_p x_p,$$

where the regression slope parameters are forced to be the same for all j. This model assumes that an explanatory variable has the same effect on the odds for every successive category comparison in much the same way the proportional odds regression model does with odds based on cumulative probabilities. Both of these adjacent-categories models can be estimated in R using the `vglm()` function. We explore these and other models further in Exercise 17.

3.6 Exercises

1. For the case of $I = J = 2$, show that $\sum_{i=1}^{I} \sum_{j=1}^{J} \frac{(n_{ij} - n_{i+}n_{+j}/n)^2}{n_{i+}n_{+j}/n}$ is equivalent to X^2 as given in Chapter 1.

2. Show that $-2\log(\Lambda)$ simplifies to $-2 \sum_{i=1}^{I} \sum_{j=1}^{J} n_{ij} \log\left(\frac{n_{ij}}{n_{i+}n_{+j}/n}\right)$ when testing for independence between two variables.

3. Consider a 2×2 contingency table structure as a special case of Table 3.1.

    (a) Discuss why the marginal distribution of n_{1+} is binomial with parameter π_{1+}.

    (b) The conditional joint distribution of n_{11}, n_{12}, and n_{21} given n_{1+} can be shown to be $f(n_{11}, n_{12}, n_{21}|n_{1+}) = f(n_{11}, n_{12}, n_{21})/f(n_{1+})$, where $f(n_{11}, n_{12}, n_{21})$ is the joint distribution of n_{11}, n_{12}, n_{21} and $f(n_{1+})$ is the marginal distribution of n_{1+}. Note that we do not need to state n_{22} within $f(n_{11}, n_{12}, n_{21})$ because $n = n_{11} + n_{12} + n_{21} + n_{22}$ and n is known. Similarly, we do not need to state n_{2+} within $f(n_{1+})$. Using Equation 3.1 for the joint distribution of $f(n_{11}, n_{12}, n_{21})$ and the binomial distribution for $f(n_{1+})$, find $f(n_{11}, n_{12}, n_{21}|n_{1+})$ and show it is equivalent to Equation 3.3 when $I = J = 2$.

    (c) In general for $I \times J$ contingency table structures, extend the result in (b) to show that Equation 3.1 leads to Equation 3.3 when conditioning on the row marginal counts.

4. Suppose that we have one continuous explanatory variable x that is to be represented in a regression model by both a linear and quadratic term.

    (a) For a c-unit increase in x in a multinomial regression model, derive the odds ratio that compares a category j ($j \neq 1$) response to a category 1 response. Show the form of the variance that would be used in a Wald confidence interval.

    (b) For a c-unit increase in x in a proportional odds regression model, derive the odds ratio that compares cumulative probabilities. Show the form of the variance that would be used in a Wald confidence interval.

5. Suppose a multinomial regression model has two continuous explanatory variables x_1 and x_2, and they are represented in the model by their linear and interaction terms.

    (a) For a c-unit increase in x_1, derive the corresponding odds ratio that compares a category j response to a category 1 response. Show the form of the variance that would be used in a Wald confidence interval.

    (b) Repeat this problem for a proportional odds regression model.

6. With the help of the CumulativeLogitModelPlot.R program discussed in Section 3.4, complete the following:

    (a) Construct plots similar to Figure 3.4 using different values of the parameters. What happens to the plots when β_1 is increased or decreased while β_{10}, β_{20}, and β_{30} remain constant? What happens to the plots when β_{10}, β_{20}, and β_{30} have more or less separation among them while β_1 remains constant?

(b) For a proportional odds regression model, find a set of regression parameter values that lead to some categories never having the largest probability at any x_1. State a mathematical reason relative to the model for why this occurs.

(c) Construct similar plots to Figure 3.6 using different values of the parameters. Find a set of parameters where nonsensical probabilities do not occur (i.e., the parameters satisfy $P(Y \leq j) < P(Y \leq j')$ for all $j < j'$).

7. Consider the non-proportional odds regression model $\text{logit}(P(Y \leq j)) = \beta_{j0} + \beta_{j1}x_1$, where $j = 1, \ldots J - 1$.

    (a) Show that $Odds_{x_1+c}(Y \leq j)/Odds_{x_1}(Y \leq j) = \exp(c\beta_{j1})$. Provide a formal interpretation of the odds ratio.

    (b) Find π_j for $j = 1, \ldots, J$. Why can a π_j be less than 0 with this model?

8. Exercise 19 of Chapter 2 examined the prevalence of hepatitis C among healthcare workers in the Glasgow area of Scotland. Convert the data for this exercise to a contingency table structure where occupational groups are located on the rows and the presence and absence of hepatitis are located on the columns. Perform Pearson chi-square and LR tests for independence using these data. Are your results here similar to what was found for the data analysis in Chapter 2? Explain.

9. Classification systems are often used by medical doctors to help make sense of patient symptoms and to guide them toward applying appropriate treatments. For this purpose, Cognard et al. (1995) proposes a new classification system with respect to dural arteriovenous fistulas (DAVFs). DAVFs are abnormal connections between arteries and veins in a person's head. If left untreated, they can lead to death due to strokes or other complications. Table 3.3 provides the data on 205 cases included in their study. The columns of the table provide the classification of each individual, where the classifications generally increase in DAVF severity from left to right. Each individual is given only one classification. The rows represent different symptoms experienced by the individuals. Assume that each individual also has only one symptom.

    (a) What would independence and dependence between the rows and columns mean in the context of DAVFs? Which conclusion do you think the researchers would prefer in order to justify their classification system? Explain.

    (b) Cognard et al. (1995) state that "the contingency table shows a statistically significant difference (P = .0001)" without mentioning the particular statistical test conducted. Perform a Pearson chi-square test for independence to produce a similar significance result.

    (c) It is unlikely that patients with a DAVF will exhibit only one symptom. While not stated in this paper, the authors may have summarized the patient's "primary" or "most extreme" symptom, so that each patient contributed to the count of only one cell in the table. However, if this was not the case, a patient could contribute to none, one, or more than one cell (e.g., a patient could have intracranial hypertension and seizures while having a level 3 classification). In fact, the total counts for the table could be greater or less than the number of patients. Would this alternative scenario invalidate the use of the Pearson chi-square test in this situation? Explain.[11]

---

[11] Section 6.4 examines situations where individuals could contribute to more than one row and/or column response combination.

Table 3.3: Classification of DAVFs by symptoms. Data source is Cognard et al. (1995).

		\multicolumn{7}{c}{Classification}						
		1	2a	2b	2a and 2b	3	4	5
Symptoms	Hemorrhage	0	0	2	1	10	19	5
	Intracranial hypertension	1	8	1	2	0	4	1
	Focal neurologic deficit	0	0	0	6	8	2	0
	Seizures	0	1	0	2	1	3	0
	Cardiac deficiency	0	1	0	1	0	0	0
	Myelopathy	0	0	0	0	0	0	6
	Non-aggressive symptoms	83	17	7	6	6	1	0

10. Section 3.3.2 discusses how the fiber-enriched crackers example can be analyzed by separating the fiber-source variable into two separate binary variables for bran and gum. This type of analysis may be preferred in order to assess separately the effects of these two sources on bloating. Using this form of the data, complete the following:

    (a) Estimate the proportional odds regression model involving bran, gum, and their interaction. Is there sufficient evidence to indicate an interaction is important?

    (b) Evaluate the proportional odds assumption by estimating the corresponding non-proportional odds model and then performing a LRT.

    (c) With the most appropriate model found through the investigations in (a) and (b), use odds ratios to interpret the effect of bran and gum on bloating.

11. For the wheat kernel data, complete the following:

    (a) Interpret the estimated odds ratios and corresponding confidence intervals for the other explanatory variables not specifically addressed in the examples of Sections 3.3.1 and 3.4.1. Relate your results back to Figure 3.2.

    (b) Compute odds ratio estimates comparing scab vs. sprout corresponding to the variables referenced in (a). Compute confidence intervals and interpret these intervals in the context of the data.

12. In order to maximize sales, items within grocery stores are strategically placed to draw customer attention. This exercise examines one type of item—breakfast cereal. Typically, in large grocery stores, boxes of cereal are placed on sets of shelves located on one side of the aisle. By placing particular boxes of cereals on specific shelves, grocery stores may better attract customers to them. To investigate this further, a random sample of size 10 was taken from each of four shelves at a Dillons grocery store in Manhattan, KS. These data are given in the cereal_dillons.csv file. The response variable is the shelf number, which is numbered from bottom (1) to top (4), and the explanatory variables are the sugar, fat, and sodium content of the cereals. Using these data, complete the following:

    (a) The explanatory variables need to be re-formatted before proceeding further. First, divide each explanatory variable by its serving size to account for the different serving sizes among the cereals. Second, re-scale each variable to be within 0 and 1.[12] Below is code we use to re-format the data after the data file is read into an object named `cereal`:

```
stand01 <- function(x) { (x - min(x))/(max(x) - min(x)) }
cereal2 <- data.frame(Shelf = cereal$Shelf, sugar =
    stand01(x = cereal$sugar_g/cereal$size_g), fat =
    stand01(x = cereal$fat_g/cereal$size_g), sodium =
    stand01(x = cereal$sodium_mg/cereal$size_g))
```

(b) Construct side-by-side box plots with dot plots overlaid for each of the explanatory variables. Below is code that can be used for plots involving sugar:

```
boxplot(formula = sugar ~ Shelf, data = cereal2, ylab =
    "Sugar", xlab = "Shelf", pars = list(outpch=NA))
stripchart(x = cereal2$sugar ~ cereal2$Shelf, lwd = 2, col
    = "red", method = "jitter", vertical = TRUE, pch = 1,
    add = TRUE)
```

Also, construct a parallel coordinates plot for the explanatory variables and the shelf number. Discuss if possible content differences exist among the shelves.

(c) The response has values of 1, 2, 3, and 4. Under what setting would it be desirable to take into account ordinality. Do you think this occurs here?

(d) Estimate a multinomial regression model with linear forms of the sugar, fat, and sodium variables. Perform LRTs to examine the importance of each explanatory variable.

(e) Show that there are no significant interactions among the explanatory variables (including an interaction among all three variables).

(f) Kellogg's Apple Jacks (http://www.applejacks.com) is a cereal marketed toward children. For a serving size of 28 grams, its sugar content is 12 grams, fat content is 0.5 grams, and sodium content is 130 milligrams. Estimate the shelf probabilities for Apple Jacks.

(g) Construct a plot similar to Figure 3.3 where the estimated probability for a shelf is on the y-axis and the sugar content is on the x-axis. Use the mean overall fat and sodium content as the corresponding variable values in the model. Interpret the plot with respect to sugar content.

(h) Estimate odds ratios and calculate corresponding confidence intervals for each explanatory variable. Relate your interpretations back to the plots constructed for this exercise.

13. Similar to Exercise 12, the cereal_supersaver.csv file contains the sugar, fat, and sodium content for cereals placed on shelves at a Super Saver grocery store in Lincoln, NE. Analyze these data to better understand the shelf placement of the cereals at this grocery store.

14. An example from Section 4.2.5 examines data from the 1991 U.S. General Social Survey that cross-classifies people according to

---

[12]The author of `multinom()` suggests this can help with the convergence of parameter estimates when using the function. We recommend the 0-1 rescaling because slow convergence occurs here when estimating the model in part (d) without re-scaling. Note that the `vglm()` function can also be used to estimate the model. When we use this function, convergence is obtained in the same number of iterations with and without the 0-1 rescaling, so `vglm()` this may be a more convenient model fitting alternative.

- Political ideology: Very liberal (VL), Slightly liberal (SL), Moderate (M), Slightly conservative (SC), and Very conservative (VC)
- Political party: Democrat (D) or Republican (R)
- Gender: Female (F) or Male (M).

Consider political ideology to be an ordinal response variable, and political party and gender to be explanatory variables. The data are available in the file pol_ideol_data.csv.

(a) Use the `factor()` function with the ideology variable to make sure that R places the levels of the ideology variable in the correct order.

(b) Because the two explanatory variables are categorical, we can view the entire data set in a three-dimensional contingency table structure. The `xtabs()` and `ftable()` functions are useful for this purpose. Below is how these functions can be used after the data file has been read into R (the data frame `set1` contains the original data):

```
> c.table <- xtabs(formula = count ~ party + ideol +
    gender, data = set1)
> ftable(x = c.table, row.vars = c("gender", "party"),
    col.vars = "ideol")
```

(c) Using multinomial and proportional odds regression models that include party, gender, and their interaction, complete the following:

  i. Estimate the models and perform LRTs to test the importance of each explanatory variable.
  ii. Compute the estimated probabilities for each ideology level given all possible combinations of the party and gender levels.
  iii. Construct a contingency table with estimated counts from the model. These estimated counts are found by taking the estimated probability for each ideology level multiplied by their corresponding number of observations for a party and gender combination. For example, there are 264 observations for `gender = "F"` and `party = "D"`. Because the multinomial regression model results in $\hat{\pi}_{\text{VL}} = 0.1667$, this model's estimated count is $0.1667 \times 264 = 44$.
  iv. Are the estimated counts the same as the observed? Explain.
  v. Use odds ratios computed from the estimated models to help understand relationships between the explanatory variables and the response.

(d) Compare the results for the two models. Discuss which model may be more appropriate for this setting.

15. Continuing Exercise 14, consider again the counts that are found using the estimated proportional odds model. Find the correct combination of counts from this table that result in estimated odds that are the same as the estimated odds found directly from the estimated model $\exp(\hat{\beta}_{j0} + \hat{\beta}_1 x_1 + \hat{\beta}_2 x_2 + \hat{\beta}_3 x_1 x_2)$, where x_1 and x_2 are the appropriate indicator variables representing `party` and `gender`, respectively.

16. Researchers at Penn State University performed a study to determine the optimal fat content for ice cream. Details of the study and corresponding data analysis are available at https://onlinecourses.science.psu.edu/stat504/node/187. In summary, 493 individuals were asked to taste and then rate a particular type of ice

cream on a 9-point scale (1 to 9 with 1 equating to not liking and 9 equating to really liking). The ice cream given to the individuals had a fat proportion level of 0, 0.04, ..., or 0.28. The data are available in ice_cream.csv, where the `count` column represents the number of individuals in a particular fat group giving a particular rating. Using these data, complete the following:

(a) Display the data in a 8 × 9 contingency table structure.

(b) Find the observed rating proportions conditional on the fat content. Construct a plot of these proportions vs. fat content. We recommend connecting the observed proportions by lines corresponding to the rating score.

(c) Discuss why Section 3.2's Pearson chi-square and LR tests for independence are not ideal analysis choices for these data.

(d) Estimate a proportional odds model using fat content to predict rating. Investigate whether or not a quadratic term is helpful. Note that the rating variable needs to be converted to a `factor` class type in order to use `polr()`.

(e) Perform a LRT to assess the proportional odds assumption.

(f) Use odds ratios and plots of the estimated probabilities to interpret the resulting model. Recommend a fat content.

17. For the fiber-enriched crackers data, an adjacent-category model can be written as $\log(\pi_{j+1}/\pi_j) = \beta_{j0} + \beta_{j1}\text{bran} + \beta_{j2}\text{gum} + \beta_{j3}\text{both}$ for $j = 1, 2,$ and 3. This model can be estimated by the `vglm()` function of the `VGAM` package in a similar manner as the non-proportional odds model discussed in Section 3.4.3.

    (a) Estimate the adjacent-category model using the following as a continuation of the code used in Fiber.R:

        ```
        > mod.fit.full <- vglm(formula = bloat ~ fiber, weights =
            count, family = acat(parallel = FALSE), data =
            diet[diet$count != 0,])
        > summary(mod.fit.full)
        ```

        The `acat()` function used in the `family` argument specifies an adjacent category model.

    (b) To better take advantage of bloating severity's ordinal response, we can estimate the model as $\log(\pi_j/\pi_{j+1}) = \beta_{j0} + \beta_1\text{bran} + \beta_2\text{gum} + \beta_3\text{both}$ where parameters are shared across the response categories for the fiber sources. Estimate this model by using `acat(parallel = TRUE)` in the `family` argument of `vglm()`.

    (c) Examine odds ratios using the fits of both models in (a) and (b). Compare their values.

    (d) State in words what is tested by $H_0 : \beta_{11} = \beta_{21} = \beta_{31}, \beta_{12} = \beta_{22} = \beta_{32}, \beta_{13} = \beta_{23} = \beta_{33}$ vs. H_a : At least one inequality. Perform the test using a LRT.

18. For the wheat kernel data, consider a model to estimate the kernel condition using the density explanatory variable as a linear term.

    (a) Write an R function that computes the log-likelihood function for the multinomial regression model. Evaluate the function at the parameter estimates produced by `multinom()`, and verify that your computed value is the same as that produced by `logLik()` (use the object saved from `multinom()` within this function).

(b) Maximize the log-likelihood function using `optim()` to obtain the MLEs and the estimated covariance matrix. Compare your answers to what is obtained by `multinom()`. Note that to obtain starting values for `optim()`, one approach is to estimate separate logistic regression models for $\log(\pi_2/\pi_1)$ and $\log(\pi_3/\pi_1)$. These models are estimated only for those observations that have the corresponding responses (e.g., a $Y = 1$ or $Y = 2$ for $\log(\pi_2/\pi_1)$).

(c) Repeat (a) and (b), but now for a proportional odds regression model. Compare your answers to what is obtained by using `polr()`. Note that to obtain starting values for `optim()`, one approach is to estimate separate logistic regression models for $\text{logit}(P(Y \leq 1))$ and $\text{logit}(P(Y \leq 2))$. The intercepts are used as initial estimates for β_{10} and β_{20}, and the two slopes are averaged to obtain an initial estimate of β_1.

19. For the wheat kernel data, consider again the estimated multinomial and proportional odds models from Sections 3.3 and 3.4 that included all of the explanatory variables as linear terms.

    (a) Construct plots of the estimated models with the estimated probability on the y-axis and `density` on the x-axis, where the remaining explanatory variables are set to their mean values. Interpret the plots and compare the models.

    (b) How do the plots from (a) change when the remaining explanatory variables are set to values other than their mean values?

    (c) Construct additional plots as in (a), but now for explanatory variables other than `density` on the x-axis.

20. Changes to habitat affects different species of wildlife differently. Birds that hunt for insects on the ground, for example, may be less affected by clearing rainforest than those that feed on plant nectars, like hummingbirds, or that extract insects from trees, like woodpeckers. Root (1967) defines a *guild* as "a group of species that exploit the same class of environmental resources in a similar way." The mist-netted birds in the example from Section 4.2.3 were classified into guilds according to their feeding habits. The guilds are labeled by food source as Air-Arth, Leaf-Arth, Wood-Arth, Ground-Arth, Carnivore, Fruit, Seeds, and Nectar, where "Arth" means arthropod (insect and spider). We wish to know whether the relative abundance of birds from different guilds is related to the amount of habitat "degradation" at the six locations. Degradation is defined here as a factor with three levels representing how much the habitat has changed from its original forested state. The two forest locations are "Low," the forest edge and forest fragment are "Moderate," and the two pasture locations are "High." The data are given in Table 3.4. Perform an analysis to determine which guilds are affected more or which are affected less by habitat degradation.

21. *Pneumoconiosis* (black lung disease) "is a lung disease that results from breathing in dust from coal, graphite, or man-made carbon over a long period of time" (http://www.ncbi.nlm.nih.gov/pubmedhealth/PMH0001187). The data frame `pneumo` in the `VGAM` package contains counts of coal miners aggregated into eight groups based on their time working in the mine (in years) and their severity of *pneumoconiosis* (three levels: normal, mild, and severe). Analyze these data with the specific goals of (1) understanding the relationship between time in the mines and resulting severity of the disease, and (2) predicting the disease status of miners after 5, 10, 15, 20, or 25 years of working in a mine.

Table 3.4: Cross-tabulation of mist-netted Ecuadorean birds by habitat degradation and food-source guild.

Guild	Degradation		
	High	Moderate	Low
Air-Arth	8	39	48
Leaf-Arth	69	139	131
Wood-Arth	2	39	50
Ground-Arth	25	57	115
Fruit	20	48	16
Seeds	7	20	19
Nectar	100	177	190

Chapter 4

Analyzing a count response

In Chapter 1 we presented the binomial distribution to model count data arising from observing a binary response on n independent Bernoulli trials. Counts observed in this manner are constrained to lie between 0 and n, with variability that decreases as the probability of success approaches 0 or 1. Count data can arise from other mechanisms that have nothing to do with Bernoulli trials. The number of cars crossing a certain bridge in a given day, the number of weeds in a plot of crop land, the number of people standing in line at a service counter, and the number of moles on a person's body are examples of counts that are not constrained by any specific number of trials. In this chapter, we use the Poisson distribution to model counts of this type. We start by exploring the model for a single count, and then extend this framework to Poisson regression models that can handle any number of explanatory variables of any type. Finally, we consider some special cases of Poisson models that are used to describe situations that do not fit cleanly into the standard Poisson regression framework. We show in all cases how inference is conducted and how R can be used to aid in the analysis.

4.1 Poisson model for count data

4.1.1 Poisson distribution

Consider a situation in which some kind of events happen over a fixed period of time, such as cars passing over a bridge. It is easy to imagine that the *rate* or *intensity* at which cars cross the bridge might vary according to time of day (e.g., more cars during rush hour, fewer at 3 a.m.) or day of the week (more on a workday, fewer on weekends and holidays). Suppose we count cars for exactly one hour at the same time every week, say Wednesday from 7:00–8:00 a.m. Even if there are no practical differences in the population intensity from week to week, we would not expect the exact same number of cars to cross the bridge each week. Some individuals who normally cross the bridge during this hour may carpool or take transit; they may be sick, on vacation, at a meeting, or tele-commuting; or they may simply be earlier or later than usual and miss the cutoffs for the hour of counting. Thus, we expect the actual count of cars during this time to vary randomly, even if the population rate underlying the process is not changing.

This is the nature of the Poisson distribution for counts. If we can assume that (1) all counts taken on some process have the same underlying intensity, and (2) the period of observation is constant for each count, then the counts follow a Poisson distribution. If Y is a Poisson random variable, then the PMF of Y is

$$P(Y = y) = \frac{e^{-\mu}\mu^y}{y!}, \ y = 0, 1, 2, \ldots,$$

where $\mu > 0$ is a parameter. We can abbreviate this distribution by writing $Y \sim Po(\mu)$. Adaptations of this model will be given later for cases where the intensity is not constant for all counts (Section 4.2) and for when the observation period is not constant for all counts (Section 4.3).

Properties of the Poisson distribution

The Poisson distribution has a number of appealing properties that make it a convenient model from which to work. First of all, the form of the distribution is fairly simple. There is only one parameter, μ, which represents both the mean and variance of the distribution:

$$E(Y) = Var(Y) = \mu.$$

Notice that, unlike the normal distribution, the variance of the Poisson distribution changes as the mean changes. This makes sense because counts are bounded below by 0. For example, suppose means for two different groups of counts are 5 and 50. Then the counts in the first group might be mostly in the 0–10 range, while those in the second group might naturally range from, say, 20–80. There is more "room" for variability with a larger mean, so we expect the variance to be increasing with the mean. The Poisson model is more specific than this: it requires that the variance *equals* the mean. We shall see in Section 5.3 that this requirement is sometimes too restrictive.

Another useful property is that sums of Poisson random variables are also Poisson random variables: If $Y_1, Y_2, \ldots, Y_m$ are independent with $Y_k \sim Po(\mu_k)$ for $k = 1, \ldots, m$, then $\sum_{k=1}^{m} Y_k \sim Po(\sum_{k=1}^{m} \mu_k)$. Thus, totals of several counts can be modeled with Poisson distributions as long as the constituent counts can.

The shape of the Poisson distribution approaches that of the normal distribution as μ grows. In fact, the normal distribution is sometimes used as a model for counts, even though it is not a discrete distribution. Exercise 22 demonstrates when the normal distribution is a reasonable approximation to the Poisson.

Example: Cars at an intersection (Stoplight.R, Stoplight.csv)

The intersection of 33rd and Holdrege Streets in Lincoln, Nebraska, is a typical north-south/east-west, 4-way intersection, except that there is a fire station located approximately 150 feet north of the intersection on the west side of the street. A back-up of vehicles waiting to go south at the intersection could block the fire station's driveway, which would prevent emergency vehicles from exiting the station. To determine the probability that this could happen, 40 consecutive stoplight cycles were watched, and the number of vehicles stopped on the north side of the intersection were counted. Note that there were no vehicles remaining in the intersection for more than one stoplight cycle.

Below we read the data into R and calculate the mean and variance for the 40 counts:

```
> stoplight <- read.csv(file = "C:\\data\\stoplight.csv")
> head(stoplight)
  Observation vehicles
1           1        4
2           2        6
3           3        1
4           4        2
5           5        3
6           6        3
```

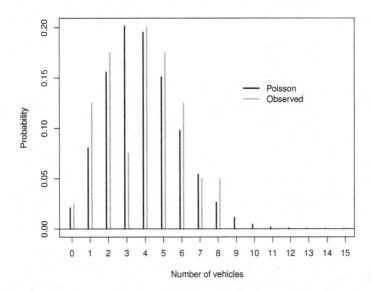

Figure 4.1: Comparison of the observed distribution of car counts and a Poisson distribution with the same mean.

```
> # Summary statistics
> mean(stoplight$vehicles)
[1] 3.875
> var(stoplight$vehicles)
[1] 4.317308
```

The sample mean is $\bar{y} = 3.9$. The sample variance, $s^2 = 4.3$, is fairly close to the mean, as is expected if the car count random variable follows a Poisson distribution. A frequency distribution and a histogram of the data, shown in Figure 4.1, are produced by the code below.

```
> # Frequencies
> table(stoplight$vehicles)   # Note that y = 0, 1, ..., 8 all
    have positive counts

0 1 2 3 4 5 6 7 8
1 5 7 3 8 7 5 2 2
> rel.freq <- table(stoplight$vehicles) /
    length(stoplight$vehicles)
> rel.freq2 <- c(rel.freq, rep(0, times = 7))

> # Poisson calculations
> y <- 0:15
> prob <- round(dpois(x = y, lambda = mean(stoplight$vehicles)),
    4)

> # Observed and Poisson
> data.frame(y, prob, rel.freq = rel.freq2)
```

```
   y    prob rel.freq
1  0  0.0208    0.025
2  1  0.0804    0.125
3  2  0.1558    0.175
4  3  0.2013    0.075
5  4  0.1950    0.200
6  5  0.1511    0.175
7  6  0.0976    0.125
8  7  0.0540    0.050
9  8  0.0262    0.050
10 9  0.0113    0.000
11 10 0.0044    0.000
12 11 0.0015    0.000
13 12 0.0005    0.000
14 13 0.0001    0.000
15 14 0.0000    0.000
16 15 0.0000    0.000
```

```
> plot(x = y-0.1, y = prob, type = "h", ylab = "Probability",
    xlab = "Number of vehicles", lwd = 2, xaxt = "n")
> axis(side = 1, at = 0:15)
> lines(x = y+0.1, y = rel.freq2, type = "h", lwd = 2, lty =
    "solid", col = "red")
> abline(h = 0)
> legend(x = 9, y = 0.15, legend = c("Poisson", "Observed"), lty
    = c("solid", "solid"), lwd = c(2,2), col = c("black", "red"),
    bty = "n")
```

The plot shows generally good agreement with the Poisson PMF assuming $\mu = \bar{y} = 3.875$. The only large deviation appears at $y = 3$. We do not believe that there is any physical reason that would prevent groups of three cars from lining up at the intersection, so we assume that this deviation is just a result of chance. Overall, it appears the Poisson distribution provides a satisfactory approximation to the distribution of cars stopped at the stoplight. We leave the use of this approximation to estimate the probability of blocking the fire station as Exercise 4.

4.1.2 Poisson likelihood and inference

The parameter μ in the Poisson model is estimated using maximum likelihood. Given a random sample $y_1 \ldots, y_n$, from a Poisson distribution with mean μ, the likelihood function is

$$L(\mu; y_1 \ldots, y_n) = \prod_{i=1}^{n} \frac{e^{-\mu} \mu^{y_i}}{y_i!}.$$

The MLE is shown in Appendix B.3.1 to be the sample mean, $\hat{\mu} = \bar{y}$. Using the second derivative of the log likelihood as described in Appendix B.3.4, the variance of $\hat{\mu}$ is easily shown to be estimated by $\widehat{Var}(\hat{\mu}) = \hat{\mu}/n$. Notice that this differs from our usual estimate for the variance of $\bar{y}$, s^2/n, because the mean and variance are the same in the Poisson distribution. In large samples from a Poisson distribution the sample mean and variance will be nearly identical.

As with the probability-of-success parameter π in a binomial distribution, there are many ways to form confidence intervals for the Poisson mean parameter μ. See Swift (2009) for a discussion of the confidence level and length properties for some of these methods. We discuss here the Wald, likelihood ratio, score, and exact methods to develop a confidence interval for μ.

In particular, the $100(1-\alpha)\%$ Wald confidence interval is

$$\hat{\mu} \pm Z_{1-\alpha/2}\sqrt{\hat{\mu}/n}.$$

The score interval is derived from the hypothesis tests of $H_0 : \mu = \mu_0$, which can be conducted using the score statistic

$$Z_0 = \frac{\hat{\mu} - \mu_0}{\sqrt{\mu_0/n}},$$

comparing Z_0 to appropriate quantiles of the standard normal distribution. Inverting this test results in the confidence interval,

$$\left(\hat{\mu} + \frac{Z_{1-\alpha/2}^2}{2n}\right) \pm Z_{1-\alpha/2}\sqrt{\frac{\hat{\mu} + Z_{1-\alpha/2}^2/4n}{n}}.$$

The LR interval is the set of values for μ_0 for which $-2\log[L(\mu_0|y_1 \ldots, y_n)/L(\hat{\mu}|y_1 \ldots, y_n)] \leq \chi_{1,1-\alpha}^2$. This does not have a closed-form solution and must be found using iterative numerical procedures. It is therefore not generally used for this simple problem. The exact interval is developed using logic similar to what was used for the Clopper-Pearson interval for the binomial probability from Section 1.1.2. Its form is

$$\chi_{2n\hat{\mu},\alpha/2}^2/2n < \mu < \chi_{2(n\hat{\mu}+1),1-\alpha/2}^2/2n.$$

Some of these methods are very similar to tests and confidence intervals used for inference on means from the normal distribution. The main advantage of these Poisson-based methods is their use of the Poisson likelihood function to obtain the relation $Var(\bar{Y}) = \mu/n$, which allows shorter confidence intervals and more powerful tests to be computed, especially in smaller samples. The use of the Poisson likelihood function can also be their disadvantage, because it is common that the processes that generate counts have intensity that does not remain rigidly constant. This causes the counts to behave as if they have more variability than the Poisson model expects.[1] Thus, $\widehat{Var}(\hat{\mu}) = \hat{\mu}/n$ often underestimates the true variability of the mean count. Although s^2 is a less precise estimate of $Var(Y)$ when the Poisson distribution holds exactly, it estimates the variability present due to *all* sources, including any variations in process intensity. Therefore, ordinary t-distribution tests and confidence intervals for a population mean are more robust to deviations from the Poisson assumption and are sometimes used with count data drawn from a single population. However, notice that the confidence intervals developed specifically for the Poisson distribution can all be used even when $n = 1$, while the t-based interval cannot.

Example: Cars at an intersection (Stoplight.R, Stoplight.csv)

We next find confidence intervals for the mean number of cars stopped at the intersection.[2]

---

[1] This is called *overdispersion* and is covered in Section 5.3.

[2] The `poisson.test()` function in the `stats` package can be used to find an exact interval. The package `epitools` also contains functions that can compute confidence intervals for the Poisson mean parameter. In particular, `pois.exact()` produces the exact interval and `pois.approx()` produces the Wald interval. Examples of their use are included in the program for this example.

```
> alpha <- 0.05
> n <- length(stoplight$vehicles)
> mu.hat <- mean(stoplight$vehicles)

> # Wald
> mu.hat + qnorm(p = c(alpha/2, 1-alpha/2))*sqrt(mu.hat/n)
[1] 3.264966 4.485034

> # Score
> (mu.hat + qnorm(p = c(alpha/2, 1-alpha/2))/(2*n)) + qnorm(p =
    c(alpha/2, 1-alpha/2)) * sqrt((mu.hat + qnorm(p =
    1-alpha/2)/(4*n))/n)
[1] 3.239503 4.510497

> # Exact
> qchisq(p = c(alpha/2, 1-alpha/2), df = c(2*n*mu.hat,
    2*(n*mu.hat+1)))/(2*n)
[1] 3.288979 4.535323

> # Usual t-distribution based interval
> t.test(x = stoplight$vehicles, conf.level = 0.95)
        One Sample t-test

data:  stoplight$vehicles
t = 11.7949, df = 39, p-value = 1.955e-14
alternative hypothesis: true mean is not equal to 0
95 percent confidence interval:
 3.210483 4.539517
sample estimates:
mean of x
    3.875
```

The 95% score confidence interval is $3.2 < \mu < 4.5$; there is a 95% chance that intervals produced in this way contain the true mean number of cars at the stop light. Other confidence intervals are fairly similar and have a similar interpretation.

Our recommendations regarding which intervals to use for estimating the mean parameter of the Poisson distribution mirror those for the probability of success parameter from the binomial. The Wald interval is not good, because its true confidence level can be very low and rarely reaches its stated level. Conversely, the exact interval can be excessively conservative, and hence wider than necessary. The score interval is a good compromise in that its true confidence level is generally better than the Wald, and it is generally shorter than the exact interval.

Swift (2009) compares numerous intervals (excluding Wald, unfortunately) and finds other intervals that he prefers to the score interval. For example, there are improvements to the exact interval, similar to those described for confidence intervals for the binomial probability parameter in Section 1.1.2, that can make it shorter while retaining the same true confidence level. The differences among the various intervals are often slight, however, as the next example shows. For more details on different possible confidence intervals and their properties, see Swift (2009).

Example: Comparison of confidence intervals for Poisson mean (PoiConf-Levels.R)

In this example, we compare the true confidence levels and mean lengths of the Wald, score, exact, and t confidence intervals for the population mean when data are sampled from a Poisson distribution. The algorithm for finding the true confidence levels for the Wald and score intervals follows that used in Section 1.1.2 for a binomial probability of success. We find the probability that a sample of size n has an observed sum equal to $0, 1, \ldots, 40n$ when the sample is drawn from $Po(\mu)$ for $\mu = 0.01, 0.02, \ldots, 20$. To do this, we use the fact the sum has a $Po(n\mu)$ distribution.[3]

Because the t-based interval depends on both the estimated mean and variance, it is much more difficult to find its true confidence level exactly. We would need to identify every possible combination of sample mean and variance for a given sample size and find the Poisson probability for each of the resulting confidence intervals. Instead, we estimate the true confidence level by using Monte Carlo simulation as in Section 1.1.3. We simulate data sets of count responses from $Po(\mu)$, use these data sets to find confidence intervals, and estimate the true confidence level with the observed proportion of these intervals that contain the μ that was used to generate the data. Estimated mean lengths are just the average length of all t intervals calculated from each μ.

From Figure 4.2 and investigations for other sample sizes, it is obvious that the true confidence level for the exact interval is always at least 95%; for small μ it is considerably higher. If strict maintenance of the confidence level is required (as in some regulatory situations), then this interval or one of its variants should be used. The confidence level for the score interval is well maintained near 95% at any sample size. The score interval only becomes erratic for very small μ (e.g., $\mu < 1$), but it is still the only interval among the three approximate intervals whose true confidence level does not drop to 0 as μ decreases toward 0. Therefore, this interval can be recommended for general use as long as the Poisson distribution is expected to hold. The t interval's estimated true confidence level seems fine—mostly within its bounds—for $n = 5$ as long as μ is large enough that the distribution of Y is reasonably approximated by a normal ($\mu > 7$ or so). It is a viable alternative to the score interval in these cases, although it creates longer intervals as Figure 4.3 shows. However, at $n = 2$ (not shown) the t interval breaks down and no longer achieves 95% for any $\mu < 20$. The score interval maintains its stated level even for $n = 1$.

4.2 Poisson regression models for count responses

As we saw in Chapter 2, binomial (binary) regression models provide a powerful framework within which a wide variety of problems can be addressed. Regression models for count data have also been developed based on the Poisson distribution. These are generically called *Poisson regression models*, and they are closely related to logistic regression models

---

[3]Technically, there are infinitely many possible values of the sum for any given μ, because the $Po(\mu)$ distribution places non-zero probability on responses $y = 0, 1, 2, \ldots$. All except a few happen with probabilities so small that they are ignorable. The program computes probabilities for the relatively small number of values of $\bar{y}$ that have non-negligible probabilities.

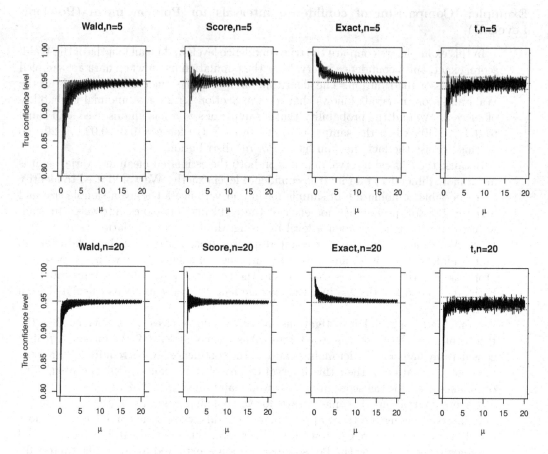

Figure 4.2: True confidence level for Wald, score, and exact 95% confidence intervals and estimated true confidence level for 95% t interval. Dotted line at 0.95 represents stated level; additional horizontal lines in t interval plot are limits within which the estimated true confidence levels should fall 95% of the time if the true confidence level is 0.95.

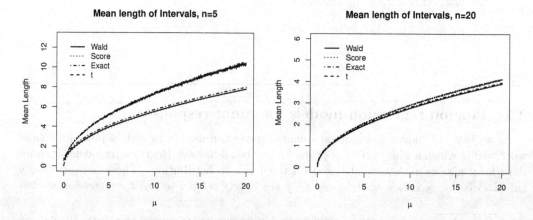

Figure 4.3: Mean 95% confidence interval length for four confidence interval methods. Note that the Wald and score intervals' lengths are nearly identical beyond about $\mu = 2$.

by the fact that both families are classes of generalized linear models that are described in Section 2.3. A special type of Poisson regression model, called a *loglinear model*, not only replicates the classical analysis of contingency tables described in Sections 1.2.3, 2.2.6, and 3.2, but can extend these analyses to any number of variables and can accommodate both continuous and ordinal variables. Poisson regression models can even be used to replicate logistic regression analyses, although their use in this manner is somewhat cumbersome.

The modeling steps with a Poisson regression are very much the same as those undertaken in any other regression modeling process. These steps include (1) specifying the model—the distribution, the parameters to be modeled (e.g., probabilities or means), and their relationship to the explanatory variables; (2) selecting explanatory variables; (3) estimating the parameters in the model; (4) assessing the model fit, and (5) performing inferences on model parameters and other quantities of interest. We assume for now that the explanatory variables to be used in the model are already known or chosen, leaving until Chapter 5 any discussion of the process by which they might have been selected. We also leave assessing the fit of the model until Chapter 5.

4.2.1 Model for mean: Log link

We have n observations of a response random variable Y, and $p \geq 1$ fixed explanatory variables, $x_1, \ldots, x_p$. We assume that for observation $i = 1, \ldots, n$,

$$Y_i \sim Po(\mu_i),$$

where

$$\mu_i = \exp(\beta_0 + \beta_1 x_{i1} + \ldots + \beta_p x_{ip}).$$

Thus, our generalized linear model has a Poisson random component, a linear systematic component $\beta_0 + \beta_1 x_1 + \ldots + \beta_p x_p$, and a log link,

$$\log(\mu) = \beta_0 + \beta_1 x_1 + \ldots + \beta_p x_p.$$

Each of these three specifications is an assumption that is open to question. For example, the Poisson distribution may provide a poor fit to the data. We will see later in this chapter and in Chapter 5 that there are different distributions that can serve as random components for count data in this case. Also, the linear predictor could conceivably be replaced with something that relates the explanatory variables to the parameters in a nonlinear fashion, at the cost of interpretative and computational complexity. Finally, the link could take an alternative form. In particular, the identity link, $\mu = \beta_0 + \beta_1 x_1 + \ldots + \beta_p x_p$, might seem a simpler form to use. However, the Poisson distribution requires that $\mu > 0$, and the identity link can lead to a non-positive value of μ for particular values of the explanatory variables. The log link function guarantees $\mu > 0$, so it is almost universally used.

A consequence of using the log link function with a linear predictor is that the explanatory variables affect the response mean multiplicatively. Consider a Poisson regression model with one explanatory variable: $\mu(x) = \exp(\beta_0 + \beta_1 x)$, where our notation emphasizes that μ changes as a function of x. When we increase the explanatory variable by c units, the result is $\mu(x + c) = \exp(\beta_0 + \beta_1(x + c)) = \mu(x) \exp(c\beta_1)$. Thus, the ratio of the means at $x + c$ and at x is

$$\frac{\mu(x + c)}{\mu(x)} = \frac{\exp(\beta_0 + \beta_1(x + c))}{\exp(\beta_0 + \beta_1 x)} = \exp(c\beta_1).$$

This leads to a convenient way to interpret the effect of x:

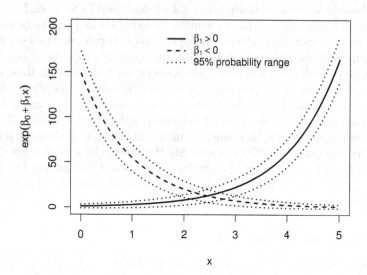

Figure 4.4: Mean functions using log links for $\beta_0 = 0.1, \beta_1 = 1$ (solid line) and $\beta_0 = 5, \beta_1 = -1$ (dashed line). Dotted lines represent approximate regions within which 95% of observations should fall according to the Poisson regression model (± 2 standard deviations).

The percentage change in the mean response that results from a c-unit change in x is $PC = 100(e^{c\beta_1} - 1)\%$.

If there were additional variables in the model, the same interpretation would apply, provided we hold the additional explanatory variables constant. When there are interaction terms or transformations involving the explanatory variable of interest, the ratio of means is more complicated, but can be derived in a similar manner as shown for odds ratios in Section 2.2.5. These derivations are the subject of Exercise 6.

Example: Poisson regression means (PoRegLinks.R)

Figure 4.4 shows what the mean function $\mu(x) = \exp(\beta_0 + \beta_1 x)$ looks like for a Poisson regression model when $\beta_1 > 0$ or $\beta_1 < 0$. It also shows the regions likely to contain most of the data. Notice in particular that because the variance and mean are equal, these regions expand as the mean grows.

4.2.2 Parameter estimation and inference

The Poisson regression model assumes that observations $y_1, y_2, \ldots, y_n$ are independent (for example, they are not serially correlated nor do they form clusters), and hence the likelihood is formed by the product of individual PMFs. Following the same steps as laid out in Equations 2.4–2.5 this leads to the log likelihood

$$\log[L(\beta_0, \ldots, \beta_p | y_1, \ldots, y_n)] = \sum_{i=1}^{n} [-\exp(\beta_0 + \beta_1 x_{i1} + \ldots + \beta_p x_{ip}) \\ + y_i(\beta_0 + \beta_1 x_{i1} + \ldots + \beta_p x_{ip}) - \log(y_i!)]. \quad (4.1)$$

As usual, we differentiate the log-likelihood with respect to each parameter β_j, set each of the resulting $p+1$ equations equal to 0, and solve the system of equations to find the MLEs $\hat{\beta}_0, \hat{\beta}_1, \ldots, \hat{\beta}_p$. As was the case with logistic regression in Section 2.2.1, these equations do not typically yield closed-form solutions, so the solutions must be found using iterative numerical procedures.

Once we have MLEs for the regression coefficients, we can compute MLEs for any function of the coefficients by taking the same function of the MLEs. For example, the *fitted value* or *prediction* is $\hat{\mu}_i = \exp(\hat{\beta}_0 + \hat{\beta}_1 x_{i1} + \ldots + \hat{\beta}_p x_{ip})$, and the estimated change in mean for a c-unit change in x_j is $\widehat{PC} = e^{c\hat{\beta}_j}$, holding the other explanatory variables constant.

Inference

The standard approaches to inference in maximum likelihood estimation are used here. We can derive Wald-type confidence intervals and tests for regression parameters and various functions of them by appealing to the large-sample normal approximation for the distribution of the MLEs $\hat{\beta}_0, \hat{\beta}_1, \ldots, \hat{\beta}_p$. While Wald methods are relatively easy to carry out, they do not always perform well. Likelihood ratio methods are better choices when the computational routines are available.

Example: Alcohol consumption (AlcoholPoRegs.R, DeHartSimplified.csv)

DeHart et al. (2008) describe a study in which "moderate to heavy drinkers" (at least 12 alcoholic drinks/week for women, 15 for men) were recruited to keep a daily record of each drink that they consumed over a 30-day study period. Participants also completed a variety of rating scales covering daily events in their lives and items related to self-esteem. Among the researchers' hypotheses was that negative events—particularly those involving romantic relationships—might be related to amount of alcohol consumed, especially among those with low self-esteem.

The repeated-measures aspect of this study (the 30 days of measurements on each participant, see Section 6.5) is beyond our current scope. We instead focus on each participant's first Saturday in the study, because Saturday is a day in which alcohol consumption might normally be high. We will model the number of drinks consumed—our count response—as a function of the variables measuring total positive and negative events. Each of these explanatory variables is a combination of several items scored on a 0–3 scale. High values of positive or negative events represent larger numbers of events and/or more extreme events.

The DeHartSimplified.csv spreadsheet contains 13 variables, measured over the first 6 days (Saturday is coded as "6" in the variable `dayweek`)[4]. Not all of the variables are needed in our current analysis, so our first step is to create a data frame that contains the required subset of these data. The variables are `id` (participant identifier), `numall` (number of alcoholic beverages, or "drinks," consumed in one day), `negevent` (an index for combining the total number and intensity of negative events experienced during the day), and `posevent` (same as `negevent`, except with positive events). Notice that only 89/100 participants had Saturday data in their first 6 days.

```
> dehart <- read.table(file = "C:\\Data\\DeHartSimplified.csv",
    header = TRUE, sep = ",", na.strings = " ")
> head(dehart)
  id studyday dayweek numall      nrel       prel     negevent
```

---

[4]Data kindly provided by Dr. Steve Armeli, School of Psychology, Fairleigh Dickinson University.

```
  1  1         1        6       9 1.000000 0.0000000 0.4000000
  2  1         2        7       1 0.000000 0.0000000 0.2500000
  3  1         3        1       1 1.000000 0.0000000 0.2666667
  4  1         4        2       2 0.000000 1.0000000 0.5333333
  5  1         5        3       2 1.333333 0.3333333 0.6633333
  6  1         6        4       1 1.000000 0.0000000 0.5900000
     posevent gender rosn      age  desired    state
  1 0.5250000      2  3.3 39.48528 5.666667 4.000000
  2 0.7000000      2  3.3 39.48528 2.000000 2.777778
  3 1.0000000      2  3.3 39.48528 3.000000 4.222222
  4 0.6083333      2  3.3 39.48528 3.666667 4.111111
  5 0.6933333      2  3.3 39.48528 3.000000 4.222222
  6 0.6800000      2  3.3 39.48528 4.000000 4.333333

> # Reduce data to what is needed for examples
> saturday <- dehart[dehart$dayweek == 6, c(1,4,7,8)]
> head(round(x = saturday, digits = 3))
   id numall negevent posevent
1   1      9    0.400    0.525
11  2      4    2.377    0.924
18  4      1    0.233    1.346
24  5      0    0.200    1.500
35  7      2    0.000    1.633
39  9      7    0.550    0.625

> dim(saturday)
[1] 89  4
```

Poisson regression models are fit using `glm()`. Details on this function are given in Section 2.2.1. The main adaptation for Poisson regression is that we must now specify `family = poisson(link = "log")`. The same follow-up functions as in logistic regression—`summary()`, `confint()`, `anova()`, `Anova()`, `predict()`, and so forth—are used to supply parameter estimates, confidence intervals, and tests.

We begin with a simplistic model using the total number of negative events to estimate the mean number of drinks:

```
> mod.neg <- glm(formula = numall ~ negevent, family =
    poisson(link = "log"), data = saturday)
> summary(mod.neg)

<OUTPUT EDITED>

Coefficients:
            Estimate Std. Error z value Pr(>|z|)
(Intercept)  1.52049    0.07524  20.208   <2e-16 ***
negevent    -0.26118    0.13597  -1.921   0.0547 .

(Dispersion parameter for poisson family taken to be 1)

    Null deviance: 250.34  on 88  degrees of freedom
Residual deviance: 246.39  on 87  degrees of freedom
AIC: 505.79

Number of Fisher Scoring iterations: 5
```

```
> 100*(exp(mod.neg$coefficients[2]) - 1)
negevent
-22.98564

> beta1.int <- confint(mod.neg, parm = "negevent", level = 0.95)
Waiting for profiling to be done...
> 100*(exp(beta1.int) - 1)
      2.5 %      97.5 %
-41.5294841  -0.3479521

> library(car)
> Anova(mod.neg)
Analysis of Deviance Table (Type II tests)
Response: numall
         LR Chisq Df Pr(>Chisq)
negevent   3.9495  1    0.04688 *

> # Matching confidence level with p-value of LRT to demonstrate
    equivalence
> confint(mod.neg, parm = "negevent", level = 1-0.04688)
Waiting for profiling to be done...
         2.3 %        97.7 %
-5.406268e-01 -5.932462e-06
```

The first model we fit is $Y_i \sim Po(\mu_i)$ with $\log(\mu_i) = \beta_0 + \beta_1 \text{negevent}_i$, where Y_i is the number of drinks consumed (numall) for person $i = 1, \ldots, 89$. The parameter estimates are $\hat{\beta}_0 = 1.52$ and $\hat{\beta}_1 = -0.26$. The negative "slope" parameter indicates that the number of drinks is decreasing as the negative events increase. Thus, a 1-unit increase in negative events leads to an estimated $\widehat{PC} = 23.0\%$ decrease in number of alcoholic beverages with 95% profile LR confidence interval $-41.5\% < PC < -0.3\%$. This change is marginally significant according to a Wald test (p-value = 0.0547) and LRT (p-value = 0.047). For illustration purposes, we re-ran the confidence interval calculations setting α to the p-value from the test, to show the equivalence between the LR confidence interval and the LRT. As expected, this yields a confidence interval for β_1 with an endpoint at zero, confirming that the two functions are performing the same calculations.

Figure 4.5 shows a plot of the data with the estimated mean curve, $\hat{\mu} = \exp(1.52 - 0.26\text{negevent}) = 4.57(0.77)^{\text{negevent}}$, using code given in the program corresponding to this example. It is apparent from the plot that the mean decreases as the negative events increase, but the change is somewhat small, decreasing from about 4.5 drinks to about 2.5 drinks across the range of negative events. The plot also shows some other interesting features. Very few of these "moderate-to-heavy" drinkers consumed no alcohol on their first Saturday in the study. One person reported consuming 21 drinks! Also, the inverse relationship that was observed between negative events and alcohol consumption is somewhat surprising. Considering that we are using a very simple model and analyzing only a fraction of the days of data from each study participant, we would view these results as preliminary, and needing further verification. Finally, a legitimate question that we do not address here is whether these people behaved differently because they were participating in a study (did recording their consumption cause them to change their drinking behavior?). If so, then this might explain why none of the people who experienced the most extreme negative events were among the

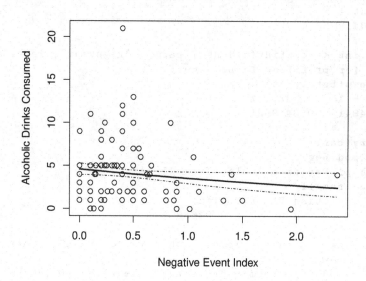

Figure 4.5: Plot of data and estimated mean (with 95% confidence interval) from Poisson regression of number of drinks consumed on total negative events for the first Saturday night in the alcohol consumption study.

heavier alcohol consumers on this day. Fuller investigation of the data (e.g., comparing the first Saturday with later Saturdays) might help to resolve this question.

Next, we consider an extended model that also takes into account the effects of positive events on drinking behavior. We now model the mean as $\log(\mu) = \beta_0 + \beta_1$negevent$+\beta_2$posevent $+\beta_3$negevent$\times$posevent. The parameter estimates and other analyses are given below:

```
> mod.negpos <- glm(formula = numall ~ negevent*posevent, family
    = poisson(link = "log"), data = saturday)
> summary(mod.negpos)

<OUTPUT EDITED>

Coefficients:
                   Estimate Std. Error z value Pr(>|z|)
(Intercept)          1.7571     0.1565  11.228  < 2e-16 ***
negevent            -1.1988     0.3231  -3.710 0.000207 ***
posevent            -0.1996     0.1199  -1.665 0.095941 .
negevent:posevent    0.7525     0.2253   3.341 0.000836 ***

(Dispersion parameter for poisson family taken to be 1)

    Null deviance: 250.34  on 88  degrees of freedom
Residual deviance: 233.07  on 85  degrees of freedom
AIC: 496.47

Number of Fisher Scoring iterations: 5
```

```
> confint(mod.negpos)
Waiting for profiling to be done...
                       2.5 %     97.5 %
(Intercept)         1.4490794  2.06217629
negevent           -1.8492097 -0.58240139
posevent           -0.4388871  0.03049091
negevent:posevent   0.3137587  1.19645560

> Anova(mod.negpos)
Analysis of Deviance Table (Type II tests)

Response: numall
                  LR Chisq Df Pr(>Chisq)
negevent            4.0153  1  0.0450894 *
posevent            1.9081  1  0.1671724
negevent:posevent  11.4129  1  0.0007293 ***
```

The estimated model is $\log(\hat{\mu}) = 1.76 - 1.20$negevent$-0.20$posevent $+0.75$negevent$\times$posevent. An interesting feature here is the strength of the interaction term. The profile LR confidence interval for β_3 from `confint()` is $0.31 < \beta_3 < 1.20$, quite clearly excluding 0. The LRT from `Anova()` places the p-value for testing $H_0 : \beta_3 = 0$ at 0.0007. The left panel of Figure 4.6 shows a plot of the data shaded according to number of drinks, with darker shades indicating more drinks. Contours of the estimated mean from the model are overlaid. We see that for small values of the positive event index (e.g., < 1), the shades seem to get lighter for larger values of negative events. However, for larger values of the positive event index (e.g., > 2), the shades tend to be lighter for smaller values of negative events. This clearly depicts the nature of the interaction. The contours show this, too: they are increasing from left to right in the upper half of the plot and decreasing in the lower half. Without interaction, the contours would have been parallel lines. Notice also that the estimated means in the upper right corner of the plot, where we have no data, are obviously absurd. This serves as a reminder of the dangers of extrapolating a regression function beyond the range of the data.

The right panel shows the estimated mean curves for drinks against negative events at the three quartiles of positive events, 0.68, 1.03, and 1.43 (a technique suggested by Milliken and Johnson, 2001 for examining interactions involving continuous variables). The three curves show that when there are low positive events, there is a more acute decrease in drinks as negative events increase than when there are higher positive events.[5] We can also calculate the ratio of means corresponding to a 1-unit increase in negative events at each of the three positive-event quartiles. For a fixed value of positive events, say a, the ratio of means is

$$\frac{\mu(\text{negevent}+1, a)}{\mu(\text{negevent}, a)} = \exp(\beta_1 + a\beta_3).$$

---

[5]The code for these plots is in the accompanying program for this example. Also in the program is code to produce an interactive 3D plot of the same points and surface. We suggest rotating this surface in different directions to see how it fits the points.

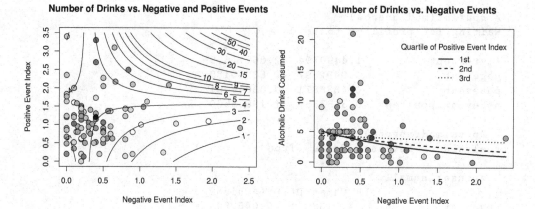

Figure 4.6: (Left) Plot of positive events vs. negative events, with the response variable, number of drinks, represented as increasingly dark shading with increasing drinks. Overlaid contours are from the estimated Poisson regression function of number of drinks against both variables and their interaction. (Right) Plot of number of drinks against negative events, with three mean curves from the same regression, estimated at the three quartiles of positive events.

These quantities are computed below, along with the corresponding percent changes, $100(\exp(\beta_1 + a\beta_3) - 1)$, and their confidence intervals:[6]

```
> posev.quart <- summary(saturday$posevent)
> posev.quart
   Min. 1st Qu.  Median    Mean 3rd Qu.    Max.
 0.0000  0.6833  1.0330  1.1580  1.4330  3.4000
> mean.ratio <- exp(mod.negpos$coefficients[2] +
+   posev.quart[c(2,3,5)]*mod.negpos$coefficients[4])
> mean.ratio
  1st Qu.    Median   3rd Qu.
0.5042493 0.6560323 0.8864268
> 100*(mean.ratio - 1)
   1st Qu.    Median   3rd Qu.
 -49.57507 -34.39677 -11.35732

> K <- matrix(data = c(0, 1, 0, 1*posev.quart[2],
                       0, 1, 0, 1*posev.quart[3],
                       0, 1, 0, 1*posev.quart[5]), nrow = 3, ncol
                    = 4, byrow = TRUE)
> linear.combo <- mcprofile(object = mod.negpos, CM = K)
> ci.beta <- confint(object = linear.combo, level = 0.95)
> 100*(exp(ci.beta$estimate) - 1)  # Verifies got same answer as
    above
    Estimate
```

---

[6]The computational process used by mcprofile() can sometimes produce slightly different confidence intervals in different runs of the same program. The differences are generally negligible, but can be eliminated by using set.seed() before mcprofile().

Table 4.1: Indicator variables for a 4-level categorical explanatory variable.

Level	x_2	x_3	x_4	Log-Mean
1	0	0	0	$\log(\mu_1) = \beta_0$
2	1	0	0	$\log(\mu_2) = \beta_0 + \beta_2$
3	0	1	0	$\log(\mu_3) = \beta_0 + \beta_3$
4	0	0	1	$\log(\mu_4) = \beta_0 + \beta_4$

```
C1 -49.57507
C2 -34.39677
C3 -11.35732
> 100*(exp(ci.beta$confint) - 1)
       lower      upper
C1 -68.15792 -23.150640
C2 -54.18113  -8.890579
C3 -36.66921  21.936802
```

We see that alcohol consumption decreases at the first quartile of positive events by nearly 50% per unit increase in negative events, and the corresponding confidence interval clearly excludes 0 ($-68\% < PC < -23\%$), while at the third quartile of positive events the increase is just 11% per unit, and the confidence interval does not exclude 0 ($-37\% < PC < 22\%$).

The next step in this analysis would be an assessment of the model fit. This is covered in Section 5.2.

4.2.3 Categorical explanatory variables

Categorical explanatory variables can be included in a Poisson regression model in a similar manner as with regression models for binary or multicategory responses. A categorical explanatory variable X with I levels is converted into $I - 1$ indicator variables, each indicating one of the levels of X. For later notational convenience, and to emulate the way that R parameterizes its indicator variables, we denote these variables as $x_2, x_3, \ldots, x_I$. Suppose for the moment that X has four levels, and that it is the only explanatory variable in the model. Then we can write the model for the mean as

$$\log(\mu) = \beta_0 + \beta_2 x_2 + \beta_3 x_3 + \beta_4 x_4.$$

This specifies a structure analogous to a 1-way ANOVA model as described by Table 4.1.

Table 4.1 also demonstrates that we can write the model as $\log(\mu_i) = \beta_0 + \beta_i x_i$ for level $i = 2, \ldots, I$ of X. This leads immediately to the interpretation of the parameters and makes clear our choice of index labels: β_0 is the log-mean for level 1 of X; β_2 is the difference in log-means between levels 2 and 1; and other parameters are interpreted similarly as β_2. The regression parameters $\beta_2, \ldots, \beta_q$ are often called *effect parameters* for the categorical variable X. Notice that $\beta_2 = \log(\mu_2) - \log(\mu_1)$, which implies that $e^{\beta_2} = \mu_2/\mu_1$, with a similar interpretation for β_3 and β_4. Thus, the effect parameters are very useful for comparing means at different levels against the mean at the first level. Furthermore, the difference between any two effect parameters also measures a ratio between two means. For example, $\log(\mu_2) - \log(\mu_3) = \beta_2 - \beta_3$, so that $\mu_2/\mu_3 = e^{\beta_2 - \beta_3}$. If we choose to write $\beta_1 = 0$ for convenience, then all ratios between pairs of means are of the form $\mu_i/\mu_{i'} = e^{\beta_i - \beta_{i'}}$.

Inferences typically focus on estimating the means for each level of X and comparing those means in some chosen way. Means are estimated as $\hat{\mu}_i = e^{\hat{\beta}_0 + \hat{\beta}_i}$, $i = 1, \ldots, q$. Confidence intervals for individual means are found by exponentiating confidence intervals for corresponding sums of regression parameters, $\beta_0 + \beta_i$.

It is common to test $H_0 : \mu_1 = \mu_2 = \ldots = \mu_q$, which is equivalent to $H_0 : \beta_2 = \ldots = \beta_q = 0$. This hypothesis is easily tested using a LRT. Further comparisons among the means are performed using tests or confidence intervals on combinations of parameters. For example, as noted above, $\mu_i/\mu_{i'} = e^{\beta_i - \beta_{i'}}$, so a confidence interval for the difference between two effect parameters can be exponentiated into a confidence interval for the ratio of their means.

Profile LR methods are preferred for confidence intervals. The `confint.glm()` method function can find confidence intervals for individual parameters, thereby allowing comparisons directly between μ_1 and any other mean. To compare other means, one can re-order levels of the categorical explanatory variable, re-estimate the model, and use `confint()` again (see Section 2.2.6). More generally, LR confidence intervals for linear combinations of parameters are available from the `mcprofile` package.

Tests for individual means or ratios are carried out using the same approach as for confidence intervals, by creating linear combinations of the corresponding model parameters. Various packages in R have the capability to produce these tests (e.g., the `deltaMethod()` function in `car`, the `glht()` function in `multcomp`, or the `summary()` function in `mcprofile`). Tests can also be carried out informally by "inverting" the confidence interval (e.g., determining whether a confidence interval for $\beta_2 - \beta_3$ contains 0), although this does not produce a p-value.

Notice that we continually work with ratios of means rather than differences. This is because the log link leads very naturally to computing differences in the log scale, which leads to ratios of the means. The regression parameter estimators are approximately normally distributed and have variances that are relatively easy to estimate as described in Appendix B.3.4. Linear combinations of normal random variables are also normal random variables, so that linear combinations of parameter estimators are also approximately normally distributed. Therefore, confidence intervals for certain functions of linear combinations of parameters, such as $e^{\beta_i - \beta_{i'}}$, are easy to develop and compute.

On the other hand, functions of means that cannot be reduced to linear combinations of parameters, such as $(\mu_2+\mu_3)/2 = (e^{\beta_0+\beta_2}+e^{\beta_0+\beta_3})/2$ and $\mu_2-\mu_3 = e^{\beta_0+\beta_2}-e^{\beta_0+\beta_3}$, are less convenient to work with. The `deltaMethod()` function in `car` allows estimation of nonlinear functions of the parameters, by applying the delta method to find an approximate standard error for the function (see Appendix B.4.2). The results can then be used to perform Wald inferences manually.

Example: Ecuadorean Bird Counts (BirdCountPoReg.R, BirdCounts.csv)

> Clearing of forests for farmland usually results in a change in the composition of plant and animal life in the area. In a study of bird species inhabiting cloud forest regions of Ecuador, researchers used mist nets to capture and identify birds at six different locations. Two locations were within undisturbed forest, one was in a fragment of forest surrounded by cleared land, one was on the edge between the forest and cleared land, and two were in pasture that had previously been forested. Several netting visits were made to each location over three years. For each bird captured, researchers recorded the species, gender, and various other measurements. See Becker et al. (2008) for details on the study. The total counts of all birds from each visit are listed in Table 4.2.[7]

---

[7] Data kindly provided by Dr. Dusti Becker, International Conservation Program Coordinator at Life Net Nature (http://www.lifenetnature.org).

Table 4.2: Data on bird counts at 6 locations in Ecuador.

Location	Counts
Forest A	155, 84
Forest B	77, 57, 38, 40, 51, 67
Edge	53, 49, 44, 47
Fragment	35, 50, 66, 51, 49, 75
Pasture A	40, 39
Pasture B	33, 31, 38, 50

We are interested in comparing overall abundance of birds across different types of habitat. To do this, we fit a Poisson regression model to estimate the mean count of bird captures at each location as an index of the overall abundance, and compare these means across different types of habitats. Thus, we use glm() to estimate the six means using 5 indicator variables indicating the last five locations in alphabetical order. The data are entered and these indicator variables are displayed below. Notice that the results from contrasts() show a form identical to Table 4.1.

```
> alldata <- read.table(file = "C:\\Data\\BirdCounts.csv", sep =
   ",", header = TRUE)
> head(alldata)
  Loc Birds
1 ForA   155
2 ForA    84
3 ForB    77
4 ForB    57
5 ForB    38
6 ForB    40
> contrasts(alldata$Loc)
     ForA ForB Frag PasA PasB
Edge    0    0    0    0    0
ForA    1    0    0    0    0
ForB    0    1    0    0    0
Frag    0    0    1    0    0
PasA    0    0    0    1    0
PasB    0    0    0    0    1
```

Next, the model is fit and the summary is obtained.

Because $\hat{\beta}_0 = 3.88$, the estimated mean count for the edge habitat is $\hat{\mu}_1 = e^{3.88} = 48.3$. The remaining $\hat{\beta}_i$'s, $i = 2, \ldots, 6$ measure the estimated difference in log-means between the given location and edge, $\log(\hat{\mu}_i) - \log(\hat{\mu}_1)$, $i = 2, \ldots, 6$. For example, the difference in log-means between fragment and edge is represented by $\hat{\beta}_4 = 0.12$. Thus, the ratio of these two means is $\hat{\mu}_4/\hat{\mu}_1 = e^{0.11874} = 1.126$.

We give code in the corresponding program for this example to show that all predicted means from this model are the same as their respective sample means. This happens because we are allowing the model to estimate a separate mean for each habitat, and the MLEs in this case can be shown to be the respective sample means. Thus, in a sense it seems that we are going to a great deal of extra effort to do something that should be easier to do. The advantage of the model-based approach lies in the numerous inference procedures that accompany the model fit. We show several of these next.

To test whether all means are equal,

$$H_0 : \mu_1 = \mu_2 = \ldots = \mu_6$$
$$H_a : \text{Not all means are equal,}$$

we use anova() to perform the LRT (Anova() could be used as well):

```
> anova(M1, test = "Chisq")
Analysis of Deviance Table
Model: poisson, link: log
Response: Birds
Terms added sequentially (first to last)
     Df Deviance Resid. Df Resid. Dev P(>|Chi|)
NULL                    23     216.944
Loc   5   149.72        18      67.224 < 2.2e-16 ***
```

This null hypothesis is strongly rejected ($-2\log(\Lambda) = 67.2$ and p-value is less than 2.2×10^{-16}), so clearly not all locations have the same mean. In order to understand the cause of this result, a plot of the means and corresponding confidence intervals can help.

We show next how to compute estimated means and corresponding LR confidence intervals using mcprofile(). We start by specifying the coefficient matrix K for estimating each $\log(\hat{\mu}_i) = \hat{\beta}_0 + \hat{\beta}_i$. The rows are arranged in order of decreasing forest habitat: forests, fragment, edge, pastures.

```
> K <- matrix(data = c(1, 1, 0, 0, 0, 0,
                       1, 0, 1, 0, 0, 0,
                       1, 0, 0, 1, 0, 0,
                       1, 0, 0, 0, 0, 0,
                       1, 0, 0, 0, 1, 0,
                       1, 0, 0, 0, 0, 1), nrow = 6, ncol = 6,
                       byrow = TRUE)
> K
     [,1] [,2] [,3] [,4] [,5] [,6]
[1,]    1    1    0    0    0    0
[2,]    1    0    1    0    0    0
[3,]    1    0    0    1    0    0
[4,]    1    0    0    0    0    0
[5,]    1    0    0    0    1    0
[6,]    1    0    0    0    0    1
> linear.combo <- mcprofile(object = M1, CM = K)
> ci.log.mu <- confint(object = linear.combo, level = 0.95,
    adjust = "none")

> mean.LR.ci1 <- data.frame(Loc = pred.data, Estimate =
    exp(ci.log.mu$estimate), Lower = exp(ci.log.mu$confint[,1]),
    Upper = exp(ci.log.mu$confint[,2]))
> mean.LR.ci1
    Loc  Estimate     Lower     Upper
C1 ForA 119.50000 104.98310 135.29716
```

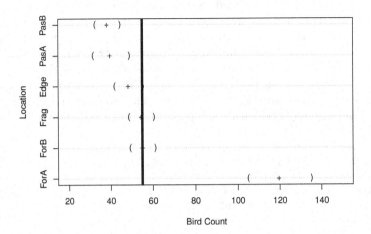

Figure 4.7: Plot of MLEs for mean bird counts, denoted by "+", and LR confidence intervals, denoted by parentheses, for mist-netted bird captures at 6 locations in Ecuador. The vertical line represents the overall mean of all counts.

```
C2  ForB  55.00000  49.27736  61.14941
C3  Frag  54.33333  48.64676  60.44668
C4  Edge  48.25000  41.75903  55.38108
C5  PasA  39.50000  31.41702  48.86316
C6  PasB  38.00000  32.27465  44.36544
>
> mean.LR.ci1$Loc2 <- factor(mean.LR.ci1$Loc, levels =
    levels(mean.LR.ci1$Loc)[c(2,3,4,1,5,6)])

> # StripChart in color
> x11(width = 7, height = 5, pointsize = 12)
> stripchart(Lower ~ Loc2, data = mean.LR.ci1, vertical = FALSE,
    xlim = c(20,150), col = "red", pch = "(", main = "", xlab =
    "Bird Count", ylab = "Location")
> stripchart(Upper ~ Loc2, data = mean.LR.ci1, vertical = FALSE,
    col = "red", pch = ")", add = TRUE)
> stripchart(Mean ~ Loc2, data = mean.LR.ci1, vertical = FALSE,
    col = "red", pch = "+", add = TRUE)
> grid(nx = NA, ny = NULL, col = "gray", lty = "dotted")
> abline(v = mean(alldata$Birds), col = "darkblue", lwd = 4)
```

The plot in Figure 4.7 shows that there is a general increasing trend in the mean counts as the amount of forestation increases. It is also apparent that one forest location has a mean that is drastically different from the other locations. Forests were expected to have higher mean counts, and those means were not necessarily expected to be identical in all forests. This shows that the variability from one location to the next can be very large.

Comparing the two forests or the two pastures to each other is actually not of interest. Instead, we want to combine the means for each pair of locations of the same habitat before comparing them to other habitats. It would be inappropriate to combine their data and estimate a single mean for the two locations of the same habitat, because the data from pairs of locations do not necessarily come from the same Poisson distribution. Instead, we estimate their means separately, and then combine the means to make comparisons. To carry out inferences, we need to develop a matrix of coefficients for the linear combinations of parameters that represent the comparisons being made. For example, to compare the two forest habitats against the fragment, we can compare the ratio for the geometric mean of the two forest means against the fragment mean, i.e., $(\mu_2\mu_3)^{1/2}/\mu_4$.[8] The log of this quantity in terms of the regression parameters is $(\beta_2 + \beta_3)/2 - \beta_4$. The coefficients are then (0, 0.5, 0.5, -1, 0, 0). Similar sets of coefficients comparing each pair of habitats are given below:

```
> contr.mat <- rbind(c(0,.5,.5,0,0,0), c(0,.5,.5,-1,0,0),
    c(0,.5,.5,0,-.5,-.5), c(0,0,0,1,0,0), c(0,0,0,1,-.5,-.5),
    c(0,0,0,0,-.5,-.5)))
> rownames(contr.mat) <- c("For-Edge", "For-Frag", "For-Past",
    "Frag-Edge", "Frag-Past", "Edge-Past")
> contr.mat
          [,1] [,2] [,3] [,4] [,5] [,6]
For-Edge     0  0.5  0.5    0  0.0  0.0
For-Frag     0  0.5  0.5   -1  0.0  0.0
For-Past     0  0.5  0.5    0 -0.5 -0.5
Frag-Edge    0  0.0  0.0    1  0.0  0.0
Frag-Past    0  0.0  0.0    1 -0.5 -0.5
Edge-Past    0  0.0  0.0    0 -0.5 -0.5
```

Tests and confidence intervals for these linear combinations are easily obtained using `mcprofile()` for LR intervals and the `multcomp` or `mcprofile` package for Wald (see Exercise 20 from Chapter 2 for an example using this package). We show the use of `mcprofile()` below. This is followed by `summary()` for tests and `confint()` for confidence intervals. The default is to adjust the results to account for multiple inferences, which ensures that the probability of *no* type I errors or coverage failures in the entire family of inferences is maintained to be *at least* $1 - \alpha$ (see the discussion on p. 84). These defaults can be skirted if desired by specifying additional arguments as shown below. However, it is considered good practice to present both multiplicity-adjusted and unadjusted results of an analysis. The adjusted results are more conservative, but any statistically significant findings are more likely to hold up against future investigations than those found only without adjustments. See Westfall and Young (1993) for a discussion of the importance of simultaneous error rates and coverages in multiple inference.

```
> linear.combo <- mcprofile(object = M1, CM = contr.mat)
> summary(linear.combo)
    mcprofile - Multiple Testing
Adjustment:     single-step
Margin:         0
Alternative:    two.sided
              Estimate Statistic Pr(>|z|)
For-Edge == 0     0.52      6.45   <0.001 ***
For-Frag == 0     0.40      5.81   <0.001 ***
For-Past == 0     0.74      9.54   <0.001 ***
```

---

[8]The geometric mean of $a_1,\ldots,a_n$ is $\left(\prod_{i=1}^{n} a_i\right)^{1/n}$, so that the log of the geometric mean is $n^{-1}\sum_{i=1}^{n}\log(a_i)$.

```
Frag-Edge == 0        0.12      1.31     0.549
Frag-Past == 0        0.34      3.86     0.001 ***
Edge-Past == 0        0.22      2.19     0.123
```

```
> exp(confint(linear.combo)$confint)
       lower     upper
1  1.3613529  2.089107
2  1.2495640  1.786597
3  1.7051334  2.586840
4  0.8943398  1.424436
5  1.1195967  1.764523
6  0.9637353  1.608817

> summary(linear.combo, adjust = "none")
   mcprofile - Multiple Testing
Adjustment:      none
Margin:          0
Alternative:     two.sided
                 Estimate  Statistic  Pr(>|z|)
For-Edge  == 0      0.52       6.45    <2e-16 ***
For-Frag  == 0      0.40       5.81    <2e-16 ***
For-Past  == 0      0.74       9.54    <2e-16 ***
Frag-Edge == 0      0.12       1.31     0.189
Frag-Past == 0      0.34       3.86    <2e-16 ***
Edge-Past == 0      0.22       2.19     0.028 *

> exp(confint(linear.combo, adjust = "none")$confint)
       lower     upper
1  1.4292924  1.983755
2  1.3023075  1.712299
3  1.7878645  2.459671
4  0.9436183  1.347467
5  1.1798823  1.671273
6  1.0235287  1.515020
```

From the adjusted tests or intervals (the top two sets of results), we see that in four of the six comparisons the ratios show differences that reflect the expectations of the researchers. Forests have a mean that is larger than edge, fragment, and pasture habitats (rows 1, 2, and 3, respectively), with an especially large ratio compared to pasture. That estimated ratio is exp(0.74) = 2.09, with a confidence interval from 1.71 to 2.59, suggesting roughly twice as many birds in the forest than in the pasture. Continuing to examine other comparisons, we see that the fragment had more birds than the pastures. The ratios of means for fragment/edge and edge/pasture are not significantly different from 1 (their logs are not significantly different from 0) according to the adjusted results. The unadjusted results suggest that the edge/pasture ratio may actually be different from 1. However, this conclusion cannot be made as strongly as the others because, without the multiplicity adjustments, it comes from a set of tests whose chance of making any type I errors is inflated relative to α.

Wald inferences provide nearly identical confidence intervals and the same interpretation. We show both the `glht()` results from `multcomp` and manual calculations in the program corresponding to this example. For the manual calculations, we need to estimate the linear combinations of the parameters, estimate the corresponding variances of these linear combinations, and combine these in the usual way to make

Table 4.3: Indicator variables and log-means from a loglinear model for a 2×2 contingency table.

X	Z	x_2	z_2	Log-Mean
1	1	0	0	$\log(\mu_{11}) = \beta_0$
1	2	0	1	$\log(\mu_{12}) = \beta_0 + \beta_2^Z$
2	1	1	0	$\log(\mu_{21}) = \beta_0 + \beta_2^X$
2	2	1	1	$\log(\mu_{22}) = \beta_0 + \beta_2^X + \beta_2^Z$

interval endpoints. The calculations are carried out easily using matrix methods. The program also shows how to use `deltaMethod()` from the `car` package to estimate the *difference*, rather than the ratio, between forest and fragment means.

4.2.4 Poisson regression for contingency tables: loglinear models

A special use for categorical explanatory variables in a Poisson regression model is to model counts from contingency tables. Practically any analysis that one might normally conduct using traditional methods for contingency tables—such as those described in Sections 1.2 and 3.2—can be carried out using Poisson regression. Tools that are available within the generalized linear model framework to perform additional analyses make it an appealing alternative to traditional table-based methods. Because Poisson regression on contingency tables is so common, it has an alternative name—the *loglinear model*—which stems from the use of the log link for the mean. Note that the loglinear model can be shown to be largely equivalent to the multinomial regression models discussed in Section 3.3.2. We will compare advantages and disadvantages of the two approaches at the end of the next section.

Using Poisson regression models for a two-dimensional contingency table structure requires nothing more than to treat both the row variable and the column variable as categorical explanatory variables. Poisson regression is then applied, using the resulting sets of indicator variables together in one model. This process is very similar to ANOVA modeling of continuous responses. However, some explanation of the details is needed to see how the model reproduces traditional statistics.

Let X represent the row variable with I levels and Z represent the column variable with J levels. We use the indicator variables $x_2, \ldots, x_I$ for X and $z_2, \ldots, z_J$ for Z within our Poisson regression model:

$$\log(\mu) = \beta_0 + \beta_2^X x_2 + \beta_3^X x_3 + \ldots + \beta_I^X x_I + \beta_2^Z z_2 + \beta_3^Z z_3 + \ldots + \beta_J^Z z_J, \quad (4.2)$$

where β_i^X, $i = 2, \ldots, I$, and β_j^Z, $j = 2, \ldots, J$ are ordinary regression effect parameters with the superscript added for convenience. Thus, the model for the mean in cell (i, j) of the contingency table can be written as

$$\log(\mu_{ij}) = \beta_0 + \beta_i^X + \beta_j^Z, \ i = 1, \ldots, I, \ j = 1, \ldots, J, \quad (4.3)$$

where is it implicit that $\beta_1^X = \beta_1^Z = 0$.

To see what these parameters measure, we evaluate Equations 4.2 and 4.3 for a 2×2 contingency table in Table 4.3. We see that

$$\log(\mu_{21}) - \log(\mu_{11}) = \log(\mu_{22}) - \log(\mu_{12}) = \beta_2^X,$$

so β_2^X measures the difference in log-mean between the two counts in the same column from rows 2 and 1. Thus, $\exp(\beta_2^X)$ measures the ratio between any two such means. Similarly,

$$\log(\mu_{12}) - \log(\mu_{11}) = \log(\mu_{22}) - \log(\mu_{21}) = \beta_2^Z,$$

so β_2^Z measures the difference in log-mean between the two counts in the same row from columns 2 and 1, and $\exp(\beta_2^Z)$ measures the ratio of those means. In larger tables, β_i^X measures the difference in log-means (i.e., the log of the ratio of means) between counts in rows i and 1 in the same column. An analogous interpretation holds for β_j^Z. It can therefore be shown that these parameters also measure differences in log-means, or log-ratios of means, in the *margins* of the table: $\beta_i^X = \log(\mu_{i+}/\mu_{1+})$ (see Exercise 8).

There are no parameters in Equation 4.2 that allow individual cells in an $I \times J$ table to deviate from what their row and column margins dictate. This corresponds to *independence* between X and Z (see Section 3.2). Another way to see this is to consider the odds ratio between any two rows and columns. For example, suppose we have a 2×2 table with cell probabilities π_{ij}, $i = 1, 2; j = 1, 2$, as in Section 3.2.1. Using the equivalence between the Poisson and multinomial models discussed in Section 4.2.5, we can denote the Poisson regression model's mean count in cell (i, j) by $\mu_{ij} = n\pi_{ij}$. It is then fairly easy to show that

$$OR = \frac{\mu_{11}\mu_{22}}{\mu_{21}\mu_{12}}.$$

Equation 4.2 implies that

$$\begin{aligned}\log(OR) &= \log(\mu_{11}) + \log(\mu_{22}) - \log(\mu_{21}) - \log(\mu_{12}) \\ &= \beta_0 + (\beta_0 + \beta_2^X + \beta_2^Z) - (\beta_0 + \beta_2^X) - (\beta_0 + \beta_2^Z) \\ &= 0\end{aligned}$$

Thus, $OR = 1$. The same thing occurs in models for larger tables regardless of which two rows or columns are chosen. That is, the loglinear model represented by Equation 4.2 represents independence between X and Z. The parameters in this model serve only to make the table of predicted counts have the same margins as the table of observed counts. In fact, the predicted counts from Equation 4.2 are exactly the same as the expected counts we computed in Section 3.2 for testing independence in a contingency table.

Because inference on margins is not typically our main concern in a contingency table, we need to augment Equation 4.2 to allow for association between X and Z. This can be accomplished by adding to the model crossproduct terms of the form $\beta_{ij}^{XZ} x_i z_j$, $i = 2, \ldots, I; j = 2, \ldots, J$, to create an interaction between X and Z. Because there are $I - 1$ indicators x_i and $J - 1$ indicators z_j, there are $(I-1)(J-1)$ new terms to be added to the model. This creates a saturated model, because the total number of parameters is equal to the number of cell means being estimated. To see this, note first that there are IJ independently sampled counts in the table. The intercept β_0 accounts for 1 degree of freedom (df). The effect parameters for X, $\beta_2^X, \ldots, \beta_I^X$, account for $I - 1$ df, while those for Z account for an additional $J - 1$. With the interaction terms, the new model consumes $1 + (I - 1) + (J - 1) + (I - 1)(J - 1) = IJ$ df.

The new model can be written as

$$\log(\mu_{ij}) = \beta_0 + \beta_i^X + \beta_j^Z + \beta_{ij}^{XZ}. \tag{4.4}$$

The effect parameters β_i^X and β_j^Z again operate on the marginal counts, but now the XZ-interaction parameters β_{ij}^{XZ}, $i = 1, \ldots, I, j = 1, \ldots, J$, allow the ratios of means between cells in two rows to change across columns and vice versa. To see this, consider again the 2×2 table, and recall that we set $\beta_{ij}^{XZ} = 0$ when either $i = 1$ or $j = 1$. Then

$$\begin{aligned}
\log(OR) &= \log(\mu_{11}) + \log(\mu_{22}) - \log(\mu_{21}) - \log(\mu_{12}) \\
&= (\beta_0 + \beta_1^X + \beta_1^Z + \beta_{11}^{XZ}) + (\beta_0 + \beta_2^X + \beta_2^Z + \beta_{22}^{XZ}) \\
&\quad - (\beta_0 + \beta_2^X + \beta_1^Z + \beta_{21}^{XZ}) - (\beta_0 + \beta_1^X + \beta_2^Z + \beta_{12}^{XZ}) \\
&= \beta_{11}^{XZ} + \beta_{22}^{XZ} - \beta_{21}^{XZ} - \beta_{12}^{XZ} \\
&= \beta_{22}^{XZ}.
\end{aligned}$$

Thus, the odds ratio in a 2×2 table is just $\exp(\beta_{22}^{XZ})$. More generally, the odds ratio between rows i and i' and columns j and j' is

$$OR_{ii',jj'} = \exp(\beta_{ij'}^{XZ} + \beta_{i'j'}^{XZ} - \beta_{i'j}^{XZ} - \beta_{ij'}^{XZ}), \tag{4.5}$$

where the extra subscripts remind us that there is more than one possible odds ratio. Because Equation 4.4 is a saturated model, its estimated means $\hat{\mu}_{ij}$ match the observed counts y_{ij} exactly. Therefore, odds ratios estimated from Equation 4.5 are exactly the same as those based on the corresponding counts in the observed table.

Tests for independence in contingency tables that were described in Section 3.2.3 can be replicated exactly from the Poisson regression model through model comparison tests. The null hypothesis of independence implies Equation 4.2, while the alternative implies the unrestricted model in Equation 4.4. Fitting both models to the tabulated counts y_{ij}, $i = 1, \ldots, I$, $j = 1, \ldots, J$, and comparing residual deviances provides the LRT statistic. In this case, the statistic is also just the residual deviance from the null model, because the alternative model is saturated, so its residual deviance is zero. This statistic can be shown to be exactly the same as the LRT statistic presented in Section 3.2 for testing independence. The degrees of freedom for the χ^2 sampling distribution, $(I-1)(J-1)$, come from the number of additional parameters in the full model compared to the null model (not counting those set to zero). Odds ratio estimates and confidence intervals are found by exponentiating estimates and confidence intervals for linear combinations of the XZ interaction parameters as determined by Equation 4.5.

It should be noted that the Pearson statistic presented in Section 3.2.3 is generally preferred to the LR statistic for testing independence in an $I \times J$ contingency table (see also the discussion in 1.2.3). However, the LR statistic is easier to adapt for tests in other models, as we discuss in Sections 4.2.5 and 4.2.6.

Example: Larry Bird's free throw shooting (LarryBirdPoReg.R)

Here we use a loglinear model to re-analyze the data on Larry Bird's free throw shooting that was previously used in Chapter 1. We show that the odds ratio estimate and Wald confidence interval are exactly the same as those given in Section 1.2.5. We further provide a LR confidence interval for the odds ratio.

Corresponding to Table 1.2, there are two variables, `First` and `Second`, each with levels "made" and "missed." For this example, we make use of the object `c.table` created in Chapter 1, and convert the object into a data frame:

```
> all.data <- as.data.frame(as.table(c.table))
> all.data
   First Second Freq
1   made   made  251
2 missed   made   48
```

```
3     made   missed     34
4     missed missed      5
```

The counts are contained in the variable Freq. We estimate the saturated model $\log(\mu_{ij}) = \beta_0 + \beta_i^F + \beta_j^S + \beta_{ij}^{FS}$, where μ_{ij} is the mean count in cell (i, j), $i = 1, 2$, $j = 1, 2$, and superscript index F stands for "First" and S for "Second." The model is estimated by using the formula argument value Freq ~ First*Second within glm(). It is important to note in the context of Table 4.3 that level 1 for both variables is represented by "made" and level 2 by "missed." Thus β_0 is the log-mean count of "made-made" sequences, β_2^F is the log-ratio of mean counts for first-missed to first-made when the second is made, and β_2^S is the log-ratio of mean counts for second-missed to second-made when the first is made. The interaction parameter β_{22}^{FS} is the log of the ratio of [the odds of *missing* the second given that the first was *missed*] to [the odds of *missing* the second given that the first was *made*]. This is equivalent to the log of the ratio of [the odds of *making* the second given that the first was *made*] to [the odds of *making* the second given that the first was *missed*]. The summary of the model fit is below:

```
> M1 <- glm(formula = Freq ~ First*Second, family = poisson(link
    = "log"), data = all.data)
> summary(M1)

<OUTPUT EDITED>

Coefficients:
                          Estimate Std. Error z value Pr(>|z|)
(Intercept)                5.52545    0.06312  87.540   <2e-16 ***
Firstmissed               -1.65425    0.15754 -10.501   <2e-16 ***
Secondmissed              -1.99909    0.18275 -10.939   <2e-16 ***
Firstmissed:Secondmissed  -0.26267    0.50421  -0.521    0.602

    Null deviance: 4.0206e+02  on 3  degrees of freedom
Residual deviance: 4.3077e-14  on 0  degrees of freedom
AIC: 29.926
```

The estimated parameters are $\hat{\beta}_0 = 5.53$, $\hat{\beta}_2^F = -1.65$, $\hat{\beta}_2^S = -2.00$, and $\hat{\beta}_{22}^{FS} = -0.26$. Table 4.3 shows how to estimate each mean count according to the model parameters. Because the model is saturated, these match the observed counts exactly. For example, $\hat{\mu}_{12} = \exp(5.53 - 2.00) = 34$ is exactly the same as the observed count for the first-made, second-missed outcome.

We follow up with the usual analysis functions to estimate the odds ratio and find confidence intervals:

```
> round(exp(M1$coefficients[4]), digits = 2)
Firstmissed:Secondmissed
                    0.77
> inter <- confint(M1, parm = "Firstmissed:Secondmissed")
Waiting for profiling to be done...
> round(exp(inter), digits = 2)
 2.5 %  97.5 %
  0.25    1.91
```

```
> round(exp(confint.default(M1, parm =
    "Firstmissed:Secondmissed")), digits = 2)
                          2.5 %  97.5 %
Firstmissed:Secondmissed   0.29   2.07
```

The estimated odds ratio is $\widehat{OR} = 0.77$, so there is a smaller estimated odds of missing the second shot after a missed first shot than after a made first shot. Equivalently, the probability of success is higher on the second shot after a missed first shot. The profile LR confidence interval from `exp(confint())` is $0.25 < OR < 1.91$, so with 95% confidence, the odds of success on the second shot are between 1/4 to almost two times as large following a made first shot compared to a missed first. For comparison purposes, the Wald confidence interval calculated by `confint.default()` is $0.29 < OR < 2.07$, which matches what was given in Section 1.2.5.

Finally, the LRT for $H_0 : \beta_{22}^{FS} = 0$, $H_a : \beta_{22}^{FS} \neq 0$, which is equivalent to $H_0 : OR = 1$, $H_a : OR \neq 1$, is provided using the `Anova()` function:

```
> library(car)
> Anova(M1)
Analysis of Deviance Table (Type II tests)

Response: Freq
              LR Chisq Df Pr(>Chisq)
First          174.958  1     <2e-16 ***
Second         226.812  1     <2e-16 ***
First:Second     0.286  1      0.593
```

The test statistic is $-2\log(\Lambda) = 0.29$ with a p-value of 0.59, indicating that there is insufficient evidence to conclude that the odds (and hence probability) of success for the second free throw depend on the result of the first. Note that this p-value is exactly the same as for the LRT performed in Section 1.2.3.

4.2.5 Large loglinear models

Loglinear models can be extended to model counts cross-classified by arbitrarily many categorical variables. A contingency table of counts corresponding to the cross-classification of p categorical variables is called a *p-way table*. For example, consider the three-way contingency table consisting of three categorical variables, X, Z, and W with I, J and K levels, respectively. We can extend our loglinear model by adding a subscript for variable W to the means and observed counts: y_{ijk} has mean μ_{ijk}, $i = 1, \ldots, I$; $j = 1, \ldots, J$; $k = 1, \ldots, K$. The independence model in Equation 4.2 is expanded to include indicator variables for W— $w_2, w_3, \ldots, w_K$—defined analogously to those for X and Z. Their corresponding regression parameters, $\beta_2^W, \ldots, \beta_K^W$, affect the marginal counts of a fitted three-way table (e.g., K different two-way tables summarizing the counts for X and Z at each value of W), so that ratios of means from levels k and k' of W are constant. Thus, $\log(\mu_{ijk}/\mu_{ijk'}) = \exp(\beta_k^W - \beta_{k'}^W)$, across all levels of X and Z. This model also represents independence among all three variables, because any odds ratio between two levels of one variable and two levels of another variable is 1. Because of this, it is sometimes called the *mutual independence* model.

Models that allow for associations among the three variables are developed depending on what associations are of interest or expected to be present. Some of these are summarized in

Table 4.4: Common models and their associated odds ratios for a three-way loglinear model with variables $X, Z,$ and W.

Model	Notation	XZ Odds Ratio at level k of W
Mutual Independence	X, Z, W	1 for all k
Homogeneous Association	XZ, XW, ZW	$\exp(\beta_{ij}^{XZ} + \beta_{i'j'}^{XZ} - \beta_{i'j}^{XZ} - \beta_{ij'}^{XZ})$ for all k
Saturated (Heterogeneous Association)	XZW	$(\beta_{ij}^{XZ} + \beta_{i'j'}^{XZ} - \beta_{i'j}^{XZ} - \beta_{ij'}^{XZ})$ $+(\beta_{ijk}^{XZW} + \beta_{i'j'k}^{XZW} - \beta_{i'jk}^{XZW} - \beta_{ij'k}^{XZW})$

Table 4.4. There are now three pairwise interactions to consider: XZ, XW, and ZW. Each is formed by adding products of the indicator variables to the regression model as described on p. 219, and each set of interaction parameters allows odds ratios for the corresponding pair of variables to be something other than 1. If all three pairwise interactions are included in the model, we have

$$\log(\mu_{ijk}) = \beta_0 + \beta_i^X + \beta_j^Z + \beta_k^W + \beta_{ij}^{XZ} + \beta_{ik}^{XW} + \beta_{jk}^{ZW}. \tag{4.6}$$

To see the corresponding odds ratios for this model, consider forming OR for levels i, i' of X and j, j' of Z at level k of W:

$$OR_{ii',jj'(k)} = \frac{\mu_{ijk}\mu_{i'j'k}}{\mu_{i'jk}\mu_{ij'k}}.$$

Applying Equation 4.6, we have

$$\log(OR_{ii',jj'(k)}) = \log(\mu_{ijk}) + \log(\mu_{i'j'k}) - \log(\mu_{i'jk}) - \log(\mu_{ij'k})$$
$$= \beta_{ij}^{XZ} + \beta_{i'j'}^{XZ} - \beta_{i'j}^{XZ} - \beta_{ij'}^{XZ} \tag{4.7}$$

All other parameters cancel except those involved in the interaction from which the odds ratio of interest is computed. The same happens with any pair of categorical variables at any pairs of levels. Therefore, this model allows all odds ratios corresponding to any pair of variables to vary freely across levels of those two variables. Notice, however, that a given odds ratio is forced to be constant across levels of the third factor. For example, the odds ratio above is $\exp(\beta_{ij}^{XZ} + \beta_{i'j'}^{XZ} - \beta_{i'j}^{XZ} - \beta_{ij'}^{XZ})$, regardless of the level k of W. Thus, this model is referred to as the *homogeneous association* model. Versions of this model can be considered in which only one or two of the interactions are present, allowing associations to be modeled between some pairs of variables and assuming independence between others. In all cases, tests and estimates are obtained using the same tools as before.

Finally, interactions of higher order are possible. The three-variable interaction allows associations between any pair of variables to vary across levels of the third. Terms of the form β_{ijk}^{XZW} are added to Equation 4.6, and these terms enter into any odds ratios. For example, the odds ratio for levels i and i' of X and levels j and j' of Z at level k of W becomes

$$\log(OR_{ii',jj'(k)}) = \log(\mu_{ijk}) + \log(\mu_{i'j'k}) - \log(\mu_{i'jk}) - \log(\mu_{ij'k})$$
$$= \beta_{ij}^{XZ} + \beta_{i'j'}^{XZ} - \beta_{i'j}^{XZ} - \beta_{ij'}^{XZ} +$$
$$\beta_{ijk}^{XZW} + \beta_{i'j'k}^{XZW} - \beta_{i'jk}^{XZW} - \beta_{ij'k}^{XZW}, \tag{4.8}$$

which shows that the XZ odds ratios have the potential to vary depending on the level k of W. While this form may seem complicated, it is predictable and not difficult to estimate in R.

In models with more variables, terms of even higher order are technically possible. Sets of interaction parameters, indexed by the names and levels of the variables, can be created to correspond to any combination of variables. However, higher-order interactions are difficult to interpret. A blanket interpretation can be developed recursively as follows: a 3-variable interaction allows 2-variable interactions (i.e., odds ratios) to vary depending on the level of the third variable; a 4-variable interaction allows 3-variable interactions to vary depending on the level of the fourth variable; and so on. Understanding what these statements imply in the context of a particular problem is a different matter. Careful tabulation or graphing of odds ratios across levels of other variables can help to extract the underlying meaning from a higher-order interaction.

Often the goal of an analysis is to determine which associations, if any, are important, and to subsequently estimate these associations. Generally it is accepted that the main-effect terms that control the marginal counts should remain in the model when 2-variable and higher-order interaction terms are the focus of study.[9] Furthermore, the principle of *marginality* dictates that all lower-order terms that are subsets of any higher-order terms should also remain in the model. For instance, the XZW interaction contains XZ, XW, and ZW, as well as X, Z, and W. If a model contains XZW then it also contains these other six terms. Following this convention, we can use abbreviated notation that identifies any model by listing only the highest-order terms in the model. Our current example is denoted as "Model XZW." This also happens to correspond to the terms that need to be specified in the `formula` argument of `glm()` in order to fit the model. For example, `formula = count ~ X*Z*W` would fit this model, where `count` denotes a cell count. As another example, "Model X, ZW" would consist of terms for X, Z, W, and ZW, specifying independence between X and W and between X and Z, but allowing for association between Z and W that is homogeneous across levels of X. This model could be fit using `formula = count ~ X + Z*W`.

In most cases, the exact nature of the association structure is not known in advance, so model terms for associations need to be selected through statistical analysis. How to select these terms is the subject of some debate. Procedures analogous to forward selection or backward elimination methods that were once popular in regression analysis (Kutner et al., 2004) are sometimes recommended (e.g., Agresti, 2007). While significance tests can tell us whether a particular hypothesis should or should not be rejected, multiple and sequential use of significance tests to repeatedly select or eliminate associations according to their estimated size violates the probability foundations behind the definition of type I error (Burnham and Anderson, 2002). This procedure should therefore be used as a last resort when no better methods are available, and the level of α used need not be the traditional 0.05. We discuss methods of model selection in Chapter 5.

---

[9]Indeed, leaving main effects out of the model in `glm()` when the corresponding interaction remains in the model results in a completely different interpretation of the interaction term. To see this, try running `M2 <- glm(formula = Freq ~ First + First:Second, family = poisson(link = "log"), data = all.data)` with the Larry Bird example, and examine the results. In particular look at `summary(M2)`, `model.matrix(M2)`, and `predict(M2)`.

Example: Political Ideology (PolIdeolNominal.R, PolIdeolData.csv)

Agresti (2007) presents a summary from the 1991 U.S. General Social Survey[10] cross-classifying people according to

- political ideology, a 5-level variable with levels 1: "Very Liberal" ("VL"), 2: "Slightly Liberal" ("SL"), 3: "Moderate" ("M"), 4: "Slightly Conservative" ("SC"), and 5: "Very Conservative" ("VC")
- political party, a 2-level variable with levels "Republican" ("R") and "Democrat" ("D")
- gender, a 2-level variable with levels "Female" ("F") and "Male" ("M").

We fit to these data various models that differ in their interpretation of which variables are associated and whether the associations are homogeneous. Notice that the ideology variable is ordinal. In the next section, we explore how that structure can be exploited to provide a more parsimonious description of the association. For now, we treat ideology as nominal. Below is part of the data:

```
> alldata <- read.table(file = "C:\\Data\\PolIdeolData.csv", sep
   = ",", header = TRUE)
> head(alldata)
  gender party ideol count
1      F     D    VL    44
2      F     D    SL    47
3      F     D     M   118
4      F     D    SC    23
5      F     D    VC    32
6      F     R    VL    18
```

The models we consider are all Poisson regressions with the variables gender (G), party (P), and ideol (I) present in all models as main effects. The models differ in how the interactions are handled. The saturated model (GPI) is

$$\log(\mu_{ijk}) = \beta_0 + \beta_i^G + \beta_j^P + \beta_k^I + \beta_{ij}^{GP} + \beta_{ik}^{GI} + \beta_{jk}^{PI} + \beta_{ijk}^{GPI},$$

where $i = F, M$, $j = D, R$, and $k = 1, \ldots, 5$. The homogeneous association model (GP, GI, PI) is given by

$$\log(\mu_{ijk}) = \beta_0 + \beta_i^G + \beta_j^P + \beta_k^I + \beta_{ij}^{GP} + \beta_{ik}^{GI} + \beta_{jk}^{PI}.$$

We consider also the model that assumes mutual independence among all variables (G, P, I),

$$\log(\mu_{ijk}) = \beta_0 + \beta_i^G + \beta_j^P + \beta_k^I.$$

One more model is considered that is intermediate to the last two. We feel fairly certain that a PI association should exist because differing ideologies are part of what defines the two parties. The real question is whether the other two associations, GP and GI,

---

[10] The source of these data is a large-scale survey with a complex probability-based sampling design. Failing to account for the survey design in the analysis could lead to biased parameter estimates and misrepresentation of the true relationships among the variables. Proper techniques for analyzing data from this type of survey are presented in Section 6.3 and should be applied here. The analyses we present here are used solely as an illustration of the present analysis techniques and as a comparison to previously published work.

exist. We therefore expect Model G, P, I to provide a poor fit, and so we consider an augmentation of it to allow an association for PI (Model G, PI). Comparing this model to Model GP, GI, PI will allow us to test whether the other two associations, whose importance is uncertain, can simultaneously be dropped from the model.

We fit all models using glm() and compare models using anova():

```
> # Saturated Model: GPI
> mod.sat <- glm(formula = count ~ gender*party*ideol, family =
    poisson(link = "log"), data = alldata)

> # Homogeneous association model in all 3 associations: GP,GI,PI
> mod.homo <- glm(formula = count ~ (gender + party + ideol)^2,
    family = poisson(link = "log"), data = alldata)
> anova(mod.homo, mod.sat, test = "Chisq")
Analysis of Deviance Table
Model 1: count ~ (gender + party + ideol)^2
Model 2: count ~ gender * party * ideol
  Resid. Df Resid. Dev Df Deviance P(>|Chi|)
1         4     3.2454
2         0     0.0000  4   3.2454    0.5176

> # Model assuming only PI association: G,PI
> mod.homo.PI <- glm(formula = count ~ gender + party*ideol,
    family = poisson(link = "log"), data = alldata)
> anova(mod.homo.PI, mod.homo, test = "Chisq")
Analysis of Deviance Table
Model 1: count ~ gender + party * ideol
Model 2: count ~ (gender + party + ideol)^2
  Resid. Df Resid. Dev Df Deviance P(>|Chi|)
1         9    17.5186
2         4     3.2454  5   14.273    0.01396 *

> # Model assuming mutual independence: G,P,I
> mod.indep <- glm(formula = count ~ gender + party + ideol,
    family = poisson(link = "log"), data = alldata)
> anova(mod.indep, mod.homo.PI, test = "Chisq")
Analysis of Deviance Table
Model 1: count ~ gender + party + ideol
Model 2: count ~ gender + party * ideol
  Resid. Df Resid. Dev Df Deviance P(>|Chi|)
1        13    79.851
2         9    17.519  4   62.333 9.376e-13 ***
```

The model comparisons are summarized in Table 4.5. For example, the comparison of the last two models tests $H_0 : \beta_{jk}^{PI} = 0, j = D, R, k = 1, \ldots, 5$ against H_a : Not all $\beta_{jk}^{PI} = 0$. Model GP, GI, PI appears to fit reasonably well relative to the saturated model ($-2\log(\Lambda) = 3.24$ and p-value $= 0.52$), indicating that all 2-variable associations seem to be reasonably consistent across levels of the third variable. The GP and GI associations do seem to contribute something to the fit of the model, as indicated by the somewhat poor fit of G, PI relative to GP, GI, PI ($-2\log(\Lambda) = 14.27$ and p-value=0.01). Clearly, the independence model G, P, I does not fit well, as expected ($-2\log(\Lambda) = 62.33$ and p-value≈ 0), showing the importance of the PI association. In view of these results, we consider GP, GI, PI as the working model from which to examine these associations further.

Table 4.5: Deviance analysis of party-ideology association. The test in each row uses that row's model as the null hypothesis and the model directly above it as the alternative.

Model	Residual Deviance	Residual df	LRT Statistic $-2\log(\Lambda)$	LRT df	p-value
GPI	0	0			
GP, GI, PI	3.25	4	3.25	4	0.52
G, PI	17.52	9	14.27	5	0.01
G, P, I	79.85	13	62.33	4	≈ 0

Our next step is to obtain a summary of the model parameter estimates so that we can see the order in which they are listed.[11] We also get tests of each term in the model using Anova():

```
> round(summary(mod.homo)$coefficients, digits = 3)
                Estimate Std. Error z value Pr(>|z|)
(Intercept)        4.740      0.089  53.345    0.000
genderM           -0.704      0.137  -5.149    0.000
partyR            -0.244      0.125  -1.962    0.050
ideolSC           -1.638      0.197  -8.302    0.000
ideolSL           -0.832      0.157  -5.308    0.000
ideolVC           -1.306      0.176  -7.438    0.000
ideolVL           -0.899      0.162  -5.534    0.000
genderM:partyR     0.276      0.147   1.878    0.060
genderM:ideolSC    0.534      0.216   2.476    0.013
genderM:ideolSL    0.236      0.216   1.093    0.274
genderM:ideolVC    0.448      0.201   2.230    0.026
genderM:ideolVL    0.372      0.227   1.636    0.102
partyR:ideolSC     0.826      0.222   3.716    0.000
partyR:ideolSL    -0.437      0.217  -2.015    0.044
partyR:ideolVC     0.702      0.203   3.458    0.001
partyR:ideolVL    -0.861      0.243  -3.548    0.000

> library(car)
> Anova(mod.homo)
Analysis of Deviance Table (Type II tests)
Response: count
             LR Chisq Df Pr(>Chisq)
gender         20.637  1  5.551e-06 ***
party           0.528  1    0.46736
ideol         154.131  4  < 2.2e-16 ***
gender:party    3.530  1    0.06026 .
gender:ideol    8.965  4    0.06198 .
party:ideol    60.555  4  2.218e-12 ***
```

Note that levels of the ideol variable have been ordered alphabetically, so that M (Moderate) is the baseline and all ideol parameters represent comparisons to this level. This could be changed using the relevel() or factor() function as described in Section 2.2.6 if desired. The LRTs for model terms show that the *PI* association is

---

[11] Although we can predict this ordering based on knowing the factor levels and the order in which they appear in the fitted model, examining them can help with forming the linear combinations that are needed for odds ratio estimation.

highly significant, but there is only moderate evidence for the existence of the other two (both p-values are 0.06).[12]

With the order of parameter estimates now clear, we can use them to estimate odds ratios for the three associations, GP, GI, and PI. Because associations are homogeneous, we need not consider a third variable when estimating an odds ratio between the other two. Each odds ratio is related to the model parameters according to Equation 4.7. For GP, we construct the ratio of the odds of being Republican for males vs. females:

$$\log(OR) = \log\left(\frac{\mu_{MR}/\mu_{MD}}{\mu_{FR}/\mu_{FD}}\right)$$
$$= \beta^{GP}_{MR} - \beta^{GP}_{MD} - \beta^{GP}_{FR} + \beta^{GP}_{FD}$$
$$= \beta^{GP}_{MR}$$

The last equality results from the fact that $\beta^{GP}_{MD} = \beta^{GP}_{FR} = \beta^{GP}_{FD} = 0$ because "F" represents level 1 of gender and "D" represents level 1 of party. Thus, the estimated odds ratio is $\exp(\hat{\beta}^{GP}_{MR}) = \exp(0.276) = 1.3$. For GI, we will construct the ratio of odds of being in a more conservative ideology relative to a less conservative ideology for males vs. females (i.e., quantities like "Odds of being Very Conservative compared to Slightly Conservative given Male, divided by Odds of being Very Conservative compared to Slightly Conservative given Female"). For example,

$$\log(OR) = \beta^{GI}_{M1} - \beta^{GI}_{M2} - \beta^{GI}_{F1} + \beta^{GI}_{F2} = \beta^{GI}_{M1} - \beta^{GI}_{M2},$$

where, as a reminder, 1=VC and 2=SC. We use a similar treatment of PI, comparing the odds of being in a more conservative ideology for Republicans vs. Democrats. Here we expect that Republicans are more conservative, so we expect positive odds ratios of increasing magnitude with increasing difference between the two ideologies being compared.

There is one odds ratio for GP, ten for GI, and ten for PI. This leads to 21 sets of linear combination coefficients on the 16 parameters listed in the output above. The first seven parameters do not contribute to the odds ratios, so their coefficients are all 0. Other coefficients are 0, 1, or -1 as determined by the $\log(OR)$ calculations. This leads to the contrast matrix shown below for the GI odds ratios:

```
contr.mat.GI <- rbind(c(rep(0,7),0,-1,0,1,0,0,0,0,0),
                     c(rep(0,7),0,0,0,1,0,0,0,0,0),
                     c(rep(0,7),0,0,-1,1,0,0,0,0,0),
                     c(rep(0,7),0,0,0,1,-1,0,0,0,0),
                     c(rep(0,7),0,1,0,0,0,0,0,0,0),
                     c(rep(0,7),0,1,-1,0,0,0,0,0,0),
                     c(rep(0,7),0,1,0,0,-1,0,0,0,0),
                     c(rep(0,7),0,0,-1,0,0,0,0,0,0),
                     c(rep(0,7),0,0,0,0,-1,0,0,0,0),
                     c(rep(0,7),0,0,1,0,-1,0,0,0,0))
```

For example, the first row multiplied by the parameter estimates (as ordered by R) gives $-\hat{\beta}^{GI}_{M2} + \hat{\beta}^{GI}_{M1}$, which is the second odds ratio described in the previous paragraph.

---

[12] It turns out that both terms *are* significant at $\alpha = 0.05$ if the other term is first removed from the model. This is an example of the difficulty with using hypothesis tests as a model-selection tool. See Chapter 5 for alternative methods of model selection.

With this matrix we can get profile LR confidence intervals using `mcprofile()`. Alternatively, we can form Wald intervals using `glht()` from the `multcomp` package or by direct calculation. Both the `mcprofile` and `multcomp` packages allow for computations with or without adjustment for simultaneous confidence levels. Note that for simultaneous inference we need to define a "family" about which we want our confidence statements to apply. While we might like to have 95% confidence that all confidence intervals in a large analysis will cover their parameters, the drawback to choosing a larger family size is that the intervals become wider. For example, if we choose all 21 intervals to be one family here, then the confidence level for each interval using the Bonferroni adjustment would be $\alpha' = 0.0024$ (hence $Z_{1-\alpha'} = 3.03$). Instead, it seems sensible for this problem to make a separate confidence statement about the odds ratios for each of the three interaction effects. Thus, the GP family contains 1 interval, while the GI and PI families each contain 10. A Bonferroni adjustment in the latter families would use $\alpha' = 0.005$ (hence $Z_{1-\alpha'} = 2.81$). The 95% adjusted LR intervals, using the preferred single-step adjustment rather than Bonferroni (see p. 110) are shown below for the PI family (details are given in the program for this example).

```
> exp(confint(LRCI))

   mcprofile - Confidence Intervals

level:               0.95
adjustment:          single-step
                  Estimate lower upper
PI VC:SC | R:D       0.884 0.439  1.76
PI VC:M  | R:D       2.019 1.169  3.54
PI VC:SL | R:D       3.124 1.597  6.25
PI VC:VL | R:D       4.775 2.314 10.25
PI SC:M  | R:D       2.284 1.260  4.24
PI SC:SL | R:D       3.535 1.736  7.40
PI SC:VL | R:D       5.403 2.522 12.08
PI M:SL  | R:D       1.548 0.864  2.82
PI M:VL  | R:D       2.365 1.244  4.68
PI SL:VL | R:D       1.528 0.712  3.35
```

Row names have been chosen to describe the odds ratio being estimated. For example, "PI VC:SC | R:D" refers to the ratio of [the odds of being very conservative vs. slightly conservative for a Republican] to [the odds of being very conservative vs. slightly conservative for a Democrat]. All odds ratios compare a more conservative ideology to a less conservative one for Republicans vs. Democrats, and hence all are expected to be ≥ 1. We see that this is generally true: seven of the ten PI odds ratios have confidence intervals entirely above 1, and the estimates and corresponding intervals tend to be farther from 1 as the ideologies being compared lie farther apart. We are 95% confident that this procedure produces a set of intervals that *all* cover their parameters.

Loglinear vs. logistic and multinomial models

We have seen that the loglinear model and the multinomial regression model in Equation 3.8 can both be applied to contingency tables. Is there a clear preference for one or the

Table 4.6: Table and parameter structures for logistic regression (top) and loglinear model (bottom) when X is a 2-level response variable and A and B are each 2-level explanatory variables.

		$B=1$					$B=2$		
		$X=0$	$X=1$				$X=0$	$X=1$	
A	1	π_{11}	$1-\pi_{11}$	1		1	π_{12}	$1-\pi_{12}$	1
	2	π_{21}	$1-\pi_{21}$	1		2	π_{22}	$1-\pi_{22}$	1

		$B=1$					$B=2$		
		$X=0$	$X=1$				$X=0$	$X=1$	
A	1	μ_{111}	μ_{112}	μ_{11+}		1	μ_{121}	μ_{122}	μ_{12+}
	2	μ_{211}	μ_{212}	μ_{21+}		2	μ_{221}	μ_{222}	μ_{22+}

other?

In many cases the answer is no. The two approaches to modeling counts in a contingency table give identical answers due to a mathematical relationship between the Poisson and multinomial distributions. On the one hand, the multinomial distribution assumes that the total count in the entire table was fixed by design (e.g., by specifying a total sample size prior to data collection). On the other hand, the Poisson model assumes that the total is a random quantity, because it is a sum of independent random variables. However, it turns out that the joint probability distribution of any set of independent Poisson random variables, conditioned on their sum, is a multinomial distribution. That is, if we treat the total count from a contingency table as a fixed quantity, the two models are identical. Furthermore, there is never any harm in treating this sum as fixed, because it is not a quantity that we attempt to model. Therefore, the two models can be used interchangeably in most settings.

A unique feature of loglinear models is that they can be used not only in cases where there is one particular variable that is considered a "response" variable, but also in cases where there are multiple response variables or none at all. Logistic regression and multinomial models require one response variable whose probabilities are to be modeled. In a loglinear model analysis with one response variable, the main interest is in modeling associations between that variable and the others. It is typical in that case to include in any model all main effects and interactions involving the explanatory variables and not explore these effects any further. For example, if the response variable is X and the explanatory variables are A, B, C, then the minimal model to be fit would be Model ABC, X.

To see why this is needed, consider the case where the response variable X has two levels, and there are two explanatory variables, A and B, each with two levels. The table structures and parameters used by the two models are laid out in Table 4.6. The saturated logistic regression model written in an ANOVA format specifies

$$\log\left(\frac{\pi_{ij}}{1-\pi_{ij}}\right) = \lambda + \lambda_i^A + \lambda_j^B + \lambda_{ij}^{AB}, \, i=1,2, \, j=1,2,$$

where we use λ's to represent the regression parameters here to avoid confusion with the saturated loglinear model, which specifies

$$\log(\mu_{ijk}) = \beta_0 + \beta_i^A + \beta_j^B + \beta_{ij}^{AB} + \beta_k^X + \beta_{ik}^{AX} + \beta_{jk}^{BX} + \beta_{ijk}^{ABX}, \, i=1,2; \, j=1,2, \, k=1,2.$$

The parameters from these models have some common interpretations:

- Recall that main effects in the logit model (λ_i^A and λ_j^B) represent associations between that variable and the response variable X. In the loglinear model, 2-variable interactions with X (β_{ik}^{AX} and β_{jk}^{BX}) represent the same associations. Odds ratios in the logistic model are found from exponentiated differences in main effect parameters. In the loglinear model the same odds ratios are found from exponentiated (1, -1, -1, 1) contrasts among the X-interaction parameters. It can be shown that *odds ratios calculated from these two models are identical*.

- In the logit model, the λ_{ij}^{AB} parameters control the heterogeneity of association. In the loglinear model, the β_{ijk}^{ABX} parameters serve the same purpose. In both models, predicted odds ratios between either A and X or B and X computed separately for each level of the other explanatory variable are the same as the corresponding observed odds ratios in the tables. Also, in both models, absence of the highest order interaction results in a model with homogeneous association.

The end result is that any logistic regression model for categorical explanatory variables model can be written as an equivalent loglinear model with (a) all main effects and interactions among the explanatory variables, (b) a main effect for the response variable, and (c) an interaction with the response for each term in the logistic model (e.g., an A effect in the logistic model produces an AX effect in the loglinear model). If this is done, then the loglinear association parameters β_{ik}^{AX} and β_{jk}^{BX}, $i = 1, \ldots, I; j = 1, \ldots, J, k = 1, 2$ are *identical* to the logistic regression parameters λ_i^A and λ_j^B, respectively, and the heterogeneity parameters β_{ijk}^{ABX} are identical to the interaction parameters λ_{ij}^{AB}.[13]

To fit a Poisson-based model as an alternative to a multinomial regression, each observed set of J multinomial counts is converted into J observations with identical values for the explanatory variables. A new categorical explanatory variable is created consisting of the levels of the response categories. The response variable for each observation is the observed count for that observations explanatory variables and response category. The model then contains all of the main effects of the explanatory variables and the new categorical variable for the response category, as well as interactions between the new variable and the explanatory variables. The main interests then lie in this set of interactions.

Because the translation to odds ratios and probabilities is a little bit simpler in the logistic regression model, we tend to prefer it over the loglinear model when there is a 2-level response variable. Multinomial regression models are not as easy to use and interpret as logistic regression, so the choice between loglinear and multinomial models when a response variable has more than two levels is not as clear. The advantage of the Poisson loglinear formulation is that much simpler model creation, diagnostic, and correction tools can be used as discussed in Chapter 5. Also, when there is an ordinal response variable, the loglinear model is more flexible than the multinomial regression model in terms of the kinds of associations that can be modeled, as is described next.

4.2.6 Ordinal categorical variables

In Section 3.4 we saw one approach to modeling counts when a response variable is ordinal and arises from a multinomial distribution. Here we provide an alternative approach using a Poisson regression model. This model also works when one or more ordinal variables are explanatory, or when there is no designated response variable.

There are two general approaches to analyzing counts at different levels of an ordinal variable: ignore the ordinality or assign numerical scores to represent the ordered levels. In

---

[13]For example, the code `logit.homo <- glm(formula = party ~ gender + ideol, family = binomial(link = "logit"), weights = count, data = alldata)` estimates the same model as `mod.homo` in the Political Ideology example.

the former case, the analyses of the previous sections are applied, but information relating the mean counts and odds ratios to the ordering of the levels is lost. This is not an efficient or informative type of analysis if the mean counts or odds ratios vary in some systematic manner with the ordering. Indeed, in the example in the previous section, we saw that analyzing the associations between the ordinal ideology variable and the nominal variables was somewhat cumbersome. We therefore try to incorporate the ordinal structure of variables into an analysis whenever we can. In a Poisson regression model, assigning numerical scores to represent levels of an ordinal variable allows us to treat the ordinal variable like a numerical variable. Once again, methods from the previous sections are used for the analysis.

Thus, the presence of ordinal variables represents no difficulty in a Poisson regression. We therefore focus on the selection of scores and the interpretation of the results. There is also a procedure available to check whether the choice of scores has failed to uncover some important aspect of the relationship between the variable and the counts.

Choosing Scores

Assigning scores is subjective, primarily because there is typically no single scoring system that is obviously better than all others. For example, ordered categorical data often arises from questionnaires that ask a respondent to rate something on a scale like "Excellent," "Very Good," "Good," "Fair," and "Poor." If one assigns numerical scores to represent these levels, what should the scores be? Should we choose 1-2-3-4-5, or 5-4-3-2-1, or some other sequence? Does it matter? The answer varies depending on the problem. Different scores can sometimes heavily influence the results of the analysis. Other times, all reasonable choices of scores lead to essentially the same results.

Scores should always be chosen in consultation with the subject-matter specialist whose questions are being answered. The important feature in any set of scores is the relative sizes of the spacings, or *gaps*, between scores. The more similar two categories are, the narrower the gap should be between their scores. For the example above, if we think that the notion behind "Fair" is closer to "Good" than to "Poor," then the gap between "Good" and "Fair" should be smaller than the gap between "Fair" and "Poor." How much closer is a matter of judgment. Whether those three scores should be 4-3-1 (gaps of 1 and 2), 5-4-1 (gaps of 1 and 3), 6-4-1 (gaps of 2 and 3), or something else entirely is often a matter of guesswork.

When there is some uncertainty in the definition of scores, it is often good practice to duplicate an analysis on several sets of reasonable scores. If the results are very similar, then the choice of scores does not matter much and conclusions can be drawn accordingly. If the results *do* change appreciably, then more exploration may be necessary to understand the cause of the difference. In any case, unless one set of scores was strongly preferred prior to seeing the analysis results, both sets of results should be reported with roughly equal consideration.

In choosing scores the *relative* sizes of the different gaps matters, not the actual numbers assigned. For example, scores 1-2-3 provide the same analysis results as do 10-20-30 or 3-2-1. Of course, estimates of ratios of means will depend on the numerical values that define a "1-unit change" in score. For example, with 1-2-3 scores, a 1-unit change equals moving to an adjacent category; with 3-2-1, the movement is to an adjacent category in the opposite direction; and with 10-20-30, a 10-unit change is required to change categories, so slope parameters relating to a 1-unit change may need to be multiplied by 10 to be meaningful. However, *test* results will be identical in all cases.

Example: Political ideology (PolIdeolOrd.R, PolIdeolData.csv)

We saw that the ideology variable in this example is ordinal, so we will attempt to use this structure to construct a more meaningful and parsimonious description of the association. As this example arises from "interesting data" rather than a problem brought to us by a political scientist, we are acting as subject-matter specialists ourselves. We choose to address two different sets of questions by using the ordering of the categories for the ideology variable in different ways. The first set of questions makes use of the fact that the five categories are naturally ordered from more liberal to less liberal (equivalently, from less conservative to more conservative). As a default, we choose 1-2-3-4-5 to represent this progression, where higher scores represent more conservative ideologies. However, we suspect that the "slightly" liberal and conservative groups might each be closer to the middle than to the extremes of their respective ideologies. We therefore consider scores with wider gaps between "very" and "slightly" than between "slightly" and "moderate." Somewhat arbitrarily, we choose to double the gap sizes, creating 0-2-3-4-6 scores. We certainly expect either set of scores to be associated with the political parties, as each party deliberately seeks to appeal to a different ideology. It would be interesting to see whether there is an association between gender and ideology (are men more liberal or conservative than women?) or whether the association between party and ideology is different for men and women.

The second set of questions relates to whether one party or gender holds more *extreme* views than the other. In other words, does one party or gender tend to have more people with "very" strong views as opposed to "slightly" strong or central views? In order to examine this, we create scores to measure extremeness of ideology: 2-1-0-1-2. This essentially creates three categories measuring strength of ideology, low, medium, and high, with 0-1-2 scores on those levels. The scores are created and added to the data as shown below:

```
> lin.score1 <- c(rep(x = c(1,2,3,4,5), times = 4))
> lin.score2 <- c(rep(x = c(0,2,3,4,6), times = 4))
> extrm.score <- c(rep(x = c(2,1,0,1,2), times = 4))
> alldata <- data.frame(alldata, lin.score1, lin.score2,
    extrm.score)
> head(alldata)
  gender party ideol count lin.score1 lin.score2 extrm.score
1      F     D    VL    44          1          0           2
2      F     D    SL    47          2          2           1
3      F     D     M   118          3          3           0
4      F     D    SC    23          4          4           1
5      F     D    VC    32          5          6           2
6      F     R    VL    18          1          0           2
```

Note that the ifelse() function could have been used instead to create the scores.

Models and interpretation

For a single ordered categorical variable X with I levels, let s_i, $i = 1, \ldots, I$ represent the set of numerical scores assigned to the levels. The ordinal Poisson regression model treats count y_i as $Po(\mu_i)$, with $\log(\mu_i) = \beta_0 + \beta_1 s_i$. Interpretation of the regression parameters is the same as with any numerical explanatory variable in a Poisson regression as described in Section 4.2.1. However, the "units" applied to scores usually have no meaning, except in those cases where the ordered categories represent intervals of a continuum. For example, if we replace categories like "Good," "Fair," and "Poor" with 3-2-1 scores, then a 1-unit change in score is merely a change in category. However, if we replace age groups "<25," "25-45," and ">45" with scores like 20-35-60, then a 1-unit change in score is approximately equivalent to a 1-year change in age.

Examining ratios of means when s changes by 1 unit may not adequately address all interests if some of the gaps between scores are more than 1. Therefore, we will often wish to examine changes for a c-unit change in s, where c represents the difference in scores between any two categories of interest. As described in Section 4.2.1, this simply means analyzing $c\beta$.

When there are two or more categorical explanatory variables, at least one of which is ordinal, then how the scores enter into the model depends on the goal of the analysis; i.e., whether the focus is on modeling and predicting means or on modeling associations between variables. If the problem is more like a typical regression and means are the primary focus, then any score sets are treated as numerical variables and the usual analyses are performed as described in Section 4.2.1. Adding interaction terms to the model is optional and depends on the goals of the analysis and the fit of the resulting models. The procedure for checking the fit of the scores that is described later in this section can be applied separately to each ordinal variable in the regression.

When associations are the primary focus, special models are needed, especially when there are additional categorical or ordinal variables. Associations between categorical variables are measured by the odds ratios, so we need to consider the assumptions that different models make regarding how the odds ratios are structured. To start, consider a problem with one nominal variable, X, and one ordinal variable, Z, with scores s_j, $j = 1, \ldots, J$, such as a contingency table with X as the row variable and Z as the column variable. The association parameters in the usual two-variable nominal model, $\log(\mu_{ij}) = \beta_0 + \beta_i^X + \beta_j^Z + \beta_{ij}^{XZ}$, assume no particular structure to the odds ratios between X and Z. Instead, we use a model that assumes that, for any pair of rows, the $\log(OR)$ changes linearly with the difference in scores between the two columns being compared. Specifically, the model is

$$\log(\mu_{ij}) = \beta_0 + \beta_i^X + \beta_j^Z + \beta_i^{XZ} s_j. \tag{4.9}$$

To see the structure that this model imposes on associations, consider again the odds ratio involving levels i and i' of X and j and j' of Z:

$$\begin{aligned}\log(OR_{ii',jj'}) &= \log(\mu_{ij}) + \log(\mu_{i'j'}) - \log(\mu_{i'j}) - \log(\mu_{ij'}) \\ &= \beta_i^{XZ} s_j + \beta_{i'}^{XZ} s_{j'} - \beta_{i'}^{XZ} s_j - \beta_i^{XZ} s_{j'} \\ &= (\beta_i^{XZ} - \beta_{i'}^{XZ})(s_j - s_{j'}).\end{aligned} \tag{4.10}$$

For example, suppose $J = 3$ and scores are 1-2-3, and we want to estimate odds ratios between two levels of Z across levels 1 and 2 of X. Then we have

$$\begin{aligned}\log(OR_{12,12}) &= -1(\beta_1^{XZ} - \beta_2^{XZ}) \\ \log(OR_{12,13}) &= -2(\beta_1^{XZ} - \beta_2^{XZ}) \\ \log(OR_{12,23}) &= -1(\beta_1^{XZ} - \beta_2^{XZ}).\end{aligned}$$

This model is sometimes called a *linear association* model, although the linearity is technically in the *log* of the odds ratios. Notice that Equation 4.9 specifies only $(I-1)$ parameters to describe the association, as opposed to the usual $(I-1)(J-1)$ parameters for association in a nominal table. Thus, using the ordering allows us to model a set of data with fewer parameters, which is always desirable, but at a cost of more assumptions about the structure of the data. Later in this section we will see how we can test whether the added assumptions about association structure fit the data well.

Notice also that the scores are *not* used in defining the effects of Z on the marginal counts: the main effect β_j^Z is retained instead of $\beta^Z s_j$. This is because we are examining whether the *associations* are affected by X and Z, and the main effect terms do not enter into the odds ratios. We therefore maintain a nominal model for each of the marginal counts. Then any difference in the overall fit of the model due to the use of linear, rather than nominal, association is caused directly by the change in the association structure, and not also by a change in the fitting of marginal means (see Exercise 9).

A test for the interaction term ($H_0 : \beta_i^{XZ} = 0$ for $i = 1, \ldots I$) for this model tests the null hypothesis of independence against the alternative of a linear change in $\log(OR)$. Failure to reject H_0 here does not mean that there is *no* association between X and Z. It means only that there is not sufficient evidence of a linear trend for $\log(OR)$. An association pattern could exist such that the odds ratios are not continually getting more extreme as the difference between their scores increases. Similarly, rejecting the null hypothesis does not imply that the association is *entirely* linear for $\log(OR)$. There can be a general increasing or decreasing trend with increasing distance between scores, but the trend does not have to be linear. Testing for goodness of fit of a model as described after the example can help to assess the possibility for alternative patterns.

Example: Political ideology (PolIdeolOrd.R, PolIdeolData.csv)

Our work with nominal models has suggested that the three-variable interaction is not needed, so we start with a homogeneous association model and consider adding scores. To begin, we consider the 1-2-3-4-5 scores contained in `lin.score.1` and apply them to the *PI* term, because we believe that the odds ratios from this association should increase as the distance between ideologies increases. The model we fit is

$$\log(\mu_{ijk}) = \beta_0 + \beta_i^G + \beta_j^P + \beta_k^I + \beta_{ij}^{GP} + \beta_{ik}^{GI} + \beta_j^{PI} s_k^I, \qquad (4.11)$$

which assumes linear association between party and ideology that is homogeneous across genders. For ideologies k and k' we have

$$\log(OR_{RD,kk'}^{PI}) = (\beta_R^{PI} - \beta_D^{PI})(s_k^I - s_{k'}^I)$$

for both genders, where the value of $s_k^I - s_{k'}^I$ is the difference in scores between the two ideologies being compared. Notice that this means the odds ratios involving any two consecutive ideologies are the same. Also, for example, the odds ratio comparing the odds of Republican for very conservative vs. slightly liberal ideologies is the same as the one comparing the odds of Republican between slightly conservative and very liberal (both score differences are 3). We fit and analyze this model in detail below. More models are fit in the accompanying program for this example.

The terms in Equation 4.11 are specified concisely in `glm()` by using only the interaction terms as shown below:

```
> mod.homo.lin1.PI <- glm(formula = count ~ gender*party +
    gender*ideol + party*lin.score1, family = poisson(link =
    "log"), data = alldata)
> summary(mod.homo.lin1.PI)

Call: glm(formula = count ~ gender * party + gender * ideol +
    party * lin.score1, family = poisson(link = "log"), data =
    alldata)

Deviance Residuals:
```

```
            Min        1Q     Median        3Q       Max
        -1.24363  -0.54155    0.02178   0.52057   0.90547

Coefficients: (1 not defined because of singularities)
                   Estimate Std. Error z value Pr(>|z|)
(Intercept)         4.72852    0.08169  57.881  < 2e-16 ***
genderM            -0.71139    0.13656  -5.209 1.90e-07 ***
partyR             -1.51082    0.20862  -7.242 4.42e-13 ***
ideolSC            -1.40587    0.14897  -9.437  < 2e-16 ***
ideolSL            -0.83121    0.13685  -6.074 1.25e-09 ***
ideolVC            -1.41136    0.15424  -9.150  < 2e-16 ***
ideolVL            -0.89370    0.14906  -5.996 2.03e-09 ***
lin.score1               NA         NA      NA       NA
genderM:partyR      0.28723    0.14555   1.973   0.0484 *
genderM:ideolSC     0.55859    0.21413   2.609   0.0091 **
genderM:ideolSL     0.23706    0.21543   1.100   0.2711
genderM:ideolVC     0.43621    0.20135   2.166   0.0303 *
genderM:ideolVL     0.37450    0.22682   1.651   0.0987 .
partyR:lin.score1   0.43058    0.06005   7.170 7.48e-13 ***
---
Signif. codes:  0 '***' 0.001 '**' 0.01 '*' 0.05 '.' 0.1 ' ' 1

(Dispersion parameter for poisson family taken to be 1)

    Null deviance: 255.1481  on 19  degrees of freedom
Residual deviance:   8.3986  on  7  degrees of freedom
AIC: 142.83

Number of Fisher Scoring iterations: 4
```

Because the linear score `lin.score1` appears only in an interaction and not as a main effect in the `formula` argument, a main effect is automatically created for it. Similarly, main effects are created for `gender`, `party`, and `ideol`. This means that the main effect of ideology is included into the model both as a series of indicator variables and as a numerical variable. This duplication leads to the note about "singularities" in the listing of estimated model coefficients and the "NA" results for the `lin.score.1` main effect.[14] This redundancy and resulting lack of estimate creates no difficulty in interpretation, because we are not modeling the main effects of ideology score. The parameter for the interaction between `party` and `lin.score.1` is properly estimated, so we can mostly ignore the issue, save for some computational details noted in the program. In particular, `glht()` and `mcprofile()` will not work on the model-fit object because of the extra parameter. We can coerce `glht()` to work by removing the NA from the vector of parameter estimates (details are in the program). Unfortunately, `mcprofile()` will still not work after this change.

We start with the tests for effects using `Anova()`.

---

[14] The exact cause of the the "singularity" is beyond the scope of this book. Readers who are familiar with *model matrices*—for example the matrix **X** in the linear model specification $\mathbf{Y} = \mathbf{X}\beta + \epsilon$—may appreciate the fact that the `lin.score1` variable can be written as a linear combination of `(Intercept)` and the four indicators representing the `ideol` variable. As a result, `lin.score1` contains no new information that hasn't already been accounted for by the `ideol` term, so the columns of the matrix representing this model are not linearly independent.

```
> library(car)
> Anova(mod.homo.lin1.PI)
Analysis of Deviance Table (Type II tests)
Response: count
                 LR Chisq Df Pr(>Chisq)
gender             20.637  1  5.551e-06 ***
party               0.528  1  0.46736
ideol             163.007  3  < 2.2e-16 ***
lin.score1                 0
gender:party        3.901  1  0.04826   *
gender:ideol        9.336  4  0.05323   .
party:lin.score1   55.402  1  9.825e-14 ***
```

We see that there is a very strong linear association between party and ideology. Tests for the other associations yield results very similar to those in the nominal case, although if one uses a rigid 0.05 significance level, GP is now significant. Practically speaking, the two tests are telling us the same thing: there is moderate evidence to suggest that GP and GI interactions are present.

Odds ratios for GP and GI are estimated as in the nominal case by constructing linear combinations of the estimated model parameters. For PI, β_R^P is estimated in the output by the partyR:lin.score1 coefficient, while $\beta_D^P = 0$, so the needed coefficients for the linear combination are just the possible differences in scores, 1, 2, 3, or 4. That is, a c-unit difference is scores results in $\widehat{OR} = \exp(c\hat{\beta}_R^P)$. Again, because Republicans are more conservative than Democrats, we expect positive odds ratios of increasing magnitude as the difference in scores increases for the two ideologies being compared. Calculations for the confidence intervals are shown in the program corresponding to this example. The results for PI are shown below.

```
> wald.ci
                     Estimate Lower CI Upper CI
PI REP | 1 Cat Ideol     1.54     1.37     1.73
PI REP | 2 Cat Ideol     2.37     1.87     2.99
PI REP | 3 Cat Ideol     3.64     2.56     5.18
PI REP | 4 Cat Ideol     5.60     3.50     8.96
```

Comparing these estimates and confidence intervals to those obtained for the nominal analysis, there is very close agreement for all GP and GI odds ratios (the latter are not shown in the output). The four odds ratios given above estimate how the odds of being Republican change as ideologies differ by progressively more categories. For adjacent categories of ideology, the odds of being Republican are 1.54 times as high for the more conservative ideology than for the more liberal one ($1.37 < OR < 1.73$). For a 2-category difference, the odds of being Republican increase to 2.37 times as high in the more conservative ideology ($1.87 < OR < 2.99$); for 3 categories to 3.64 ($2.56 < OR < 5.18$); and finally the odds of being Republican are 5.6 times as high ($3.50 < OR < 8.96$) for very conservative ideology compared to very liberal ideology.

Notice that both the estimation and the interpretation of the PI association are much easier when the ordinality is used. This highlights the appeal of using ordinality when it is available and when the association can reasonably be described by the linear trend across the log odds ratios. When viewed nominally, the estimates based on consecutive ideologies were 0.88, 2.28, 1.55, and 1.53. Our estimate of 1.54 lies in the middle of these figures. However, notice that the first two of these estimates

(`VC:SC` and `SC:M`) have confidence intervals that do not overlap in the nominal analysis (0.54 to 1.46 for `VC:SC`, and 1.48 to 3.52 for `SC:M`). This suggests that the true odds ratios that these two intervals are estimating may not both have the same value. Thus, it remains to be seen whether the linear pattern for $\log(OR)$ really fits these data.

These confidence intervals also demonstrate the other big advantage of using ordinality explicitly: confidence intervals for comparable odds ratios are generally much shorter in the ordinal analysis than in the nominal analysis. For example the confidence interval comparing the two most extreme ideologies was ($2.3 < OR < 10.3$) in the nominal analysis, which is considerably longer than the corresponding confidence interval from the ordinal analysis. The ability to represent the PI association using fewer parameters allows those parameters to be estimated more precisely. This precision is transferred to inferences about the odds ratios.

When both variables in a two-variable model are ordinal we can define scores $s_i^X, i = 1, \ldots, I$ for X and $s_j^Z, j = 1, \ldots, J$ for Z. The $I - 1$ association parameters from Equation 4.9 are replaced with a single parameter by using the model

$$\log(\mu_{ij}) = \beta_0 + \beta_i^X + \beta_j^Z + \beta^{XZ} s_i^X s_j^Z.$$

It is easy to see that for this model $\log(OR_{ii',jj'}) = \beta^{XZ}(s_i^X - s_{i'}^X)(s_j^Z - s_{j'}^Z)$ (see Exercise 11). This structure is sometimes called the *linear-by-linear association*. The reduction of association parameters comes at the cost of more-structured assumptions regarding the nature of the log-odds ratios. Therefore, when a test for $H_0 : \beta^{XZ} = 0$ is rejected, it does not imply that the association between X and Z is entirely linear; there may be additional structure not captured by β^{XZ}. Similarly, failing to reject $H_0 : \beta^{XZ} = 0$ means only that there is insufficient evidence of a *linear-by-linear* association, not that there is no association at all.

With more than two variables, associations continue to be defined by two-variable and higher-order interactions. Nominal, linear, and linear-by-linear association terms for different pairs of variables can appear together in any combination—indeed, the political ideology example contained two nominal association terms and one linear association term. Placing linear structures on *three*-variable interactions assumes that the $\log(OR)$'s between two variables change linearly across the scores of the third variable. Modeling interactions involving more than three variables can be done. The main difficulty then is interpreting the resulting models. Carrying out assessments as shown in Equations 4.8 and 4.10 provides the key to interpreting models.

Checking goodness of fit for a score set

For a single ordered categorical variable X with I levels, let s_i represent the set of scores assigned to the levels. As in Section 4.2.3 let $x_2, \ldots, x_I$ be the indicator variables that would be used to represent the levels of X if the ordering is not considered. We can consider the nominal model $\log(\mu_i) = \beta_0 + \beta_2 x_2 + \ldots + \beta_I x_I$ and the ordinal model $\log(\mu_i) = \gamma_0 + \gamma_1 s_i$ as two alternative explanations for the changes in mean count. The nominal model creates a separate parameter for each mean, and thus represents a saturated model, so that the residual deviance is zero when there is one observation per level of X. The ordinal model assumes that the population means change log-linearly with the scores. When this is true, the data ought to fit the model reasonably closely, and hence the residual deviance should be relatively small. If the ordinal model is wrong, then the means predicted by it may lie far from the observed means, and the residual deviance will be large. A comparison of the residual deviances of the nominal and ordinal models represents a test for the assumption that the means follow a linear pattern with the chosen scores.

Formally, this is a LRT. The test statistic $-2\log(\Lambda)$ is the difference between the nominal- and ordinal-model deviances. The degrees of freedom for the test are derived from the difference in the number of parameters in the two models. In the nominal model, the variable X uses $I-1$ df (the parameters $\beta_2, \ldots, \beta_I$), while in the ordinal model X uses only one parameter, γ_1. Therefore, the test statistic is compared to a χ^2_{I-2} distribution. Notice that using different sets of scores can affect the residual deviance of the ordinal model, and hence the test statistic and conclusion. Therefore, this test is often used to help determine whether a particular set of scores is appropriate. This particular type of test is referred to as a *goodness-of-fit* test.

The hypotheses for the goodness-of-fit test are $H_0 : \log(\mu_i) = \gamma_0 + \gamma_1 s_i$ vs. $H_a : \log(\mu_i) = \beta_0 + \beta_2 x_2 + \ldots + \beta_I x_I$. In words, we are testing the null hypothesis that the scores provide a reasonable explanation for the pattern of mean counts, against the alternative that the pattern of means does *not* fit the shape implied by the scores. Failing to reject H_0 implies that the scores seem to do a reasonably good job of capturing the trend in means across levels of X. Note that this is not the same as saying that the scores are *correct*. It is quite possible that models based on several different sets of scores can provide "adequate" fits based on the lack of significance in the test. Note also that this test is different from the LRT for the significance of the score variable s in the ordinal model, for which the hypotheses are $H_0 : \log(\mu_i) = \gamma_0$ vs. $H_a : \log(\mu_i) = \gamma_0 + \gamma_1 s_i$. These two questions are different from each other; their null hypotheses can be true or false in any combination.

This procedure extends to association models for two or more variables. The null hypothesis for the goodness-of-fit test is always represented by the model that uses scores, while the alternative is the model in which one or more sets of scores have been replaced with indicator variables. Rejecting H_0 means that the linear structure imposed by the scores is not an adequate description of the association for which they are used.

When there are multiple ordinal variables, it is possible to create a series of models by successively loosening the assumptions in a given association. For example, if X and Z are both ordinal, then in each variable we have a choice of whether or not to use scores in a linear structure. Suppose we use a subscript L to denote that a variable appears in an interaction in its linear scores form. In the interaction term, we could use scores for both variables $(X_L Z_L)$, scores for X only $(X_L Z)$, scores for Z only $(X Z_L)$, or leave both variables nominal (XZ). The middle two models which use scores for only one variable are both generalizations of the linear-by-linear model, and are both restricted versions of the fully nominal model. Thus, model-comparison tests can be used in several different ways, comparing $H_0 : X_L Z_L$ to one of several possible alternatives, $H_a : X_L Z$ or $X Z_L$ or XZ. We can also test $H_0 : X_L Z$ or $H_0 : X Z_L$ against $H_A : XZ$. Only $X_L Z$ and $X Z_L$ are not directly comparable using LRTs.

Testing pairs of models at a time can end with ambiguous results. It could happen, for example, that $X_L Z_L$ is rejected as a reduced model for $X_L Z$ and/or $X Z_L$ but not as a reduced model for XZ. Then one is left wondering whether or not $X_L Z_L$ is an adequate model. This is a problem any time series of tests are used to try to select a single model. An alternative is to use information criteria or some other measure to assess the fit of each model individually, rather than in comparison to some other model. These measures are discussed in greater detail in Chapter 5.

Example: Political ideology (PolIdeolOrd.R, PolIdeolData.csv)

We test the fit of the model that assumes a linear association in the PI interaction against the one that assumes a nominal association. The null hypothesis is H_0 : *The PI association is adequately described by the 1-2-3-4-5 scores*, and the alternative is simply H_a : Model GP, GI, PI. The test is shown below.

```
> anova(mod.homo.lin1.PI, mod.homo, test = "Chisq")
Analysis of Deviance Table
Model 1: count ~ gender * party + gender * ideol + party *
    lin.score1
Model 2: count ~ (gender + party + ideol)^2
  Resid. Df Resid. Dev Df Deviance P(>|Chi|)
1         7     8.3986
2         4     3.2454  3   5.1532    0.1609
```

The test statistic is $-2\log(\Lambda) = 5.15$ and the p-value=0.16, so using any common level of α we would not reject H_0. We can conclude that there is insufficient evidence that the extra parameters given by the nominal variable version of the model are needed. Less formally, this means that the assumption of loglinear odds ratios using the 1-2-3-4-5 score set provides a reasonable explanation for the association between party and ideology.

Further simplifications of this model can be considered. We can try using the same scores to describe the *GI* association; that is, we can specify that the odds formed from two ideologies must have a ratio for males vs. females that changes in a log-linear fashion with the distance between the ideologies (as measured by the scores). We can try the alternative 0-2-3-4-6 scores for either or both of the ordinal associations. And finally, and perhaps most interestingly, we can use the "extremeness" scores, 2-1-0-1-2, in the associations with party, gender, or both. In fact, we can even use these scores *in addition to* the ones that measure ideologies linearly. We leave these pursuits as exercises.

4.3 Poisson rate regression

Consider the following claim: "Most automobile accidents occur within five miles of home." If this claim is true, it would seem to imply that the most dangerous roads are close to your home—and close to our homes, and close to everyone else's home! An explanation that one might manufacture for this claim is that we become so familiar with the roads near our own homes that we relax too much when we drive near home, and thus open ourselves up to grave danger.

A much simpler explanation for this result is that perhaps most *driving* occurs within 5 miles of home! If this is so, then the *rate* at which we collide with others may be no different, or even lower, near home than elsewhere, but the *exposure* to potential accidents is much greater. In this context, counting the *number* of accidents at different distances from home may be a less meaningful measure of risk than estimating the *rate* of accidents *per mile driven*.

This highlights an important point that was made at the start of this chapter: an assumption behind the use of the Poisson model for counts is that both the intensity (or rate) of event occurrence and the opportunity (or exposure) for counting are constant for all available observations. In Section 4.2 we investigated models that allow a mean count to change. Implicitly we were assuming that the exposure was constant for all observations. For example, in counting the drinks consumed on a Saturday, the length of observation ("Saturday") was constant for all participants, so that only their rate or intensity of drink consumption varied. We then allowed mean counts (i.e., rate of consumption) to vary by using a Poisson regression model to relate them to explanatory variables.

In problems where exposure is not constant among all observations, modeling the counts directly leaves interpretations unclear. For example, if we measured some participants in the alcohol study for one hour and others for one day and others for a week or a month, we should expect that the number of drinks consumed among participants would differ simply because of the way we conducted our measurements. This might obscure any effects that the explanatory variables might have. We therefore would need to take the exposure effect out of consideration, so that we may get to the heart of the issue relating the explanatory variables to the *rate* of consumption.

Very generally, "rate" is defined as mean count per unit exposure (e.g., drinks per day, animals trapped per visit, accidents per mile driven, trees per hectare, and so forth). That is, $R = \mu/t$, where R is the rate, t is the exposure, and μ is the mean count over an exposure duration of t. When our observed counts all have the same exposure, modeling the mean count μ as a function of explanatory variables $x_1, \ldots, x_p$ is the same as modeling the rate. When exposures vary, we can still use a Poisson regression model for the means, but we need to account for the exposure in the model. This is done using a *Poisson rate regression* model by writing the model as:

$$\log(r_i) = \log(\mu_i/t_i) = \beta_0 + \beta_1 x_{1i} + \ldots + \beta_p x_{pi}$$

or

$$\log(\mu_i) = \log(t_i) + \beta_0 + \beta_1 x_{1i} + \ldots + \beta_p x_{pi}, \ i = 1, \ldots, n. \quad (4.12)$$

Notice that the $\log(t_i)$ term is known and is not multiplied by a parameter. It simply adjusts the mean for each different observation to account for its exposure. The term $\log(t)$ is called an *offset*, so this model is sometimes called *Poisson regression with offsets*. A direct consequence of Equation 4.12 is that $\mu = t \times \exp(\beta_0 + \beta_1 x_1 + \ldots + \beta_p x_p)$, which shows the direct influence of exposure on the mean count.

There is a similarity between rates and binomial proportions that sometimes causes confusion. Both are expressed as {counts of events}/{opportunities for events}. One might incorrectly use the number of trials in a binomial model as an exposure and apply a Poisson rate regression, or conversely might incorrectly think of the exposure in a Poisson rate problem as a number of trials and apply a logistic regression. However, the two situations are different, and the distinction is easily made by asking the following question: Can each unit of exposure conceivably produce more than one event, or must each unit of exposure produce exactly 0 or 1 event? If more than one event is possible in a single unit of exposure, then this violates the binomial model assumptions and the Poisson rate model is more appropriate. For example, there is nothing that prevents multiple accidents from happening in a single mile driven, and indeed a very drunk or careless driver may experience this misfortune. A binomial model would be inappropriate in this case. On the other hand, each placekick or free throw must be either a success or a failure, and one cannot count more than one successful event per trial. For these problems the logistic regression model should be used.

Parameters $\beta_0, \ldots, \beta_p$ from Equation 4.12 are estimated using maximum likelihood estimation as before, but their interpretation is in terms of the *rate* rather than the mean. For example, β_0 is the true event rate when all explanatory variables are set to 0, and β_1 is the change in rate of occurrence per unit increase in x_1. Note that the units of measurement for the rate are count-units per exposure-unit (e.g., accidents per mile driven).

It is sometimes helpful to convert the exposure to other units to create a more interpretable rate. For example, if we are analyzing accidents, and exposure is given as miles driven, the average number of accidents in 1 mile should be extremely small. If we prefer, we can choose to rewrite the rate as a more interpretable quantity like "accidents per 1000 miles" by dividing the exposure miles by 1000 before applying the model.

Given a set of estimates $\hat{\beta}_0, \ldots, \hat{\beta}_p$, predicted rates for given values $x_1, \ldots, x_p$ are found from $\hat{R} = \exp(\hat{\beta}_0 + \hat{\beta}_1 x_1 + \ldots + \hat{\beta}_p x_p)$. A predicted count for a given exposure t is found simply by multiplying the rate by the exposure, $\hat{\mu} = t \times \hat{R}$.

Example: Beetle egg-laying response to crowding (BeetleEggCrowding.R, BeetleEggCrowding.txt)[15]

Tauber et al. (1996) describe an experiment examining the effects of crowding on reproductive properties of a certain species of leaf beetle, *Galerucella nymphaeae*. Cages of a fixed size containing either 1 female and 1 male or 5 females and 5 males were kept in temperature chambers at either 21°C or 24°C. Among the measurements taken on each cage is the number of egg masses produced by the females in the cage over the lifetime of the females. We use Poisson regression to examine whether crowding and/or temperature influence the number of egg masses produced by each female. A complicating feature is that in cages with 5 females, there is no easy way to identify which females laid a given egg mass, so we must consider "egg masses per female" as the *rate* response of interest.

The data are entered using the code below. The temperature variable `Temp` takes values 21 and 24; `TRT`, which is the crowding treatment variable, takes values "I" for cages with individual females and "G" for cages with groups of 5; `Unit` identifies the cage within a combination of temperature and crowding level (there are 35 individual cages and 7 group cages at each temperature); `NumEggs` counts the egg masses laid in the box over the female's lifetime. We create a variable, "`females`" to represent the number of females in the cage that will serve as our exposure variable.

```
> eggdata <- read.table(file = "C:\\Data\\BeetleEggCrowding.txt",
    header = TRUE)
> eggdata$females <- ifelse(test = eggdata$TRT == "I", yes = 1,
    no = 5)
> head(eggdata)
  Temp TRT Unit NumEggs females
1   21   I    1       8       1
2   21   I    2       0       1
3   21   I    3       3       1
4   21   I    4       5       1
5   21   I    5       1       1
6   21   I    6       3       1
```

As an initial summary, the sample mean rates (egg masses per female) for each group were calculated within the program, yielding the results 3.54 for TRT=I, Temp=21, 4.34 for TRT=I, Temp=24, 0.51 for TRT=G, Temp=21, and 1.46 for TRT=G, Temp=24.

The model that we will fit is $\log(\mu) = \log(t) + \beta_0 + \beta_1 \text{Temp} + \beta_2 \text{TRTI} + \beta_3 \text{Temp} \times \text{TRTI}$, where TRTI is the indicator for TRT=I, and t is the number of females in the cage (1 or 5). In glm(), the offset is incorporated into the model using the `offset` argument as shown below:

---

[15] Data kindly provided by Dr. Jim Nechols, Department of Entomology, Kansas State University.

```
> eggmod1 <- glm(formula = NumEggs ~ Temp*TRT, family =
    poisson(link = "log"), offset = log(females), data = eggdata)
> round(summary(eggmod1)$coefficients, digits = 3)
            Estimate Std. Error z value Pr(>|z|)
(Intercept)   -7.955      2.125  -3.743    0.000
Temp           0.347      0.091   3.799    0.000
TRTI           7.795      2.314   3.369    0.001
Temp:TRTI     -0.279      0.100  -2.796    0.005
```

From the positive coefficients on the main effect terms, it is apparent that the egg-laying rate is higher at the higher temperature and in the individual cages. The negative interaction coefficient indicates that the ratio of rates between individual and group cages is higher at the lower temperature. Rates can be calculated from these coefficients. For example, the mean rate for TRT=I, Temp=21 is $\exp[-7.955 + 0.347(21) + 7.795(1) - 0.279(21)(1)] = 3.54$, matching the previous estimate. An easy way to get predicted rates is to use predict() on the original data frame. This produces one prediction for each cage, which we have abbreviated to one for each TRT and Temp combination in the output below. Note that this returns the predicted mean *count*, which we must divide by the exposure in order to form a rate. Mean counts for other levels of exposure can be estimated by multiplying these rates by the new exposure.

```
> newdata <- data.frame(TRT = c("G", "I", "G", "I"), Temp = c(21,
    21, 24, 24), females = c(5, 1, 5, 1))
> mu.hat <- round(predict(object = eggmod1, newdata = newdata,
    type = "response"), 2)
> data.frame(newdata, mu.hat, rate = mu.hat/newdata$females)
  TRT Temp females mu.hat  rate
1   G   21       5   2.57 0.514
2   I   21       1   3.54 3.540
3   G   24       5   7.29 1.458
4   I   24       1   4.34 4.340
```

The reason that the estimated egg-laying rates (rate) are identical to our sample estimates is that the model we are using is saturated. The four observed rates are fully explained by the 4 df accounted for by $\beta_0, \beta_1, \beta_2$, and β_3. These predicted rates would not usually match the observed mean rates in a reduced model.

Below is a summary of the LRTs for the model terms:

```
> library(car)
> Anova(eggmod1)
Analysis of Deviance Table (Type II tests)
Response: NumEggs
         LR Chisq Df Pr(>Chisq)
Temp       10.842  1   0.000992 ***
TRT       132.994  1   < 2.2e-16 ***
Temp:TRT    8.450  1   0.003650 **
```

The Temp:TRT interaction term is significant, indicating that the ratios of means between the two crowding treatments are not the same at both temperatures ($-2\log(\Lambda) = 8.4, \mathrm{df} = 1$, p-value=.004). Wald confidence intervals for the true egg-laying rate per female are produced using the usual coding. This is given in the

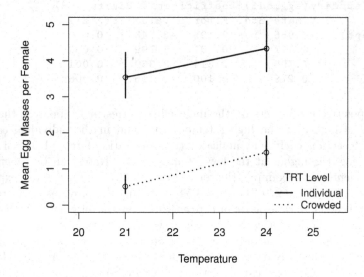

Figure 4.8: Mean rates of egg mass production and confidence intervals for all treatment-temperature combinations in the beetle egg-laying example. Individual cages have one female and male; crowded cages have five females and males. The vertical lines represent 95% Wald confidence intervals for the rate.

accompanying program for this example. They are depicted along with the estimated rates in Figure 4.8. The effect of crowding is apparent: fewer egg masses are laid per female in the cages with the crowded condition. The interaction effect is explored in more detail in Exercise 20.

Note that it sometimes happens that a rate may depend on the amount of exposure. For example, crime rates (e.g., crimes per 10,000 population) are often higher in cities with larger populations. This was also implicit in the beetle egg example above, where the crowding treatment related to the number of females in each box, which was also the exposure for the eggs/female rate. As is clear from the example, it is certainly possible to use the exposure variable as another explanatory variable in the regression, in addition to using it as an offset. That is, we can define $x_{p+1} = t$ (or $\log(t)$ if preferred) and add $\beta_{p+1} x_{p+1}$ to Equation 4.12. The dual use of the offset causes no problems for the model parameter estimation like those seen in the Political Ideology example in Section 4.2.6, because no parameter is estimated for the offset (in essence, the parameter is fixed at 1).

4.4 Zero inflation

Many populations consist of subgroups of different types of units. A particularly common situation is that a population consists of two classes: "susceptible" and "immune" individuals. For example, magnetic resonance imaging (MRI) machines are large and very expensive pieces of medical equipment. Some, but not all, hospitals have an MRI machine. If a hospital has one, it surely does not sit idle for long periods of time. Suppose that hospital administrators are asked in a survey to report the number of MRIs they perform in a month. Hospitals with no MRI machine are not performing any MRIs, while the rest will report some count other than 0.

As a different example, suppose that we record the number of fish caught on various lakes in 4-hour fishing trips in Minnesota. Some lakes in Minnesota are too shallow for fish to survive the winter, so fishing on those lakes will yield no catch. On the other hand, even on a lake where fish are plentiful, we may or may not catch any fish due to conditions or our own competence. Thus, the number of fish caught will be zero if the lake does not support fish, and will be zero, one, or more if it does. These situations are examples where the responses come from *mixtures* of distributions, and in particular are cases where one of the distributions places all of its mass on zero. That is, the "susceptibles" in the population return a count according to some distribution, while the "immunes" all return a zero.

Data that result from this type of mixture have the property that there are more zero counts than would be predicted by a single distributional model. This is referred to as *zero-inflation*. In general, let π be the proportion of immunes in the population, and let the counts for susceptibles have a $Po(\mu)$ distribution. This mixture is called the *zero-inflated Poisson* model. Figure 4.9 shows histograms of the PMFs from zero-inflated Poisson distributions with varying values of π and μ (note that $\pi = 0$ is the ordinary Poisson distribution). The zero inflation is obvious when $\mu >> 0$ as is the case in the top row. One would hardly be tempted to use a single Poisson distribution as a model for data with a histogram like those shown for $\pi > 0$. However, when the mean count among susceptibles can place appreciable mass on zero or one (bottom row), then the presence of zero inflation can go unnoticed if the nature of the problem does not suggest it in advance. The presence of explanatory variables can also blur the distinction between the classes, because then mean counts for susceptibles and probabilities of immunity can vary greatly among members of the population. In practice, zero inflation is often discovered only upon fitting a model and finding the fit poor using techniques as described in Chapter 5.

Corrective models

When a population is known or believed to consist of two subgroups as described above, or when an excess of zero counts appears unexpectedly, then ordinary Poisson models will likely yield poor fits to the data. Several models have been proposed to account for the excess zeroes. The main model is the zero-inflated Poisson (ZIP) model of Lambert (1992). The ZIP model specifies that immunity is a binary random variable that may depend on explanatory variables collectively denoted by z, while the counts among susceptibles follow a Poisson distribution whose mean may depend on explanatory variables collectively denoted by x. Thus,

$$Y = 0 \quad \text{with probability} \quad \pi(z)$$
$$Y \sim Po(\mu(x)) \text{ with probability } 1 - \pi(z)$$

Typically a logistic model is assumed for $\pi(z)$,

$$\text{logit}(\pi(z)) = \gamma_0 + \gamma_1 z_1 + \ldots + \gamma_r z_r.$$

while the usual log link is used for $\mu(x)$,

$$\log(\mu(x)) = \beta_0 + \beta_1 x_1 + \ldots + \beta_p x_p,$$

where $\gamma_0, \gamma_1, \ldots, \gamma_r$ and $\beta_0, \beta_1, \ldots, \beta_p$ are unknown regression parameters. It is permissible that $x = z$ so that the same explanatory variables are used both to distinguish the classes

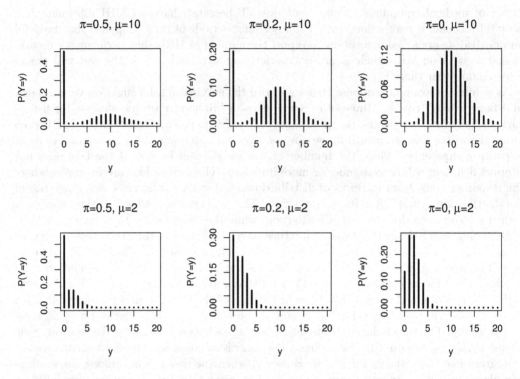

Figure 4.9: Probability mass histograms from zero-inflated Poisson models with varying means and levels of zero inflation. From program ZeroInflPoiDistPlots.R.

and to model the mean for susceptibles, but this is not required. Generally, x and z can be arbitrarily similar or different sets of variables. One can show that

1. $P(Y = 0|x, z) = \pi(z) + (1 - \pi(z))e^{-\mu(x)}$,

2. $E(Y|x, z) = \mu(x)(1 - \pi(z))$, and

3. $Var(Y|x, z) = \mu(x)(1 - \pi(z))(1 + \mu(x)\pi(z))$.

Thus, the mean of Y when $\pi(z) > 0$ is always less than $\mu(x)$, but the variance of Y is greater than its mean by a factor of $1 + \mu(x)\pi(z)$.

A relative of the ZIP model, called the *hurdle model* (Mullahy, 1986), differs in that a count is always greater than zero whenever some real or conceptual hurdle is crossed. The MRI example that started this section is of this type, because if a hospital has gone to the expense of installing an MRI machine (the "hurdle"), then they surely use it. In this case *all* zeroes correspond to immunes (those units that have not crossed the hurdle of installing a machine), while the count model for susceptibles is assumed to take values starting at 1 rather than 0.[16] The logistic model for probability of immune and the log link for the mean of the left-truncated Poisson distribution of counts for susceptibles are used in the hurdle model as in the ZIP model.

---

[16]This is done by imposing 0 probability at $Y = 0$ and then increasing all other probabilities proportionately so that they once again sum to 1. The resulting distribution is called a "left-truncated Poisson," because the left tail of the distribution ($Y = 0$) is cut off from the usual Poisson distribution.

Both models can be estimated by maximum likelihood using functions in the `pscl` package. The syntax for fitting both models is very similar and is described in detail by Zeileis et al. (2008). The model fitting can be easier with the hurdle model, because it is immediately clear that all zeroes are immunes. Then the modeling of the mean function is based on the subset of data with $Y > 0$ and is independent of the probability of immunes. However, the ZIP model seems to be preferred in most areas. Our preference is generally determined by which model for the immunes better agrees with the source of the data. If a zero count defines an immune, as with the MRI example, a hurdle makes better sense. If susceptibles can sometimes provide zero counts, as with the fishing example, then this points toward a ZIP model. If it is not clear whether susceptibles can produce zero counts, then empirical considerations such as measures of fit can preside (see Chapter 5 for details).

Example: Beetle egg-laying response to crowding (BeetleEggCrowding.ZIP.R, BeetleEggCrowding.txt)

One of the goals of the experiment in Tauber et al. (1996) was to observe a particular response called *diapause* in the females. Diapause occurs in insects when they temporarily cease biological development in some way as a response to environmental conditions. In this case, the researchers suspected that females' reproductive cycles might stop in response to temperature or perceived crowding. Therefore, they expected that females might enter diapause and produce no eggs with higher probability for some conditions than for others. The fact that some females in the group cages might enter diapause while others do not creates a complication because those cages of mixed females will still produce eggs, but at a reduced rate. Thus, estimating the probability that a particular female produces zero egg masses in a group cage for the entire study duration is an example of a group testing problem (see Bilder, 2009 for an introduction to group testing). This is beyond our current scope, so for this example we will limit ourselves to the individual cages. We will use a ZIP model to determine whether temperature affects either the probability of diapause or the mean number of eggs laid by females not in diapause. We leave the fitting of a hurdle model as Exercise 15.

First we examine the extent of the zero inflation. We use the following code to reduce the data to only the individual cages and to plot these data:

```
> eggdata2 <- eggdata[eggdata$TRT == "I",]
> stripchart(NumEggs ~ Temp, method = "stack", data = eggdata2,
    ylab = "Temperature", xlab = "Number of egg masses")
```

In our previous example, the estimated means were 3.54 for TRT=I, Temp=21 and 4.34 for TRT=I, Temp=24. Figure 4.10 shows that there are considerably more zeroes than would be expected from Poisson distributions with these means (the estimated probabilities are $P(Y = 0) = e^{-\hat{\mu}} = 0.029$ and 0.013, respectively). Furthermore, there are a few more zeroes at Temp=24 than at Temp=21 (16 vs. 11), and the means of the non-zero counts appear to differ between temperatures. Overall, the ZIP model is preferred for this setting because a female not in diapause could nonetheless produce a zero count (e.g., due to some other biological process such as infertility or early death).

Our ZIP model uses a logistic regression to describe the probability of "immunity" to egg-laying (i.e., diapause) and a Poisson regression on the counts for all females who are not "immune." In both models, we will allow the probabilities or the mean counts to vary according to temperature. That is, we are fitting

$$\log\left(\pi/(1+\pi)\right) = \gamma_0 + \gamma_1 \text{Temp}$$

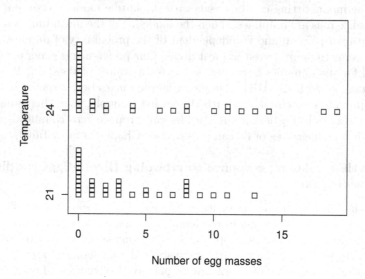

Figure 4.10: Histograms of beetle egg masses per female, showing zero inflation.

and
$$\log(\mu) = \beta_0 + \beta_1 \text{Temp}.$$

The ZIP model is fit using the function zeroinfl() from the pscl package. The main argument required for zeroinfl() is the model formula. In general, the syntax formula = y ~ x1 + x2 +...|z1 + z2 +... fits the count model to the x-variables and the probability-of-immune model to the z-variables. If the same variables are to be used in both models, as is the case for the beetles, the shortcut y ~ x1 + x2 ... can be used.

We first fit the model in which both the probability of immune and the means are allowed to vary by temperature:

```
> library(pscl)
> zip.mod.tt <- zeroinfl(NumEggs ~ Temp | Temp, dist = "poisson",
    data = eggdata2)
> summary(zip.mod.tt)

Call: zeroinfl(formula = NumEggs ~ Temp | Temp, data = eggdata2,
    dist = "poisson")

Pearson residuals:
    Min      1Q  Median      3Q     Max
-1.1688 -0.9659 -0.5211  0.8134  3.2601

Count model coefficients (poisson with log link):
            Estimate Std. Error z value Pr(>|z|)
(Intercept)  -1.4631     0.9249  -1.582 0.113672
Temp          0.1476     0.0407   3.626 0.000288 ***

Zero-inflation model coefficients (binomial with logit link):
```

```
              Estimate Std. Error z value Pr(>|z|)
(Intercept)   -5.1843     3.7897  -1.368   0.171
Temp           0.2088     0.1671   1.249   0.212

Number of iterations in BFGS optimization: 18
Log-likelihood: -187.2 on 4 Df
```

The output from the `zeroinfl()` call gives the parameter estimates for the Poisson mean model first, followed by those for the logistic probability model. The estimated mean function is $\hat{\mu} = \exp(-1.46 + 0.15 \text{ Temp})$, and the estimated probability of immune is $\hat{\pi} = \exp(-5.20 + 0.21 \text{ Temp})/(1 + \exp(-5.20 + 0.21 \text{ Temp}))$. The Wald tests for temperature effects suggest significance in the mean model but not in the probability model. We explore this more later.

There is no `anova()` method for objects produced by `zeroinfl()`, so we must instead use other tools for LRTs. The package `lmtest` provides a variety of testing and diagnostic tools for linear models, some of which have been extended to handle `zeroinfl`-class objects. In particular, the `lrtest()` function from the `lmtest` package can produce LRTs for comparing nested models. We use it here to compare the full model given in `zip.mod.tt` to the fits of three reduced models: (1) no temperature effect on π, (2) no temperature effect on μ, and (3) no temperature effect anywhere.

```
> # Fit ZIP Models: Temp in mean only
> zip.mod.t0 <- zeroinfl(NumEggs ~ Temp | 1, dist = "poisson",
    data = eggdata2)
> # Fit ZIP Models: Temp in probability only
> zip.mod.0t <- zeroinfl(NumEggs ~ 1 | Temp, dist = "poisson",
    data = eggdata2)
> # Fit ZIP Models: no explanatories
> zip.mod.00 <- zeroinfl(NumEggs ~ 1 | 1, dist = "poisson", data
    = eggdata2)

> # LRTs of each reduced model against largest model.
> library(lmtest)
> lrtest(zip.mod.t0, zip.mod.tt)
Likelihood ratio test

Model 1: NumEggs ~ Temp | 1
Model 2: NumEggs ~ Temp | Temp
  #Df  LogLik Df  Chisq Pr(>Chisq)
1   3 -187.97
2   4 -187.18  1 1.5866     0.2078

> lrtest(zip.mod.0t, zip.mod.tt)
Likelihood ratio test

Model 1: NumEggs ~ 1 | Temp
Model 2: NumEggs ~ Temp | Temp
  #Df  LogLik Df  Chisq Pr(>Chisq)
1   3 -193.82
2   4 -187.18  1 13.294  0.0002662 ***

> lrtest(zip.mod.00, zip.mod.tt)
Likelihood ratio test
```

```
Model 1: NumEggs ~ 1 | 1
Model 2: NumEggs ~ Temp | Temp
  #Df  LogLik  Df  Chisq  Pr(>Chisq)
1  2  -194.58
2  4  -187.18  2  14.808  0.0006087 ***
```

The first LRT shows that reducing from the full model zip.mod.tt, with temperature effects on both mean and probability, to zip.mod.t0, with temperature effects only on the mean, does not cause a significant drop in the observed value of the likelihood function ($-2\log(\Lambda) = 1.59$, df=1, p-value=0.21). However, removing the temperature effect from the mean does result in a significantly worse fit ($-2\log(\Lambda) = 13.29$, df=1, p-value=0.0002). We therefore would consider zip.mod.t0 to be a good working model, pending diagnostic evaluation as discussed in Chapter 5.

We summarize this model in several ways using code given in the program for this example. First, we estimate the effect of temperature as the ratio of mean egg-mass counts for each 1°C increase in temperature. We also find a 95% confidence interval for this quantity. The estimated ratio is $\exp(0.1476) = 1.16$, and the confidence interval runs from 1.07 to 1.25. Thus, with 95% confidence we estimate that the mean number of egg masses laid by females not in diapause increases between 7% and 25% for each 1°C increase in temperature.

Next, we calculate estimates and individual 95% Wald confidence intervals for the conditional mean egg-mass count in each temperature group (conditioned on insects that are susceptible) and for the overall probability of immunity (diapause). The code uses the multcomp package and is given in the program for this example. The results are: $\hat{\mu} = 5.14$ ($4.31 < \mu < 6.15$) at 21°C; $\hat{\mu} = 8.00$ ($6.82 < \mu < 9.38$) at 24°C; and $\hat{\pi} = 0.38$ ($0.28 < \pi < 0.50$). Thus, more eggs are laid by the non-diapause females at 24°C than at 21°C, and approximately 38% of females enter diapause regardless of temperature.

To compare the handling of zeroes in these ZIP models with the Poisson model fit in Section 4.3, we calculated the number of zero counts predicted by all four ZIP models and by the Poisson model. This is done by computing the estimated probability of a zero response for each observation based on its explanatory variables, and then summing these probabilities across the full data set. The code is given in the program for this example. The observed number of zeroes was 27. Each ZIP model comes very close to this total, with expected totals ranging from 27.04 to 27.11. The Poisson model that does not account for zero inflation predicts only 1.47 zeroes. Clearly, the Poisson model is not adequate for these data. However, diagnostics done in Exercise 33 from Chapter 5 reveal that the final ZIP model given in zip.mod.t0 has problems as well!

Example: Simulation comparing ZIP and Poisson models (ZIPSim.R)

A potential difficulty with the ZIP model is the fact that zero counts can arise from two separate sources: immune members of the population and susceptible members who happen to record a zero. Individual observations of zero generally do not provide any direct indication as to whether they are immune or susceptible. It is therefore difficult to separate the two sources precisely using only data on observed counts, particularly when the mean for susceptibles is close to zero, so that an appreciable fraction of the susceptible population yields counts of zero.

Table 4.7: Simulation results of ZIP parameter estimation from 1000 simulated data sets of $n = 35$ generated from a ZIP model. True confidence level was 95% and $\mu = 1$.

	$\hat{\pi}$				$\hat{\mu}$			
		Conf.	Median C.I. limit			Conf.	Median C.I. limit	
π	Mean	Level	Lower	Upper	Mean	Level	Lower	Upper
0	0.06	0	0^b	1	1.07	94.2	0.72	1.59
0.1	0.13	88.5	0.00	0.91	1.05	92.4	0.65	1.70
0.2	0.19	93.0	0.02	0.77	1.01	91.7	0.58	1.76
0.5	0.44	96.5	0.19	0.81	0.99	90.4	0.47	2.01
0.8^a	0.64	98.4	0.45	0.95	0.96	80.1	0.30	3.08
0.9^a	0.58	89.1	0.47	0.99	0.91	57.0	0.14	3.96

[a] Summaries based on 991 ($\pi = 0.8$) and 899 ($\pi = 0.9$) data sets.
[b] Rounded. Actual median is 7.2×10^{-99}. No lower confidence limits contain 0.

The purpose of this Monte Carlo simulation is to explore the properties of the estimates of π and μ under various conditions when their true values are known. We generate a large number of data sets from a ZIP model with fixed values for π and μ. We then use the ZIP model to estimate these parameters and find Wald confidence intervals for them. We calculate the average value of the parameter estimates so that we can check for bias in the estimates. We estimate the true confidence level of the Wald intervals and compute the median endpoints to see what a "typical" interval might look like.

The data generation is performed as follows. Given the sample size n for a single data set and the number of data sets we wish to create, say R, we generate nR independent observations of two independent random variables. The first is a Bernoulli random variable, say S, that creates the immune and susceptible members of the sample. We set $S = 1$ for susceptible for reasons that will be apparent momentarily. Then $P(S = 1) = 1 - \pi$. The second random variable is a count variable, say C, from $Po(\mu)$. Finally, we create ZIP observations by taking $Y = S \times C$. This way, immunes are generated with probability π and their counts are set to 0 because $S = 0$, while susceptibles are generated with probability $1 - \pi$ and their counts are distributed as $C \sim Po(\mu)$. We form a data matrix of these counts and apply the ZIP analysis function to them, one set of n at a time. We then compute confidence intervals for both μ and π.

We have chosen $n = 35$ because this was the sample size in the beetle egg-laying example above. We set $\mu = 1$ as a case where the count distribution will naturally produce a fairly large number of zeroes ($P(C = 0) = \exp(-1) \approx 0.37$) and explore values of $\pi = 0, 0.1, 0.2, 0.5, 0.8, 0.9$. Finally, we use $R = 1000$ so that we can obtain a reasonably good estimate of the distributions of $\hat{\pi}$ and $\hat{\mu}$ and for the true level of the corresponding confidence intervals.

Pilot runs highlighted a problem with the use of large values of π: some data sets were generated that contained 35 counts of 0. This created problems that are described in the program for this example. We therefore added code to omit these cases from the calculation of the summary statistics. The results from the simulations are presented in Table 4.7 and Figure 4.11.

It is evident from the results that it can sometimes be difficult to estimate the parameters from a ZIP model. When the probability of immune is very small, there is great uncertainty in the estimates of π, as indicated by the excessively wide confidence intervals. The estimated true confidence levels of the corresponding confidence intervals are too low, despite their vast widths. With $\mu = 1$ the susceptibles generate

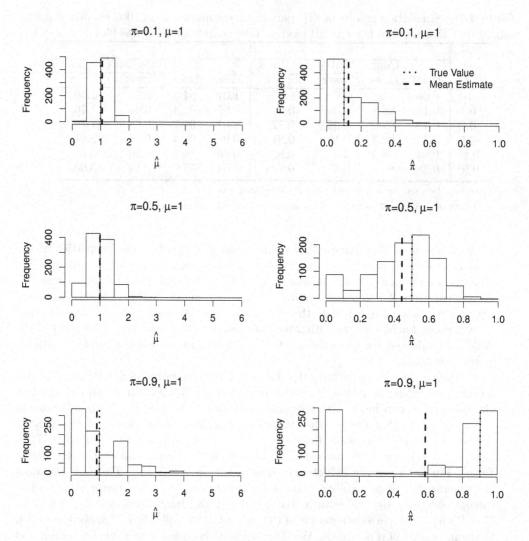

Figure 4.11: Simulated sampling distributions of estimates from ZIP model with various parameter values.

an average of 37% zeroes, and the immunes add very little to the total. The model estimates a near-zero probability of immune, but with an immense standard error on logit(π). Thus, the interval endpoints get transformed to nearly (but not quite) 0 and 1.

On the other hand, with very large values of π there are too few non-zero counts to provide reliable estimates of μ. The estimates $\hat{\mu}$ and $\hat{\pi}$ are mostly too small, indicating that the procedure tends to prefer attributing the many zeroes to a small μ rather than a large π. Most of this difficulty is associated with the small sample size. There is not enough information with which to reliably estimate a small mean and a large probability of immune with $n = 35$. Additional simulations using $n = 100$ and $n = 500$ with $\pi = 0.9$ provide mean estimates of 0.99 and 0.98 for μ and 0.83 and 0.90 for π, respectively.

From this example we can conclude that some caution is needed in using the zero-inflated Poisson models with small sample sizes, especially when there are very few non-zero observations. In Exercise 21 the simulations can be repeated for the hurdle model.

4.5 Exercises

1. There are many simple data-collection exercises that can be performed with the goal of estimating a population mean for a particular problem. We encourage you to perform your own. The parts below discuss some possible data collection ideas that would result in count data. For each part, outline how to take a sample and discuss any simplifying assumptions that may be needed in order to apply the statistical methods from this chapter. Collect the data and examine the appropriateness of the Poisson distribution by comparing the sample mean to the sample variance and comparing observed relative frequencies and the corresponding probabilities from a $Po(\hat{\mu})$ distribution.

    (a) Observe the number of people standing in line at a grocery store checkout(s).
    (b) Examine the number of Ph.D. graduates from a university's academic department.
    (c) Count the number of cars passing through an intersection of streets over a period of time.

2. Suppose you count 225 cars crossing a particular bridge between 8:00-9:00 a.m. on a "typical" Monday.

    (a) Assume that a Poisson distribution is a reasonable model for the process of cars crossing the bridge between 8:00 and 9:00 a.m. on "typical" Mondays. Find a confidence interval for the true mean car count for this time period.
    (b) If the intensity of cars crossing this bridge is variable from week to week, what impact will this have on your calculation? What impact would it have on the interval's true confidence level or even your interpretation?

3. The Great Plains region of the United States usually has a number of tornadoes during the non-winter months. Groups of scientists, called *storm chasers*, travel throughout the region when thunderstorms are expected in the hopes of encountering a tornado and taking various measurements on it. The file Stormchaser.csv contains the number of tornadoes encountered on 69 12-hour long storm chases during 1999-2004 for one particular storm chaser. Using these data, do the following:

    (a) Make a plot comparing the observed counts of tornadoes against the Poisson PMF. Comment on the fit of the Poisson model; does zero inflation appear to be a problem?
    (b) What different assumptions do the hurdle and ZIP models make regarding the cause of zero counts of tornadoes?

(c) There are many different kinds of thunderstorms, but storm chasers specifically seek those that seem to have the potential for tornadic development. With this in mind, describe what the parameter π would represent in a zero-inflated model.

(d) Fit both the ZIP and hurdle models, using only an intercept for both the mean and probability portions of the models. In each model, estimate both parameters with 95% confidence intervals and interpret the results.

4. Refer to the "Cars at an intersection" example in Section 4.1.1 to answer the questions below.

   (a) The data for this problem were collected between 3:25p.m. and 4:05p.m. on a non-holiday weekday. Ideally, it would also be important to understand the traffic patterns for other time periods. The number of cars stopped at the intersection is likely to vary due to factors such as time of day, day of the week, whether school is in session, and so forth. Discuss how these factors could have been accounted for in designing this study.

   (b) Are the observations truly independent? If not, discuss what assumptions must be made to use the Poisson distribution for this problem.

   (c) The lengths of vehicles and the distances between stopped vehicles vary. For the purpose of this problem, suppose all vehicles are 14 feet long with a distance between cars of 4 feet when stopped at the intersection. This suggests that 9 vehicles ($150/18 = 8.3$) or more will at least partially block the fire station's driveway. Using a Poisson distribution, estimate the probability this will happen for one stoplight cycle. Considering part (a), what caveats need to be placed on the interpretation of this probability?

   (d) Using the probability from part (c), estimate the probability that the fire station's driveway is at least partially blocked at least once over 60 cycles of the light (roughly one hour). Use the binomial distribution to help answer this problem.

5. Once each week for a year, you drive around a county and observe the number of houses being built. You record the number each week. Is it reasonable to think of these 52 counts as a random sample from a Poisson distribution with mean μ? Why or why not? (Hint: it can take several months, or even longer, to build a house. Is construction cyclic?)

6. Consider the Poisson regression models below involving x_1, x_2, and x_3. Derive the multiplicative change in the mean response for a c-unit change in x_2.

   (a) $\mu = \exp(\beta_0 + \beta_1 x_1 + \beta_2 x_2 + \beta_3 x_3)$
   (b) $\mu = \exp(\beta_0 + \beta_1 x_1 + \beta_2 x_2 + \beta_3 x_1 x_2 + \beta_4 x_3)$
   (c) $\mu = \exp(\beta_0 + \beta_1 x_1 + \beta_2 x_2 + \beta_3 x_2^2 + \beta_4 x_3)$

7. For a Poisson regression with $\mu_i = \exp(\beta_0 + \beta_1 x_{i1} + \ldots + \beta_p x_{ip})$, let $x_{i1}, x_{i2}, \ldots, x_{ip}$ be the explanatory variables corresponding to response y_i, $i = 1, \ldots, n$. Derive the log-likelihood in Equation 4.1.

8. In Equation 4.3, show that $\beta_i^X = \log(\mu_{i+}/\mu_{1+})$.

9. Consider a two-way contingency table with counts y_{ij}, $i = 1, \ldots, I$, $j = 1, \ldots, J$. Following Section 4.2.4, use Equation 4.2 to model the mean counts.

(a) Write out the log-likelihood $\log[L(\beta_0, \beta_2^X, \ldots, \beta_I^X, \beta_2^Z, \ldots, \beta_J^Z | y_{11}, \ldots, y_{IJ})]$ analogously to Equation 4.1.

(b) Differentiate the log-likelihood with respect to β_2^X, and set the result equal to zero. Show that this equation implies that $\hat{\mu}_{2+} = \sum_{j=1}^{J} y_{ij}$; i.e., the MLE for the row-marginal total is the sum of counts in that row. (This is also true for $i = 3, \ldots, I$.)

(c) Repeat this calculation for β_2^Z and hence infer that the MLE for a column total is the observed column total.

(d) Now consider adding interactions as shown on p. 219. How does this affect the results of parts (b) and (c)?

(e) Suppose that ordinal forms of interaction were added instead, but the main effects remained nominal. Would this affect the results from (b) and (c)? Explain.

10. For a $2 \times 2 \times 2$ contingency table, assume that the model is given by Equation 4.6.

(a) Write out the corresponding regression model in terms of indicator variables x_2, z_2, and w_2.

(b) Write out the log-means in a format like Table 4.3.

11. For the linear-by-linear association model given on p. 238, show that $\log(OR_{ii',jj'}) = \beta^{XZ}(s_i^X - s_{i'}^X)(s_j^Z - s_{j'}^Z)$.

12. The purpose of this problem is to further investigate the political ideology data of Section 4.2.6.

(a) Refit the model in Equation 4.11 by trying different score sets for I in the PI interaction as suggested in the example (0-2-3-4-6 and 2-1-0-1-2). Do any of them fit as well as 1-2-3-4-5?

(b) Consider the ordinality of I in the GI interaction.

   i. Starting from Equation 4.11 consider all three score sets for GI. Compare deviances for these models to the one for Equation 4.11 to see whether the reduced number of parameters in each still provides an adequate explanation for the association.

   ii. Is the association significant in each case?

(c) Use the GI linear association model with the smallest residual deviance from part (b) as the final model.

   i. Estimate the GI odds ratio for a 1-unit difference in scores. Interpret what this measures and find the Wald confidence interval for the estimate.

   ii. Compare the results of the ordinally estimated odds ratio to the nominal estimates obtained in the example on p. 237. In particular, compare the confidence intervals for each odds ratio. Do they give similar interpretations?

(d) Write a brief conclusion for the results of this analysis. Explain the nature of any associations that were found. Do not use mathematical symbols in this discussion.

13. In the downtown areas of very large cities, it is common for Starbucks locations to be within a block of one another. Why does Starbucks decide to put their locations so close together? One would expect that it has something to do with how busy a

current location is. If an order line is long, a potential customer may not even get into line, and instead leave without making a purchase, which is lost business for the store.

Using this as motivation, a Starbucks location in downtown Lincoln, NE, was visited between 8:00a.m. and 8:30a.m. every weekday for five weeks. The number of customers waiting in line was counted at the start of each visit. The collected data are stored within the file starbucks.csv, where Count (number of customers) is the response variable and Day (day of week) is the explanatory variable. Using these data, complete the following.

(a) What is the population of inference? In other words, define the setting to which one would like to extend inferences based on this sample.

(b) Construct side-by-side dot plots of the data where the y-axis gives the number of customers and the x-axis is for the day of the week. Describe what information this plot provides regarding the mean number of customers per day. In particular, does it seem plausible that the true mean count is constant across the days? (We recommend putting the factor values given within Day in their chronological order using the factor() function before completing this plot.)

(c) Using a Poisson regression model that allows different mean counts on different days, complete the following:

   i. Estimate the model.
   ii. Perform a LRT to determine if there is evidence that day of the week affects the number of customers waiting in line.
   iii. Estimate the ratio of means comparing each pair of the days of the week and compute 95% confidence intervals for these same comparisons. Do this both with and without control for the familywise confidence level for the family of intervals. Interpret the results.
   iv. Compute the estimated mean number of customers for each day of the week using the model. Compare these estimates to the observed means. Also, compute 95% confidence intervals for the mean number of customers for each day of the week.

(d) The hypotheses for the LRT in part (c) can be written as $H_0: \beta_2 = \beta_3 = \beta_4 = \beta_5 = 0$ vs. H_a: At least one $\beta_r \neq 0$. These hypotheses can be equivalently expressed as $H_0: \mu_{\text{Monday}} = \mu_{\text{Tuesday}} = \mu_{\text{Wednesday}} = \mu_{\text{Thursday}} = \mu_{\text{Friday}}$ vs. H_a: At least one pair of means is unequal, where μ_i represents the mean number of customers in line on day i. Discuss why these two ways of writing the hypotheses are equivalent. Write out the proper forms of the Poisson regression model to support your result.

14. Often the word "rate" is used informally to refer to something that would better be described as a proportion. For each of the situations below, indicate whether a logistic regression or a Poisson rate regression would be more appropriate:

(a) Crime rate: in different cities, measure counts of violent crimes relative to population sizes.

(b) Product defect rate for a particular product: on different days of production, measure the number of defective products produced (or sampled) relative to total number produced (or sampled).

(c) Egg-laying rate: on different plants, measure the number of insect eggs laid relative to the number of adult females placed on the plants.

(d) Alcohol consumption rate: at different soccer games, measure the number of beers sold relative to number of fans.

(e) Illegal parking rate: in different parking lots, measure the number of cars illegally parked relative to the total number of cars in the lot.

(f) Medical resource usage rate: for different patients, measure the number of doctor visits since last heart attack relative to time since last heart attack.

(g) Caesarean delivery rate: at different hospitals, measure the number of babies delivered by Caesarean section relative to total number of babies delivered at the hospital.

15. Use the `hurdle()` function in `pscl` to repeat the analysis of the beetle egg example from Section 4.4 with a hurdle model. Compare the results to those from the zero-inflated Poisson model. Are there any differences that lead you to prefer one model over the other? In terms of the beetles, what is the difference in assumptions between the two models regarding the source of zero counts? (Note that the coefficients in the probability model estimate $P(Y > 0|z)$, but `predict(..., type = "prob")` returns $P(Y = 0|z)$.)

16. Deb and Trivedi (1997) and Zeileis et al. (2008) examine the relationship between the number of physician office visits for a person (`ofp`) and a set of explanatory variables for individuals on Medicare. Their data are contained in the file dt.csv. The explanatory variables are number of hospital stays (`hosp`), number of chronic conditions (`numchron`), gender (`gender`; male = 1, female = 0), number of years of education (`school`), and private insurance (`privins`; yes = 1, no = 0). Two additional explanatory variables given by the authors are denoted as `health_excellent` and `health_poor` in the data file. These are self-perceived health status indicators that take on a value of yes = 1 or no = 0, and they cannot both be 1 (both equal to 0 indicates "average" health). Using these data, complete the following:

    (a) Estimate the Poisson regression model to predict the number of physician office visits. Use all of the explanatory variables in a linear form without any transformations.

    (b) Interpret the effect that each explanatory variable has on the number of physician office visits.

    (c) Compare the number of zero-visit counts in the data to the number predicted by the model and comment. Can you think of a possible explanation for why there are so many zeroes in the data?

    (d) Estimate the zero-inflated Poisson regression model to predict the number of physician office visits. Use all of the explanatory variables in a linear form without any transformations for the $\log(\mu_i)$ part of the model and no explanatory variables in the π_i part of the model. Interpret the model fit results.

    (e) Complete part (d) again, but now use all of the explanatory variables in a linear form to estimate π_i. Interpret the model fit results and compare this model to the previous ZIP model using a LRT.

    (f) Examine how well each model estimates the number of 0 counts.

17. Explore the effectiveness of ML estimation in the ZIP model using ZIPSim.R from Section 4.4. For example,

Table 4.8: Salamander count data.

Years after burn	12	12	32	20	20	27	23	19	23	26
Salamanders	3	4	8	6	10	5	4	7	2	8
Years after burn	21	3	8	35	2	19	8	25	33	35
Salamanders	6	0	2	6	1	5	1	5	4	10

(a) Examine how well it estimates the probability of diapause and mean number egg masses for the beetle egg-laying example at 21°C. This can be done using the function call `save.beetle <- sim.zip(n = 35, pi = 0.38, mu = 5.14, sets = 1000, seed = 1200201)`. Include the following in your examination.

  i. Recall that the Wald interval is based on the large-sample normality of $\log(\hat{\mu})$. Make a histogram and a normal quantile (QQ) plot (`qqnorm()`) of the estimated log-mean (the estimated means are stored in the first row of `save.beetle`) and indicate how much faith you have in the use of a normal approximation for the distribution of $\log(\hat{\mu})$. What does this suggest about the general quality of the Wald intervals for μ here?

  ii. Repeat for $\log(\hat{\pi}/(1 - \hat{\pi}))$ (from the fourth row of `save.beetle`).

(b) Rerun simulations as in the example from Section 4.4, but this time hold π constant at 0.8 and let μ vary across 0.5, 1, 2, 5, and 10. Summarize the results and draw conclusions about the effect of different population means on the quality of estimation of π when it is relatively large and $n = 35$. As an example, `save.05 <- sim.zip(n = 35, pi = 0.8, mu = 0.5, sets = 1000, seed = 29950288)` will generate the first results.

(c) Explore the effect of sample size on these conclusions. Repeat the simulations from this exercise using $n = 100$. How do the conclusions change?

18. The data in Table 4.8 are from a survey of salamander counts taken on plots of fixed size sampled from prairie regions in Manitoba, Canada.[17] The prairies are subject to periodic controlled burning to reduce the chance of uncontrolled wildfires. The research question being investigated is how salamander populations react to these controlled burns. Use a Poisson regression model to address this question. Also comment on the apparent fit of the model.

19. The file BladderCancer.csv contains data concerning recurrence of bladder cancer tumors in patients who had received previous treatment to remove a primary tumor (Seeber, 2005). Patients were observed for different periods of time post-surgery, which were measured in months, and have different sizes of primary tumor, classified as 0 if the tumor was small (< 3 cm) and 1 if large (> 3 cm). The question to be answered is whether the size of the initial primary tumor relates to the number of tumors (Tumors in the data file) discovered in follow-up. The different lengths of follow-up time represent different exposures to potential tumor development. Analyze these data.

20. Use the data from the egg-laying example from Section 4.3 to find confidence intervals for the ratio of egg-laying rates between individual and group cages separately at 21°C and 24°C. Use these results to explain the interaction. Interpret exactly what the interaction coefficient, −0.279, measures in terms of egg-laying rates.

---

[17] Data kindly provided by Dr. Carl Schwarz, Department of Statistics and Actuarial Science, Simon Fraser University.

21. Repeat the simulations from Section 4.4 for the hurdle model. How does its performance compare to the ZIP model? Draw some conclusions about the relative performance of the two models, making sure to indicate the limitations of your conclusions. (Considering that the data were actually generated from a ZIP model, this is a demonstration that sometimes one gets better answers from an approximate model than from the correct one!)

22. Count data are often modeled using the usual linear regression and ANOVA models that assume normally distributed responses. While such models ignore the discrete nature of count data and the potential for unequal variances, they are supported by the fact that the Poisson distribution is increasingly well approximated by the normal distribution $N(\mu, \mu)$ as μ increases. The program PoissonNormalDistPlots.R generates a probability mass histogram for a Poisson distribution with $\mu = 1$ and a probability density curve for the corresponding normal. Try different values of μ with this program and observe the relationship between the two distributions. Above what value of μ do you feel that the normal approximation seems excellent? Below what value does it seem poor?

23. McCullagh and Nelder (1989, p. 205) provide data on the number of reported incidents of wave damage incurred by ships of five different types (ShipType), four different construction periods (ConstYear), and two different operation periods. An amended version of these data, consisting of just one of the two periods of operation, are in the file ships.csv. There are different numbers of ships of each type, and different years of construction, and hence different potential for damage incidence in each combination of year and ship type. There is an additional measure, "Aggregate months of service" (Months), that represents the exposure.

    (a) Use a Poisson rate regression with variables ShipType and ConstYear to test whether each affects the rate of damage incidents per month. Interpret all tests using $\alpha = 0.05$ and draw conclusions. Note that there is one observation for which there was no exposure! This will cause an error if left in the data. It can be removed by adding a condition to the data in the glm() function call, such as dataframename[Months>0,].

    (b) Use a 95% confidence interval to estimate the geometric mean incident rate per 1000 months of service for each ship type across all years of construction. Also perform pairwise comparisons at $\alpha = 0.05$. Which ships appear to be least or most prone to damage? Make a plot to support your conclusions.

    (c) Now include in the model an interaction between ship type and construction period.
        i. Explain what such an interaction term is attempting to measure.
        ii. Does there appear to be a significant interaction between ship type and construction year? Explain.

    (d) Draw conclusions about the problem without referring to mathematical symbols.

24. In Section 4.2 we analyzed alcohol consumption as a function of number of positive and negative events for a sample of moderate-to-heavy drinkers during their first Saturday on the study. Repeat the analysis for Friday and draw conclusions. Are the results similar to those for Saturday?

25. The researchers in the alcohol consumption study proposed the following hypothesis (DeHart et al., 2008, p. 529): "We hypothesized that negative interactions with romantic partners would be associated with alcohol consumption (and an increased desire to

drink). We predicted that people with low trait self-esteem would drink more on days they experienced more negative relationship interactions compared with days during which they experienced fewer negative relationship interactions. The relation between drinking and negative relationship interactions should not be evident for individuals with high trait self-esteem." In DeHartSimplified.csv, trait self-esteem (a long-term view of self-worth) is measured by the variable `rosn`, while the measure of negative relationship interactions is `nrel`. Conduct an analysis to address this hypothesis, using the data for the first Saturday in the study.

26. Consider an expanded regression model for the alcohol consumption study from Section 4.2. We will use a model that regresses the number of drinks consumed (`numall`) against positive romantic-relationship events (`prel`), negative romantic-relationship events (`nrel`), age (`age`), trait (long-term) self-esteem (`rosn`), state (short-term) self-esteem (`state`), and two other variables that we will create below. We will again use Saturday data only.

The `negevent` variable is the average of the ratings across 10 different types of "life events," one of which is romantic relationships. We want to isolate the relationship events from other events, so create a new variable, `negother`, as 10*`negevent` − `nrel`. Do the same with positive events to create the variable `posother`.

   (a) Construct plots of the number of drinks consumed against the explanatory variables `prel`, `nrel`, `posother`, `negother`, `age`, `rosn`, and `state`. Comment on the results: which variables seem to have any relationship with the response?

   (b) Fit the full model with each of the variables in a linear form. Report the regression parameter estimates, standard errors, and confidence intervals. Do these estimates make sense, considering the plots from part (a)?

   (c) Conduct LRTs on the regression parameters to determine which corresponding variables make a significant contribution to the model. State the hypotheses, test statistic, p-value, and use the results to draw conclusions regarding the contributions of each variable to the model.

   (d) Determine whether any variables except the two negative events variables are needed in the model. To do this, refit the model with only `nrel` and `negother`. Perform a LRT comparing the full model above with this reduced model. State the hypotheses, test statistic and df, p-value, and conclusions.

27. Suppose you take a random sample, $y_{11}, y_{12}, \ldots, y_{1n_1}$, from a $Po(\mu_1)$ distribution and an independent random sample, $y_{21}, y_{22}, \ldots, y_{2n_2}$, from a $Po(\mu_2)$ distribution.

   (a) Write out a single likelihood function for both parameters using both samples.

   (b) Find the MLEs for μ_1 and μ_2 from this likelihood. Do the estimates make sense?

   (c) Now consider the test $H_0 : \mu_1 = \mu_2$. Under H_0, both samples are drawn from the same Poisson distribution, say $Po(\mu_0)$. Show that the MLE for μ_0 is $(n_1\bar{y}_1 + n_2\bar{y}_2)/(n_1 + n_2)$, where $\bar{y}_j = (1/n_j)\sum_{i=1}^{n_j} y_{ji}$.

   (d) Derive the LRT statistic for this test.

28. Prove the three properties of the zero-inflated Poisson model listed on p. 246.

29. Refer to Table 1.7 and Exercise 17 in Chapter 1. Re-analyze these data using a Poisson log-linear model.

(a) Fit the saturated loglinear model and report the parameter estimates.

(b) Explain what independence of the two variables means in the context of the problem.

(c) Use a LRT to test for independence between strategy and field goal outcome. Report all parts of the test and draw conclusions.

(d) Explain what the population odds ratio measures in the context of this problem.

(e) Estimate the odds ratio between strategy and field goal outcome. Include a 95% confidence interval for the population odds ratio.

(f) Draw conclusions from the analysis: Does calling a time-out appear to reduce the kicker's field goal success rate?

30. Use Poisson regression methods to perform the analysis suggested in Exercise 20 in Chapter 3.

31. Recall the fiber-enriched cracker example from Table 3.2 in Section 3.2.

   (a) Consider at least three different possible sets of scores for the ordinal response variable, bloating severity. Explain why each might be relevant.

   (b) Re-analyze these data using loglinear models. Draw conclusions about the effect of the fiber source on bloating.

32. The data file pregnancy.csv gives results on a study of *pre-eclampsia* (also known as *toxaemia*), a medical condition affecting women in pregnancy. It is indicated by the presence of both high blood pressure, or *hypertension* (HT), and protein in the urine, or *proteinurea* (PU). The cause of pre-eclampsia is not immediately known, and other medical conditions may cause either of these symptoms to appear without the other. It is therefore of interest to understand the factors that relate to HT, PU, and their association with each other. Brown et al. (1983) report on a study in which women in their first pregnancy were classified according to smoking status (Smoke, with levels 0, 1–19, 20+, coded as 1, 2, 3), social class (Social, with levels I, II, III, IV, V, coded as 1, 2, 3, 4, 5), and presence of symptoms HT (yes/no) and PU (yes/no). The goal was to see whether smoking and social class relate to either symptom's frequency and/or whether they are associated with the symptoms' interaction. Use these data to perform the following analyses.

   (a) Examine the data through the following steps:

   i. Convert the explanatory variables Smoke and Social to variables Smokef and Socialf using as.factor().

   ii. Use these new factors to create a cross tabulation of the data (e.g., xtabs(formula = Count ~ HT + PU + Smokef + Socialf, ...)).

   iii. Use ftable(..., col.vars = c("HT", "PU")) to create a nicer-looking table and print out both the counts and the proportions (use prop.table() for the latter).

   iv. Comment on these table summaries: What effects do the explanatory variables have on the two symptoms?

   (b) Treat social class and smoking status levels as nominal for now. Consider a loglinear model using formula = Count ~ (Smokef + Socialf + HT + PU)^3.

i. Assign the letters S, C, H, P to represent the four variables in this model, respectively. Use the notation given in Section 4.2.5 to describe the model being fit and identify all of its terms.
ii. Write out the estimated model.
iii. Considering the study's objectives, why is this a reasonable starting point?
iv. Which terms in the model address whether the explanatory variables relate to the symptoms' frequency?
v. Which terms address whether they relate to the symptoms' association?

(c) Fit the model and perform the following analyses:

i. Use the residual deviance to test whether the excluded four-variable interaction is needed in the model. Report results and draw conclusions.
ii. Use `Anova()` from the `car` package to test the significance of terms identified in part (b) as being important for the analysis. Draw conclusions.

(d) Fit the simpler model that omits all 3-variable interactions and perform the following analyses:

i. Test whether the simpler model provides an adequate fit relative to the model that includes all 3-variable interactions. State the hypotheses, test statistic, and p-value. Draw conclusions.
ii. According to these results, are smoking or social class related to the association between proteinurea and hypertension? Explain.
iii. Use odds ratios based on the SH term to examine the relationship between smoking and hypertension. Compare all pairs of smoking levels. List the estimated odds ratios and 95% confidence intervals and draw conclusions.
iv. Repeat this for the relationship between social class and proteinurea.
v. Analyze the HP odds ratio in the same way. Draw conclusions.

33. Using the data described in Exercise 32, consider treating smoking as an ordinal variable.

    (a) Discuss possible choices of scores.
    (b) Starting from the model with all two-variable interactions, use 1-2-3 scores to analyze smoking effects on HT and PU. (Leave the smoking main effect nominal as well as its interaction with social class.)
    (c) Compare the fit of the ordinal model relative to the nominal model from Exercise 32d. Draw conclusions about the ordinal model.

34. The `datasets` package in R contains a `table` object called `HairEyeColor` consisting of a cross-classification of 592 people by hair color (black, brown, red, blond) and eye color (brown, blue, hazel, green). According to Snee (1974), the sample was taken as a class project by students in a Statistics class at the University of Delaware. It is not clear how the sampling was conducted.

    (a) The data were later artificially split out by sex by Friendly (1992), so the `HairEyeColor` table is actually a 3-way table with variables Sex, Hair, and Eye. We first need to remove the artificial sex-split using `margin.table(x = HairEyeColor, margin = c(1,2))`. The result should be a 4×4 table. Print out the resulting table and comment on the results. Do there appear to be any associations between the hair color and eye color?

(b) Fit the model in Equation 4.4 to the data. Obtain the summary of the fit. Explain what the symbols in Equation 4.4 mean in terms of the hair and eye colors.

(c) Write out the log-odds ratio comparing the odds of blue vs. brown eyes for someone with blond hair vs. brown hair as in Equation 4.5. Estimate the odds ratio with a 95% confidence interval and interpret the result.

35. Continuing Exercise 34, consider the hypothesis that brown and black hair have the same eye-color distribution and that red and blond hair have the same eye-color distribution, but that these two distributions differ.

    (a) Create a new variable (say Hair2) to represent these two groups ("dark" vs. "light" hair), and re-fit the model using this new variable instead of the variable Hair. Perform a LRT of the proposed hypothesis. Provide the test statistic and p-value and draw conclusions about the hypothesis.

    (b) A second hypothesis is that eye color can be reduced to two groups: "dark"=brown, and "light"=(blue, hazel, green). Starting from the model on the full 4 × 4 hair-color data, repeat the LRT analysis on the eye-color hypothesis.

36. Long (1990) gathered data comparing the publication success of recent male and female Ph.D. recipients in biochemistry. The data are available in the data frame bioChemists in the pscl package. The response is art, the number of publications produced in the three-year period ranging from one year before the year in which the Ph.D. was granted to one year after. Explanatory variables available include fem, an indicator for gender of the subject, mar, an indicator for marital status in the year of the Ph.D., kid5, indicating whether the subject had children aged 5 or less, phd, a numerical index of the Ph.D. department's prestige, and ment, the number of journal articles published by the student's mentor over the same period. The main research hypothesis is that the various factors, other than gender, have a more serious effect on females' publication rates than on males'. That is, gender likely interacts with the other factors in such a way that other explanatory variables have more of an effect on females, especially in reducing publications, than on males.

    (a) Make a histogram of art and note the large number of zeroes.
    (b) Analyze these data, making sure to address the hypotheses.

Chapter 5

Model selection and evaluation

In Chapters 1-4 we developed a variety of probability models for analyzing different types of categorical data. In each case we started with a fixed set of explanatory variables and explored techniques of model fitting and inference assuming that the model and the chosen variables were correct. However, each of these probability models is an *assumption* that may or may not be satisfied by the data from a particular problem. Also, in practice there is often uncertainty regarding which explanatory variables are needed in a model. Indeed, the goal of many categorical regression analyses is to identify which variables from a large number of candidates are associated with a response or with one another, and which are not.

"Model selection" consists of both (1) identifying an appropriate probability model for a problem, and (2) identifying an appropriate set of explanatory variables to be used in this model. In this chapter we first present techniques that can be used to select an appropriate set of explanatory variables from among a larger pool of candidate variables. We show both classical techniques and more recent developments that have distinct advantages over their older counterparts. Once the variable-selection process is addressed, we then explore methods for assessing whether the assumptions that surround our probability model are satisfied. We introduce residuals and show how they can be used in plots and tests to identify model assumptions that may be violated. In particular, we identify a very common model violation—*overdispersion*—and discuss its causes, its impacts on inferences, and adjustments that can be made to a model that can reduce its adverse effects. We conclude the chapter with two worked examples.

5.1 Variable selection

Variable selection refers to the process of reducing the size of the model from a potentially large number of variables to a more manageable and interpretable set. There are many approaches to selecting a subset of variables from a larger pool. All have strengths and weaknesses, and new approaches are being developed each year. We will first present the methods that are historically the most often used, and indeed are still in common use today. However, these methods are no longer recommended by experts on variable selection. We will present critiques of these methods and offer alternatives that are generally more reliable.

5.1.1 Overview of variable selection

Why do variable selection at all? Why not simply use the model with all available variables in it? There are two answers to these questions. First, in many modern applications the number of available variables may be too large to form an interpretable model. Part of the goal of the analysis may be to try to understand which variables are actually related to the response. Furthermore, in areas like genetics and medicine, the number of available variables may be larger than the number of observations, a problem referred to as "$p > n$." In these applications, the full model cannot actually be fit, which means that some sort of variable selection is needed.

The second answer has to do with something called the *bias-variance tradeoff*. Errors made by a model's predictions are a result of two sources: bias and variance. Bias in the specification of a model includes the inability to guess the correct relationship between response and the variables in the model, as well as errors in the choice of variables used in the model. Using all available variables would seem to be a remedy for the second of these two problems, and indeed minimizes the bias due to variable selection. However, the process of estimating parameters adds variability to any predicted values, because the estimate does not usually match the true parameter value exactly. To see this, consider fitting a model to, say, 10 variables when only one is really important. Although the true parameter values should be 0 for the other nine variables, they will likely be estimated by some random non-zero amount. Even if we were to guess the important variable's parameter *exactly*, the predicted values would vary randomly from their true means due to the noise added by the parameter estimates. The more unimportant variables we add to a model, the more random noise gets added to the predictions. It can even be the case that a variable's true effect on the response is not zero, but is so small that the increase in bias caused by leaving it out of a model is smaller than the variability added from having to estimate its parameter. When the main goal of an analysis is predicting new observations, it may be worthwhile to use a model that has *too few* variables in it rather than one that has all of the right ones. This is known as *sparsity* or *parsimony* in the variable-selection literature. On the other hand, in problems where understanding the relationships between the response and the explanatory variables is the main goal, we do not want to overlook small contributions. Both goals are worthwhile in different contexts. Exercise 1 explores the bias-variance tradeoff in the context of logistic regression.

Variable selection is therefore a common and important part of many statistical analyses. All of the variable-selection techniques we discuss apply to any of the probability models from the previous chapters. We therefore begin by assuming that we have selected a probability model that we believe is appropriate for our problem (e.g., Poisson, binomial, or multinomial), and that there is a "pool" of P explanatory variables, $x_1, \ldots, x_P$, that are candidates for inclusion into a regression model. In accordance with generalized linear model notation, we assume that the parameters of the model (e.g., the mean μ for a Poisson model or the probability π for a binomial) are linked to the regression parameters $\beta_0, \beta_1, \ldots, \beta_P$ via an appropriate link function $g(\cdot)$ (e.g., log or logit; see Section 2.3). We therefore express all of our models in terms of the linear predictor $g(\cdot) = \beta_0 + \beta_1 x_1 + \ldots + \beta_P x_P$.

Obviously, we want the final model to contain all of the important variables and none of the unimportant ones. If knowledge or some theory about the problem tell us that certain variables are definitely important, then these should be included in the model at the start and are no longer considered for selection. Similarly, if any variables are somehow known to be unrelated to the response, then these can be immediately excluded from the model.

The P remaining variables form the pool upon which the selection process should focus. This pool of variables may include transformations of other variables in the pool or interactions among variables. We need to somehow create models out of these variables, compare them using some criterion, and select variables or models that give the best results. We first discuss popular criteria for comparing models. Then we describe algorithms for creating models that can be compared using these criteria. We finally explain how the selected models should (and should not) be used.

5.1.2 Model comparison criteria

Comparison of two models can be done using a hypothesis test as long as one of the models is *nested* within the other. That is, the larger model must contain all of the same variables as in the smaller model, plus at least one additional variable. For example, models $g(\cdot) = \beta_0 + \beta_1 x_1$ and $g(\cdot) = \beta_0 + \beta_1 x_1 + \beta_2 x_2$ can be compared using a hypothesis test, while models $g(\cdot) = \beta_0 + \beta_1 x_1$ and $g(\cdot) = \beta_0 + \beta_2 x_2$ cannot. This severely limits the use of hypothesis tests for variable selection. Instead, a criterion is needed that can assess how well *any* model explains the data.

When models are fit using ML estimation, the residual deviance provides an aggregate measure of how far the model's predictions lie from the observed data, with smaller values indicating a closer fit. It might seem that deviance could be used as a variable-selection criterion in the sense that a model with a smaller deviance would be preferred over one with a larger deviance. However, like the sum of squared errors in linear regression, residual deviance cannot increase when a new variable is added to a model. The model with all available variables always has the smallest deviance, or equivalently the largest value for its evaluated likelihood function.

The residual deviance or log-likelihood value needs to be adjusted somehow so that adding a variable does not automatically improve the measure. *Information criteria* are measures based on the log likelihood that include a "penalty" for each parameter estimated by the model (see, e.g., Burnham and Anderson, 2002). If the penalty is sufficiently large, then adding variables to a model improves the likelihood but also increases the penalty, and the combination can result in either a better or a worse value of the criterion.

Numerous versions of information criteria have been proposed that use different penalties for model size. Let n be the sample size, r be the number of parameters in the model (including regression parameters, intercepts, and any other parameters in the model, such as the variance in normal linear regression) and let $\log(L(\hat{\boldsymbol{\beta}}|y_1,....y_n))$ be the log likelihood of an estimated model, evaluated at the MLEs for the parameters. The general form of most information criteria is

$$IC(k) = -2\log(L(\hat{\boldsymbol{\beta}}|y_1,....y_n)) + kr,$$

where k is a penalty coefficient that is chosen in advance and used for all models. For a given k, smaller values of $IC(k)$ indicate models that have a large log-likelihood relative to their penalty, and hence are better models according to the criterion.

The three most common information criteria are:

1. Akaike's Information Criterion:

$$\text{AIC} = IC(2) = -2\log(L(\hat{\boldsymbol{\beta}}|y_1,....y_n)) + 2r$$

2. Corrected AIC:

$$\text{AIC}_c = IC(2n/(n-r-1)) = -2\log(L(\hat{\boldsymbol{\beta}}|y_1,....y_n)) + \frac{2n}{n-r-1}r = \text{AIC} + \frac{2r(r+1)}{n-r-1}$$

3. Bayesian Information Criterion

$$\text{BIC} = IC(\log(n)) = -2\log(L(\hat{\boldsymbol{\beta}}|y_1,....y_n)) + \log(n)r$$

The penalty coefficient k for AIC_c is always larger than that for AIC. It increases as the model size increases, but becomes more similar to the AIC penalty when the sample size is large relative to the model size. The BIC penalty becomes more severe when n is large. It is larger than the AIC penalty for all $n \geq 8$, and larger than the AIC_c penalty whenever $n - 1 - (2n/\log(n)) > r$, which is the case in many problems.

Information criteria are used for variable selection as follows. Suppose we have a group of models that differ in the regression variables that they use. We select a k and compute $IC(k)$ on all models. The model with the smallest $IC(k)$ value is the one preferred by that information criterion. Notice that this does not require the nesting of models, which gives information criteria a distinct advantage over hypothesis tests for variable selection.

It is possible that many models have values of $IC(k)$ that are near that of the best model. Roughly, models whose $IC(k)$ values are within about 2 units of the best are considered to have similar fit to the best model (Burnham and Anderson, 2002, p. 70; see also Section 5.1.6). The more models that are near the best, the less definitive the selection of the best model is, and the greater the chance that a small change in the data could result in a different model being selected. For this reason, *model averaging* is becoming more popular in variable selection. We discuss this topic in Section 5.1.6.

Notice that using larger values of k results in a criterion that favors smaller models. Thus, AIC tends to choose larger models than BIC and AIC_c, while BIC will often choose smaller models than AIC_c. Opinions vary as to which among these is best, as there are theoretical results that support the use of each of these. BIC has a property called *consistency*, which means that as n increases, it chooses the "right" model (i.e., the one with the same explanatory variables as the model from which the data arises) with probability approaching 1, assuming that the true model is among those being examined. AIC enjoys a property called *efficiency*, meaning that, asymptotically, it selects models that minimize the mean squared error of prediction, which accounts for both the error in estimating the coefficients and the variability of the data. AIC_c was developed for linear models to extend the efficiency property to smaller samples. It is used in generalized linear models to approximately achieve the same goal, although no theory exists to show that the correction succeeds in this purpose.

Given that parsimonious models are generally preferred over models with too many variables, we generally prefer to use AIC_c or BIC for variable selection. Both AIC and BIC are generally easily computed in R using the generic function `AIC()`, which can be applied to any model-fit object that produces a log-likelihood that can be accessed through the `logLik()` generic function. Setting the argument value k = 2 in `AIC()` gives AIC, while k = log(n) gives BIC, where n is an object containing the sample size of the data set. There is no automatic way to calculate AIC_c, but it is typically not hard to obtain from an AIC calculation using the second equation in point 2 above.

5.1.3 All-subsets regression

Given a criterion for comparing and selecting models, the next step is deciding which models to compare. Perhaps the most obvious thing to consider is to make every possible combination of variables and select the one that "fits best." Indeed, this approach—called *all-subsets regression*—is popular where it can be used and we describe it in more detail below.

When there are P candidate explanatory variables, including any transformations and interactions that might be of interest, then there are 2^P different models that be formed. We refer to this set of all possible models as the *model space*. All-subsets regression for generalized linear models is limited to problems in which P is not too large, because each model must be fitted using iterative numerical techniques. While modern computers can be very fast, the sheer scope of the problem can be overwhelming. For example, if $P = 10$ then there are just over 1000 models to be fit, and this can be done fairly quickly in most cases. If $P = 20$, the number of models is over 1 million, while if $P = 30$, the number is over

1 billion, and time or memory requirements may make it infeasible to exhaustively search the entire model space.

Assuming that the task is possible, then for a chosen k an information criterion $IC(k)$ is computed for each of the 2^P models. The model with the smallest $IC(k)$ is considered "best," although it is quite common that there are numerous models with similar values of $IC(k)$. This is especially likely when P is large, so that there are many possible variables whose importance is borderline.

When an exhaustive search is not possible, an alternative search algorithm can be used to explore the model space without evaluating every model. These algorithms can be either deterministic (for a given data set, they explore a specific set of models) or stochastic (they can explore different sets of models every time they are run). Several deterministic algorithms are described in Sections 5.1.4 and 5.1.5. An example of a stochastic search algorithm is the *genetic algorithm* (Michalewicz, 1996). This algorithm makes a first "generation" of models out of randomly selected combinations of variables, and then makes a new generation of models by splicing together parts of those models from the previous generation that perform the best. This step is iterated many times with occasional random additions and deletions of variables (called *mutations*). The combining structure allows the algorithm to focus on combinations of variables that work well together, while the mutations allow it to explore the model space more broadly. The algorithm eventually converges to a "best" model according to $IC(k)$. This model is not guaranteed to be the one with the smallest $IC(k)$ among all possible models, but in many problems it does find the best model or one very close to it. Because it is a random search process and the optimal selection is not guaranteed, we recommend running the algorithm a few times to see whether better models can be found.

Example: Placekicking (AllSubsetsPlacekick.R, Placekick.csv)

We use the S4-class `glmulti()` function from the `glmulti` package to perform all-subsets regression on the placekicking data (see page 65 for a detailed description of the variables, and page 184 for tips on using S4 objects).[1] The package is described in more detail by Calcagno and de Mazancourt (2010). This package has several appealing features that most other similar packages do not have: (1) it can fit a wide variety of model families, including all of those from Chapters 2–4; (2) it can be set to automatically include two-variable interactions without having to specify them individually; (3) when an interaction involving a categorical explanatory variable is included in the pool, it includes or excludes the entire set of interaction variables that are created from the products involving the categorical variable's indicator variables[2]; and (4) for very large search problems, it can use the genetic search algorithm to attempt to find a best model.

---

[1] The `glmulti` package depends on the `rJava` package, which is installed and loaded together with `glmulti`. We have encountered numerous difficulties with installing and loading `rJava` on our Windows-based computers. In particular, `rJava` requires an installation of Java software for the same architecture on which R is running (i.e., 32- or 64- bit; see http://java.com/en/download/manual.jsp). Also, at least on Windows machines, `rJava` may have difficulty finding the Java installation. As a precaution, running `Sys.setenv("JAVA_HOME" = "")` prior to `library(glmulti)` will usually solve the problem. Numerous resources on the Internet, such as those listed in Appendix A.7.4, can help to resolve other issues that may arise.

[2] Recall that categorical variables are coded as groups of indicator variables for regression. Some alternative packages treat the individual indicator variables and subsequent interaction products as separate variables that are selected individually. This can drastically increase the number of variables in a pool, and hence result in greater difficulty in selecting an appropriate model.

The argument y in `glmulti()` can contain a formula, as shown below, or an object like a previous `glm()` fit of the full model. The `fitfunction` argument takes the name of any R function that can evaluate a log likelihood from a `formula`-class object, including "glm", "lm", or similar user-defined functions. Other important arguments include `level` (1 for main effects only, 2 to include two-way interactions) and `method` ("h" for exhaustive search, "g" for genetic algorithm). When `level = 2` is chosen, adding `marginality = TRUE` limits the interactions to only those based on variables whose main effects are also in the model. The `print()` generic function is used instead of `summary()` to give a summary of the results, while `weightable()` lists the models in order of their $IC(k)$ values. By default, a status report is given after every 50 models.

```
> library(glmulti)

> # Using "good ~ ." to include all explanatory variables
> search.1.aicc <- glmulti(y = good ~ ., data = placekick,
    fitfunction = "glm", level = 1, method = "h", crit = "aicc",
    family = binomial(link="logit"))
Initialization...
TASK: Exhaustive screening of candidate set.
Fitting...

After 50 models:
Best model: good~1+week+distance+change+PAT
Crit= 767.299644897567
Mean crit= 861.376317807727

After 100 models:
Best model: good~1+week+distance+change+PAT
Crit= 767.299644897567
Mean crit= 849.706367688597

<OUTPUT EDITED>

After 250 models:
Best model: good~1+distance+change+PAT+wind
Crit= 766.728784139471
Mean crit= 774.167191225549
Completed.

> print(search.1.aicc)
glmulti.analysis
Method: h / Fitting: glm / IC used: aicc
Level: 1 / Marginality: FALSE
From 100 models:
Best IC: 766.728784139471
Best model:
[1] "good ~ 1 + distance + change + PAT + wind"
Evidence weight: 0.0669551501665866
Worst IC: 780.47950528365
12 models within 2 IC units.
51 models to reach 95% of evidence weight.

> aa <- weightable(search.1.aicc)
> cbind(model = aa[1:5,1], round(aa[1:5,2:3], digits = 3))
```

```
                                              model    aicc weights
1             good ~ 1 + distance + change + PAT + wind 766.729  0.067
2 good ~ 1 + week + distance + change + PAT + wind      767.133  0.055
3      good ~ 1 + week + distance + change + PAT        767.300  0.050
4              good ~ 1 + distance + change + PAT       767.361  0.049
5                     good ~ 1 + distance + PAT + wind  767.690  0.041
```

The status updates show the best model found so far, its $IC(k)$ value, and the mean $IC(k)$ from among the top 100 models fit so far. The results of print() indicate that the best model contains variables distance, change, PAT, and wind, with an AIC_c=766.7. It is interesting to see that 12 models have AIC_c within 2 units of the best model, which is an indication that the declaration of this model as "best" is not very definitive. The worst of the top 100 models had AIC_c=780.5.

The top five models are shown next along with their AIC_c values. It is meaningful to note that all of these top models include distance and PAT, while change appears in four of these models, wind in three, and week in two. This gives an informal indication of which variables are the most important for explaining the probability of a made placekick. In Section 5.1.6 we will describe a more formal approach to this assessment and discuss the notes about Evidence weight and weights within the output.

Adding pairwise interactions to the set of variables could be a useful thing to do. However, this hugely increases the number of possible models—from $2^8 = 256$ to $2^{36} = 68.7$ billion—and makes exhaustive search no longer feasible (we estimate that it would have taken about 13 years to complete on one of our computers). This is where the genetic search algorithm is useful. The program accompanying this example includes code to complete this search (please see Exercise 6). It took about seven minutes to complete, averaged over four runs of the algorithm, and found the same "best" model in three of the four runs.

The bestglm and BMA packages also do all-subsets regression, but they are restricted to families of models available in glm(). This excludes, for example, proportional odds and zero-inflated Poisson models (Sections 3.4 and 4.4, respectively). Examples using bestglm and BMA on the placekicking data are included in the AllSubsetsPlacekick.R program. The search procedure used in the glmulti package is considerably slower than those used in the bestglm or BMA packages. For larger problems that can be modeled using a family available in glm(), these other packages might be better choices.

Burnham and Anderson (2002) strongly urge caution in using automated variable-selection routines because, in the presence of so many candidate models, finding the true "best" model becomes a needle-in-a-haystack problem. Many models may have similar $IC(k)$ values simply by luck. For example, picture having a world-champion darts player throw one dart at a dart board, and then randomly scattering another 64 billion darts around the board. The chances are extremely good that some of the random darts will land closer to the bulls-eye than the champion's, no matter how good the champion is. The list of candidate explanatory variables should therefore be made as small as possible using subject-matter knowledge before beginning all-subsets regression. Indeed, the wisdom of routinely adding pairwise interactions and transformations to a variable search is questionable in this light. Instead, it often makes better sense to include only certain interactions that are of specific interest and to use subject-matter knowledge and diagnostic tools from Section 5.2.

5.1.4 Stepwise variable selection

As noted in the previous section, unless the number of variables, P, is fairly small, it is not feasible to complete an exhaustive search of the entire model space to locate the "best" model. In such cases we must use some kind of algorithm to explore models that can quickly be identified as "promising." *Stepwise search algorithms* are simple and fast methods for selecting a relatively small number of potentially good models from the model space. These algorithms have shortfalls that are not widely known among practitioners, and so while we describe them below, we do not generally recommend them unless no better approach is available.

Stepwise variable selection uses a structured algorithm to examine models in only a small portion of the model space. The key to the algorithm is to evaluate the models in a specific sequence, where each model differs from the previous model by only a single variable. Once a sequence of models is chosen, an information criterion is used to compare them and select a "best" model. There are several different versions of the algorithm, as described below. In each version, let x_j, $j = 1, \ldots, P$ represent either a single variable or a set of variables that must be included or excluded together (e.g., a categorical explanatory variable that is represented by a group of indicator variables that are included in the model or excluded from it as a group).

Forward Selection

Forward selection starts from a model with no variables in it and then adds one variable at a time until either all of the variables are in the model or, optionally, some stopping rule has been reached. Specifically:

1. Select a k for the information criterion.

2. Start with an empty model, $g(\cdot) = \beta_0$, and compute $IC(k)$ on this model. All variables are in the "selection pool."

3. Add variables to the model one at a time as follows:

   (a) Compute $IC(k)$ for each model that can be formed by adding a single variable from the selection pool into the current model. In other words, we consider models that contain all of the variables in the current model plus one additional variable.

   (b) Choose the variable whose $IC(k)$ is smallest and add it to the model.

   (c) (Optional) If the model with the added variable has a larger $IC(k)$ than the previous model, stop. Report the model with the smaller $IC(k)$ value as the final model.

4. Repeat step 3 until the termination rule in part (c) is used or until all variables have been added into the model.

5. If no stopping rule is used, then a sequence of $P + 1$ models is created, each with a different number of variables. Report the model with the smallest $IC(k)$ as the final model.

Notice that the model size at each step increases by at most one variable. The logic here is that, at each step, the variable added is the one that improves the model most. Therefore, including it into the model yields the best model of its size, given the previous model. The next example illustrates this algorithm.

Example: Placekicking (StepwisePlacekick.R, Placekick.csv)

Stepwise variable selection methods are implemented in several R functions. We use the `step()` function because it is part of the `stats` package and it is flexible and fairly easy to use. To begin, the smallest and largest models to be considered are first fit using `glm()` and saved as objects. One of these becomes the value for the `object` argument in `step()`, and the other is specified in the `scope` argument, depending on the form of stepwise algorithm being used. If interactions are to be considered, they must be included in the largest model, either by listing them explicitly or by using a "~2" around the main-effect model. The `anova()` function summarizes the selection process.

Here we show forward selection (`direction = "forward"`) for a logistic regression model using the placekicking data. We start from a model with no variables (`empty.mod`) and add one variable at a time until no variable can be added that improves the criterion or the model with all variables (`full.mod`) is reached. Unfortunately, `step()` requires a constant penalty coefficient for all models, so AIC_c is not available. Instead, we base the selection on BIC, $IC(\log(n))$, by specifying the argument value `k = log(nrow(placekick))`.

```
> empty.mod <- glm(formula = good ~ 1, family = binomial(link =
    logit), data = placekick)
> full.mod <- glm(formula = good ~ ., family = binomial(link =
    logit), data = placekick)
> forw.sel <- step(object = empty.mod, scope = list(upper =
    full.mod), direction = "forward", k = log(nrow(placekick)),
    trace = TRUE)
Start:  AIC=1020.69
good ~ 1
```

	Df	Deviance	AIC
+ distance	1	775.75	790.27
+ PAT	1	834.41	848.93
+ change	1	989.15	1003.67
<none>		1013.43	1020.69
+ elap30	1	1007.71	1022.23
+ wind	1	1010.59	1025.11
+ week	1	1011.24	1025.76
+ type	1	1011.39	1025.92
+ field	1	1012.98	1027.50

```
Step:  AIC=790.27
good ~ distance
```

	Df	Deviance	AIC
+ PAT	1	762.41	784.20
<none>		775.75	790.27
+ change	1	770.50	792.29
+ wind	1	772.53	794.32
+ week	1	773.86	795.64
+ type	1	775.67	797.45
+ elap30	1	775.68	797.47
+ field	1	775.74	797.53

```
Step:  AIC=784.2
```

```
            good ~ distance + PAT

            Df Deviance    AIC
<none>         762.41   784.20
+ change    1  759.33   788.38
+ wind      1  759.66   788.71
+ week      1  760.57   789.62
+ type      1  762.25   791.30
+ elap30    1  762.31   791.36
+ field     1  762.41   791.46

> anova(forw.sel)
Analysis of Deviance Table
Model: binomial, link: logit
Response: good
Terms added sequentially (first to last)
           Df Deviance Resid. Df Resid. Dev
NULL                       1424     1013.43
distance    1  237.681     1423      775.75
PAT         1   13.335     1422      762.41
```

The `trace = TRUE` argument value allows one to see the full forward selection process at each step. In the first step, the algorithm starts with an intercept only and gives the $IC(k)$ values—listed as AIC, regardless of the value specified for k—for each model that arises based on adding one of the eight explanatory variables. The models with each added variable and the model with no added variables ("<none>") are sorted by their $IC(k)$ values. We see that `distance` has the smallest BIC, and hence is the first variable entered. The next step begins with `distance` in the model and finds that adding `PAT` results in a smaller BIC (notice that BIC is different now than it was for `PAT` in the first step because `distance` is now in the model). Thus, `PAT` is added to the model and the next step begins. Because no added variable improves BIC, the algorithm terminates and declares that our final model uses variables `distance` and `PAT`. Had we used AIC instead of BIC by specifying k = 2, the model chosen would have included `change` and `wind` as well.

Backward elimination

A different stepwise algorithm is sometimes used that approaches the problem from the opposite direction. Backward elimination starts from a model that has *all* available variables in it (assuming that this model can be fit) and then removes one variable at a time as long as doing so improves the $IC(k)$ for the model. Specifically,

1. Select a k for the information criterion.

2. Start with a full model, $g(\cdot) = \beta_0 + \beta_1 x_1 + \ldots + \beta_P x_P$, and compute $IC(k)$ on this model.

3. Remove variables from the model one at a time as follows:

    (a) Compute $IC(k)$ for each model that can be formed by deleting a single variable from the current model. In other words, we consider models that contain all but one of the variables in the current model.

    (b) Choose the variable whose deletion results in the smallest $IC(k)$ and remove it from the model.

(c) (Optional) If the new model with the variable deleted has a larger $IC(k)$ than the previous model, stop. Report the model with the smaller $IC(k)$ value as the final model.

4. Repeat step 3 until the termination rule in part (c) is used or until all variables have been removed from the model.

5. If no stopping rule is used, then a sequence of $P+1$ models is created, each with a different number of variables. Report the model with the smallest $IC(k)$ as the final model.

Notice that at each step the current model is compared against all models that are exactly one variable smaller. Thus, model size decreases by at most one variable. The logic here is that, at each step, the variable is removed that is contributing the least to the current model.

Example: Placekicking (StepwisePlacekick.R, Placekick.csv)

Only minor modifications to the program from the previous example are needed in order to perform backward elimination. The starting object is now the full model, the ending object is listed in scope, and we set direction = "backward". The output for this example is long, so we show an abridged version of it:

```
> back.sel <- step(object = full.mod, scope = list(lower =
    empty.mod), direction = "backward", k = log(nrow(placekick)),
    trace = TRUE)
Start:  AIC=818.87
good ~ week + distance + change + elap30 + PAT + type + field +
    wind

             Df Deviance    AIC
- elap30      1   753.72  811.82
- field       1   754.25  812.35
- type        1   754.76  812.86
- week        1   755.13  813.23
- change      1   756.69  814.79
- wind        1   756.74  814.84
<none>            753.52  818.87
- PAT         1   764.63  822.73
- distance    1   822.64  880.73

Step:  AIC=811.82
good ~ week + distance + change + PAT + type + field + wind

             Df Deviance    AIC
- field       1   754.49  805.32
- type        1   755.04  805.87
- week        1   755.32  806.16
- change      1   756.78  807.61
- wind        1   756.96  807.79
<none>            753.72  811.82
- PAT         1   764.81  815.64
- distance    1   824.69  875.53

<OUTPUT EDITED>
```

```
Step:  AIC=788.38
good ~ distance + change + PAT

          Df Deviance    AIC
- change   1   762.41 784.20
<none>         759.33 788.38
- PAT      1   770.50 792.29
- distance 1   831.75 853.53

Step:  AIC=784.2
good ~ distance + PAT

          Df Deviance    AIC
<none>         762.41 784.20
- PAT      1   775.75 790.27
- distance 1   834.41 848.93
```

Starting from a model with all eight variables in it, removing `elap30` results in the most improvement in BIC. At the next step `field` is removed. Steps continue to remove variables until only `PAT` and `distance` remain. Neither of these can be removed from the model without increasing BIC, so this is the final model.

Alternating stepwise selection

One of the criticisms of the previous two algorithms is that once a variable is added to (or removed from) the model, this decision can never be reversed. But in all forms of regression, relationships among the explanatory variables can cause the perceived importance of a variable to change depending on which other variables are considered in the same model. For example, a person's average daily calorie intake may seem very important by itself in a model for the probability of high blood pressure, but that relationship might disappear once we also account for their weight and exercise level. Therefore, it can be useful in some cases to have an algorithm that allows past decisions to be reconsidered as the model structure changes. The standard such algorithm is based on forward selection, but with the amendment that each time a variable is added, a round of backward elimination is carried out so that variables that have been rendered unimportant by the new addition can be removed from the model.

Example: Placekicking (StepwisePlacekick.R, Placekick.csv)

Because this algorithm is the default in `step()`, the forward-selection algorithm can be modified to run stepwise selection simply by deleting the `direction` argument (`"both"` is its default value). We leave running the code and interpreting the results as Exercise 7. In this case, the results are identical to forward selection, because there are no steps at which any previously added variables become non-significant and can be removed.

Stepwise procedures have historically been applied using hypotheses tests to determine the sequence of models to be considered. However, these tests are not actually valid hypothesis tests (Miller, 1984), and so it is becoming more common to apply stepwise methods using an information criterion as we have shown above. The program corresponding to the example from this section provides code showing how to manipulate the `step()` function to use hypothesis tests for variable addition and deletion. It is important to note that both hypothesis tests and information criteria order the variables the same way at each step, and thus produce exactly the same sequence of models if allowed to run until all variables are added or removed. The only possible difference between using a test and an information criterion is the selection of a final model from the sequence.

For the same reasons as just noted, significance tests for variables in a model selected by an information criterion are not valid tests. Therefore, it is not appropriate to follow a stepwise procedure with further tests of significance to see whether the selected model can be simplified further. Doing so would typically cause the value of $IC(k)$ to increase, resulting in a *worse* model according to the chosen criterion.

Although the forward, backward, and stepwise procedures all pointed to the same "best" model in the placekicking example, this does not generally have to happen. It is quite possible that the three procedures select three different models. In fact, Miller (1984) discusses a real example with eleven explanatory variables in which the first variable entered into the model by forward selection is also the first one removed by backward elimination. See Miller (1984), Grechanovsky (1987), Shtatland et al. (2003), and Kutner et al. (2004) for more details and critiques of stepwise procedures.

5.1.5 Modern variable selection methods

The past two decades have seen many new algorithms proposed for searching a model space. The most popular among these is the *least absolute shrinkage and selection operator* (LASSO) proposed by Tibshirani (1996). The LASSO has been refined, improved, and redeveloped in many different ways. We next describe the original LASSO and mention several improvements to it.

The LASSO

A variable selection procedure is more likely to select a particular variable when random chance in the data causes the variable to seem more important than it really is, as compared to those times when random chance makes it seem less important than it really is. As a result, parameter estimates for the selected variables tend to be biased: they are frequently estimated farther away from zero than they should be. This is the main motivation for the LASSO, which attempts to simultaneously select variables and shrink their estimates back toward zero to counteract this bias.

The LASSO estimates parameters using a procedure that adds a penalty to the log-likelihood to prevent the parameter estimates from being too large. Specifically, for a model with p explanatory variables, the LASSO parameter estimates $\hat{\beta}_{0,\text{LASSO}}, \hat{\beta}_{1,\text{LASSO}}, \ldots, \hat{\beta}_{p,\text{LASSO}}$ maximize

$$\log(L(\beta_0, \beta_1, \ldots, \beta_p | y, \ldots, y_n)) - \lambda \sum_{j=1}^{p} |\beta_j|, \qquad (5.1)$$

where λ is a *tuning parameter* that needs to be determined. For a given value of λ, the penalty term $\sum_{j=1}^{p} |\beta_j|$ in the likelihood criterion has two effects. It causes some of the regression parameter estimates to remain at zero, so that their corresponding variables are considered to be excluded from the model. Also, it discourages other parameter estimates from becoming too large unless increasing their magnitudes provides a sufficient increase in the likelihood to overcome the additional penalty. Thus, the procedure simultaneously selects variables and shrinks their estimates toward zero, which can help to overcome the bias noted above.

As the penalty parameter λ grows larger, more shrinkage occurs and smaller models are chosen. This parameter is usually chosen by *crossvalidation* (CV), which randomly splits the data into several groups and predicts responses within each group based on models fit to the rest of the data. Comparing the predicted and actual responses in some way provides an estimate of the prediction error of the model. The prediction error is computed for a sequence of different values for λ, and the one that has the smallest estimated prediction error) is a possible choice for the penalty. Note that CV is a randomized procedure. It is likely that slightly different prediction errors would result from repeated runs of the CV algorithm, leading to different chosen values for λ and resulting in different models and parameter estimates. Furthermore, there are often many values of λ that lead to similar values of prediction error. In practice it is common to use the smallest model whose prediction error is within 1 standard error of the lowest CV error, where the standard error refers to the variability of the CV-prediction-error estimates and can be calculated in several ways. See Hastie et al. (2009) for details.

Although the LASSO estimates often result in better prediction than the ordinary MLEs, subsequent inference procedures have not yet been fully developed. For example, confidence intervals for regression parameters or predicted values do not yet exist. Therefore the LASSO is used primarily as a variable selection tool or for making predictions where interval estimates are not required.

Example: Placekicking (LASSOPlacekick.R, Placekick.csv)

The `glmnet` package includes functions that can compute LASSO parameter estimates and predicted values for binomial (logit link), Poisson (log link), and multinomial (nominal response) regression models. The `glmnet()` function computes LASSO estimates for a sequence of up to 100 values of λ. The `cv.glmnet()` function uses crossvalidation to choose a value of λ. Both `glmnet()` and `cv.glmnet()` require that the observations for the explanatory variables and response be separated into matrix-class objects, entered as argument values for x and y, respectively. While both of these functions have various additional control parameters, the default levels serve fairly well for a relatively easy analysis. The objects created by these functions can be accessed using `coef()`, `predict()`, and `plot()` generic functions. By default, `coef()` and `predict()` produce output for every value of λ for objects from `glmnet()`; alternatively, values of λ can be specified using the s argument.

We apply these functions to the placekicking data below. After creating the required matrices (stored as objects yy and xx), we run the LASSO algorithm and examine the results.

```
> yy <- as.matrix(placekick[,9])
> xx <- as.matrix(placekick[,1:8])
> lasso.fit <- glmnet(y = yy, x = xx, family = "binomial")
> round(coef(lasso.fit), digits = 3)
9 x 63 sparse Matrix of class "dgCMatrix"
   [[ suppressing 63 column names 's0', 's1', 's2' ... ]]

(Intercept) 2.047  2.359  2.632  2.875  3.095  3.297  3.482
week         .      .      .      .      .      .      .
distance     .     -0.011 -0.021 -0.029 -0.036 -0.042 -0.048
change       .      .      .      .      .      .      .
elap30       .      .      .      .      .      .      .
PAT          .      .      .      .      .      .      .
type         .      .      .      .      .      .      .
```

```
field           .       .       .       .       .       .       .
wind            .       .       .       .       .       .       .

<OUTPUT EDITED>

(Intercept)   4.789   4.788   4.788   4.787   4.786   4.786   4.785
week         -0.023  -0.023  -0.023  -0.023  -0.023  -0.023  -0.024
distance     -0.086  -0.086  -0.086  -0.086  -0.086  -0.086  -0.086
change       -0.337  -0.338  -0.339  -0.340  -0.341  -0.342  -0.342
elap30        0.004   0.004   0.004   0.004   0.004   0.004   0.004
PAT           1.195   1.199   1.203   1.207   1.210   1.214   1.216
type          0.238   0.245   0.250   0.256   0.260   0.265   0.269
field        -0.151  -0.156  -0.162  -0.166  -0.171  -0.175  -0.178
wind         -0.566  -0.572  -0.577  -0.581  -0.586  -0.590  -0.593
```

The algorithm produced parameter estimates that maximize Equation 5.1 at 63 values of λ, of which we display only the first and last few. The estimates are listed in order of *decreasing* λ, so that models progress from intercept-only to a nearly unpenalized fit. In the program corresponding to this example, we provide additional coding to get the λ values printed in each column along with the estimates.

We next use cv.glmnet() to select a value for λ using the default value of 10 CV subsets and then plot and print the results. We set a seed so that the CV procedure will always give the same results.

```
> set.seed(27498272)
> cv.lasso.fit <- cv.glmnet(y = yy, x = xx, family = "binomial")

> # Default plot method for cv.lambda() produces CV errors +/- 1
    SE  at each lambda.
> plot(cv.lasso.fit)
> # Print out coefficients at optimal lambda
> coef(cv.lasso.fit)
9 x 1 sparse Matrix of class "dgCMatrix"
                    1
(Intercept)  4.45567259
week         .
distance    -0.07745453
change       .
elap30       .
PAT          .
type         .
field        .
wind         .
> predict.las <- predict(cv.lasso.fit, newx = xx, type =
    "response")
> head(cbind(placekick$distance, round(predict.las, digits = 3)))
             1
[1,]  21  0.944
[2,]  21  0.944
[3,]  20  0.948
[4,]  28  0.908
[5,]  20  0.948
[6,]  25  0.925
```

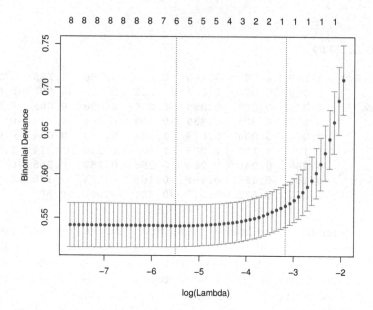

Figure 5.1: Cross-validation estimates of prediction error (binomial residual deviance) ±1 standard error for each value of λ. Numbers at the top of the plot are the number of variables at the corresponding λ. The first dotted vertical line is where the smallest CV error occurs; the second is the smallest model where the CV error is within 1 standard error of the smallest.

The CV estimates of error at each λ are shown in Figure 5.1; note that the binomial residual deviance is the function evaluated to compare the predicted and observed data in each CV subset. The minimum CV estimate of deviance occurs with a 6-variable model, but there are many models with practically the same deviance. The smallest model whose deviance is within 1 standard error of the minimum has only one variable. The print of the coefficients for this model shows that the variable is `distance`, with a parameter estimate of -0.077. Comparing this to the MLE, -0.115, we see that some shrinkage has taken place. The predicted probabilities of a successful field goal are printed out for the first few observations in the data set.

The LASSO has been improved in several ways since its introduction. The original formulation for the LASSO does not handle categorical explanatory variables, so a revised version, called the *grouped LASSO* was developed to correct this (Yuan and Lin, 2006). The grouped LASSO is available for logistic regression in the `SGL` package, and for logistic and Poisson regression in the `grplasso` package. Also, it has been noted in several places that the LASSO may perform poorly when faced with a problem in which there are many variables in the pool but few are important (see, e.g., Meinshausen, 2007). It has a tendency to include many more variables than are required, albeit with very small parameter estimates. Improvements such as the *relaxed LASSO* (Meinshausen, 2007) and the *adaptive LASSO* (Zou, 2006) both correct this tendency by separating the task of shrinkage from the task of variable selection. Unfortunately, these improvements have not yet been implemented into standard R packages for generalized linear models, as of this writing. They may be available as functions or packages offered privately, such as on the authors' websites.

We suggest that readers check periodically for new packages offering these techniques.

5.1.6 Model averaging

As noted in the previous section, it is often the case that there are many models with values of $IC(k)$ very close to the smallest value. This is an indication that there is some uncertainty regarding which model is truly best. In such cases, changing the data slightly could result in selecting a different model as best. In Exercise 2 we illustrate this uncertainty by splitting the placekick data approximately in half and showing that there are considerable differences in the top 5 models selected by the two halves. This uncertainty is difficult to measure if variable selection is applied using the traditional approach of selecting a single model and using it for all further inference. In particular, if a model is selected and then fit to the data, any variables not in that model implicitly have their regression parameters estimated as zero, *with a standard error of zero*. That is, we act as if we *know* that these variables do not belong in the model, when in fact the data cannot determine this for certain. It would be useful to be able to better estimate the uncertainty in each of our parameter estimates.

When all-subsets variable selection is feasible, then *model averaging* can account for model-selection uncertainty in subsequent inferences (see, e.g., Burnham and Anderson, 2002). In particular, *Bayesian model averaging* (BMA, Hoeting et al., 1999) uses Bayesian theory to compute the probability that each possible model is the correct model (assuming that one of them is, indeed, correct). It turns out that these probabilities can be approximated by relatively simple functions of the BIC value for each model. Suppose that a total of M models are fit and denote the BIC for model m by BIC_m, $m = 1, \ldots, M$. Denote the smallest value of BIC among all models by BIC_0, and define $\Delta_m = BIC_m - BIC_0 \geq 0$. Assuming that all models were considered equally likely before the data analysis began, then the estimated probability that model m is correct, τ_m, is approximately

$$\hat{\tau}_m = \frac{\exp(-\frac{1}{2}\Delta_m)}{\sum_{a=1}^{M} \exp(-\frac{1}{2}\Delta_a)}. \tag{5.2}$$

For the model with the smallest BIC, $\exp(-\Delta_m/2) = 1$; for all other models, $\exp(-\Delta_m/2) < 1$, and this number shrinks very quickly as Δ_m grows. Notice that when $\Delta_m = 2$, $\exp(-\Delta_m/2) = 0.37$, so that the probability of model m is about a third as large as that for the best-fitting model. This is not a big difference, considering that variable-selection procedures that select a single model essentially declare one model to have probability 1 and the rest to have probability zero. This is also the meaning behind the reporting of the number of models within 2 IC units of the best in `glmulti()`.

Now, let θ be any quantity that we want to estimate from the models (such as a regression parameter, an odds ratio, or a predicted value). Denote its estimate in model m by $\hat{\theta}_m$ with corresponding variance estimated by $\widehat{Var}(\hat{\theta}_m)$. Then the model-averaged estimate of the parameter is

$$\hat{\theta}_{\text{MA}} = \sum_{m=1}^{M} \hat{\tau}_m \hat{\theta}_m,$$

and the variance of this estimate is estimated by

$$\widehat{Var}(\hat{\theta}_{\text{MA}}) = \sum_{m=1}^{M} \hat{\tau}_m [(\hat{\theta}_m - \hat{\theta}_{\text{MA}})^2 + \widehat{Var}(\hat{\theta}_m)]. \tag{5.3}$$

If one model is clearly better than all others, then its $\hat{\tau}_m \approx 1$. In this case, the model-averaged estimate, $\hat{\theta}_{\text{MA}}$, is not much different from this model's estimate, and the variance of $\hat{\theta}_{MA}$ is essentially just the estimated variance of $\hat{\theta}_m$ from its model. On the other hand, if many models have comparable probabilities, then the variance of the estimate $\hat{\theta}_{MA}$ depends both on how variable the parameter estimates are from model to model through $(\hat{\theta}_m - \hat{\theta}_{MA})^2$ and on the variance of each model's parameter estimates.

This process can be applied to any number of parameters associated with the model. As an example, consider the case of a single regression parameter, $\theta = \beta_j$. There is an estimate, $\hat{\beta}_{j,m}$, in each model. The model-averaged estimate of β_j is $\hat{\beta}_{j,MA} = \sum_{m=1}^{M} \hat{\tau}_m \hat{\beta}_{j,m}$. Notice here that half of the models in all-subsets selection exclude X_j, so $\hat{\beta}_{j,m} = 0$ for these models. In models where X_j does appear, and that also have high probability attached to them, β_j is likely to have its magnitude overestimated. These facts have several implications:

1. Model averaging generally results in a non-zero estimate for *all* regression parameters.

2. Variables that are truly unimportant are unlikely to appear in the models that have high probability. Thus, their model-averaged estimates are close to zero, because the probabilities for the models in which they appear are all small, and there is little variability in these estimates. Practically speaking, they can be taken to be zero, because their impact on the model-averaged predictions is negligible.

3. On the other hand, variables that appear in some, but not all, of the models with the highest probability typically have their estimates shrunk by being averaged with zeroes from those models in which they do not appear. This reduces the post-selection bias noted in Section 5.1.5.

Example: Placekicking (BMAPlacekick.R, Placekick.csv)

The main R package that performs Bayesian model averaging, BMA, focuses on variable selection and assessment (i.e., taking $\theta = \beta_j, j = 1, \ldots, p$). Computations for model-averaged inference on other parameters, such as predicted values, must be programmed manually using results from this package. On the other hand, the glmulti package has functions that can easily produce model-averaged predicted values and corresponding confidence intervals. We therefore demonstrate model averaging here using glmulti(), and we include an example using bic.glm() from the BMA package in the program corresponding to this example.

The glmulti() syntax for the variable search is essentially the same as that used in the example from Section 5.1.3, except that we choose crit = "bic" so that we can carry out Bayesian model averaging. The top 6 models are shown below

```
> search.1.bic <- glmulti(y = good ~ ., data = placekick,
    fitfunction = "glm", level = 1, method = "h", crit = "bic",
    family = binomial(link = "logit"))

<OUTPUT EDITED>

> print(search.1.bic)
glmulti.analysis
Method: h / Fitting: glm / IC used: bic
Level: 1 / Marginality: FALSE
From 100 models:
Best IC: 784.195620372141
Best model:
[1] "good ~ 1 + distance + PAT"
Evidence weight: 0.656889992065609
```

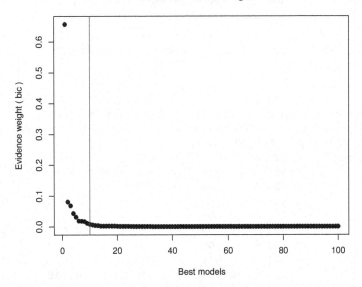

Figure 5.2: Estimated model probabilities from Bayesian model averaging applied to the placekicking data. The models to the left of the vertical line account for 95% of the cumulative probability.

```
Worst IC: 810.039529019811
1 models within 2 IC units.
9 models to reach 95% of evidence weight.

> head(weighttable(search.1.bic))
                             model      bic    weights
1          good ~ 1 + distance + PAT 784.1956 0.65688999
2 good ~ 1 + distance + change + PAT 788.3802 0.08106306
3   good ~ 1 + distance + PAT + wind 788.7094 0.06875864
4   good ~ 1 + week + distance + PAT 789.6186 0.04364175
5                good ~ 1 + distance 790.2689 0.03152804
6   good ~ 1 + distance + PAT + type 791.3004 0.01882426

> plot(search.1.bic, type = "w")
```

Compared to the AIC_c results from page 271, we see that BIC prefers smaller models, as expected. The results from `print()` show that the best model uses variables `distance` and `PAT` and that it is the only model within two units of its BIC value. The notes about `evidence weight` refer to model probabilities: the top model has probability 0.66 and the top nine out of $2^8 = 256$ models account for 95% of the probability. The last column in the `weighttable()` output shows the model probabilities, where we see that the top model is rather dominant. The second best model has 0.08 probability. A plot of these model probabilities for the top 100 models is produced by the `plot` method function for `glmulti` objects and is shown in Figure 5.2.

Additional programming statements create model-averaged parameter estimates and confidence intervals, shown below. The `coef()` function produces the BMA parameter estimates and variances computed as per Equation 5.3, along with the number of models in which the variable was included, the total probability accounted for by those models, and the half-width of a confidence interval for the parameter. The variables are printed in order from least- to most-probable, so we reverse this order and compute the confidence intervals. Exponentiating the estimates and interval endpoints allows them to be interpreted in terms of odds ratios per 1-unit change in the corresponding explanatory variable.

```
> parms <- coef(search.1.bic)
> # Renaming columns to fit in book output display
> colnames(parms) <- c("Estimate", "Variance", "n.Models",
    "Probability", "95%CI +/-")
> round(parms, digits = 3)
            Estimate Variance n.Models Probability 95%CI +/-
field          0.000    0.000       29       0.026     0.010
set            0.001    0.000       28       0.026     0.012
elap30         0.000    0.000       29       0.027     0.001
type           0.003    0.000       30       0.028     0.017
week          -0.002    0.000       35       0.062     0.007
wind          -0.051    0.010       35       0.095     0.197
change        -0.042    0.007       39       0.119     0.159
PAT            1.252    0.191       62       0.944     0.857
(Intercept)    4.624    0.271      100       1.000     1.021
distance      -0.088    0.000      100       1.000     0.024
> parms.ord <- parms[order(parms[,4], decreasing = TRUE),]
> ci.parms <- cbind(lower = parms.ord[,1] - parms.ord[,5], upper
    = parms.ord[,1] + parms.ord[,5])
> round(cbind(parms.ord[,1], ci.parms), digits = 3)
                       lower   upper
(Intercept)    4.626   3.609   5.644
distance      -0.088  -0.111  -0.064
PAT            1.252   0.394   2.109
change        -0.042  -0.201   0.117
wind          -0.051  -0.248   0.147
week          -0.002  -0.008   0.005
type           0.003  -0.015   0.020
elap30         0.000  -0.001   0.001
field          0.000  -0.011   0.010
> round(exp(cbind(OR = parms.ord[,1], ci.parms))[-1,], digits = 2)
            OR lower upper
distance  0.92  0.89  0.94
PAT       3.50  1.48  8.24
change    0.96  0.82  1.12
wind      0.95  0.78  1.16
week      1.00  0.99  1.01
type      1.00  0.99  1.02
elap30    1.00  1.00  1.00
field     1.00  0.99  1.01
```

As one might expect, distance is the most important variable, appearing in all of the top 100 models. Next was PAT, appearing in models accounting for 0.94 probability. No other variables were found to be particularly important. This is reflected in the 95% t-based confidence intervals for both the parameters and the odds ratios, where the default setting uses degrees of freedom averaged across models. The parameter intervals cover 0 and the corresponding OR intervals cover 1 for all variables except for distance and PAT. The program corresponding to this example gives the results from fitting just the best model, good ~ 1 + distance + PAT, and ignoring all others. The parameter estimates from the two methods are quite similar—not surprising, considering that this model was fairly dominant among all possible subsets—but the confidence intervals from the single-model fit are slightly narrower, and we get no confidence intervals for the other variables.

Our program also shows how to compute model-averaged predicted probabilities of a successful placekick, along with confidence intervals for the true probabilities based on the variance in Equation 5.3. Unfortunately, it appears that no packages in R can easily carry out the necessary computations to apply Equation 5.3 to more general statistics that are functions of multiple regression parameters.

Although the theory for BMA applies only to BIC, the methods of this section are often used with other $IC(k)$ as well. Burnham and Anderson (2002) refer to the model and variable probabilities calculated in this section as *evidence weights* rather than posterior probabilities when alternative $IC(k)$ are used.

Model averaging as a variable-selection tool

As noted in Section 5.1.3, it is common in an exhaustive search that there are many models with $IC(k)$ values very similar to the best model. These models may contain many of the same explanatory variables, thereby providing evidence regarding which variables are most important. We can formalize this evidence using model averaging.

Specifically, using the BIC criterion we calculate the posterior probability that each explanatory variable belongs in the model by summing the probabilities for all models in which the variable appears (we could alternatively work with evidence weights produced by a different information criterion). If a variable is important, then the models with that variable tend to be among those with the highest probabilities and models without that variable all have very small probabilities. Thus, the models that contain the variable have probabilities that add to nearly 1. Similarly, an unimportant variable tends not to be among the top models, so its probabilities sum to something fairly small.

Raftery (1995) suggests guidelines for interpreting the posterior probability that a variable belongs in the model. First, any variable with posterior probability below 0.5 is more likely not to belong in the model than to belong. Evidence that a variable belongs in the model is "weak" when its posterior probability is between 0.50 and 0.75, "positive" when it lies between 0.75 and 0.95, "strong" when it lies between 0.95 and 0.99, and "very strong" when its posterior probability is above 0.99. Of course, these are not the only interpretations possible. For example, using the plot() function with type = "s" on an object from glmulti() produces a bar-chart of the posterior probabilities for each variable, with a guideline drawn at 0.80.

See Section 5.4.2 for an example that uses model averaging for variable selection.

5.2 Tools to assess model fit

When we fit a generalized linear model (GLM), we assume that the following three things are true: (1) the distribution, or random component, is the one we specify (e.g., binomial or Poisson); (2) the mean of that distribution is linked to the explanatory variables by the function g that we specify (e.g., $g(\pi) = \beta_0 + \beta_1 x$, where π is the mean of a binary response variable Y); and (3) the link relates to the specified explanatory variables in a linear fashion. Any of these assumptions could be violated; indeed, we expect that a model is only an approximation to the truth. In this section we present tools that can help us to identify when a model is not a good approximation.

The methods we use are largely analogous to those used in normal linear regression, particularly in their reliance on *residuals*. For a refresher on the use of residuals in diagnosing problems with linear models with normally distributed errors, see Kutner et al. (2004). In this section we will show how to compute and interpret residuals that are appropriate for diagnosing problems in distributions where the mean and variance are related. We will also offer some methods for testing the fit of a model. These techniques should always be used as part of an iterative model-building process. A model is proposed, fit, and checked. If it is found lacking, a new model is proposed, fit, and checked, and this process is repeated until a satisfactory model is found.

This section discusses tools that are used primarily with models for a single count response per observation, such as binomial and Poisson models. Historically, more effort has been spent developing diagnostic tools for binomial regression models—particularly, logistic regression—than for other models. Also, binomial models require special consideration because the counts that are modeled may be based on different numbers of trials for different subjects. We therefore focus our discussion on binomial models, and generalize to other models for counts as appropriate. Specialized diagnostics for the multinomial regression models of Sections 3.3 and 3.4, which model multiple responses simultaneously, are less well developed and not readily available in R. We discuss them briefly at the end of the section.

5.2.1 Residuals

Generally, a residual is a comparison between an observed and a predicted value. With count data in regression models, the observed data are counts, and the predicted values are estimated mean counts produced by the model. For the Poisson model, this is straightforward, because the mean count is modeled and predicted directly. However, some other cases, such as binomial, multinomial, and Poisson rate regressions, models are written in terms of parameters other than the mean—the probability of success or the rate. These parameters need to be converted into counts before computing residuals. For a Poisson rate regression, this means simply multiplying the rate by the exposure for each observation. For binomial regression, this requires a little more care.

Consider a binomial regression problem in which there are M unique combinations of the explanatory variable(s). We refer to these combinations as *explanatory variable patterns* (EVPs). For EVP $m = 1, \ldots, M$, let n_m be the number of trials observed for variable(s) x_m. For example, in the context of the placekicking example with distance as the only explanatory variable, there are 3 placekicks at a distance of 18 yards, 7 at 19 yards, 789 at 20 yards, and so forth, leading to $n_1 = 3, n_2 = 7, n_3 = 789, \ldots$. We have a choice between fitting the model to the individual binary responses or first aggregating the trials and fitting the model to the summarized counts of successes, w_m, as shown in Section 2.2.1. As noted in that section, the parameter estimates are the same either way and lead to the same estimated probability of success, $\hat{\pi}_m$. But modeling the binary trials individually produces n_m residuals for each unique x, of which w_m are $-\hat{\pi}_m$ and the remaining $n_m - w_m$ are $1 - \hat{\pi}_m$, regardless of whether $\hat{\pi}_m$ is at all close to the observed proportion of success, w_m/n_m. On the other hand, modeling the aggregated count results in a single residual that provides a better summary of the model fit. Therefore, prior to computing residuals or any of the quantities that we later derive from them, it is important to aggregate all trials for each

EVP into a single count as shown in the example on page 74. We refer to this aggregation as *explanatory variable pattern form*.

Henceforth, we assume that all binomial data have been aggregated in this way into counts y_m, $m = 1, \ldots, M$. We use y instead of w for consistency with other models, where the count response is typically denoted by y. Suppose that EVP m has n_m trials. Then the corresponding predicted count is $\hat{y}_m = n_m \hat{\pi}_m$, $m = 1, \ldots, M$. We use similar notation for counts from other models that are not based on aggregations of trials: observed responses on M units are denoted by y_m, with corresponding predicted counts $\hat{y}_m$, $m = 1, \ldots, M$.

Often the fits of several possible models are to be compared, for example through likelihood ratio tests or some other comparison of residual deviance. Likelihoods based on different EVP structures behave as though they were calculated on different data sets, which means that they are not comparable. Aggregation of binomial counts should therefore be done at the level of the most complex model being compared before fitting all models that are being compared.

Counts from the distributional models discussed in Chapters 2–4 all have variances that depend on their respective means. Therefore, the "raw residual" that is commonly used in normal linear regression, $y_m - \hat{y}_m$, $m = 1, \ldots, M$, is of limited use, because a difference of a given size may be either negligible or extraordinary depending on the size of the mean. We must instead use quantities that adjust for the variability of the count. The basic residual used with count data is the *Pearson residual*,

$$e_m = \frac{y_m - \hat{y}_m}{\sqrt{\widehat{Var}(Y_m)}}, \quad m = 1, \ldots, M$$

where $\widehat{Var}(Y_m)$ is the estimated variance of the count Y_m based on the model (e.g. $\widehat{Var}(Y_m) = \hat{y}_m$ for Poisson, $\widehat{Var}(Y_m) = n_m \hat{\pi}_m (1 - \hat{\pi}_m)$ for binomial). Pearson residuals behave approximately like a sample from a normal distribution, particularly when the $\hat{y}_m$ (and $n_m - \hat{y}_m$ in the binomial case) are large. However, the denominator overestimates the standard deviation of $y_m - \hat{y}_m$, so that the variance of e_m is smaller than 1. To correct this, we compute the *standardized Pearson residual*,

$$r_m = \frac{y_m - \hat{y}_m}{\sqrt{\widehat{Var}(Y_m - \hat{Y}_m)}} = \frac{y_m - \hat{y}_m}{\sqrt{\widehat{Var}(Y_m)(1 - h_m)}},$$

where h_m is the mth diagonal element of the hat matrix described in Sections 2.2.7 and 5.2.3.

A different kind of residual is sometimes used that is based on the model's residual deviance. As first discussed in Section 2.2.2, the residual deviance is a measure of how far the model lies from the data and can be written in the form $\sum_{m=1}^{M} d_m$, where the exact nature of d_m depends on the model for the distribution (for example, Equation 2.10 shows that $d_m = -2y_m \log[\hat{y}_m / y_m] - 2(n_m - y_m) \log[(n_m - \hat{y}_m)/(n_m - y_m)]$ in the binomial case). The quantity $e_m^D = \text{sign}(y_m - \hat{y}_m)\sqrt{d_m}$ is called the *deviance residual*, and a standardized version of it, $r_m^D = e_m^D / \sqrt{(1 - h_m)}$ has approximately a standard normal distribution. Among the different types of residuals, we prefer to use the standardized Pearson residuals when they are available. They are generally easy to interpret and they are also not difficult to compute from many common fitted model objects.

Standardized residuals can be interpreted approximately as observations from a standard normal distribution. We use convenient thresholds based on the standard normal distribution as rough guidelines for identifying unusual residuals. For example, values beyond ± 2 should happen only for about 5% of the sample when the model is correct, values beyond ± 3 are extremely rare, and values beyond ± 4 are not expected at all.

For EVPs with small n_m in binomial models, the standard normal approximation is poor, and these guidelines should be viewed as tentative. For example, binary responses ($n_m = 1$) can produce only two possible raw residual values, $0 - \hat{\pi}$ and $1 - \hat{\pi}$, and hence also only two possible Pearson, deviance, or standardized residual values. This discrete situation is not at all well-approximated by a normal distribution, which is for continuous random variables. More generally, there are $n_m + 1$ possible residual values for an EVP with n_m trials, so with small n_m there are not enough different possible residual values to allow meaningful interpretation. The normal approximation can also be poor when $\hat{\pi}_m$ is near 0 or 1. Again, these cases tend to generate only a few possible responses unless the number of trials is large.

A similar phenomenon can happen with Poisson models where the estimated means are very small, for example $\hat{y}_m < .5$. Then most observed y_m are either 0 or 1, with only occasional larger values, so that there are very few possible residual values. Surprisingly, however, the rough guidelines are still reasonably useful for identifying extreme residuals even when $\hat{y}_m$ is somewhat smaller than 1. See Exercise 17.

In cases where the normal approximation is not reliable, part of the interpretation of any seemingly extreme residuals is calculating the probability of such an extreme residual, assuming that the model-estimated count $\hat{y}_m$ is the correct expected value. For example, this means computing the probability of a count *at least as extreme* as the observed y_m under the binomial model with n_m trials and probability $\hat{\pi}_m$. A similar probability calculation should be done for Poisson residuals with small $\hat{y}_m$.

Computing residuals in R

Many model-fitting functions have method functions associated with them that can compute various kinds of residuals for models shown in Chapters 2–4. Most notably, the generic function `residuals()` has method functions that work on model-fit objects produced by `glm()`, `vglm()`, `zeroinfl()`, and `hurdle()`. All of these functions use the `type` argument to select the types of residuals computed. Pearson or raw residuals are produced with the argument values `"pearson"` or `"response"`, respectively. The `residuals.glm()` method also has a `"deviance"` argument value to produce deviance residuals. For models fit with `logistf()`, there are currently no functions available that produce useful residuals automatically. For these objects, relatively simple manual calculation can create Pearson or deviance residuals.

Standardized Pearson and deviance residuals are available for `glm`-class objects using `rstandard()`. Standardizing residuals for fits from other models entails calculating the hat matrix manually. Finding the diagonal elements of the hat matrix is aided by the code for computing $(\mathbf{X'VX})^{-1}$ shown in the example on page 70, although not all model-fit objects readily provide all of the required elements to complete these calculations easily.

Graphical assessment of residuals

The information content in a set of residuals is best understood in graphical displays. Most of these displays mirror the residual plots used in linear regression (see, e.g., Kutner et al., 2004). However, their interpretation in generalized linear models can be slightly different. In particular, whereas residual plots in linear regression mainly diagnose problems with the mean model and possible unusual observations, in generalized linear models the plots also can diagnose when other model assumptions do not fit the data well, including the choice of distributional family. Of course, interpretation of any residual plot for a count-data model is subject to the caveats noted in the previous section regarding the interpretation of residuals when there are limited numbers of possible responses.

A plot of the standardized residuals against each explanatory variable can show whether the form for that explanatory variable is appropriate. The plot should show roughly the same variance throughout the range of the explanatory variable of interest, x_j, and should show no serious fluctuations in the mean value. Curvature in the plot suggests that a transformation or additional polynomial terms are needed for that variable, or possibly that the link needs to be changed. We recommend adding a smooth curve to the plot, such as a *loess* smoother, representing the average residual over the range of the x-axis. This makes changes in the mean residuals more prominent. The curve should wiggle randomly around zero when the model is a reasonable fit. Clear patterns of curvature, such as a "smile" or a "frown," are indications of a problem. In binomial models, the loess curve can be weighted by n_m, so that EVPs based on larger numbers of trials contribute relatively more to the curve placement than those with relatively few. Note that loess curves are highly variable where data are sparse or near the extreme values of the variable on the x-axis. Care should be taken not to "overinterpret" apparent changes in the curve at the edges of the plot.

Similarly, a plot of residuals against the fitted values $\hat{y}$ ($\hat{\pi}$ in binomial models) is useful for assessing when the link function is not fitting well. The points should again have roughly constant variance and show no clear curvature. Alternatively, a plot of the residuals against the linear predictor, $g(\hat{y})$, may show patterns of change in the mean residuals more clearly and can help to diagnose how the link $g(\cdot)$ should be changed to better fit the data.

Any of these plots can be read to check for extreme residuals. As noted above, only about 5% of standardized residuals should be beyond ± 2 and typically none should lie beyond ± 3. Presence of one or two very extreme residuals could indicate *outliers*—unusual observations that should be checked for possible errors. However, outliers are rare by definition. If there are numerous residuals whose magnitudes are larger than expected, and these values are randomly scattered across the range of the plot's x-axis, then this may be a sign of *overdispersion*, meaning that there is more variability to the counts than what the model assumes there should be. This is an indication that there may be important explanatory variables missing from the model, or that a different distribution may be needed to model the data. See Section 5.3 for details regarding overdispersion.

A slightly different interpretation applies to these plots in binomial models. Hosmer and Lemeshow (2000) show that the opportunity for extreme residuals to occur is greater in regions of high or low estimated probability of success (e.g., $\hat{\pi} < 0.1$ or > 0.9). There are two main reasons for this. First, hat values h_m tend to be larger for more extreme values of the explanatory variables, for which the probability-of-success estimates may also be extreme. Second, the largest possible raw residual value is $0 - n_m \hat{\pi}_m$ or $n_m - n_m \hat{\pi}_m$, and these are at their largest if $\hat{\pi}_m$ is near 0 or 1. Therefore, part of the interpretation of any seemingly extreme residuals in a binomial regression is calculating the binomial probability that such an extreme value of y_m could occur in n_m trials with probability $\hat{\pi}_m$.

Also, when the response counts take on relatively few unique values, residual plots often show odd-looking bands of points. As the most extreme example, consider a plot of raw residuals against $\hat{\pi}$ in a logistic regression with binary responses. As noted above, for each value of $\hat{\pi}$ there are only two possible residuals. The plot will therefore show just two bands of points: one at $-\hat{\pi}$ and the other at $1 - \hat{\pi}$. This lack of "randomness" in the plot is an indication that the response counts on which the model is based are rather small and limited in range. In this case, the approximate normality of the residuals is in doubt, thus weakening the potential for the residual plots to detect violations of model assumptions.

Example: Placekicking (PlacekickDiagnostics.R, Placekick.csv)

We demonstrate the interpretation of residual plots in a logistic regression using the placekicking data with a model containing only the `distance` variable. Aggregation of these data into EVP form for this model was shown in Section 2.2.1. The model fit and functions calling the residuals are shown below, along with the code for creating the plot of standardized Pearson residuals against the explanatory variable. This produces the first plot in Figure 5.3. The other two plots in Figure 5.3 use very similar code. Notice the use of `loess()` to create an object whose predicted values add the smooth trend for the mean residual. The `weights = trials` argument value causes `loess()` to use the number of trials in the EVP—denoted in the `w.n` data frame as `trials`—to weight the mean trend. Thus, EVPs with many trials influence the trend curve more than those with fewer.

```
> mod.fit.bin <- glm(formula = success/trials ~ distance, weights
    = trials, family = binomial(link = logit), data = w.n)
> pi.hat <- predict(mod.fit.bin, type = "response")
> p.res <- residuals(mod.fit.bin, type = "pearson")
> s.res <- rstandard(mod.fit.bin, type = "pearson")
> lin.pred <- mod.fit.bin$linear.predictors
> w.n <- data.frame(w.n, pi.hat, p.res, s.res, lin.pred)
> round(head(w.n), digits = 3)
  distance success trials prop pi.hat  p.res  s.res lin.pred
1       18       2      3 0.667  0.977 -3.571 -3.575    3.742
2       19       7      7 1.000  0.974  0.432  0.433    3.627
3       20     776    789 0.984  0.971  2.094  3.628    3.512
4       21      19     20 0.950  0.968 -0.444 -0.448    3.397
5       22      12     14 0.857  0.964 -2.136 -2.149    3.281
6       23      26     27 0.963  0.960  0.090  0.091    3.166

> # Standardized Pearson residual vs X plot
> plot(x = w.n$distance, y = w.n$s.res, xlab = "Distance", ylab =
    "Standardized Pearson residuals", main = "Standardized
    residuals vs. \n X")
> abline(h = c(3, 2, 0, -2, -3), lty = "dotted", col = "blue")

> smooth.stand <- loess(formula = s.res ~ distance, data = w.n,
    weights = trials)
> # Make sure that loess estimates are ordered by "X" for the
    plots, so that they are displayed properly
> order.dist <- order(w.n$distance)
> lines(x = w.n$distance[order.dist], y =
    predict(smooth.stand)[order.dist], lty = "solid", col = "red",
    lwd = 1)
```

Because `distance` is the only variable in the model and it has a negative coefficient, the plot of the residuals against the linear predictor is just a mirror image of the plot against `distance`. All three plots show the same general features. First, note the general trend of increasing variability as $\hat{\pi}$ approaches 1, which is most evident in the center plot. This is expected to some degree even when the model fits well. Next, there are two very extreme residuals—magnitudes beyond 3—and both occur for very low distances. Whether these EVPs contain errors or have some other explanation is

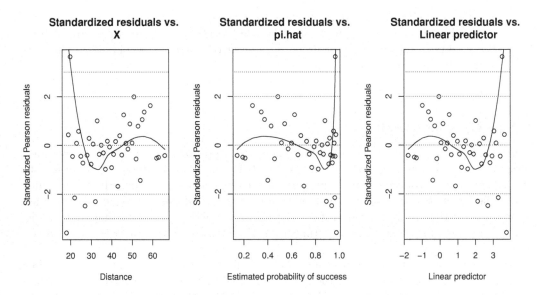

Figure 5.3: Residual plots for modeling of the probability of success vs. distance.

not clear just from the plot. These two points need to be investigated further. Also, there are 6 residuals that have magnitudes of 2 or more, which is a little more than the 5% we would expect to see from a sample of 43 observations from a standard normal distribution. These occur mostly for small distances, and four of those five are negative. A possible explanation could be that the extreme positive value is an outlier of some type (i.e., a special case), in which case the remaining residuals follow a general increasing pattern, which would disappear upon refitting to the data with the questionable value removed or fixed. Finally, the shape of the loess curve is interesting. It suggests that the residuals for long distances (lower probabilities) tend to be positive, while the residuals for shorter distances tend to be negative, except for one clear point with a large positive residual. However, we do not take this curve too seriously until we determine the cause of the two extreme residuals. If nothing about them is found to be unusual, this may suggest that either the form of `distance` in the model or the link function may be wrong.

The two most extreme residuals are seen in EVP's 1 and 3 in the output, at distances of 18 and 20 yards. The extreme residual at 18 yards is caused by observing only 2 successes in 3 trials when the probability of success is estimated to be 0.977. The bounds at ±2 and ±3 shown on the plot are not very accurate for so few trials. In fact we can easily compute that the probability is 0.067 that one would observe only two successes in three trials when the true probability of success is 0.977 as predicted by the model.[3] So, this residual is not actually as extreme as it would seem to be based on its position relative to the bounds. On the other hand, the other extreme residual occurs at 20 yards, where there are 789 trials, and more observed successes than expected. In most cases, a 20-yard placekick takes place as a point-after-touchdown (PAT). Unlike ordinary field goals, these kicks are placed exactly in front of the center of the goalposts, creating the largest possible angle for a successful kick, and they are worth only one point rather than three. It might be possible that these kicks have a

---

[3] The 0.067 probability is computed from `pbinom(q = 2, size = 3, prob = 0.977)`.

higher probability of success than would be expected for an ordinary field goal at a comparable distance. Perhaps the model could be improved by adding a variable that allows PATs to have a different probability of success than other placekicks. We will explore this possibility in a further example in Section 5.4.1

5.2.2 Goodness of fit

Examining residual plots is essential for understanding problems with the fit of any generalized linear model, but interpretation of these plots is somewhat subjective. Goodness-of-fit (GOF) statistics are often computed as more objective measures of the overall fit of a model. While these statistics are useful, they tend to provide little information regarding the cause of any poor fit that they might detect. We therefore strongly recommend that these statistics be used as an adjunct to interpreting residual plots, rather than as a replacement for them.

Statistical assessment of the fit of a GLM typically begins with looking at the residual deviance and Pearson statistics for the model. The residual deviance, which we can denote here by D, is a comparison of the model-estimated counts to the observed counts and has a form that varies depending on the distribution used as the model (see Appendix B.5.3). The Pearson statistic has a more generic form, but can also be calculated in several different ways, depending on the distribution being modeled. For distributions with a univariate response, such as the Poisson or binomial, the Pearson statistic is calculated as

$$X^2 = \sum_{m=1}^{M} e_m^2 = \sum_{m=1}^{M} \frac{(y_m - \hat{y}_m)^2}{\widehat{Var}(y_m)}.$$

For example, $\widehat{Var}(y_m) = \hat{y}_m$ for a Poisson model and $\widehat{Var}(y_m) = n_m \hat{\pi}_m (1 - \hat{\pi}_m)$ for a binomial model. For binomial models, it is sometimes more convenient to use an equivalent form of the statistic based on both the success and failure counts,

$$X^2 = \sum_{m=1}^{M} \frac{(y_m - \hat{y}_m)^2}{\hat{y}_m} + \sum_{m=1}^{M} \frac{((n_m - y_m) - (n_m - \hat{y}_m))^2}{n_m - \hat{y}_m}. \tag{5.4}$$

Both D and X^2 are summary measures of the distance between the model-based estimates and the observed data. As such, they are often used in formal tests of goodness of fit, which means testing the null hypothesis that the model is correct[4] against the alternative that it is not. Both statistics have large-sample $\chi^2_{M-\tilde{p}}$ distributions, where $\tilde{p}$ is the total number of regression parameters estimated in the model. However, this result is only valid under the assumption that set of EVPs in the data is fixed and would not change with additional sampling. This is true if, for example, all explanatory variables are categorical, and all possible categories have already been observed, or if numerical values are fixed by the design of an experiment. However, in most cases with continuous explanatory variables, additional sampling would surely yield new EVPs, and so the distributions of D and X^2 may not be approximated well by a $\chi^2_{M-\tilde{p}}$ distribution even if the model is correct. Furthermore, the $\chi^2_{M-\tilde{p}}$ approximation holds reasonably well only if all predicted counts $\hat{y}_m$—and $n_m - \hat{y}_m$ in binomial models—are large enough (e.g., all are at least 1 and many are at least 5). Thus, the models in which these statistics can be used in formal tests are limited primarily to

---

[4]Of course, the model is never really considered "correct" because we never formally accept the null hypothesis. If the null is not rejected, we merely conclude that the model is a "reasonable" fit to the data.

those with only categorical predictors, and not too many of them, so that all combinations have reasonably high estimated counts.

As an informal alternative to formal hypothesis testing, it is common to compute the ratio of residual deviance to residual degrees of freedom, which we denote in the text as "deviance/df," or symbolically by $D/(M-\tilde{p})$. Because the expected value of a $\chi^2_{M-\tilde{p}}$ random variable is $M - \tilde{p}$, we can expect that this ratio ought not be too far from 1 when the model is correct. There are no clear guidelines for how far is "too far," owing in part to the uncertainty surrounding whether the $\chi^2_{M-\tilde{p}}$ distribution is a reasonable approximation to the distribution of D. When it is, and when $M - \tilde{p}$ is not small, then values of D that are more than, say, two or three standard deviations above their expected value would be unusual when the model is correct.[5] Because the variance of a chi-square random variable is twice its degrees of freedom, then this suggests a rough guideline of $D/(M-\tilde{p}) > 1 + 2\sqrt{2/(M-\tilde{p})}$ to indicate a potential problem, and $D/(M-\tilde{p}) > 1 + 3\sqrt{2/(M-\tilde{p})}$ to indicate a poor fit. We do not intend these thresholds to be taken as firm rules, but rather to provide the user with some way to reduce the subjectivity in interpreting this statistic.

Goodness-of-fit tests

There are several alternative GOF tests that can be used even when the number of EVPs may increase with additional sampling. These are especially common in logistic regression, but variants of them can be used in other regression models as well. Chapter 5 of Hosmer and Lemeshow, 2000 provides a very nice discussion of these tests for logistic regression. We summarize their discussion and provide our own insight.

The best-known GOF test for EVPs with continuous variables is the Hosmer and Lemeshow test for logistic regression (Hosmer and Lemeshow, 1980). The idea is to aggregate similar observations into groups that have large enough samples so that a Pearson statistic computed on the observed and predicted counts from the groups has approximately a chi-square distribution. As an example, consider a single explanatory variable, such as distance in the placekicking data. We might expect that placekicks from similar distances have similar enough success probabilities that their observed successes and failures could be grouped into a single pair of counts. This could be done before fitting a model (i.e., "binning" the data into intervals), but doing so can sometimes create a poor-looking fit out of a good model. Instead, Hosmer and Lemeshow suggested fitting the model to the data first, and then forming g groups of "similar" EVPs according to their estimated probabilities. The aggregated counts of successes and failures within each group are then combined with the aggregated model-estimated counts within the same group into a Pearson statistic as shown in Equation 5.4, summing over both successes and failures for the g groups. The test statistic, X^2_{HL}, is compared to a χ^2_{g-2} distribution.

Failing to reject the null hypothesis of a good fit does not mean that the fit is, indeed, good. The power of the test can be affected by sample size, and the process of aggregation can sometimes mask potential problems. For example, a large positive outlier and a large negative outlier in the same group will somewhat "cancel" each other out, resulting in a smaller value of X^2_{HL} than some other grouping might have created. This points to another difficulty with the test: there are numerous ways to form the group of similar EVPs, and the outcome of the test can depend on the chosen grouping pattern. In the most common implementation of the test, the model-estimated probabilities are divided into $g = 10$ groups of approximately equal size based on the 10th percentiles of the values of $\hat{\pi}$. However, there are even several ways to do *this*—for example, making the count of

---

[5] As ν gets large, the χ^2_ν behaves more and more like a normal distribution with mean ν and variance 2ν.

EVPs in each group approximately constant or making the number of trials in each group approximately constant—and these can give different groupings when some EVPs have multiple trials. Test results may vary with the grouping, and no one algorithm is clearly better than others. Thus, it is a good idea to try several groupings, for example by setting g to be several different values (sample size permitting), and to ensure that the test results from the multiple groupings are in substantive agreement before drawing conclusions from this test. In general the more groups that are used, the less chance that very dissimilar EVPs are grouped together, so the test can be more sensitive to poor fits. However, it is also important that the expected counts—i.e., aggregated predicted successes and failures from the model—are large enough to justify using the large-sample chi-square approximation.

Hosmer et al. (1997) compare several GOF tests for logistic regression and recommend two other tests that can be used in addition to the Hosmer-Lemeshow test. The first of these is the Osius-Rojek test, which is based on the original model's Pearson GOF statistic. Osius and Rojek (1992) derived the large-sample mean and variance for X^2 in such a way that allows the number of sampled categories to grow as the sample size grows, enabling it to be used as a test statistic even when there are continuous explanatory variables. Then a standardized version of X^2 is compared against a standard normal distribution, with extreme values representing evidence that the fitted model is not correct. A second test is the Stukel test (Stukel, 1988), which is based on an expanded logistic regression model that allows the upper and lower halves of the logistic curves to differ in shape. Two additional explanatory variables are calculated from the logits of original model fit, and these are added to the original model. A test of the null hypothesis that their regression coefficients are both zero indicates whether the original model shape is adequate—which is implied by the null hypothesis—or whether a different link or a transformation of continuous variables might be needed. We forgo the calculation details here and direct interested readers to refer to Hosmer et al. (1997) and to our R code described in the example below.

Example: Placekicking (PlacekickDiagnostics.R, AllGOFTests.R, Placekick.csv)

None of the Hosmer-Lemeshow, Osius-Rojek, or Stukel tests are included in the default distribution of R, so we have written functions to perform them: `HLTest()`, `o.r.test()`, and `stukel.test()`, respectively. All three are contained in the file AllGOFTests.R. In all three functions, the `obj` argument is a `glm` object fit using `family = binomial`. The `HLTest()` function includes an additional argument for g, which is set to `g = 10` by default. The code for these functions is not shown below, but is given separately on the website for this book. Each function's R code needs to be run once before carrying out the tests (e.g., using the `source("AllGOFTests.R")`). For each function, the test statistic and p-value are produced. For the Hosmer-Lemeshow test, the Pearson residuals upon which the test's Pearson statistic is based are available for examination, as are the observed and expected counts in each group. We run this test here using $g = 10$.

```
> HL <- HLTest(obj = mod.fit.bin, g = 10)
Warning message:
In HLTest(obj = mod.fit.bin, g = 10) :
  Some expected counts are less than 5. Use smaller number of
      groups
> cbind(HL$observed, round(HL$expect, digits = 1))
               Y0    Y1  Y0hat  Y1hat
[0.144,0.353]   3     2    3.8    1.2
(0.353,0.469]  19    13   18.3   13.7
```

```
(0.469,0.589]   25   39   30.1   33.9
(0.589,0.699]   24   49   26.4   46.6
(0.699,0.79]    32   74   26.6   79.4
(0.79,0.859]    18   75   15.9   77.1
(0.859,0.908]   12   69    9.2   71.8
(0.908,0.941]   10   72    5.9   76.1
(0.941,0.963]    3   53    2.6   53.4
(0.963,0.977]   17  816   24.3  808.7
> HL
        Hosmer and Lemeshow goodness-of-fit test with 10 bins
data:  mod.fit.bin
X2 = 11.0282, df = 8, p-value = 0.2001

> round(HL$pear, digits = 1)   # Pearson residuals for each group
interval            Y0    Y1
  [0.144,0.353]   -0.4   0.8
  (0.353,0.469]    0.2  -0.2
  (0.469,0.589]   -0.9   0.9
  (0.589,0.699]   -0.5   0.4
  (0.699,0.79]     1.0  -0.6
  (0.79,0.859]     0.5  -0.2
  (0.859,0.908]    0.9  -0.3
  (0.908,0.941]    1.7  -0.5
  (0.941,0.963]    0.3  -0.1
  (0.963,0.977]   -1.5   0.3

> o.r.test(obj = mod.fit.bin)
z = 1.563042 with p-value = 0.1180426

> stukel.test(obj = mod.fit.bin)
Stukel Test Stat = 6.977164 with p-value = 0.03054416
```

The Hosmer-Lemeshow test function first reports a warning that some expected counts are less than 5 for the cells of the 10 × 2 table of expected counts for the groups. We print out the observed successes and failures (labeled Y1 and Y0, respectively) and estimated expected number of successes and failures (Y1hat and Y0hat, respectively) in each group. For example, among all $\hat{\pi}$ in the interval [0.144, 0.353][6] there are only 5 observations, whose expected counts are split as 3.8 failures and 1.2 successes according to the fitted model. Also, the interval (0.941,0.963] has only 2.6 estimated expected failures. These expected counts are not drastically low (e.g., below 1), and there are not too many of them (only 15% of cells), so the χ_8^2 approximation may not be too bad here. The test statistic is $X_{HL}^2 = 11.0$ with a p-value of 0.20, which suggests that the fit is not bad. However, the Pearson residuals for the grouped counts show a potential pattern in the signs. In each row, there must be a positive and a negative residual, and these should ideally be equally likely to occur with either response. Notice that five intervals in a row have the same pattern of positive residual for the $Y = 0$ response. Although none of these values is large (e.g., none larger than 2), the pattern suggests potential for a specific problem of overestimating the true success probability when

---

[6]Square brackets, "[" or "]", indicate that the corresponding interval endpoint is included in the interval, while round parentheses, "(" or ")", indicate that the corresponding interval endpoint is *not* in that interval. For example, the interval (0.353,0.469] includes all values of $\hat{\pi}$ such that $0.353 < \hat{\pi} \leq 0.469$. Using this notation, there is no ambiguity regarding which interval contains the endpoints.

the probability is large. We have already seen evidence of this in the residual plots in Figure 5.3.

The Osius-Rojek test yields a test statistic of $z = 1.56$ with a p-value of 0.12 which offers no serious evidence of a poor fit. However, the Stukel test provides a p-value of 0.03, which provides some evidence that the model is not fitting the data well. In particular, because this test targets deviations from the model separately for large and small probabilities, it may be more sensitive to the possible model flaw suggested by Figure 5.3 and the Pearson residuals from the Hosmer-Lemeshow grouping. We therefore should investigate further to determine what may be causing this problem.

For response variables from other distributions, there is no formal goodness-of-fit test that is recognized as standard when continuous explanatory variables are involved. For models with a univariate response, an analog to the Hosmer-Lemeshow test can be conducted, wherein groupings are formed based on fitted values from the model, such as predicted counts from a Poisson model or rates from a Poisson rate model. This idea was suggested on p. 90 of Agresti (1996), but it does not appear to be well studied. We offer a function that can perform this test on a set of counts and predicted values, `PostFitGOFTest()` in the file PostFitGOFTest.R. However, this test should be checked for accuracy before using it more broadly (e.g., by using a simulation to ensure that the test holds its size for the data set on which it is used; see Exercise 27).

Note that the p-values from any goodness-of-fit test are valid only if the model being tested could have been chosen before beginning the data analysis. The reasons are the same as those discussed in Section 5.1.4. In particular, using the data to select the variables to include into a model has the effect of making the model look better for that data set than it might on another set of data. Applying a goodness-of-fit test to such a model tends to result in larger p-values that would be found if the fit were assessed more fairly on an independent data set.

This leads to two important points. First, whenever possible, models should be tested on independent data before they are adopted as valid for some broader purpose than describing the data on which they are based. Second, the p-value from a goodness-of-fit test needs to be interpreted carefully when the test is done on a model that results from a variable-selection process. If the p-value is "small," then this is, indeed, evidence that the model does not fit well. However, it is not clear what interpretation can be applied when the p-value is not small.

5.2.3 Influence

The fit of a regression model is influenced by each of the observations in the sample. Removing or changing any observation (or equivalently an EVP in a binomial regression) can result in a change to the parameter estimates, predicted values, and GOF statistics. An observation is considered to be *influential* if the regression results change a lot when it is temporarily removed from the data set. Obviously, we do not want our conclusions to depend too strongly on a single observation, so it is wise to check for influential observations. *Influence analysis* consists of computing and interpreting quantities that measure different ways in which a model changes when each observation is changed or removed.

Removing observations and refitting models to measure how they change may seem cumbersome. Fortunately, the quantities most commonly computed to measure influence in linear regression models all have "shortcut" formulas that allow them to be computed from a single model fit. The same quantities are not so easily computed from a GLM model fit, but they can be approximated using formulas analogous to those used in linear regression. We present these approximations here, and we discuss another quantity, leverage, that is useful for identifying potentially influential observations. For more details on influence analysis for linear models, see Belsley et al. (1980) or Kutner et al. (2004). Pregibon (1981) and Fox (2008) show how to apply some of these measures to GLMs.

Leverage

In linear regression, *leverage* values are often computed as a way to measure the potential for an observation to influence the regression model. In Section 2.2.7 we defined the *hat matrix* in a logistic regression as $\mathbf{H} = \mathbf{V}^{1/2}\mathbf{X}(\mathbf{X'VX})^{-1}\mathbf{X'V}^{1/2}$, where $\mathbf{X}$ is the $M \times (p+1)$ matrix whose rows contain the explanatory variables $\mathbf{x}'_m$, $m = 1, \ldots, M$, and $\mathbf{V}$ is a diagonal matrix whose mth diagonal element is $v_m = \widehat{Var}(Y_m)$, the variance of the response variable calculated using the regression estimate of the mean or probability at $\mathbf{x}'_m$. This definition comes about through an analog to weighted least squares estimation used for normal linear regression models and applies to all GLMs (Pregibon, 1981). The diagonal elements of $\mathbf{H}$—h_m, $m = 1, \ldots, M$—are the leverage values for the GLM. Pregibon (1981) and Hosmer and Lemeshow (2000) show that h_m measures the potential for observation m to have large influence on the model fit. Because the average value of h_m in any model is p/M, observations with $h_m > 2p/M$ are considered to have moderately high leverage, and those with $h_m > 3p/M$ to have high leverage. Whether or not influence is actually exerted depends on whether an extreme *response* occurs, and this can be measured by residuals and by other measures described below. Thus leverage values are important tools, but cannot stand alone as measures of influence. We discuss more useful tools to examine influence next.

"Leave-one-out" measures of influence

Cook's distance from linear regression is a standardized summary of the change in *all* of the parameter estimates simultaneously when observation m is temporarily deleted. An approximate version of Cook's distance for GLMs is found as

$$CD_m = \frac{r_m^2 h_m}{(p+1)(1-h_m)^2}, \quad m = 1, \ldots, M.$$

In general, points with values of CD_m that stand out above the rest are candidates for further investigation. Alternatively, observations with $CD_m > 1$ might be considered to have high influence on the regression parameter estimates, while those with $CD_m > 4/M$ have moderately high influence.

The change in the Pearson goodness-of-fit statistic X^2 caused by the deletion of observation m is measured approximately by

$$\Delta X_m^2 = r_m^2,$$

while the change in the residual deviance statistic is approximately

$$\Delta D_m = (e_m^D)^2 + h_m r_m^2.$$

The latter approximation is especially excellent, in many cases replicating the actual change in the residual deviance almost perfectly. For both statistics, we use thresholds of 4 and 9 to suggest observations that have moderate or strong influence on their respective fit statistics. These numbers are derived from squaring the ± 2 and ± 3 suggested thresholds that we use for the standardized residuals upon which they are based. The two statistics can suggest possible observations that may have high influence on the overall model fit.

Plots

Plotting the four measures—h_m, CD_m, ΔX_m^2, and ΔD_m—against the fitted values is recommended by Hosmer and Lemeshow (2000) for helping to locate influential observations for logistic regression models. These plots can be used for other GLMs as well. Hosmer and Lemeshow (2000) also recommend plotting the latter three statistics against the leverage values, as a way to simultaneously judge whether an observation with *potential* influence is causing an *actual* change in the model fit. However, all three of these statistics are direct functions of h_m, so large leverage can affect their values. Also, all of them are functions of residuals, so they may be large because of large residuals associated with observations that are not particularly influential. This is especially true for ΔX_m^2 and ΔD_m.

Once potentially influential observations have been identified, we are left to decide what to do about them. We discuss this after the next example.

Example: Placekicking (PlacekickDiagnostics.R, glmDiagnostics.R, Placekick.csv)

For `glm`-class model objects, the influence measures are all easy to calculate: `hatvalues()` produces h_m; `cooks.distance()` produces CD_m; `residuals()` produces e_m and e_m^D with a `type = "pearson"` or `type = "deviance"` argument value, respectively; and `rstandard()` produces r_m and r_m^D, again with a `type = "pearson"` or `type = "deviance"` argument value, respectively. The leave-one-out measures of influence ΔX_m^2 and ΔD_m can be subsequently calculated using r_m, e_m^D, and h_m. For convenience, we have created a function, `glmInflDiag()`, to produce the plots and form a list of all observations that are identified as potentially influential by at least one statistic. This function expects a `glm`-class object for its first argument, but will also work on any other object for which there are `residuals()` and `hatvalues()` method functions and for which `summary()$dispersion` is a legitimate element.

The output from this function is given below. We include the `print.output` and `which.plots` arguments in the function to control the output to be printed and sets of plots to be constructed; both are set here to their default values to maximize the information given. The observations printed by `glmInflDiag()` correspond to those where $h_m > 3p/M$, $CD_m > 4/M$, $\Delta X_m^2 > 9$, or $\Delta D_m > 9$, and those with both $\Delta X_m^2 > 4$ and $h_m > 2p/M$ or $\Delta D_m > 4$ and $h_m > 2p/M$. Figures 5.4 and 5.5 show the plots.

```
> source("glmDiagnostics.R")
> save.diag <- glmInflDiag(mod.fit = mod.fit.bin, print.output =
    TRUE, which.plots = 1:2)
Potentially influential observations by any measures
      h hflag Del.X2 xflag Del.dev dflag Cooks.D cflag
1  0.00        12.78    **    3.84         0.015
3  0.67    **  13.16    **   13.95    **  13.163    **
34 0.08         3.98         4.11     *   0.174     *

   Data for potentially influential observations
     distance success trials   prop  yhat
1          18       2      3 0.6667 0.977
3          20     776    789 0.9835 0.971
34         51      11     15 0.7333 0.486
```

The function identifies three potentially influential observations: #1, #3, and #34. Observation #1 has large ΔX_1^2 without a large h_1 value, indicating that the large

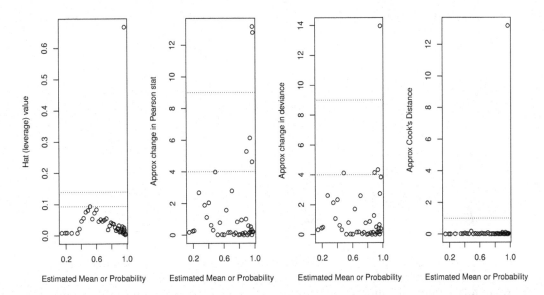

Figure 5.4: Plots of influence measures vs. estimated probabilities of success for the placekicking data. Dotted lines represent thresholds for potentially high influence.

influence measure value is due to the size of its standardized residual r_1. This case was addressed in Section 5.2.1, where we saw that the large residual was merely a result of observing 1 failure out of 3 trials when the estimated probability of success is very large. There is no cause for alarm with this observation. Observation #3 has large ΔX_3^2, ΔD_3, and CD_3 values along with a large h_3. This observation was also identified as having an extreme residual in Section 5.2.1. Its high leverage is undoubtedly due to the fact that it accounts for over half of the total 1425 trials in the data set. Thus, we expect this observation to be influential, although its high influence is not necessarily a bad thing. However, because it also has an extreme CD_3, its influence is clearly affecting the parameter estimates and the overall fit of the model. We will find out in Section 5.4.1 that simply adding the PAT explanatory variable to the model eliminates the extreme standardized residual, but still leaves an influential observation corresponding to this same EVP.

Observation #34 is also potentially influential, but the evidence is mild. Both CD_{34} and ΔD_{34} are only slightly above their respective thresholds. It has a somewhat smaller estimated probability of success than the observed proportion. A quick calculation shows that the probability of 11 or more successes out of 15 trials when $\hat{\pi}_{34} = 0.486$ is 0.047, so this observation is not extremely atypical and hence is not really of great concern.

Another useful measure not covered here is the *studentized residual*. It is a standardized version of the deviance residual using a denominator that excludes its own count. A large value indicates that the observation lies far from where the model suggests it should be. The influencePlot() function in the car package plots the studentized residual against the leverage value, with bubbles proportional to the Cook's distance. This is another useful tool for identifying points that have a combination of extremeness relative to the explanatory

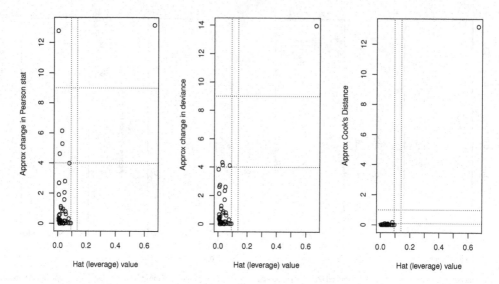

Figure 5.5: Plots of influence measures vs. leverage values for the placekicking data. Dotted lines represent thresholds for potentially high influence.

variables and extremeness relative to the response.

Model adjustments for influence and outliers

How one handles outliers and influential values depends on the cause of the extreme results and the goals of the analysis. In general, any flagged observations should be investigated for other features that might cause their outlying and/or influential status. If there is reason to believe that an error has been made in measuring or recording their data, then this should obviously be fixed if possible, and the analysis rerun. If the observations are fundamentally different in some way, such as a goat in a study of sheep, or a professional athlete in a study of regular people, then it is reasonable to exclude the observation(s), provided that the reason for doing so is made clear and the results of the analysis are explicitly applied to this narrower population (e.g., people who are not professional athletes). In the context of our placekicking data, there are differences between PATs and other field goals that might cause their probabilities of success at the same distance to be different. We could choose to exclude the PATs from the data and rerun the analysis, in which case our inferences would apply only to field goals and not to all placekicks.

When there is no clear reason for the influence or extremeness of an observation, then removing it from the study should not be automatic. When there are not too many flagged values to consider, it is possible to rerun the analysis both with and without the extreme values. If the main inferences from the analysis do not change in any substantive way, then there is no worry that these values are interfering with the analysis, and there is no problem with presenting the original results. On the other hand, if the inferences do change meaningfully, then there is uncertainty regarding what conclusions should be drawn. Reporting of the results should reflect this uncertainty by clearly indicating the differences and explaining that it is not clear which results are closer to the truth.

5.2.4 Diagnostics for multicategory response models

As noted in the introduction to this section, diagnostic tools for multicategory response models are not readily available. To begin the discussion of what can be used, we assume that the data have been aggregated into EVP form as in a binomial regression model. Then there are M EVPs, where EVP m consists of n_m trials for $m = 1, \ldots, M$. If there are J categories in the response variable, then the response is a list of J counts rather than the single count of successes the binomial models use (for example, if $n_m = 1$, one of the J counts is 1 and the rest are 0). Represent the counts as y_{mj}, $m = 1, \ldots, M$; $j = 1, \ldots, J$.

Model fitting provides estimated probabilities $\hat{\pi}_{mj}$ for all J categories for each EVP, and hence estimated expected counts $\hat{y}_{mj} = n_m \hat{\pi}_{mj}$. Then there are J residuals, $y_{mj} - \hat{y}_{mj}$, $j = 1, \ldots, J$ per EVP, although these are not independent, because both the observed counts and the expected counts must sum to n_m. A set of Pearson residuals can be defined for each observation as

$$e_{mj} = \frac{y_{mj} - \hat{y}_{mj}}{\sqrt{\hat{y}_{mj}}}, \ j = 1, \ldots, J.$$

Each EVP's contribution to the model-fit Pearson statistic is the sum of its squared Pearson residuals, $X_m^2 = \sum_{j=1}^J e_{mj}^2$. Similarly, each EVP's contribution to the residual deviance can be computed.

From the Pearson residuals a number of further quantities can be computed similar to those in Sections 5.2.1 and 5.2.3. Defining the hat matrix, and hence the leverage values h_m, requires some care due to the multivariate nature of the response. Mathematical details are given in Lesaffre and Albert (1989), who extend the measures from Pregibon (1981) to the multicategory case.

We have written R functions that calculate statistics and create plots that can help to identify outlying or influential EVPs. These are contained in the R program multinomDiagnostics.R for models fit using `multinom()`. The statistics calculated on each EVP include Pearson residuals, the contributions to the Pearson and residual deviance goodness-of-fit statistics, leverage, Cook's distance, and the two case-deletion statistics, ΔD_m and ΔX_m^2.

As an alternative to performing specialized diagnostics for multinomial models, the problem can be broken down into a series of ordinary logistic regressions and each of these can be assessed for fit separately. For example, the multinomial regression model for a nominal response,

$$\log(\pi_j / \pi_1) = \beta_{j0} + \beta_{j1} x_1 + \cdots + \beta_{jp} x_p,$$

uses $J - 1$ logits, each comparing response j to response 1, $j = 2, \ldots, J$. A series of $J - 1$ logistic regressions can be fit, each considering only the data from category 1 and category j, and the fit for each of these models can be assessed using standard tools described earlier in this section. The appeal of this approach is that the diagnostic tools are easily accessed and well understood. However, each model is only a portion of the total model.

5.3 Overdispersion

In most linear regression problems, we assume that errors are normally distributed with constant variance. Under this assumption, the variance of the data is not affected by the model we pose for the mean; the mean and variance are controlled independently by separate parameters. That is, suppose we write the mean response in a conditional form as $E(Y|x) =$

$\mu(x)$ to emphasize the relationship between the mean and the explanatory variables x, with analogous notation for $Var(Y|x)$. In the normal model we have $Var(Y|x) = \sigma^2$, a constant that does not depend on x.

In most GLM settings, such as logistic and Poisson regressions, the mean and variance are related. For Poisson regression with $E(Y|x) = \mu(x)$, we have $Var(Y|x) = \mu(x)$. In a logistic regression, an EVP with n_m trials has $E(Y|x) = \mu(x) = n_m \pi(x)$, so that $Var(Y|x) = n_m \pi(x)(1 - \pi(x))$. Thus, with most GLMs, when we specify a model for the relationship between the mean and x, we are implicitly imposing a model on the relationship between the variance and x.

In linear regression, it is somewhat common that the variance does not follow its assumed model; that is, the variance is *not* constant for all values of x. This situation is known as *heteroscedasticity*, or more simply *non-constant variance*, and it is often detected through analysis of the residuals (see, e.g., Kutner et al., 2004). Similarly, in GLMs the variance may not follow the structure that is imposed on it by the distributional model. In particular, it is fairly common to have counts or proportions that exhibit more variability that what the models indicate there should be. This indicates a bad fit of the model, even when the estimated mean or probability seems to fit the data well. This phenomenon is called *overdispersion*. We explain in the next sections what causes overdispersion, how to detect it, and how to correct models to account for it.

5.3.1 Causes and implications

Overdispersion is a failure of the *model*, not a failure of the *data*. It is a symptom of another problem rather than a problem by itself.

In linear regression, when an important variable is added to the model, the variability that it explains is taken out of the error sum of squares and put into the model sum of squares. Looking at this another way, when an important variable is *left out* of the model, the error sum of squares is much larger than it should be. As a result, the mean squared error becomes inflated and overestimates the true variance σ^2. In a sense, this is an example of overdispersion, except that the model has a built-in parameter, σ^2, that adapts to the model's failure. Inferences in linear regression are all based on using the mean squared error to measure variance, so inferences are automatically adjusted for the model failure by making test statistics less extreme and confidence intervals wider.

Models such as the Poisson and binomial have no separate variance parameter to allow them to adapt to the absence of important variables. When an important variable is left out of the model, then observations that have the same mean according to the model may actually have different means if they have different values of the missing variable. Thus, the responses from this common estimated mean may exhibit more variability than what the model expects. The next example shows that data are more variable when they are generated from Poisson distributions with variable means than when they are generated from a Poisson distribution with a constant mean. Thus, the response mean that is estimated by a Poisson regression model underestimates the amount of variability in the data around that mean. Exactly the same phenomenon occurs with other distributional models that lack a separate parameter for variance, such as the binomial.

Unlike the case with a normal distribution, there is no mechanism within these models to adjust the inferences to make them less certain in consideration of this extra variability. The inferences assume that the model-based estimates of variance are correct, when in fact they are too small. Thus, confidence intervals are too narrow—that is, they do not achieve their stated confidence levels—and p-values are smaller than they should be, meaning that type I error rates are higher than stated. The end result is that inferences tend to detect effects that are not really there.

Example: Simulation of overdispersion (OverdispSim.R)

In this example, we simulate data under a Poisson model with and without overdispersion. First, we simulate data directly from a Poisson distribution with mean 100, and compute the sample mean and variance of the data (called mean0 and var0, respectively). Then we repeat the simulation, except that the mean is allowed to vary by simulating it from a normal distribution with mean 100 and standard deviation 20. The sample mean and variance is again computed (called mean20 and var20, respectively). Both simulations use sample sizes of 50, and the simulations are repeated 10 times so that the patterns are apparent. The program and results are given below:

```
> set.seed(389201892)
> poi0 <- matrix(rpois(n = 500, lambda = rep(x = 100, times =
    500)), nrow = 50, ncol = 10)
> poi20 <- matrix(rpois(n = 500, lambda = rnorm(n = 500, mean =
    100, sd = 20)), nrow = 50, ncol = 10)

> mean0 <- apply(X = poi0, MARGIN = 2, FUN = mean)
> var0 <- apply(X = poi0, MARGIN = 2, FUN = var)
> mean20 <- apply(X = poi20, MARGIN = 2, FUN = mean)
> var20 <- apply(X = poi20, MARGIN = 2, FUN =v ar)

> all <- cbind(mean0, var0, mean20, var20)
> round(all, digits=1)
       mean0   var0 mean20  var20
 [1,]   99.9   85.9  104.3  456.0
 [2,]  101.5   91.7   97.0  533.5
 [3,]   99.1   83.3  104.5  534.0
 [4,]   99.9  121.1  100.4  452.2
 [5,]   99.5  102.4  101.0  452.3
 [6,]   99.7   73.8  104.0  408.8
 [7,]  101.0   94.5   96.9  509.0
 [8,]  100.4  110.9  102.1  292.8
 [9,]   98.0   92.4   98.1  303.8
[10,]   99.8   96.1  104.3  524.6
> round(apply(X = all, MARGIN = 2, FUN = mean), digits = 1)
  mean0    var0  mean20   var20
   99.9    95.2   101.3   446.7
```

The sample means for both cases vary randomly around the true value of 100, although there is more variability to the means from the case where the Poisson means are variable. The variances for the standard Poisson also vary randomly around 100, but those from the case with variable means are considerably larger, averaging 447. Thus, it is clear that generating data from a situation where the means may vary causes the resulting data to exhibit much more variability than is expected under a Poisson model.[7]

The program corresponding to this example also contains code to simulate data sets from the same model used above, except that the amount of overdispersion is controlled more finely by setting the standard deviation on the random means to 0, 1, ..., 20. For each standard deviation we generated 100 data sets of size 20. These data

---

[7] It can be shown that the true mean and variance of the response Y under the second model are 100 and 500, respectively. See Exercise 32.

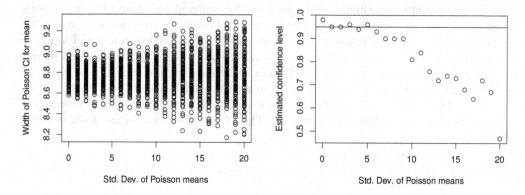

Figure 5.6: Likelihood ratio confidence interval results from simulations using overdispersed Poisson data. Confidence interval widths (left) and estimated confidence levels (right) are plotted against the standard deviation of the means used to generate the Poisson data. The horizontal line in the confidence level plot shows the stated confidence level of 95%.

sets are then analyzed by a Poisson regression model using only an intercept, resulting in maximum likelihood estimation of the constant mean for each data set. Model-based LR confidence intervals for μ are computed, and the average confidence interval widths and estimated true confidence levels—the fraction of data sets whose confidence interval correctly contains μ—are calculated. These widths and estimated confidence levels are plotted against the standard deviations used to generate the Poisson means in Figure 5.6.

The plot of confidence interval widths shows that the average width is not affected by the increased levels of overdispersion. There does appear to be a little more variability to the widths as overdispersion increases, which reflects the observation made in the first simulation that the means from the overdispersed case are more variable than those from the standard Poisson. The major problem we see is in the estimated confidence levels. These levels decrease as the overdispersion increases, and drop substantially below the stated 95% level for larger standard deviations. This is an indication that inferences are unreliable when ordinary Poisson regression models are fit to overdispersed data, even when the model for the mean is correct.

Other causes for overdispersion have been given in the literature. For example, McCullagh and Nelder (1989), among others, point out that the standard binomial and Poisson models assume that each trial or observation is independent of all others. However, sometimes the sampling process is such that this assumption is not satisfied. A common example is sampling data in *clusters*, meaning that groups of observations are sampled together. Typically, observations within a cluster respond more similarly to one another than observations in different clusters. For example, married couples may hold opinions that are more similar to their spouse's opinions than to those of some other random person. Cattle in the same pen or field tend to have more similar health issues than cattle selected from different locations. Products built at similar times on a given assembly line may have more similar defects than those from different lines or different production periods. If sampling is done on couples, pens of cattle, or batches of consecutive products produced—i.e., in clusters—then the resulting individuals will tend to be positively correlated within their clusters. The positive correlation within the clusters causes means or probabilities to be more variable than their respective models expect. Thus, treating units that were gathered in clusters as if they were independent is likely to lead to overdispersion. This phenomenon is discussed in more detail in Section 6.5.

5.3.2 Detection

The main symptom of overdispersion is a poor model fit without any obvious cause. In particular, large goodness-of-fit statistics described in Section 5.2.2, such as the deviance/df statistic or formal goodness-of-fit tests, indicate a problem with the model. However, this is not evidence of overdispersion by itself. Examination of standardized residual plots is required in order to rule out other issues. The plots must not show a poor fit of the mean model, nor identify one or perhaps two particular outliers that might be causing the excessively large deviance and/or Pearson statistic. If these other problems *are* present, then they must be addressed before considering whether there is overdispersion. Typically, overdispersion causes too many Pearson or standardized residuals to lie near or beyond the expected boundaries for "extreme" values—considerably more than 5% of them beyond ± 2 and often several beyond ± 3. These extreme residuals generally occur fairly uniformly across all values of the linear predictor, the estimated mean or probability, or any explanatory variable.

Example: Ecuadorean Bird Counts (BirdQuasiPoi.R, BirdCounts.csv)

In this example, we revisit the Ecuadorean Bird Count data originally analyzed in Section 4.2.3. We refit the Poisson model used for these data and examine the results for signs of overdispersion. This is done using the deviance/df statistic, and a plot of the standardized residuals against the estimated means. The relevant results are given below:

```
> Mpoi <- glm(formula = Birds ~ Loc, family = poisson(link =
    "log"), data = alldata)
> summary(Mpoi)

<OUTPUT EDITED>

(Dispersion parameter for poisson family taken to be 1)

    Null deviance: 216.944  on 23  degrees of freedom
Residual deviance:  67.224  on 18  degrees of freedom
AIC: 217.87

Number of Fisher Scoring iterations: 4
> pred <- predict(Mpoi, type = "response")
> # Standardized Pearson residuals
> stand.resid <- rstandard(model = Mpoi, type = "pearson")
> plot(x = pred, y = stand.resid, xlab = "Predicted count", ylab
    = "Standardized Pearson residuals", main = "Residuals from
    regular likelihood", ylim = c(-5,5))
> abline(h = c(qnorm(0.995), 0, qnorm(0.005)), lty = "dotted",
    col = "red")
```

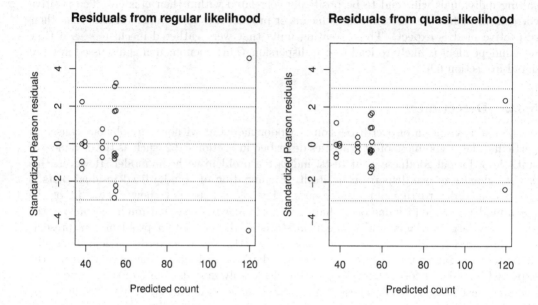

Figure 5.7: Standardized residuals plotted against the mean for the Ecuadorean Bird Count data. Left: Residuals from the fit of the original Poisson model. Right: Residuals from the fit of a quasi-Poisson model. In both cases, upper and lower dotted lines are given at ±2 and ±3 to aid interpretation of the plots.

The residual deviance/df ratio is $67.2/18 = 3.73$, which is much larger than the suggested guidelines in Section 5.2.2: $1 + 3\sqrt{2/18} = 2.0$. The residual plot on the left in Figure 5.7 shows that 4 out of 24 residuals lie beyond ±3, and 2 even lie beyond ±4. Because the explanatory variable is categorical, there is no concern with nonlinearity in this plot. Furthermore, there are no other variables to consider adding to this model, and there is no clustering to the sampling process. Therefore, we conclude that the poor fit of the model is likely caused by overdispersion. We examine solutions to the problem in the next section.

Note that overdispersion occurs when the apparent variance of response counts is larger than the model suggests. In binomial (or multinomial) regressions where all EVPs have $n_m = 1$, overdispersion cannot be detected, even when it might be expected to occur (e.g., if there are clusters of highly correlated data that are mistakenly being handled as independent observations). This is because when $n_m = 1$, the responses can take on only values 0 or 1 and cannot express extra variability, whereas when $n_m > 1$, counts can be distributed farther from $n_m \hat{\pi}_m$ than expected.

5.3.3 Solutions

Because overdispersion is a result of an inadequacy in the model, the obvious way to fix the overdispersion problem is to fix the model. How one does this depends on the source of the overdispersion.

The easiest case to deal with is when there are additional variables that have been measured but not included in the model. Then adding one or more of these into the model will create predictions that are closer to the observed counts, especially if the added variables are strongly related to the response. Residuals from the new model will therefore be generally smaller than those from the original model, and the residual deviance will be reduced. If augmenting the model in this way is successful in removing the symptoms of overdispersion, then inferences can be carried out in the usual manner using the augmented model.

Example: Placekicking (OverdispAddVariable.R, Placekick.csv)

We have seen in previous examples that the distance at which a placekick is attempted is the most important predictor of the kick's probability of success. This result makes sense given an understanding of placekicking in football. In many other problems, however, there is insufficient intuition to lead to a good guess as to which explanatory variables will turn out to be important. In this example, we demonstrate what can happen when an important explanatory variable—represented by distance in the placekicking data—is left out of the model.

We fit a logistic regression model using only wind as a predictor and assess the fit of this model. Then we add distance to the model, refit, and show the revised assessment of fit. Prior to the model fitting and assessment, the data are aggregated into EVP form corresponding to both wind and distance. This allows us to compare fit statistics between models with and without the distance variable.

```
> w <- aggregate(formula = good ~ distance + wind, data =
    placekick, FUN = sum)
> n <- aggregate(formula = good ~ distance + wind, data =
    placekick, FUN = length)
> w.n <- data.frame(success = w$good, trials = n$good, distance =
    w$distance, wind = w$wind)

> mod.fit.nodist <- glm(formula = success/trials ~ wind, weights
    = trials, family = binomial(link = logit), data = w.n)
> summary(mod.fit.nodist)

<OUTPUT EDITED>

    Null deviance: 317.35  on 70  degrees of freedom
Residual deviance: 314.51  on 69  degrees of freedom
AIC: 423.81

> pred <- predict(mod.fit.nodist)
> stand.resid <- rstandard(model = mod.fit.nodist, type =
    "pearson")
> par(mfrow=c(1,2))

> plot(x = pred, y = stand.resid, xlab = "Estimated
    logit(success)", ylab = "Standardized Pearson residuals", main
    = "Standardized residuals vs. Estimated logit", ylim = c(-6,
    13))
> abline(h = c(qnorm(0.995), 0, qnorm(0.005)), lty = "dotted",
    col = "red")

> plot(w.n$distance, y = stand.resid, xlab = "Distance", ylab =
    "Standardized Pearson residuals", main = "Standardized
    residuals vs. Distance", ylim = c(-6, 13))
```

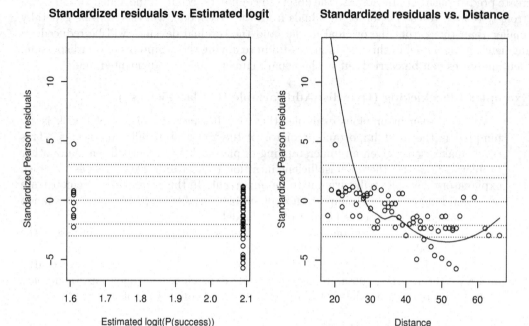

Figure 5.8: Standardized residual plots for placekick data when the logistic regression model contains only wind. Lines at ±2 and ±3 are approximate thresholds for standardized residuals. Left: Residuals versus the logit of the estimated probability of success. Right: Residuals versus the omitted variable `distance`, with the curve from a weighted loess smoother added to accentuate the mean trend.

```
> abline(h = c(qnorm(0.995), 0, qnorm(0.005)), lty = "dotted",
    col = "red")
> dist.ord <- order(w.n$distance)
> lines(x = w.n$distance[dist.ord], y = predict(loess(stand.resid
    ~ w.n$distance))[dist.ord], col = "blue")
```

The model fit without `distance` is shown above. The deviance/df ratio is $314.51/69 = 4.5$, well above the expected upper limit for this statistic, $1 + 3\sqrt{(2/69)} = 1.5$. This result leaves no doubt that the model does not fit, but does not explain why. Plotting the standardized residuals against the linear predictor (Figure 5.8, left plot) shows a generally overdispersed pattern, with numerous residuals lying well outside the expected range. This plot does nothing to explain the cause of the extreme residuals, and might generically be thought to indicate overdispersion. However, the plot on the right shows that the residuals have a very clear decreasing pattern as the omitted variable `distance` increases. This is because the model assumes a constant probability of success for all distances at a given level of `wind`. Therefore, observed success probabilities are higher than the model expects for short placekicks, leading to generally positive residuals, and the observed success probabilities are lower than the model expects for long placekicks, leading to negative residuals.

Next, we add `distance` to the model and repeat the analysis:

```
> mod.fit.dist <- glm(formula = success/trials ~ distance + wind,
    weights = trials, family = binomial(link = logit), data = w.n)
```

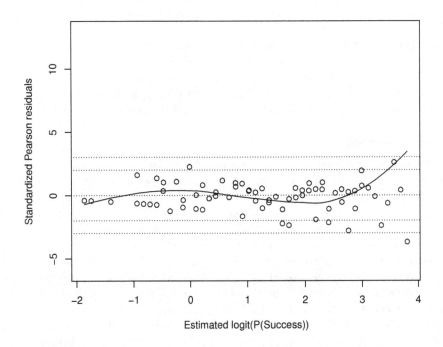

Figure 5.9: Standardized residuals against linear predictor for placekick data when the logistic regression model for success contains wind speed and distance. Upper and lower red lines at ±2 and ±3 are approximate thresholds for standardized residuals. The solid curve is from a weighted loess smoother to highlight any mean trend.

```
> summary(mod.fit.dist)

<OUTPUT EDITED>

    Null deviance: 317.345  on 70  degrees of freedom
Residual deviance:  76.453  on 68  degrees of freedom
AIC: 187.76
```

The added variable reduces the deviance/df ratio to $76.453/68 = 1.1$, which is well within the lower threshold of $1 + 2\sqrt{(2/68)} = 1.3$. The residuals are shown in Figure 5.9 on the same scale as the previous plots to make comparison easier. The weighted loess mean-trend curve shows a pattern very similar to that observed in Figure 5.3, and the interpretation offered for that plot in the corresponding example applies here. Although there are still a few points outside the ±2 and ±3 thresholds, none are as extreme as when distance was left out of the model.

Measured variables with a large enough effect to create noteworthy overdispersion are rarely omitted from a model when an appropriate variable-selection procedure has been applied to the data. However, variable selection often does not consider interactions until a poor model fit suggests that they might be needed. Thus, if overdispersion appears and no other variables are available that have not already been considered, it is possible that adding one or more interactions to the model may alleviate the problem.

Other causes of overdispersion require different model amendments. For example, if the apparent overdispersion is caused by analyzing clustered data as if they were independent, the problem can often be addressed by adding a random-effect term to the model with a different level for each cluster. Then a separate parameter is fit which accounts for the extra variability that is created by the within-cluster correlation. The result is a *generalized linear mixed model*, meaning a model containing both fixed- and random-effect variables. Details on how to formulate and account for random effects are given in Section 6.5.

Quasi-likelihood models

When there is no apparent cause for the overdispersion, then it may be due to the omission of important variables that are unknown and not measured, sometimes called *lurking variables* (Moore and Notz, 2009). Adding these variables to the model is obviously not an option. Instead, the solution is to change the distributional family to one with an additional parameter to measure the extra variance that is not explained by the original model. There are several different approaches to doing this, each of which provides a more flexible model for the variance than the Poisson or binomial families allow. *We strongly recommend against the routine use of these methods unless all other avenues for model improvement have been exhausted.* These methods are often just a patch that covers up a symptom of a problem, rather than a cure for the problem.

One approach for doing this, called *quasi-likelihood*, can be used for both binomial and Poisson analyses. In the current context, a quasi-likelihood function is a likelihood function that has extra parameters added to it that are not a part of the distribution on which the likelihood is based. As an example, suppose we model counts as Poisson with mean $\mu(x)$, where x represents one or more explanatory variables. Suppose that the overdispersion stems from a constant inflation of the variance; that is, suppose that $Var(Y|x) = \gamma\mu(x)$, where $\gamma \geq 1$ is an unknown constant called a *dispersion parameter*. A quasi-likelihood can be created to account for this by dividing the usual Poisson log-likelihood by γ.[8] This has no effect on the regression-parameter estimates, but after they are computed, an estimate of γ is found by dividing the Pearson goodness-of-fit statistic for the model by its residual degrees of freedom, $M - \tilde{p}$, where $\tilde{p}$ is the number of parameters estimated in the model. See Wedderburn (1974) for details.

Parameter estimates and test statistics computed from a quasi-likelihood function have the same properties as those computed from a regular likelihood, so inferences are carried out using the same basic tools, with minor changes. First, the dispersion parameter estimate, $\hat{\gamma}$, which is greater than 1 when overdispersion exists, is used to adjust standard errors and test statistics for subsequent inferences. Second, the sampling distributions used for tests and confidence intervals need to reflect the fact that the dispersion is being estimated with a statistic based on the model Pearson statistic, which has approximately a chi-square with $M - \tilde{p}$ degrees of freedom in large samples.

Specifically, consider a LRT statistic for a model effect that would normally be compared to a chi-square distribution with q df. The statistic is divided by $\hat{\gamma}$, which not only makes it smaller, but also changes its distribution. Further dividing the statistic by q creates a new statistic that, in large samples, can be compared to an F distribution with $(q, M - \tilde{p})$

---

[8]Quasi-likelihood is more rigorously defined in Wedderburn (1974) according to properties of the first derivative of the log-likelihood with respect to μ. Dividing either the log-likelihood or its first derivative by γ provides equivalent results. See McCullagh and Nelder (1989) for a more comprehensive treatment of quasi-likelihood.

degrees of freedom. Similarly, standard errors for regression parameters are multiplied by $\sqrt{\hat{\gamma}}$, which changes the large-sample distribution on which confidence intervals are based from normal to $t_{M-\tilde{p}}$. Diagnostic statistics like standardized residuals and Cook's distance are also appropriately adjusted by $\hat{\gamma}$.

Finally, it is often recommended that $\hat{\gamma}$ be computed from a "maximal model," meaning one that contains all available variables, even when the final model to which quasi-likelihood is applied is a smaller model. This is done to ensure that there can be no artificial inflation of the dispersion parameter estimate due to missing explanatory variables. However, in modern studies there are often huge numbers of explanatory variables available, many of which are expected to be unimportant. In these cases, basing $\hat{\gamma}$ on a maximal model can be excessively conservative, if not impossible. If the residual degrees of freedom from the full model are much smaller than the residual degrees of freedom from a model containing just the important variables, power for tests and widths of confidence intervals may be adversely affected. Estimating γ from a full model is "safe," but if one can be fairly confident that no important measured variables have been left out of a particular smaller model, then that model should be able to provide a reasonable estimate of γ that can result in more powerful inferences.

When quasi-likelihood is applied to a Poisson model as described above, the result is sometimes referred to as a *quasi-Poisson* model. The same quasi-likelihood approach can be applied to a binomial model, in which case we model $Var(Y|x_m) = \gamma n_m \pi(x_m)(1 - \pi(x_m))$, where $\pi(x_m)$ is the probability of success and n_m is the number of trials for EVP m. The result is sometimes called a *quasi-binomial* model. This model is only relevant when at least some $n_m > 1$; otherwise, the Pearson statistic cannot be used to measure overdispersion.

Example: Ecuadorean Bird Counts (BirdOverdisp.R, BirdCounts.csv)

In this example, we refit the Ecuadorean Bird Count data using a quasi-Poisson model. Quasi-Poisson and quasi-binomial model fitting are available in glm() using family = quasipoisson(link = "log") or quasibinomial(link = "logit"). Alternative links can be specified. Here we apply a quasi-Poisson model to the Ecuadorean bird counts:

```
> Mqp <- glm(formula = Birds~Loc, family = quasipoisson(link =
    "log"), data = alldata)
> summary(Mqp)

<OUTPUT EDITED>

(Dispersion parameter for quasipoisson family taken to be
    3.731881)

    Null deviance: 216.944  on 23  degrees of freedom
Residual deviance:  67.224  on 18  degrees of freedom
AIC: NA

Number of Fisher Scoring iterations: 4

> # Demonstrate calculation of dispersion parameter
> pearson <- residuals(Mpoi, type = "pearson")
> sum(pearson^2)/Mpoi$df.residual
[1] 3.731881
```

Notice several things about these results. First, the residual deviance is identical to that from the Poisson model fit. This is an indication that the portion of the likelihood function that is used to estimate the regression parameters has not been changed. The dispersion parameter is now estimated to be 3.73, and we show code demonstrating that this is the Pearson statistic divided by its degrees of freedom. A plot of the standardized residuals is given in the plot on the right in Figure 5.7 (code is given in the program corresponding to this example). All residuals are now contained within ±3 and only two remain outside ±2.

To compare the inferences from the Poisson and quasi-Poisson models, we compute LRTs for the effect of locations on the means using the anova() function, and then compute LR confidence intervals for the mean bird counts. The coding involves a reparameterization to simplify the calculations: the intercept parameter is removed from the model, allowing log-means to be estimated directly by the model. We can then apply confint() to the model objects and exponentiate the results to make confidence intervals for the means.

```
> anova(Mpoi, test = "Chisq")
Analysis of Deviance Table
Model: poisson, link: log
Response: Birds
Terms added sequentially (first to last)
     Df Deviance Resid. Df Resid. Dev  Pr(>Chi)
NULL                   23     216.944
Loc   5   149.72       18      67.224 < 2.2e-16 ***
---
Signif. codes:  0 '***' 0.001 '**' 0.01 '*' 0.05 '.' 0.1 ' ' 1

> anova(Mqp, test = "F")
Analysis of Deviance Table
Model: quasipoisson, link: log
Response: Birds
Terms added sequentially (first to last)
     Df Deviance Resid. Df Resid. Dev      F    Pr(>F)
NULL                   23     216.944
Loc   5   149.72       18      67.224 8.0238 0.0003964 ***
---
Signif. codes:  0 '***' 0.001 '**' 0.01 '*' 0.05 '.' 0.1 ' ' 1

> Mpoi.rp <- glm(formula = Birds ~ Loc - 1, family = poisson(link
    = "log"), data = alldata)
> Mqp.rp <- glm(formula = Birds ~ Loc - 1, family =
    quasipoisson(link = "log"), data = alldata)
> round(exp(cbind(mean = coef(Mpoi.rp), confint(Mpoi.rp))), 2)
Waiting for profiling to be done...
           mean    2.5 %  97.5 %
LocEdge   48.25   41.76   55.38
LocForA  119.50  104.98  135.30
LocForB   55.00   49.28   61.15
LocFrag   54.33   48.65   60.45
LocPasA   39.50   31.42   48.86
LocPasB   38.00   32.27   44.37
> round(exp(cbind(mean = coef(Mqp.rp), confint(Mqp.rp))), 2)
      Waiting for profiling to be done...
           mean   2.5 %  97.5 %
LocEdge   48.25  36.27   62.62
LocForA  119.50  92.57  151.20
```

```
LocForB    55.00  44.32  67.27
LocFrag    54.33  43.72  66.54
LocPasA    39.50  24.97  58.79
LocPasB    38.00  27.49  50.89
```

The test for $H_0 : \mu_1 = \mu_2 = \ldots = \mu_6$ rejects this null hypothesis in both models, although overdispersion correction does alter the extremeness of the test statistic, as measured by the p-values. The LRT statistic for the Poisson model is $-2\log(\Lambda) = 149.72$, which has a very small p-value using a χ_5^2 approximation. On the other hand, the quasi-Poisson model uses $F = (149.72/5)/(67.22/18) = 8.33$, which has a p-value of 0.0004 using a $F_{5,18}$ distribution approximation.

The means estimated by the two models are identical as expected, because both models use the same likelihood to estimate the regression parameters. However, the confidence intervals for the quasi-Poisson model are wider than those for the Poisson. Again, the main conclusion is not changed: Forest A has a considerably higher mean than all other locations. However, with the quasi-Poisson there is more overlap between the confidence intervals for the two pastures and those for the edge, fragment, and Forest B.

Notice that $\hat{\gamma}$ is not found by maximum likelihood—the computation is performed *after* the MLEs are found for the regression parameters. Thus, the value of $\hat{\gamma}$ does not impact the value of the log-likelihood. In particular, this means that the standard information criteria described in Section 5.1.2 cannot be computed on quasi-likelihood estimation results. There is instead a series of quasi-information criteria, which we can denote by $QIC(k, \gamma)$, that can be used to compare different regression models within the same quasi-likelihood family. These are obtained simply by dividing the log-likelihood $\log(L(\hat{\beta}|y_1,....y_n))$ by $\hat{\gamma}$ in the formulas for any $IC(k)$, and adding 1 to the number of parameters. For example, QAIC = $-2\log(L(\hat{\beta}|y_1,....y_n))/\hat{\gamma}+2(k+1)$, where k is the number of regression parameters estimated, including the intercept; $QAIC_c$ and QBIC are similarly defined.

Using a $QIC(k, \gamma)$ to compare different regression models must be done carefully. Because $\hat{\gamma}$ is computed externally to the MLE calculations, the effect of changing its value from model to model is not properly accounted for in the calculation of $QIC(k, \gamma)$. It is therefore important that all regression models be compared using the same value of $\hat{\gamma}$. An appropriate strategy is to estimate γ from the largest model being considered, and then use the same estimate in the computations of $QIC(k, \gamma)$ for all models being compared. As noted before the Bird Count example, this also ensures that the Pearson goodness-of-fit statistic on which γ is based is least likely to be inflated due to omitted variables. For details regarding computing $QIC(k)$ values in R, see Bolker (2009).

Negative binomial and beta-binomial models

There are other models that can be used as alternatives to the standard models or their quasi-counterparts. The *negative binomial* distribution is most often used when an alternative to the Poisson distribution is needed. The negative binomial distribution results from allowing the mean μ in a Poisson distribution to be a random variable with a gamma distribution.[9] This is very similar to the problem we studied in our simulation example on page 303, except that we used a normal distribution for the mean rather than a gamma

---

[9] The negative binomial was originally derived from properties of Bernoulli trials. It is the distribution of the number of successes that are observed before the rth failure, which gives rise to the name "negative binomial." See p. 95 of Casella and Berger (2002).

distribution. When a gamma distribution is used instead, then a specific relationship arises between the variance of the count random variable Y and its mean μ:

$$Var(Y) = \mu + \theta\mu^2,$$

where $\theta \geq 0$ is an unknown parameter. Notice that $\theta = 0$ returns the same mean-variance relationship as in the Poisson model. Also notice that this relationship is different from the one assumed by the quasi-Poisson model, $Var(Y) = \gamma\mu$. Therefore, these two models are distinct and there can be instances where one model is preferred over the other.

Ver Hoef and Boveng (2007) provide some guidance for helping to decide between these two models. In particular, a plot of $(y_i - \hat{\mu}_i)^2$ vs. $\hat{\mu}_i$ from the Poisson model can help to identify the variance-mean relationship. If a smoothed version of this plot follows a mostly linear trend, then a quasi-Poisson model is appropriate. If the trend is more of an increasing quadratic, then the negative binomial is preferred. Ver Hoef and Boveng (2007) suggest that the trend in the plot can be made clearer by first grouping the data according to similar values of $\hat{\mu}_i$ (similar to what is done in the Hosmer-Lemeshow test discussed in Section 5.2.2), then plotting the average squared residual in each group against the average $\hat{\mu}_i$ in the group.

Unlike the quasi-Poisson model, all parameters of the negative binomial model, including θ, are estimated using MLEs. Therefore, information criteria can be used without further adjustment to compare different negative binomial regression models. The value of θ counts as one additional parameter in the penalty calculations of $IC(k)$. Also, because these are the usual information criteria, they can be used to compare models across the negative binomial and Poisson families. However, no direct comparison to a quasi-Poisson model is possible, because $QIC(k)$ is a different criterion.

Example: Ecuadorean Bird Counts (BirdOverdisp.R, BirdCounts.csv)

We now refit the Ecuadorean Bird Count data using a negative binomial distribution. The MASS package has a function glm.nb() that fits the negative binomial distribution and estimates θ. A log link is the default, but alternative links can be specified using a link argument. We use this function below to fit the model and examine the results. We then refit the model using the reparameterization that removes the intercept and compute LR confidence intervals for the means.

```
> library(MASS)

> M.nb <- glm.nb(formula = Birds ~ Loc, data = alldata)
> summary(M.nb)
Call: glm.nb(formula = Birds ~ Loc, data = alldata, init.theta =
    33.38468149, link = log)
Deviance Residuals:
     Min       1Q   Median       3Q      Max
 -1.80758  -0.48773  -0.08495   0.55106   1.65702

Coefficients:
            Estimate Std. Error z value Pr(>|z|)
(Intercept)   3.8764     0.1126  34.438  < 2e-16 ***
LocForA       0.9069     0.1784   5.083 3.71e-07 ***
LocForB       0.1309     0.1438   0.910    0.363
LocFrag       0.1187     0.1440   0.825    0.410
LocPasA      -0.2001     0.2008  -0.997    0.319
LocPasB      -0.2388     0.1635  -1.460    0.144
```

```
---
Signif. codes:  0 '***' 0.001 '**' 0.01 '*' 0.05 '.' 0.1 ' ' 1

(Dispersion parameter for Negative Binomial(33.3847) family taken
    to be 1)

    Null deviance: 73.370  on 23  degrees of freedom
Residual deviance: 22.705  on 18  degrees of freedom
AIC: 198.06

Number of Fisher Scoring iterations: 1

              Theta:  33.4
          Std. Err.:  14.9

 2 x log-likelihood:  -184.061

> anova(M.nb, test = "Chisq")
Analysis of Deviance Table
Model: Negative Binomial(33.3847), link: log
Response: Birds
Terms added sequentially (first to last)
     Df Deviance Resid. Df Resid. Dev  Pr(>Chi)
NULL                    23     73.370
Loc   5   50.665        18     22.705 1.013e-09 ***
---
Signif. codes:  0 '***' 0.001 '**' 0.01 '*' 0.05 '.' 0.1 ' ' 1
Warning message:
In anova.negbin(M.nb, test = "Chisq") :
  tests made without re-estimating 'theta'

> M.nb.rp <- glm.nb(formula = Birds ~ Loc - 1, data = alldata)
> round(exp(cbind(mean=coef(M.nb.rp),confint(M.nb.rp))), 2)
Waiting for profiling to be done...
          mean  2.5 %  97.5 %
LocEdge  48.25  38.75   60.26
LocForA 119.50  91.70  157.91
LocForB  55.00  46.20   65.64
LocFrag  54.33  45.62   64.87
LocPasA  39.50  28.54   54.84
LocPasB  38.00  30.13   47.99
```

The goodness-of-fit results from **summary()** show that the negative binomial model fits somewhat better than the Poisson model: the residual deviance is 22.7 on 18 df. The extra dispersion parameter θ is estimated to be 33.4 with a standard error of 14.9. This estimate is more than two standard errors from 0, indicating that the overdispersion correction is important. The LRT for equality of the means at the six locations is given in the **anova()** results. The deviance of 50.7 is highly significant, although less extreme than the uncorrected Poisson model. The estimated means are the same in this model as in the previous models (this does not have to be so, but happens here because we are using only a nominal explanatory variable). The confidence intervals are similar to those from the quasi-Poisson model, except wider for larger means and narrower for smaller means. This is consistent with the variance-mean relationship of the negative binomial model.

Which of the two models to correct for overdispersion is better for these data? To answer this, we make a plot of $(y_i - \hat{\mu}_i)^2$ vs. $\hat{\mu}_i$. We plot a linear and a quadratic fit to examine trends and see which one fits better.

```
> res.sq <- residuals(object = Mpoi, type = "response")^2
> set1 <- data.frame(res.sq, mu.hat = Mpoi$fitted.values)

> fit.lin <- lm(formula = res.sq ~ mu.hat, data = set1)
> fit.quad <- lm(formula = res.sq ~ mu.hat + I(mu.hat^2), data =
    set1)
> summary(fit.quad)

Call: lm(formula = res.sq ~ mu.hat + I(mu.hat^2), data = set1)

Residuals:
    Min      1Q  Median      3Q     Max
-159.49  -95.80  -19.95   41.66  320.51

Coefficients:
              Estimate  Std. Error  t value  Pr(>|t|)
(Intercept)  -108.88160   331.02260   -0.329    0.745
mu.hat         -0.61133     9.69505   -0.063    0.950
I(mu.hat^2)     0.10115     0.05993    1.688    0.106

Residual standard error: 137.5 on 21 degrees of freedom
Multiple R-squared: 0.8638,    Adjusted R-squared: 0.8508
F-statistic:  66.6 on 2 and 21 DF,  p-value: 8.1e-10

> plot(x = set1$mu.hat, y = set1$res.sq, xlab = "Predicted
    count", ylab = "Squared Residual")
> curve(expr = predict(object = fit.lin, newdata =
    data.frame(mu.hat = x), type = "response"), col = "blue", add
    = TRUE, lty = "solid")
> curve(expr = predict(object = fit.quad, newdata =
    data.frame(mu.hat = x), type = "response"), col = "red", add =
    TRUE, lty = "dashed")
> legend(x = 50, y = 1000, legend = c("Linear", "Quadratic"), col
    = c("red", "blue"), lty = c("solid", "dashed"), bty = "n")
```

The results are given in Figure 5.10. The quadratic trend seems to describe the relationship slightly better than the straight line, although the improvement is not vast. The coefficients of the quadratic fit of the squared residuals against the predicted mean are shown above. The quadratic coefficient has a p-value of 0.1, which is not significant but not very large. The results in this case are somewhat inconclusive, so there is justification for either model.

Similar to the negative binomial model, the *beta-binomial* model (Williams, 1975) can serve as an alternative to quasi-binomial as a model for overdispersed counts from fixed numbers of trials. Suppose that we observe binomial counts from M EVPs, $y_1, y_2, \ldots, y_M$, with y_m resulting from n_m trials with probability π_m, $m = 1, \ldots, M$. The beta-binomial model assumes that each π_m follows a beta distribution, the mean of which may depend on the explanatory variables. Reparameterizing the beta distribution for this purpose leads to the result that the expected count for EVP m is $E(Y_m) = n_m \pi_m$, while the variance

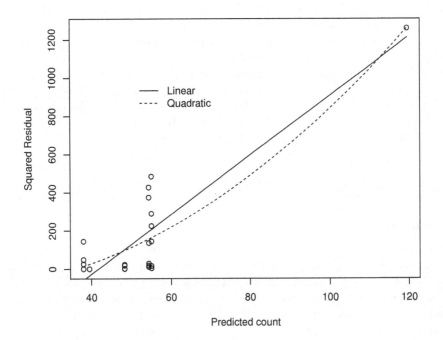

Figure 5.10: Plot of squared residuals vs. predicted values for choosing between quasi-Poisson and negative binomial models. Blue line is the straight line fit for quasi-Poisson model; red line is quadratic fit for negative binomial.

is $Var(Y_m) = n_m \pi_m (1 - \pi_m)(1 + \phi(n_m - 1))$, where $0 \leq \phi < 1$ is an unknown dispersion parameter. Notice that the mean of this distribution is the same as in the binomial distribution, as is the variance as long as either $\phi = 0$ or $n_m = 1$. When $\phi > 0$ and $n_m > 1$, the variance of the count is strictly greater than the usual variance specified by the binomial distribution, thereby allowing for overdispersion. Also, when only individual binary trials are observed, each with its own explanatory variables, all $n_m = 1$, and the beta-binomial model cannot be fit. Finally, notice that when all n_m are equal, the factor $(1+\phi(n_m-1))$ is the same for all observations, so that the beta-binomial assumes the same mean-variance relationship as the quasi-binomial model. In this case, the simpler quasi-binomial is generally preferred.

Parameters of the beta-binomial model are estimated using MLEs, so that standard inference procedures can be used. Also, information criteria can be computed, so this model can be compared to an ordinary binomial regression model to see whether the overdispersion is severe enough to warrant the fitting of a model with an extra parameter. The fitting is done in R using the `betabin()` function of the package `aod` (short for "Analysis of Overdispersed Data"). The function has capacity to further allow ϕ to depend on explanatory variables. An example is given in Exercise 37d.

Overdispersion can affect multivariate regression models like Equation 3.4. Unfortunately, the complexity of a multi-category regression model makes detecting and resolving overdispersion somewhat more complicated than in single-response models. Generally, it can present itself as an inflation of goodness-of-fit statistics that are not explained by other diagnostic investigations. The simplest solution is a quasi-likelihood-type formulation, in which the variances and covariances among the counts for different categories are all multiplied by a constant $\gamma > 1$. This process is discussed in McCullagh and Nelder (1989, p. 174). An alternative method of formulating and fitting the model is discussed in Mebane and Sekhon (2004) and is carried out using the `multinomRob` package. An extension of the beta-binomial model to multivariate response, called the Dirichlet-multinomial model, can be used as an alternative to an ordinary multinomial regression model as a way to introduce extra variability into the model. Unfortunately these models are difficult to fit; we are not aware of any R packages that can fit general Dirichlet-multinomial regression models.

As a simpler alternative to fitting complex models, a Poisson-based formulation for the multivariate counts could be used as discussed in Section 4.2.5. This allows the use of simpler tools for model diagnostics and for correcting overdispersion.

5.4 Examples

In this section we re-analyze two examples that have thus far been presented only in pieces throughout the book: the placekicking data using logistic regression and the alcohol consumption data using Poisson regression. Our goal is to demonstrate the process of undertaking a complete analysis, starting with variable selection, continuing through model fitting and assessment, and concluding with inferences. We present R code and graphics liberally and explain our rationale for each decision made during the analyses.

5.4.1 Logistic regression - placekicking data set

We now examine the placekicking data set more closely to find models for the probability of a successful placekick that fit the data well. The data used here include 13 additional observations that were not part of any previous example. The reason these additional observations are included (and then eventually excluded) is made clear from our analysis. The data used here also include these additional variables:

- `altitude`: Official altitude in feet for the city where the placekick is attempted (not necessarily the exact altitude of the stadium)
- `home`: Binary variable denoting placekicks attempted at the placekicker's home (1) vs. away (0) stadium
- `precip`: Binary variable indicating whether precipitation is falling at game time (1) vs. no precipitation (0)
- `temp72`: Fahrenheit temperature at game time, where 72° is assigned to placekicks attempted within domed stadiums

The data are stored in the placekick.mb.csv file, and we include all of our R code in Placekick-FindBestModel.R.

Variable selection

While we could use `glmulti()` and the genetic algorithm to help us search among the main effects and all of their two-way interactions, we choose instead to limit the interactions to those that make sense in the context of the problem. Based on our football experience, we believe that the following two-way interactions are plausible:

- `distance` with `altitude`, `precip`, `wind`, `change`, `elap30`, `PAT`, `field`, and `temp72`; distance clearly has the largest effect on the probability of success, and all of these other factors could conceivably have a larger effect on longer kicks than on shorter ones.

- `home` with `wind`; a placekicker on his home field may be more accustomed to local wind effects than a visiting kicker.

- `precip` with `type`, `field`, and `temp72`; precipitation cannot directly affect a game in a domed stadium, can have a bigger effect on grass than on artificial turf, and can be snow or otherwise make players miserable if the temperature is cold enough.

Ideally, one would like to consider all main effects and these 12 interactions at one time using `glmulti()`, allowing its `exclude` argument to remove those interactions that are not plausible. Unfortunately, we encountered a potential error in the function that prevented us from doing so. As an alternative, we first use the `glmulti()` function to search among all possible main effects to find the model with the smallest AIC_c. Similar to the results from Section 5.1.2, the model with `distance`, `wind`, `change`, and `PAT` has the smallest AIC_c. Next, starting from this base model, we perform forward selection among the 12 interactions. We wrote our own function to calculate AIC_c in order to do the selection one step at a time (`step()` does not calculate AIC_c). This process involved manually adding each interaction to the base model in a sequence of `glm()` runs. Note that for any interaction that involves a main effect not among `distance`, `wind`, `change`, and `PAT`, we include this main effect in the model along with its interaction. For example, we add an `altitude` main effect to the model when evaluating `distance:altitude`. Below is the first step of forward selection:

```
> # Data has been read in and it is stored in placekick.mb

> AICc <- function(object) {
    n <- length(object$y)
    r <- length(object$coefficients)
    AICc <- AIC(object) + 2*r*(r+1)/(n-r-1)
    list(AICc = AICc, BIC = BIC(object))
  }

> # Main effects model
> mod.fit <- glm(formula = good ~ distance + wind + change + PAT,
    family = binomial(link = logit), data = placekick.mb)
> AICc(object = mod.fit)
$AICc
[1] 780.8063

$BIC
[1] 807.1195

> # Models with one interaction included
> mod.fit1 <- glm(formula = good ~ distance + wind + change + PAT
    + altitude + distance:altitude, family = binomial(link =
    logit), data = placekick.mb)

<OUTPUT EDITED>

> mod.fit12 <- glm(formula = good ~ distance + wind + change +
    PAT + precip + temp72 + precip:temp72, family = binomial(link
    = logit), data = placekick.mb)
```

```
> inter <- c("distance:altitude", "distance:precip",
    "distance:wind", "distance:change", "distance:elap30",
    "distance:PAT", "distance:field", "distance:temp72",
    "home:wind", "type:precip", "precip:field", "precip:temp72")
> AICc.vec <- c(AICc(mod.fit1)$AICc, AICc(mod.fit2)$AICc,
    AICc(mod.fit3)$AICc, AICc(mod.fit4)$AICc, AICc(mod.fit5)$AICc,
    AICc(mod.fit6)$AICc, AICc(mod.fit7)$AICc, AICc(mod.fit8)$AICc,
    AICc(mod.fit9)$AICc, AICc(mod.fit10)$AICc,
    AICc(mod.fit11)$AICc, AICc(mod.fit12)$AICc)
> all.AICc1 <- data.frame(inter = inter, AICc.vec)
> all.AICc1[order(all.AICc1[,2]), ]
                inter AICc.vec
3       distance:wind 777.2594
6        distance:PAT 777.3321
7      distance:field 782.4092
4     distance:change 782.6573
9           home:wind 783.6260
1   distance:altitude 784.4997
10         type:precip 784.6068
2      distance:precip 784.6492
8      distance:temp72 784.6507
5      distance:elap30 784.6822
11        precip:field 785.7221
12       precip:temp72 786.5106
```

The base model has $AIC_c = 780.8$. The results of adding each interaction individually to this model are sorted by increasing AIC_c. Adding `distance:wind` improves the model the most ($AIC_c = 777.25$), so the first step of forward selection adds `distance:wind` to the model.

The code for subsequent steps is included in the corresponding program. In summary, the second step adds `distance:PAT` ($AIC_c = 773.80$), and the third step results in no model with a smaller AIC_c. Thus, forward selection finds that the "best" model adds `distance:wind` and `distance:PAT` to the base model. As an alternative method for evaluating the importance of interactions, we also used `glmulti()` with the main effects of `distance`, `PAT`, `wind`, `change` along with the pairwise interactions among only these variables. This identified the same model as with our previous forward selection.

We next consider whether the current model can be improved by considering transformations of the explanatory variables. Because `PAT`, `wind`, and `change` are binary explanatory variables, no transformations for these variables need to be considered. To evaluate `distance`, we convert the data into EVP form, re-fit the model, obtain the standardized Pearson residuals, and then plot these standardized Pearson residuals vs. `distance`.

```
> # Convert data to EVP form; interactions are not needed in
    aggregate() because they do not change the number of unique
    combinations of explanatory variables.
> w <- aggregate(formula = good ~ distance + wind + change + PAT,
    data = placekick.mb, FUN = sum)
> n <- aggregate(formula = good ~ distance + wind + change + PAT,
    data = placekick.mb, FUN = length)
> w.n <- data.frame(w, trials = n$good, prop =
    round(w$good/n$good, 4))
> head(w.n)
```

```
  distance wind change PAT good trials prop
1       18    0      0   0    1      1 1.00
2       19    0      0   0    3      3 1.00
3       20    0      0   0   15     15 1.00
4       21    0      0   0   11     12 0.92
5       22    0      0   0    7      8 0.88
6       23    0      0   0   15     15 1.00
> nrow(w.n)  # Number of EVPs (M)
[1] 124
>  sum(w.n$trials)  # Number of observations
[1] 1438

> # Estimates here match those had before converting data to EVP
    form
>  mod.prelim1 <- glm(formula = good/trials ~ distance + wind +
    change + PAT + distance:wind + distance:PAT, family =
    binomial(link = logit), data = w.n, weights = trials)
> round(summary(mod.prelim1)$coefficients, digits = 4)
              Estimate Std. Error z value Pr(>|z|)
(Intercept)     4.4964     0.4814  9.3399   0.0000
distance       -0.0807     0.0114 -7.0620   0.0000
wind            2.9248     1.7850  1.6385   0.1013
change         -0.3320     0.1945 -1.7068   0.0879
PAT             6.7119     2.1137  3.1754   0.0015
distance:wind  -0.0918     0.0457 -2.0095   0.0445
distance:PAT   -0.2717     0.0980 -2.7726   0.0056

> # Plot of standardized Pearson residuals vs. distance
> stand.resid <- rstandard(model = mod.prelim1, type = "pearson")
> plot(x = w.n$distance, y = stand.resid, ylim = c(min(-3,
    stand.resid), max(3, stand.resid)), ylab = "Standardized
    Pearson residuals", xlab = "Distance")
> abline(h = c(3, 2, 0, -2, -3), lty = "dotted", col = "blue")
> smooth.stand <- loess(formula = stand.resid ~ distance, data =
    w.n, weights = trials)
> ord.dist <- order(w.n$distance)
> lines(x = w.n$distance[ord.dist], y =
    predict(smooth.stand)[ord.dist], lty = "solid", col = "red")
```

Figure 5.11 shows the plot along with a loess curve. The curve undulates gently around 0, showing no apparent trends to the mean value of the residuals. This suggests that no transformations of distance are needed in the model. For illustration purposes, we also temporarily added a quadratic term for distance to the model. This increases AIC_c to 775.72, which agrees with our findings from the plot. Therefore, a preliminary model for our data includes distance, PAT, wind, change, distance:wind, and distance:PAT.

Assessing the model fit - preliminary model

To expedite the process of assessing the fit of the preliminary model, we have written a function that implements most of the methods outlined in Section 5.2 in a simplified manner specifically for logistic regression. This function, called examine.logistic.reg(), is available from the textbook's website in the program Examine.logistic.reg.R. The only required argument is mod.fit.obj, which corresponds to the glm-class object containing the model whose fit is to be assessed. Optional arguments include identify.points that

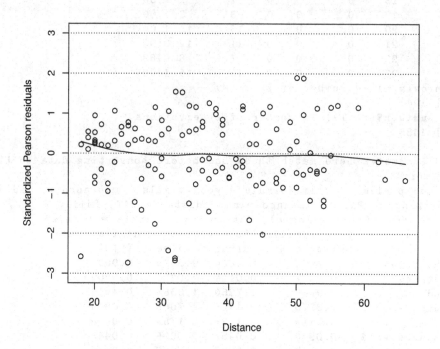

Figure 5.11: Standardized Pearson residuals vs. distance of the placekick.

allows users to interactively identify points on a plot, scale.n and scale.cookd which allow users to rescale numerical values used as the circle size in bubble plots, and pearson.dev to denote whether standardized Pearson or standardized deviance residuals are plotted (default is Pearson). These arguments are explained in more detail shortly. The function's use is demonstrated below, and the resulting plots are in Figure 5.12.

```
> # Used with examine.logistic.reg() for rescaling numerical
    values
> one.fourth.root <- function(x) {
    x^0.25
  }
> one.fourth.root(16)   # Example
[1] 2

> # Read in file containing examine.logistic.reg() and run
    function
> source(file = "C:\\Rprograms\\Examine.logistic.model.R")
> save.info1 <- examine.logistic.reg(mod.fit.obj = mod.prelim1,
    identify.points = TRUE, scale.n = one.fourth.root,
    scale.cookd = sqrt)
> names(save.info1)
 [1] "pearson"          "stand.resid"      "stand.dev.resid"
 [4] "deltaXsq"         "deltaD"           "cookd"
 [7] "pear.stat"        "dev"              "dev.df"
[10] "gof.threshold"    "pi.hat"           "h"
```

Model selection and evaluation 323

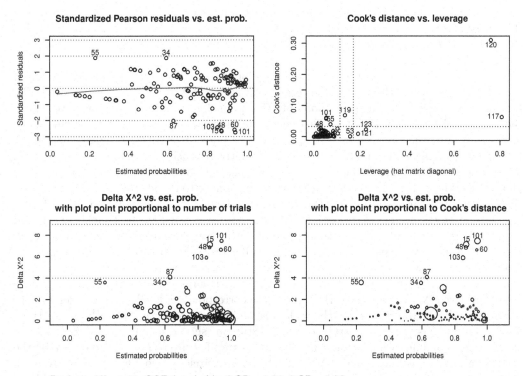

Figure 5.12: Diagnostic plots for the placekicking logistic regression model; all observations are included in the data set.

After each plot is created, R prompts the user to identify points in that plot. This identification is done by left-clicking on the points of interest, and these are subsequently labeled by the row name in `placekick.mb`.[10] For the (row, column) = (1,1) plot in Figure 5.12, EVPs 15, 34, 48, 55, 60, 87, 101, and 103 are all identified in this manner. To end the identification process for a particular plot, a user can right-click the mouse and select Stop from the resulting window. By default, the `identify.points` argument is set to TRUE, but can be changed to FALSE in the function call if identification is not of interest.

The (1,1) and (1,2) plots are similar to those shown in Section 5.2 for the standardized Pearson residuals, Cook's distances, and leverages. Horizontal lines in the (1,1) plot are drawn at $-3, -2, 2$, and 3 to help judge outlying EVPs. There are a few EVPs outside or close to these lines that we may want to investigate further. A loess curve included on the plot shows no clear relationship between the residuals and estimated probabilities. A horizontal line is drawn at $4/M$ in the (1,2) plot to help determine which EVPs may be potentially influential as judged by Cook's distance. The plot shows that EVP 120 has a very large value in comparison to the other EVPs, so we definitely want to investigate this EVP further. There are a few other EVPs with Cook's distances which stand out from the rest that we may want to examine more closely as well. Vertical lines on the (1,2) plot are

---

[10]Every row of a data frame has a "name." These row names are shown in the left-most column of a printed data frame. Typically, the row number of the data frame is the row name. However, these row names can be changed to be more descriptive. For example, if `set1` is a two-row data frame, then `row.names(set1) <- c("row 1", "row 2")` replaces the default row names for `set1` with the names row 1 and row 2.

drawn at $2p/M$ and $3p/M$ to judge potential for influence as indicated by the leverage. For this measure, we see that both EVP 117 and 120 have very high leverage. Two other EVPs, 121 and 123, also have leverage values to the right of the $3p/M$ vertical line indicating the potential for influence.

The (2,1) and (2,2) plots both give bubble plots of ΔX_m^2 against the model-estimated probability of success. Horizontal lines are drawn at 4 and 9 to help point out those EVPs that may be influential. The (2,1) plot's plotting point has a radius that is proportional to the number of trials for the EVP. This helps one judge whether or not an outlying value indicates a potential problem or is simply due to a poor distributional approximation. A few EVPs involving PATs in the data set have a relatively much larger number of trials than others (e.g., EVP 117 has the largest number of trials with 614), which can distort a bubble plot of this type. This led us to use the `scale.n` argument to rescale the number of trials so that relative differences in radius between 1 and 614 trials are less pronounced. After trying `sqrt()` and `log()`, we decided that a fourth-root scaling created the best-looking plot, and because there is no built-in R function to compute this transformation, we created our own. For the (2,2) plot, we encountered somewhat similar problems due to EVP 120's very large Cook's distance value, and we decided that `sqrt()` worked the best for rescaling.

The (2,1) bubble plot shows that a few EVPs with large estimated probabilities have somewhat large ΔX_m^2 values too. However, these corresponding EVPs have a small number of trials as shown by the size of their plotting point (except perhaps EVP 15), so these EVPs may not necessarily be truly unusual. Still, we should examine these EVPs more closely before making a final judgment. An interesting feature of the (2,2) plot is that the very large Cook's distance value (EVP 120, the largest circle on plot) has a relatively small squared standardized residual. This could indicate a very influential EVP that is "pulling" the model toward it to better its fit.

Below the plots generated by `examine.logistic.reg()`, the deviance/df statistic is shown to be $D/(M - \tilde{p}) = 1.01$. The thresholds of $1 + 2\sqrt{2/(M - \tilde{p})} = 1.26$ and $1 + 3\sqrt{2/(M - \tilde{p})} = 1.39$ are also given at the bottom of the plot. Because the deviance/df is well below these thresholds, there is no evidence of any problems with the overall fit of the model. We also applied the Hosmer-Lemeshow test ($g = 10$; output not shown here) and obtained a p-value of 0.97, indicating again there is insufficient evidence of a problem with the model's overall fit.

Next, we examine the EVPs identified in the plots more closely. Our preferred method for this is to examine all EVPs with their estimated probabilities, standardized Pearson residuals, Cook's distances, and leverage values together in one data set. With only 124 EVPs, this is manageable to do; however, in order to save space here and to illustrate what can be done when the number of EVPs is much larger, we print only those EVPs with standardized Pearson residuals greater than 2 in absolute value, Cook's distances greater than $4/124 = 0.0323$, or leverage values greater than $3 \times 7/124 = 0.17$. The code below illustrates the process of extracting these EVPs from the main `w.n` data frame:

```
> w.n.diag1 <- data.frame(w.n, pi.hat = round(save.info1$pi.hat,
    2), std.res = round(save.info1$stand.resid, 2), cookd =
    round(save.info1$cookd, 2), h = round(save.info1$h, 2))
> p <- length(mod.prelim1$coefficients)
> ck.out <- abs(w.n.diag1$std.res) > 2 | w.n.diag1$cookd >
    4/nrow(w.n) | w.n.diag1$h > 3*p/nrow(w.n)    # "|" means "or"
> extract.EVPs <- w.n.diag1[ck.out,]
> extract.EVPs[order(extract.EVPs$distance),]    # Order by distance
  distance wind change PAT good trials prop pi.hat
60      18      0      1    0      1      2 0.50   0.94
```

117	20	0	0	1	605	614	0.99	0.98
121	20	1	0	1	42	42	1.00	0.99
123	20	0	1	1	94	97	0.97	0.98
101	25	1	1	0	1	2	0.50	0.94
119	29	0	0	1	0	1	0.00	0.73
120	30	0	0	1	3	4	0.75	0.65
103	31	1	1	0	0	1	0.00	0.85
15	32	0	0	0	12	18	0.67	0.87
48	32	1	0	0	0	1	0.00	0.87
87	45	0	1	0	1	5	0.20	0.63
55	50	1	0	0	1	1	1.00	0.23

	std.res	cookd	h
60	-2.57	0.01	0.01
117	0.32	0.06	0.81
121	0.52	0.01	0.19
123	-0.75	0.02	0.23
101	-2.73	0.06	0.05
119	-1.76	0.07	0.13
120	0.83	0.31	0.76
103	-2.43	0.03	0.03
15	-2.67	0.06	0.05
48	-2.62	0.02	0.02
87	-2.03	0.02	0.03
55	1.90	0.04	0.07

Most of these EVPs are not truly of concern because they consist of a very small number of trials. As detailed in Section 5.2.1, thresholds of 2 and 3 based on a standard normal distribution approximation are not very useful in this situation. The large leverage (h) values for EVPs 121 and 123 are likely because of their relatively large numbers of trials, which automatically gives them the potential for influence.

There are three EVPs that require further discussion. EVP 117 has one of the larger Cook's distances and the largest leverage. However, this EVP consists of 614 out of the 1,438 total trials in the data set. We would expect one EVP with such a large percentage of the total observations to be influential. The main concern with such an influential observation is that it might "pull" the regression model toward it, which would create a large Cook's distance and may cause larger residuals (in absolute value) for similar EVPs. We see only a slightly elevated Cook's distance, and no apparent effects on other residuals, so EVP 117 does not cause us further concern. EVP 15 appears simply to have an unusually low number of successes compared to its estimated probability of success. This causes a somewhat large standardized Pearson residual (still less than 3 in absolute value) and Cook's distance. Finally, EVP 120 has the largest Cook's distance by far and the second largest leverage. This suggests that it is likely influential. Its observed explanatory variables correspond to a very unusual type of placekick—a 30-yard PAT, rather than the customary 20-yard PAT—that occurs only as a result of a penalty on an initial placekick attempt. In fact, we note that EVP 119 is a PAT at a non-standard distance too. This leads us to wonder whether these unusual types of placekicks are somehow different than the more typical types of placekicks and perhaps need separate attention.

To examine EVPs 119 and 120 and their effect on the model more closely, we temporarily remove them from the data set and re-estimate the model:

```
> mod.prelim1.wo119.120 <- glm(formula = good/trials ~ distance +
    wind + change + PAT + distance:wind + distance:PAT, family =
```

```
              binomial(link = logit), data = w.n[-c(119, 120),], weights =
              trials)
> round(summary(mod.prelim1.wo119.120)$coefficients, digits = 4)
              Estimate  Std. Error   z value  Pr(>|z|)
(Intercept)     4.4985      0.4816    9.3400    0.0000
distance       -0.0807      0.0114   -7.0640    0.0000
wind            2.8770      1.7866    1.6103    0.1073
change         -0.3308      0.1945   -1.7010    0.0889
PAT           -12.0703     49.2169   -0.2452    0.8063
distance:wind  -0.0907      0.0457   -1.9851    0.0471
distance:PAT    0.6666      2.4607    0.2709    0.7865
```

The output shows a dramatic change in the estimate corresponding to `distance:PAT`. When EVP 119 and 120 are in the data set, the estimate is -0.2717 with a Wald test p-value of 0.0056. Now, the estimate is 0.6666 with a Wald test p-value of 0.7865. Therefore, it appears the presence of this interaction was solely due to the five placekicks corresponding to these EVPs.

There are 5 EVPs containing a total of 13 observations which are non-20 yard PATs in the data set, and 11 of these placekicks are successes (the two failures are included in EVPs 119 and 120). Due to the small number of these types of placekicks, it is difficult to determine if what we observe here is a real trend or an anomaly in the data. This leads then to three possible options:

1. Return EVPs 119 and 120 to the data set and acknowledge that they are responsible for making the `distance:PAT` interaction appear important

2. Remove all non-20 yard PATs from the data set, which subsequently reduces the population of inference

3. Remove all PATs from the data set and find separate models for field goals and PATs

We decided that option #2 was a slightly better choice than the other two. Non-20-yard PATs are somewhat unusual events and always follow penalties. If they truly have different probabilities of success, we do not have a good way to make this determination. We therefore believe that non-20-yard PATs may be fundamentally different from other placekicks and should be removed from the data. In turn, this leads to a corresponding reduction to the population of inference, albeit extremely small, for the analysis.

Whether to keep the 20-yard PATs in the data is a separate judgment. They, too, are different from field goals because they are worth one point instead of three, and they are always kicked from the middle of the field. Thus, there is reason to remove them and develop separate models for the two types of placekicks. However, we surmise that the effects of other variables, such as wind and its interactions, are the same for either type of placekick, and the loss of the 614 20-yard PATs from the data would reduce precision for estimating these effects. We therefore choose option 2: we remove the non-20-yard PATs and continue the analysis.

Assessing the model fit - revised model

Below is the code to remove the non-20 yard PATs:

```
> # Remove non-20 yard PATs - "!" negates and "&" means "and"
> placekick.mb2 <- placekick.mb[!(placekick.mb$distance!=20 &
    placekick.mb$PAT==1),]
```

```
> nrow(placekick.mb2)  # Number of observations after 13 were
    removed
[1] 1425
```

We now return to the variable selection step with this revised data set containing 1425 total placekicks. Details are given in our corresponding program, and the resulting model is the same as before, except that `distance:PAT` is no longer selected. Below is the corresponding output from fitting the new model.

```
> # w.n2 contains 119 EVPs
> mod.prelim2 <- glm(formula = good/trials ~ distance + wind +
    change + PAT + distance:wind, family = binomial(link = logit),
    data = w.n2, weights = trials)
> summary(mod.prelim2)

Call:
glm(formula = good/trials ~ distance + wind + change + PAT +
    distance:wind, family = binomial(link = logit), data = w.n2,
    weights = trials)

Deviance Residuals:
    Min       1Q   Median       3Q      Max
-2.2386  -0.5836   0.1965   0.8736   2.2822

Coefficients:
               Estimate Std. Error z value Pr(>|z|)
(Intercept)     4.49835    0.48163   9.340  < 2e-16 ***
distance       -0.08074    0.01143  -7.064 1.62e-12 ***
wind            2.87783    1.78643   1.611  0.10719
change         -0.33056    0.19445  -1.700  0.08914 .
PAT             1.25916    0.38714   3.252  0.00114 **
distance:wind  -0.09074    0.04570  -1.986  0.04706 *
---
Signif. codes:  0 '***' 0.001 '**' 0.01 '*' 0.05 '.' 0.1 ' ' 1

(Dispersion parameter for binomial family taken to be 1)

    Null deviance: 376.01  on 118  degrees of freedom
Residual deviance: 113.86  on 113  degrees of freedom
AIC: 260.69

Number of Fisher Scoring iterations: 5
```

We again use `examine.logistic.reg()` to evaluate the fit of the model. The corresponding plots are not given here because they essentially show the same results as before, excluding those EVPs that we removed. EVP 116 (formerly EVP 117) has the largest Cook's distance now. This EVP contains the 614 PATs at 20 yards, so we would again expect this to be influential. None of the other EVPs with relatively large Cook's distances have any identifiable characteristic about them that may lead to their size. We took the extra step of temporarily removing a few of these EVPs one at a time and refitting the model to determine if the regression parameter estimates changed substantially. These estimates did not, so we make no further changes to the data or the model.

Our final model is

$$\text{logit}(\hat{\pi}) = 4.4983 - 0.08074\text{distance} + 2.8778\text{wind} - 0.3306\text{change} + 1.2592\text{PAT}$$
$$- 0.09074\text{distance} \times \text{wind}.$$

Interpreting the model

We calculate odds ratios and corresponding 90% profile LR confidence intervals to interpret the explanatory variables in the model:

```
> library(package = mcprofile)
> OR.name <- c("Change", "PAT", "Distance, 10-yard decrease,
    windy", "Distance, 10-yard decrease, not windy", "Wind,
    distance = 20", "Wind, distance = 30", "Wind, distance = 40",
    "Wind, distance = 50", "Wind, distance = 60")
> var.name <- c("int", "distance", "wind", "change", "PAT",
    "distance:wind")
> K <- matrix(data = c(0,    0, 0, 1, 0,    0,
                       0,    0, 0, 0, 1,    0,
                       0,  -10, 0, 0, 0,  -10,
                       0,  -10, 0, 0, 0,    0,
                       0,    0, 1, 0, 0,   20,
                       0,    0, 1, 0, 0,   30,
                       0,    0, 1, 0, 0,   40,
                       0,    0, 1, 0, 0,   50,
                       0,    0, 1, 0, 0,   60),
    nrow = 9, ncol = 6, byrow = TRUE, dimnames = list(OR.name,
        var.name))
> # K  # Check matrix - excluded to save space
> linear.combo <- mcprofile(object = mod.prelim2, CM = K)
> ci.log.OR <- confint(object = linear.combo, level = 0.90,
    adjust = "none")
> exp(ci.log.OR)

   mcprofile - Confidence Intervals

level:          0.9
adjustment:     none

                                        Estimate  lower   upper
Change                                    0.7185  0.5223  0.991
PAT                                       3.5225  1.8857  6.785
Distance, 10-yard decrease, windy         5.5557  2.8871 12.977
Distance, 10-yard decrease, not windy     2.2421  1.8646  2.717
Wind, distance = 20                       2.8950  0.7764 16.242
Wind, distance = 30                       1.1683  0.5392  3.094
Wind, distance = 40                       0.4715  0.2546  0.869
Wind, distance = 50                       0.1903  0.0598  0.515
Wind, distance = 60                       0.0768  0.0111  0.377
```

The **change** and **PAT** explanatory variable interpretations are the simplest because there are no interactions involving them in the model. With 90% confidence, the odds of success are between 0.52 and 0.99 times as large for lead-change placekicks than for non-lead change placekicks, when holding the other variables constant. This indicates moderate evidence of a pressure effect for placekicks; namely, the probability of success is lower for lead-change kicks. With respect to the PAT variable, the odds of a success are much larger for 20-yard PATs than for field goals of the same distance, when holding the other variables constant.

The `distance` and `wind` explanatory variable interpretations need to take into account their interaction. For a 10-yard decrease in `distance`, the 90% confidence interval for the odds ratio is (2.89, 12.98) when there are windy conditions and (1.86, 2.72) when there are not windy conditions. This means that decreasing the distance of a placekick is even more important when conditions are windy than when they are not. We also calculated the estimated odds ratios and corresponding confidence intervals for windy vs. not windy conditions for placekicks of 20, 30, 40, 50, and 60 yards. Again, we see that the effects of windy conditions are minimal for short placekicks, but pronounced for longer placekicks.

The estimated probability of success along with profile LR intervals can be calculated for specific combinations of explanatory variables. For example, we examine the probabilities of success for PATs and field goals at a distance of 20-yards with `wind` = 0 and `change` = 0 as follows:

```
> K <- matrix(data = c(1, 20, 0, 0, 1, 0,
                       1, 20, 0, 0, 0, 0), nrow = 2, ncol = 6,
              byrow = TRUE, dimnames = list(c("PAT",
              "FG"), var.name))
> linear.combo <- mcprofile(object = mod.prelim2, CM = K)
> ci.lin.pred <- confint(object = linear.combo, level = 0.90,
    adjust = "none")
> # as.matrix() is needed to obtain the proper class for plogis()
> round(plogis(q = as.matrix(ci.lin.pred$estimate)), digits = 3)
      Estimate
PAT      0.984
FG       0.947
> round(plogis(q = as.matrix(ci.lin.pred$confint)), digits = 3)
     lower upper
[1,] 0.976 0.991
[2,] 0.921 0.966
```

The results here show that the probability of success is larger for PATs than for field goals, as we demonstrated before with using odds ratios.

Figure 5.13 plots the estimated model as a function of `distance` (code is included in the program). With the addition of "risk" factors, windy conditions and lead-change attempts, the estimated probability of success generally decreases. Figure 5.14 further emphasizes this point by plotting the estimated probability success for the least risky (`change` = 0 and `wind` = 0) and the most risky (`change` = 1 and `wind` = 1) field goals along with 90% Wald confidence interval bands. Notice that the 90% confidence intervals no longer overlap after a distance of approximately 38 yards.

It is important to note that our confidence intervals here are all at stated 90% levels and do not control an overall familywise error rate. For a set number of odds ratios and probability of success calculations, the `confint()` function provides a convenient way for this control by using the `adjust` argument (see Section 2.2.6 for examples).

5.4.2 Poisson regression - alcohol consumption data set

In this example, we return to Poisson regression analysis of the alcohol consumption data introduced in Section 4.2.2. We again focus on the subjects' consumption during their first Saturday in the study. We now perform a more complete analysis of the data,

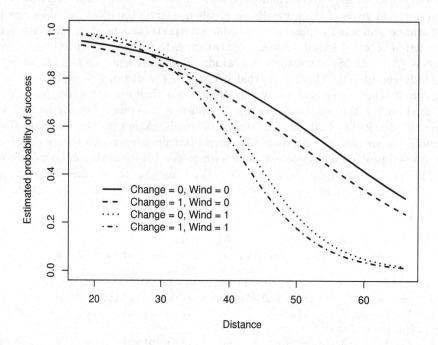

Figure 5.13: Estimated probability of success vs. the field goal (`PAT` = 0) distance for four combinations of change and wind.

including initial examination of the data, variable selection, assessment of fit, correction for overdispersion, and analysis and interpretation of the results. The code for the analysis is contained in AlcoholDetailed.R; the data are in DeHartSimplified.csv.

Examining the data

First, we introduce new variables that have not been used before. The example in Section 4.2.2 regressed the number of drinks consumed (`numall`) against an index for the positive events (`posevent`) and negative events (`negevent`) experienced by the subject each day. In Exercise 26 of Chapter 4, we considered additional explanatory variables for positive romantic-relationship events (`prel`), negative romantic-relationship events (`nrel`), age (`age`), trait (long-term) self-esteem (`rosn`), and state (short-term) self-esteem (`state`). In this example we add two more variables: `gender` (1=male, 2=female) and `desired`, which measures a subject's desire to drink each day, with a higher score meaning greater desire. Note that trait self-esteem was measured once at the start of the study, while state self-esteem was measured daily. Also note that, although `gender` is a numerical representation of a categorical variable, there is no harm in treating it as numerical because there are only two levels. The "slope" coefficient is exactly the same as the "difference" parameter for a two-level categorical variable when it is coded with numbers that are 1 unit apart.

The example from Section 4.2.2 shows the initial creation of the data set from a larger data file. We continue next with boxplots and a scatterplot matrix of the nine numerical variables. The boxplots show the extent of skewness or outlying data. The scatterplots help to show whether there are substantial issues with multicollinearity to be concerned with,

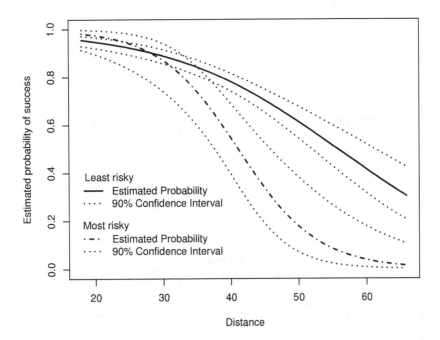

Figure 5.14: Estimated probability of success vs. the field goal (PAT = 0) distance for the most and least risky field goal attempts.

and they also give a preliminary impression regarding which explanatory variables might be related to the response. Using the enhanced scatterplot matrix function spm() from the car package, we get a smooth fit and a linear fit added to each scatterplot, as well as an estimate of the density function or, optionally, a boxplot for each variable. The results are shown in Figures 5.15 and 5.16.

The boxplots and the density estimates show some marked right skewness in nrel and negevent, which may be a potential source of high influence in a regression. We also see somewhat milder skewness in their positive counterparts, and a bit of left skewness in rosn. In the response variable numall the extreme drink count of 21 that has been mentioned in previous examples stands out. This represents a possible outlier that may have a large residual, or it could be influential by forcing variables into the model to try to explain it or by altering regression parameters on some variables.

The scatterplot matrix reveals no substantial correlations among the explanatory variables. The most linear-looking relationship seems to be prel vs. posevent, but there is considerable variability to the relationship, so it is not likely to be an important source of multicollinearity. Of course, multicollinearity can occur due to combinations of more than two variables, and this plot cannot depict such relationships.

Among associations between response and explanatory variables, the strongest positive relationship seems to be with desired, which suggests that a strong desire to drink alcohol may be followed by increased consumption. It is also interesting to note that the subject with the extreme numall value reported the maximum possible desire to drink (desired = 8, tied with a few others) and a zero value for negative relationships (tied with many others), but was otherwise not extreme in any other measurement.

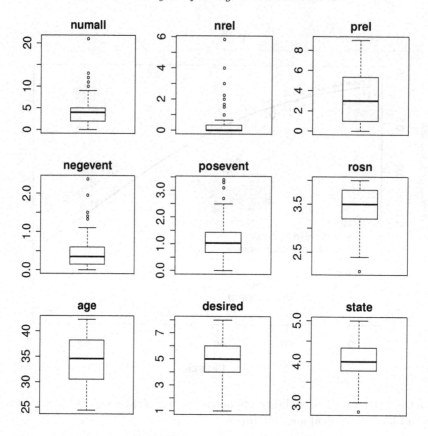

Figure 5.15: Boxplots of response (numall) and explanatory variables for alcohol consumption data.

Selecting variables

We use model averaging (Section 5.1.6) to select variables for a model. There are nine explanatory variables, so an exhaustive search of the main effects is possible as a first step to see which variables seem most strongly related to the drink count. We use AIC_c as the criterion for assessing each model. The code and results are below and in Figure 5.17.

```
> library(glmulti)
> search.1.aicc <- glmulti(y = numall ~ ., data =
    saturday[,-c(1:3)], fitfunction = "glm", level = 1, method =
    "h", crit = "aicc", family = poisson(link = "log"))

<OUTPUT EDITED>

After 500 models:
Best model: numall~1+nrel+negevent+age+desired+state
Crit= 430.109908874369
Mean crit= 441.121806562024
Completed.

> print(search.1.aicc)
glmulti.analysis
```

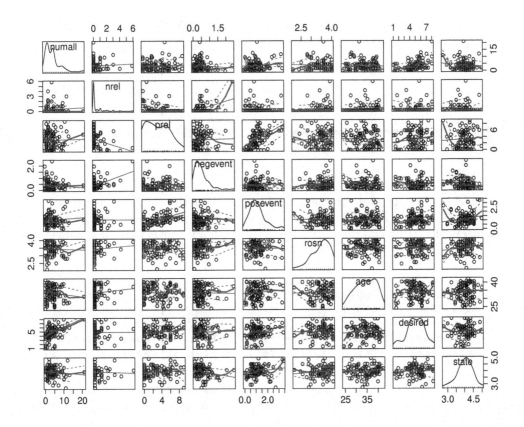

Figure 5.16: Scatterplot matrix of response (`numall`) and explanatory variables for alcohol consumption data.

```
Method: h / Fitting: glm / IC used: aicc
Level: 1 / Marginality: FALSE
From 100 models:
Best IC: 430.109908874369
Best model:
[1] "numall ~ 1 + nrel + negevent + age + desired + state"
Evidence weight: 0.258040276400709
Worst IC: 444.502983808218
2 models within 2 IC units.
37 models to reach 95% of evidence weight.
> head(weightable(search.1.aicc))
                                                              model
1                 numall~1 + nrel + negevent + age + desired + state
2          numall~1 + nrel + negevent + rosn + age + desired + state
3          numall~1 + nrel + prel + negevent + age + desired + state
4      numall~1 + nrel + negevent + posevent + age + desired + state
5        numall~1 + nrel + negevent + gender + age + desired + state
6 numall~1 + nrel + prel + negevent + rosn + age + desired + state
       aicc    weights
1 430.1099 0.25804028
2 432.0628 0.09718959
3 432.2393 0.08898058
```

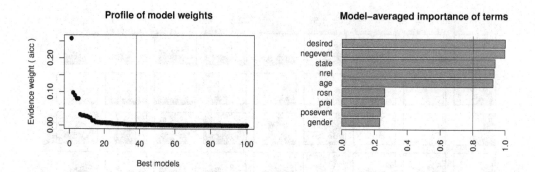

Figure 5.17: Results of main-effect variable search for the alcohol consumption example using model averaging with evidence weights: Profile of evidence weights on top 100 individual models (left) and total evidence weight from all models for each variable (right).

```
4  432.4503  0.08007105
5  432.4645  0.07950429
6  434.2853  0.03198980

> plot(search.1.aicc, type = "w")   # Plot of model weights
> plot(search.1.aicc, type = "s")   # Plot of variable weights
```

There are $2^9 = 512$ models in the search. The top model contains 5 variables and has approximately 0.25 weight, so support for the top model is not overwhelmingly large. However the next-best models all have weights between 0.05 and 0.10, so there is at least a 2.5-to-1 preference for the top model relative to any other model. The plot of model-averaged variable weights clearly indicates that there are five variables that are important—desired, negevent, state, nrel, and age—with weights of at least 0.9. The remaining variables have weights between 0.2 and 0.3, so they are not completely useless, but they are not well supported by the data.

As described in DeHart et al. (2008), the researchers were interested in possible interactions among certain variables. We can extend the search to consider only those interactions, or we can consider interactions more broadly between pairs of variables. For the sake of demonstration, we choose the latter approach here. However, note that among the nine main effects, there are $9*8/2 = 36$ possible pairwise interactions that can be formed, leading to a potential variable-selection problem consisting of 45 variables. This is much too large to be carried out exhaustively in glmulti(). Even imposing marginality on the search—i.e., requiring main effects to be included in any model that contains their interactions—still leaves too large a set of possible models to complete an exhaustive search (note that running glmulti() with method = "d" runs several minutes and returns, "Your candidate set contains more than 1 billion (1e9) models."). One way around this problem is to consider interactions only among the main effects that the previous search has chosen as important. Then there are only $5*4/2 = 10$ interactions, yielding a set of "only" $2^{5+10} = 32{,}768$ models. Imposing marginality reduces this set further. Running this selection takes just a few seconds and results in a best model with the five main effects, plus state:negevent and age:desired. The AIC_c for this model is 421.2, compared to 430.1 for the best main-effects-only model, so including these interactions does improve the model.

An alternative approach is to not limit the variables considered, and instead to use a genetic algorithm (GA) to explore the model space randomly but intelligently. Because this is a randomized sampling process, it is not guaranteed to find the best model every time. It is possible for a given run of the GA to become "stuck" searching around slightly inferior models and never find the ones that perform the best. In order to increase chances of finding the best models, we run the GA search several times. We choose to run the search four times, which is a compromise between increasing the chance of finding the best models and decreasing execution time. We retain the top 100 models from each run. These four sets of 100 models are not identical, but contain considerable overlap. Many models appear in more than one search result, but not all of the best results appear in any one search. Combining results of the multiple GA runs gives a "consensus" in which the best models from each run are brought together into a final set of 100 models. Calculations of evidence weights are carried out from this consensus set. The process completed in around 10 minutes on a dual-core 2.7 GHz processor with 8 GB of RAM. Abridged results are shown below and depicted in Figure 5.18.

```
> set.seed(129981872)
> search.2g0.aicc <- glmulti(y = numall ~ ., data =
    saturday[,-c(1:3)], fitfunction = "glm", level = 2,
    marginality = TRUE, method = "g", crit = "aicc", family =
    poisson(link = "log"))
After 730 generations:
Best model: Crit= 407.129916150943
Mean crit= 425.066862696667
Algorithm is declared to have converged.

> search.2g1.aicc <- glmulti(y = numall ~ ., data =
    saturday[,-c(1:3)], fitfunction = "glm", level = 2,
    marginality = TRUE, method = "g", crit = "aicc", family =
    poisson(link = "log"))
After 590 generations:
Best model: Crit= 407.129916150943
Mean crit= 423.983364416536
Algorithm is declared to have converged.

> search.2g2.aicc <- glmulti(y = numall ~ ., data =
    saturday[,-c(1:3)], fitfunction = "glm", level = 2,
    marginality = TRUE, method = "g", crit = "aicc", family =
    poisson(link = "log"))
After 610 generations:
Best model: Crit= 406.09300880825
Mean crit= 420.313095864784
Algorithm is declared to have converged.

> search.2g3.aicc <- glmulti(y = numall ~ ., data =
    saturday[,-c(1:3)], fitfunction = "glm", level = 2,
    marginality = TRUE, method = "g", crit = "aicc", family =
    poisson(link = "log"))
After 760 generations:
Best model: Crit= 406.09300880825
Mean crit= 420.043673540324
Algorithm is declared to have converged.

> search.2allg.aicc <- consensus(xs = list(search.2g0.aicc,
    search.2g1.aicc, search.2g2.aicc, search.2g3.aicc),
    confsetsize = 100)
```

```
> print(search.2allg.aicc)
consensus of 4-glmulti.analysis
Method: g / Fitting: glm / IC used: aicc
Level: 2 / Marginality: TRUE
From 100 models:
Best IC: 406.09300880825
Best model:
[1] "numall ~ 1 + prel + negevent + gender + rosn + age + desired
    + "
[2] "     state + rosn:prel + age:rosn + desired:gender +
    desired:age + "
[3] "     state:negevent"
Evidence weight: 0.0664972544357641
Worst IC: 414.234599845948
6 models within 2 IC units.
74 models to reach 95% of evidence weight.
```

It is perhaps disconcerting to note that the best model is not always the same in the four GA runs, as indicated by the different AIC_c values (407.1, 407.1, 406.1, 406.1). Based on this variability, we cannot be certain than the best among these, 406.1, is truly the best value among all models. This might lead to a lack of faith in the GA. However, from a practical point of view, all four runs had best models that were substantially better than what was found in the first analysis using a reduced set of variables. In fact, the *worst* of the 100 models in the consensus top 100 has $AIC_c = 414.2$, which is still much better than the best model from the search on the restricted set of variables. This indicates clearly that (1) interactions among variables may be important, even when their main effects appear not to be, and (2) the GA is a useful, although not perfect, tool for finding these interactions.

The variables preferred in this unrestricted search include all of those from the exhaustive search on the restricted set of variables, except `nrel`. They also include interactions `desired:gender`, `prel:rosn`, and `age:rosn`, as well as the additional main effects required by these interactions, `rosn`, `gender`, and `prel`. We use this model as an initial working model.

Assessing the model

To assess the fit of the chosen model, we first fit the Poisson regression using `glm()` and then use the tools from Section 5.2 on the model object. As an initial check, we look at the deviance/df:

```
> mod.fit <- glm(formula = numall ~ 1 + prel + negevent + gender
    + rosn + age + desired + state + rosn:prel + age:rosn +
    desired:gender + desired:age + state:negevent, family =
    poisson(link = "log"), data = saturday)
> deviance(mod.fit)/mod.fit$df.residual
[1] 1.576825
> 1 + 2*sqrt(2/mod.fit$df.residual)
[1] 1.324443
> 1 + 3*sqrt(2/mod.fit$df.residual)
[1] 1.486664
```

The model's deviance/df ratio is 1.57, which is somewhat larger than the rough thresholds. We therefore conclude that this model may have some problems. To explore this further, we first check for zero inflation; that is, we check whether there are more subjects consuming

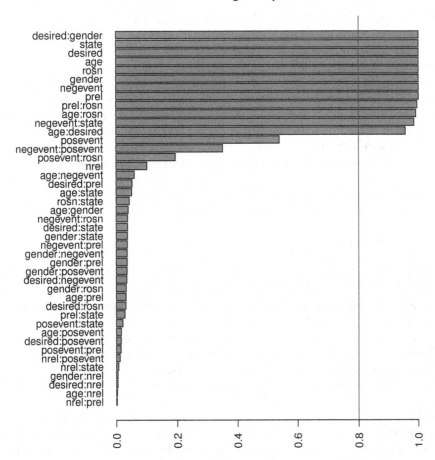

Figure 5.18: Results of genetic algorithm variable search for the alcohol consumption example using model averaging with evidence weights on all variables and 2-way interactions.

no alcoholic beverages than the model expects. We compute the observed number of zero counts, n_0, along with the expected number based on the model. For a subject with estimated expected value $\hat{\mu}_i$, the probability that they consume zero drinks according to the Poisson model is $e^{-\hat{\mu}_i}$. The expected number of zeroes is therefore $\hat{n}_0 = \sum_{i=1}^{M} e^{-\hat{\mu}_i}$. A Pearson statistic can then be used to measure the discrepancy between n_0 and $\hat{n}_0$, $X_0^2 = (n_0 - \hat{n}_0)^2/\hat{n}_0$.

```
> mu.poi <- exp(predict(mod.fit))
> zero.poi <- sum(exp(-mu.poi))   # Expected number of zeroes for
    Poisson
> zero.obs <- sum(saturday$numall==0)   # Total zeroes in data
> zero.pear <- (zero.poi - zero.obs)^2/zero.poi   # Pearson Stat
    comparing them
>
> c(zero.poi,zero.obs, zero.pear)
```

```
[1] 7.075870834 7.000000000 0.000813523
```

There were 7 zero counts in the data, while the model estimates that there should be about 7.07. These numbers are so close that there is clearly no evidence to suggest a problem with zero inflation. We could further confirm this by fitting the zero-inflated Poisson model and comparing its AIC_c to that of the Poisson fit. We do not pursue this here.

There are other aspects of a model fit that can cause inflated deviance/df value. In particular, we can check whether the estimated mean values produced by the model deviate from the observed counts in other ways. A standard Pearson or deviance GOF test cannot be used because many of the observed and expected counts are small. Instead, we apply the omnibus fit test described at the end of Section 5.2.2 that is contained in our program PostFitGOFTest.R. Using the default number of groups yields the following results:

```
> source("PostFitGOFTest.R")
> GOFtest <- PostFitGOFTest(obs = saturday$numall, pred = mu.poi,
    g = 0)
Post-Fit Goodness-of-Fit test with 18 bins
Pearson Stat = 13.88894
p= 0.6069871
```

Because this model was determined using data-based variable selection, we cannot interpret the p-value literally as a probability. However, the fact that it is so large is somewhat comforting, if not conclusive evidence of a good fit. Rerunning the test with different numbers of groups (not shown) does not substantively change this result. Additional output (not shown) shows no particularly large Pearson residuals, nor any long strings of positive or negative Pearson residuals. From this we conclude that the log link appears to be reasonable.

Next, we check residuals and influence for individual observations. Standardized Pearson residuals are plotted against each explanatory variable's main effect and against the estimated means and linear predictors using code given in the program for this example. The results are shown in Figure 5.19. Keeping in mind that loess smoothing is unreliable near the edges of the range of the x-axis, the plots do not show any clear problems with poor choice of link or clear curvature in any predictors. The loess curves generally wiggle around the horizontal line plotted at 0. The plot for **negevent** may show some curvature, but the loess curve in the right half of the plot is estimated using very few data points. It may be distorted by these points that attract the curve toward their residual values. There is one very large standardized residual (> 4), and a small number between 2 and 3. In all, 6/89 lie beyond 2, with another three at nearly -2. This number is perhaps a few more than what might be expected on average to happen by chance (5% of $M = 89$ is 4.45), and there are several negative residuals just inside the threshold at -2. Overall, what we see in these plots is not necessarily unusual for a Poisson regression model fit.

Finally, we perform influence diagnostics using our function glmInfDiag() in the program glmDiagnostics.R. The results are plotted in Figure 5.20 and shown below.

```
> source("glmDiagnostics.R")
> save.info <- glmInflDiag(mod.fit = mod.fit)
Potentially influential observations by any measures
      h hflag Del.X2 xflag Del.dev dflag Cooks.D cflag
2  0.46   **    0.01         0.01         0.000
5  0.22         2.26         4.01    *    0.050    *
39 0.14         3.83         5.11    *    0.047    *
```

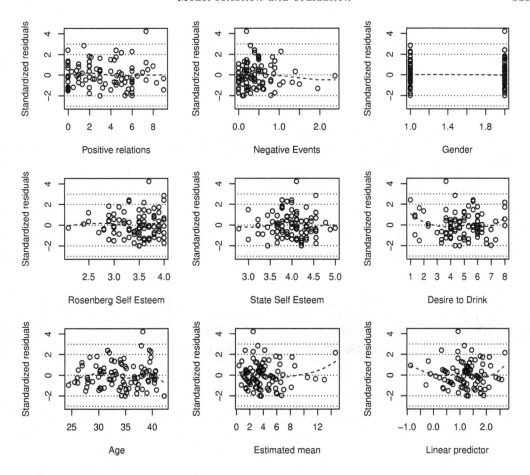

Figure 5.19: Residual plots from the working Poisson regression model for the alcohol consumption data.

```
     52  0.11          5.78   *    3.98        0.052   *
     57  0.12          8.14   *    6.01   *    0.085   *
     58  0.41    *     4.72   *    4.39   *    0.257   *
     61  0.35    *     3.02        2.73        0.123   *
    108  0.32    *     1.92        2.16        0.069   *
    113  0.56   **     0.18        0.19        0.018
    118  0.66   **     0.12        0.12        0.018
    129  0.44   **     0.05        0.05        0.003
    137  0.08         17.88  **   11.25   **   0.114   *

<OUTPUT EDITED>
```

From the plots and the listings, 6 points stand out: four have large leverage values (observation 2, 113, 118, and 129); one has large ΔX_m^2 and ΔD_m^2 and moderately large Cook's distance, but low leverage (observation 137); and one has a combination of moderately large values of all measures, including a Cook's distance that stands out from the rest (observation 58). Further examination of the first four reveals that observation 2 has the largest value of **negevent** but with only a moderate estimated mean; observation 113 has a large **prel**, an odd combination of very low trait self-esteem but very high state self-esteem, and

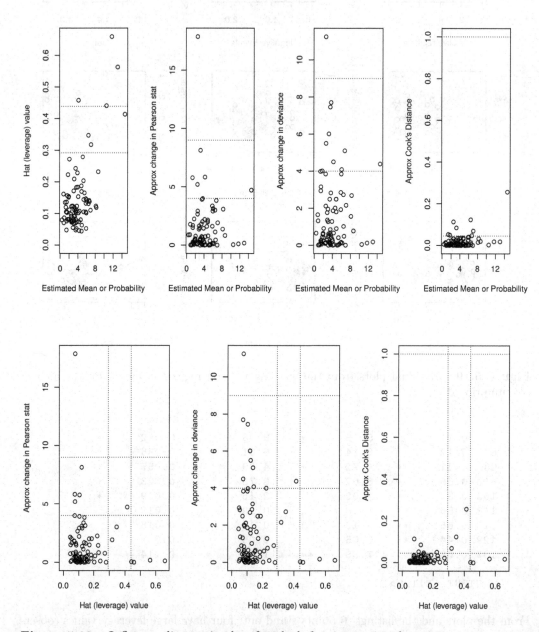

Figure 5.20: Influence diagnostic plots for alcohol consumption data.

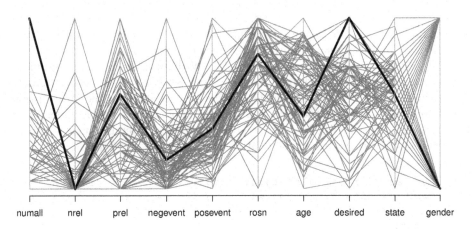

Figure 5.21: Parallel coordinate plot highlighting observation 58's position relative to the rest of the data.

a very high estimated mean; observation 118 has the largest prel, the smallest state, and a high estimated mean; and observation 129 has moderate values of all variables but a high estimated mean. In the cases of the high estimates, the actual count is very close to the estimated mean and the actual influence on the model parameters is not large according to Cook's distance. We therefore are not especially worried about these four subjects. The subject with the very large ΔX_m^2 and ΔD_m^2 has a very large prel, a somewhat low negevent and desired, and is very poorly predicted by the model (observed 9 drinks but predicted 2.5). This subject appears to influence the regression mainly by having a very unexpected count. This person's data could be checked for accuracy. The same is true for subject 58, whose 21 drinks is by far the highest observed. This subject has the maximum values of prel and desired, and the largest estimated mean at 14.6. To highlight these findings, we created a parallel coordinates plots (see Section 3.3) showing each observation's position relative to the rest of the data. An example for observation 58 is shown in Figure 5.21. The code to make these plots is in the program for this example.

Because we do not have access to the original log-books or to the subjects themselves, we cannot check the veracity of their data. We therefore have two options: (1) ignore the problems and proceed, or (2) rerun the analysis without one or both of the questionable observations and compare the results to the present analysis. Following the second option, we reran the genetic algorithm without observation 58 and found that the gender and desired:gender interaction were no longer very important. Their evidence weight was reduced from nearly 1 to about 0.4. All other variables remained important and no new ones rose to be considered. We therefore conclude that this one observation has considerable influence on the importance of the desired:gender interaction. We now have additional uncertainty regarding the choice of model that would ideally be resolved in conversation with the researchers. For the remainder of this example, we assume that observation 58 is a legitimate value. We note there were recorded drink counts of 18, 15, 14, and 13 on other days not included in this analysis, so the 21 drinks consumed by this person on this Saturday is not as extreme relative to the larger data set.

Addressing overdispersion

Our investigation of the model fit has provided no concrete evidence of any specific flaw in our model. But the deviance/df indicator still suggests a problem with the model fit. Lacking any other explanation, we conclude that there is overdispersion for some reason. Perhaps there are additional important variables that have not been measured, or maybe three-way or higher interactions could be useful. Without additional data or guidance from the researchers regarding which (of very many) three-way or higher interactions we should consider, we are left to try adjusting the inferences to account for the overdispersion.

First, we assess whether a quasi-Poisson or negative binomial model is preferred. Applying the analysis from Section 5.3.3, we fit a regression of the squared residuals against the fitted means and look for curvature in the relationship:

```
> res.sq <- residuals(object = mod.fit, type = "response")^2
> set1 <- data.frame(res.sq, mu.hat = saturday$mu.hat)
> fit.quad <- lm(formula = res.sq ~ mu.hat + I(mu.hat^2), data =
   set1)
> anova(fit.quad)
Analysis of Variance Table

Response: res.sq
            Df  Sum Sq  Mean Sq  F value    Pr(>F)
mu.hat       1   681.9   681.93  11.9460  0.0008532 ***
I(mu.hat^2)  1    20.1    20.10   0.3521  0.5544753
Residuals   86  4909.2    57.08
```

The quadratic term is clearly not important, indicating that a quasi-Poisson model is preferred. *Our next step should be to return to the variable selection stage and repeat the model-averaging assessment of variables using a quasi-Poisson model.* We have done this in the program for this example, and found that the same variables are considered important as before, although with somewhat smaller evidence weight in some cases. We therefore proceed to the analysis of the chosen quasi-Poisson model.

Analyzing and interpreting parameter estimates

Our final step is to quantify the explanatory variables' effects on the number of drinks consumed. This analysis is complicated by the fact that every variable we consider is involved in at least one interaction. We therefore need to estimate the effects of a given explanatory variable separately at different levels of its interacting variables. This involves computing certain linear combinations of parameter estimates and computing their variances accordingly. Unfortunately, variances cannot be computed accurately using the model-averaged parameter estimates, because `glmulti()` does not provide the full variance-covariance matrix of the parameter estimates. We could compute the estimates, but would have no reliable estimate of their variability. We show the model-averaged analysis of individual parameters in the program for this example. Here, we focus our analysis on the single best model, refit by `glm()` using a quasi-Poisson model below.

```
> mq <- glm(formula = numall ~ prel + negevent + gender + rosn +
    age + desired + state + rosn:prel + age:rosn + desired:gender
    + desired:age + state:negevent, family = quasipoisson(link =
    "log"), data = saturday)
> round(summary(mq)$coefficients, 3)
```

	Estimate	Std. Error	t value	Pr(>|t|)
(Intercept)	8.016	5.548	1.445	0.153
prel	0.474	0.210	2.255	0.027
negevent	-4.961	2.189	-2.266	0.026
gender	1.645	0.593	2.775	0.007
rosn	-2.869	1.462	-1.962	0.053
age	-0.260	0.160	-1.624	0.109
desired	1.482	0.375	3.951	0.000
state	-0.925	0.266	-3.473	0.001
prel:rosn	-0.156	0.063	-2.489	0.015
rosn:age	0.102	0.043	2.379	0.020
gender:desired	-0.294	0.106	-2.775	0.007
age:desired	-0.021	0.011	-1.802	0.075
negevent:state	1.143	0.550	2.079	0.041

The estimated model is $\log(\hat{\mu}) = 8.016 + 0.474\text{prel} - 4.961\text{negevent} + \ldots - 0.021(\text{age} \times \text{desired}) + 1.143(\text{negevent} \times \text{state})$. Seeing the signs and magnitudes of the coefficients helps us to understand their relationships between the response and the explanatory variables. For example, prel has a coefficient of 0.474, while that for prel:rosn is -0.156. Thus, the effect of rosn on the number of drinks is estimated to be

$$\frac{\hat{\mu}(\text{prel}+1, \text{rosn})}{\hat{\mu}(\text{prel}, \text{rosn})} = \exp(0.474 - 0.156\text{rosn})$$

Furthermore, from Figure 5.15, we see that the range of rosn is 2–4. Thus, when we hold all other explanatory variables constant, drinking increases as the number of positive relationship events increases for lower values of trait self-esteem (e.g., $\exp(0.474-0.156*2) = 1.18$). However, when trait self-esteem is high, the effect of positive relationship events on drinking is reversed (e.g., $\exp(0.474 - 4*(0.156)) = 0.86$). Code in the program for this example shows how to make a 3D plot of this surface.

To further this analysis, we compute the ratio of estimated means corresponding to a 1-unit increase for an explanatory variable at each of the quartiles of the other variables (in the case of desired:gender, at the two levels of gender). For example, to further assess the effects of prel, we first compute the three quartiles of rosn, then compute $\exp(0.474 - 0.156\text{rosn})$. The code for doing this manually is shown below.

```
> bhat <- mq$coefficients
> rosn.quart <- summary(saturday$rosn)[c(2,3,5)]
> rosn.quart
1st Qu.   Median  3rd Qu.
    3.2      3.5      3.8
> mprel.rosn <- exp(bhat["prel"] + rosn.quart * bhat["prel:rosn"])
> mprel.rosn
    1st Qu.     Median    3rd Qu.
  0.9753995  0.9308308  0.8882985
> 100*(mprel.rosn - 1)
    1st Qu.     Median    3rd Qu.
  -2.460046  -6.916922 -11.170151
```

Note that for convenience we can specify the element of the coefficient vector by its name, as in bhat["prel"]. At the first quartile of rosn, the effect of prel is already negative, reducing the mean number of drinks consumed by about $100(1 - 0.975) = 2.5\%$ for each

1-unit increase in `prel`. The decrease in number of drinks reaches about 11% per 1-unit increase in `prel` by the third quartile of `rosn`.

We can use these quantities to obtain profile (quasi-)likelihood confidence intervals for the true parameter values using the `mcprofile` package. See Sections 2.2.4 and 4.2.2 for previous examples using this package. The matrix of coefficients contains a row for each estimate and a column for each parameter from the model. An entry of 1 is placed in the position for the main effect for the variable whose slope we are estimating. The value of the quartile of the interacting variable goes into the position for the interaction. All other positions are set to coefficients of 0. For confidence interval calculation, we have not specified a method for multiplicity adjustment in the code below (no `adjust` argument), so the default single-step method is used (see the description in the example on page 110). Finally, we re-express the parameter estimates and confidence intervals as percentage changes.

```
> library(mcprofile)
> # Create coefficient matrices as 1*target coefficient +
    quartile*interaction(s)
> # Look at bhat for ordering of coefficients
> # CI for prel, controlling for rosn
> K.prel <- matrix(data =
 c(0, 1, 0, 0, 0, 0, 0, 0, rosn.quart[1], 0, 0, 0, 0,
   0, 1, 0, 0, 0, 0, 0, 0, rosn.quart[2], 0, 0, 0, 0,
   0, 1, 0, 0, 0, 0, 0, 0, rosn.quart[3], 0, 0, 0, 0),
              nrow = 3, byrow = TRUE)
> # Produces wider intervals than with regular Poisson regression
    model
> profile.prel <- mcprofile(object = mq, CM = K.prel)
> ci.prel <- confint(object = profile.prel, level = 0.95)

> 100*(exp(ci.prel$estimate) - 1)   # Verifies got same answer as
    above
      Estimate
 C1   -2.460046
 C2   -6.916922
 C3  -11.170151
> 100*(exp(ci.prel$confint) - 1)
       lower      upper
 1   -8.491358   3.843081
 2  -12.856058  -0.717001
 3  -18.980590  -2.814959
```

The estimates match with what we previously calculated manually. The confidence interval at the first quartile of `rosn` contains zero, so the effect of positive relationships on drinking is not clear. However, the association is decreasing for the higher quartiles of `rosn`. The code for finding confidence intervals for the other variables' effects is given in the program for this example, including how to find the confidence intervals for variables that are involved in two interactions.

5.5 Exercises

1. The bias-variance tradeoff described on page 266 is sometimes difficult to picture. The program LogisticRegBiasVarianceSim.R contains a simulation designed to demonstrate that it can sometimes be better to use a model that is too small rather than one that is too large, or even the correct model! The program simulates binary data, $Y_1, \ldots, Y_n$, with probability of success π_i defined by $\text{logit}(\pi_i) = \beta_0 + \beta_1 x_{i1} + \beta_2 x_{i2} + \beta_3 x_{i3}$, where each x_{ij} is selected from a uniform distribution between 0 and 1. The values of the regression parameters are initially set to $\beta_0 = -1, \beta_1 = 1, \beta_2 = 1, \beta_3 = 0$, so that the correct model has just x_1 and x_2. The program fits three models with variable sets $\{x_1\}, \{x_1, x_2\}$, and $\{x_1, x_2, x_3\}$, and compares the models' estimated probabilities to the true probabilities over a grid of (x_1, x_2, x_3) combinations. The average bias, variance, and mean squared error (MSE = bias^2 + variance) for the predictions are calculated at each x-combination over 200 simulated data sets. The closer to 0 that each of these three summary quantities is, the better the model.

   (a) Run the model using the default settings, which includes $n = 20$. Examine the plots of bias, variance, and MSE as well as the averages for these quantities printed out at the end of the program. Which models perform best and worst with respect to each quantity? Is the correct model ("Model 2" in the program) always the best? Explain.

   (b) Increase the sample size to 100 and repeat the analysis. How do the bias, variance, and MSE change for the three models? Do the best and worst models change?

   (c) Increase the sample size to 200 and repeat the analysis again. Answer the same questions as given in (b).

   (d) Using $n = 20$, try increasing and decreasing the values of β_1 and β_2, maintaining $\beta_0 = -(\beta_1 + \beta_2)/2$ so that probabilities remain centered at 0.5 (the `logit` summary results should remain symmetric around zero). Can you draw conclusions about the effects of each of these two parameters on bias, variance, and MSE?

2. Split the placekick data set approximately in half and repeat variable selection on each half. Determine whether results are similar in the two halves. Partial code for doing this is shown below (assuming that data are contained in the `data.frame` called `placekick`).

```
set.seed(32876582)
placekick$set <- ifelse(runif(n = nrow(placekick)) > 0.5, yes
    = 2, no = 1)
library(glmulti)
search.1.aicc <- glmulti(y = good ~ ., data =
    placekick[placekick$set==1, -10], fitfunction = "glm",
    level = 1, method = "h", crit = "aicc", family =
    binomial(link = "logit"))
a1 <- weightable(search.1.aicc)
cbind(model = a1[c(1:5),1], round(a1[c(1:5),c(2,3)], digits =
    3))
```

3. Refer to the definitions of AIC, AIC_c, and BIC in Section 5.1.2. Construct a plot comparing the penalty coefficients (k) against p for the three criteria with $n = 10$ and $p = 1, 2, \ldots, n/2$. Considering that a larger penalty coefficient causes an information criterion to prefer a smaller model, what do these results imply for models selected by these three criteria? Repeat for $n = 100$ and $n = 1000$. Do the conclusions change?

4. Refer to Exercise 7 in Chapter 2 for the placekicking data collected by Berry and Wood (2004). Using `Distance`, `Weather`, `Wind15`, `Temperature`, `Grass`, `Pressure`, and `Ice` as explanatory variables and `Good` as the response, perform model selection in the following ways:

    (a) Compare which variables are selected under forward selection, backward elimination, and stepwise selection using BIC.

    (b) Perform BMA using BIC. Which variables seem important and which seem clearly unimportant?

    (c) Estimate the regression parameters for the stepwise analyses. Compare these estimates to those of BMA. How are they different?

    (d) Explain why the regression parameter estimate for `Grass` using BMA is somewhat closer to 0 than it is using stepwise.

5. Continuing Exercise 4 in Chapter 2, consider the model using temperature as the only explanatory variable.

    (a) Refit the model using a probit and a complementary log-log link instead of a logit link. Plot all three curves with the data, using a temperature range of 31° to 81° on the x-axis. Does one appear to fit better than the others? Do they provide similar probabilities of failure at 31°?

    (b) Compare BIC values for the three different links. Which link appears to fit best?

    (c) Compute the BMA model probabilities for the three links. Comment on the uncertainty of the model selection among these three links.

6. Refer to the example and program on all-subsets regression from Section 5.1.3 with the placekicking data set. Run the genetic algorithm using AIC_c and include pairwise interactions in the model with `marginality = TRUE` (no pairwise interactions between variables are considered unless both variables are in the model as well). Use `set.seed(837719911)` prior to using `glmulti()`. Note that the function will take a few minutes to run.

    (a) Report the best model and its AIC_c. Compare this to the AIC_c from the best model found without interactions. Do the interactions improve the model?

    (b) How many models are within 2 AIC_c units of the best model? What does this imply about our confidence in having found the best model?

    (c) How many generations did the algorithm produce?

    (d) Rerun the algorithm, but *do not rerun the* `set.seed()` code. This will cause the algorithm to use different random numbers. Re-examine the same items as requested in (a) – (c) and compare to what you obtained initially.

    (e) Rerun the algorithm *and the original* `set.seed()` code, increasing the mutation probability to 0.01 (add the argument value `mutrate = 0.01`; the default is 0.001). Report the results as requested in parts (a) – (c). Are the results any different from those in (a) – (c)? Note that in this context, increasing the mutation probability increases the probability that a variable is randomly added to or removed from a model. In general, increasing the mutation probability can make the algorithm better at finding the top models, but can increase run time.

7. Section 5.1.4 includes an example involving the placekicking data set and alternating stepwise selection of its variables for a logistic regression model. Run the alternating stepwise selection program referred to in the example and report the results. For the final model, compute how much BIC would increase with each possible change that could be made to the model.

8. Refer again to the example referenced in Exercise 7 and its program using alternating stepwise selection. Run the alternating stepwise selection code using k = 0.

    (a) Notice that all variables are in the final model. Why must this happen?
    (b) In the final model, the variables are in order relative to how much $IC(0)$ would change if the variable were dropped. Notice that this is not the same as the order in which the variables were entered. Explain why this can happen.

9. Refer to the hospital-visit data described in Exercise 16 of Chapter 4. Consider the problem of trying to predict whether a person has private insurance based on their pattern of health-care usage. This suggests a logistic regression model for privins, a binary variable. Use the LASSO to identify which of the remaining variables relate to the probability that a person has private insurance. Interpret the results; in particular, estimate the effect of each explanatory variable on this probability.

10. A criminologist studying capital punishment was interested in identifying whether certain social, economic, and political attributes of a country related to its use of the death penalty. She gathered data from public sources on 194 countries, recording the following variables:

    - COUNTRY: The name of the country;
    - STATUS: The country's laws on death penalties, coded as a = Public executions; b = Private executions; c = Executions allowed but none carried out in last 10 years; d = No death penalty;
    - LEGAL: The basis for the country's legal system, coded as a = Islamic; b = Civil; c = Common; d = Civil/Common; e = Socialist; f = Other;
    - HDI: Human Development Index, a numerical measure ranging from 0 (very low human development) to 1 (very high human development);
    - GINI: GINI Index of income inequality, a numerical measure ranging from 0 (perfect equality) to 1 (perfect inequality);
    - GNI: Gross National Income per capita (US$);
    - LITERACY: Literacy rate (% of adults aged 15 and above who are literate);
    - URBAN: Percentage of total population living in urban areas;
    - POL: Political Instability Index for the level of threat posed to governments by social protest, a numerical measure ranging from 0 (no risk) to 10 (high risk);
    - CONFLICT: Level of conflict experienced in the country, coded as a = No conflict, b = latent (non-violent) conflict, c = manifest conflict, d = crisis, e = severe crisis, f = war.

The data for the 141 countries with complete records are available in the DeathPenalty.csv file.[11] The response variable is the country's death penalty status, while all other variables except COUNTRY are explanatory.

(a) Fit a multinomial (nominal response) regression model to these data using all available explanatory variables as linear terms. Compute the AIC and the BIC for this model. Also compute AIC_c by using the formula on page 267. Note that the model degrees of freedom that are needed for computing AIC_c can be found from the extractAIC() function.

(b) Explain why STATUS can be viewed as an ordinal variable.

(c) Fit a proportional odds regression model (see Section 3.4) to these data using all available explanatory variables. Compute AIC, BIC, and AIC_c as in the multinomial regression model.

(d) Note that the proportional odds model assumes both the ordinality of the response and equal coefficients across different logits. Based on the information criteria, is there evidence to suggest that these assumptions lead to a poor model fit? Explain.

11. Continuing Exercise 10, the researcher's main interest was to determine which, if any, of the explanatory variables are associated with the country's use of the death penalty. Complete the items below with this in mind.

(a) Explain why a BMA approach would be more appropriate than a stepwise approach given the context of this problem.

(b) Treating the response as nominal multinomial, carry out the BMA analysis using main effects only. Report the results and draw conclusions. Note that this will require programming a model-fitting function for use in glmulti(), because the multinomial model is not normally fit using glm(). The code for doing this is available in the program glmultiFORmultinom.R. This code needs to be run before glmulti() is called on the multinom-class model fit.

12. Repeat the analysis from Exercise 11 using a proportional odds regression model. Note that this will require programming a model-fitting function for use in glmulti(), because the proportional odds model is not normally fit using glm(). The code for doing this is available in the program glmultiFORpolr.R. This code needs to be run before glmulti() is called on the polr-class model fit.

(a) For each variable, report the posterior probability that it belongs in the model and interpret the results.

(b) Obtain model-averaged estimates of the odds ratios corresponding to the effect of each explanatory variable. Interpret these values.

(c) Compute 95% confidence intervals for each odds ratio and interpret. In particular, which variables appear to be most important in describing countries' use of the death penalty?

(d) Repeat this analysis on the odds ratios from a single proportional odds model using all explanatory variables. Compare the results to the model-averaged values.

---

[11] Data courtesy of Diana Peel, Department of Criminology, Simon Fraser University.

13. Refer to the death penalty data given in Exercise 10. Identify which variables are related to the binary response "Allows Death Penalty" (STATUS is a, b, or c) versus "No Death Penalty" (STATUS is d). Form a model and use the model to estimate the probabilities that the United States and Canada would allow the death penalty.

14. Refer to the hospital visit data described in Exercise 16 of Chapter 4. Use BMA with a BIC criterion on a zero-inflated Poisson model to identify which of the explanatory variables are related to the number of physician office visits for a person (ofp) and the probability of zero visits. Note that health_excellent and health_poor are really just two levels of a three-level factor. Combine them into a single new variable with three levels, e.g., by taking health = as.factor(health_excellent - health_poor).

   For each variable, report:

   (a) the estimated probability that it belongs in the model,

   (b) its parameter estimate and confidence interval in both parts of the model.

   Draw conclusions from your investigation. Because this problem requires programming a model-fitting function to use in glmulti() (a ZIP model is not normally fit using glm()), we have provided some of the necessary code in the program glmulti-FORzeroinfl.R. This code needs to be run before glmulti() is called on the zeroinfl-class model fit. Please see the comments in the program file for help on how to use glmulti() for this model.[12]

15. Repeat Exercise 14 including pairwise interactions into the list of variables, maintaining marginality of the model. Note that this will require using the genetic algorithm. We suggest turning off the plotting (plotty = FALSE) because it produces an enormous amount of graphical output that may crash some systems. Also, note that this may take several hours to run.

16. Using the placekick data from the examples in Section 5.1, fit a logistic regression model with variables distance and PAT. Save the results as an object, say mod.fit.

    (a) Execute summary(mod.fit) and notice that there is a value called AIC in the output. Report this value. Report also the residual deviance.

    (b) Obtain the AIC and residual deviance by extracting them from the model fit object using mod.fit$aic and mod.fit$deviance. Extract the number of parameters in the model using mod.fit$rank. Using the resulting values, compute the AIC and the AIC_c.

    (c) The functions AIC() and extractAIC() can compute values of $IC(k)$ for given values of k, where fit is the model object. Use these functions to compute the model deviance, the AIC, the BIC, and the AIC_c. Note that for AIC_c you will need to use the value of mod.fit$rank from the previous part in the computation of the penalty parameter k.

---

[12]Note that variable selection is applied only to the variables listed in the model for the mean. The model for the probability must either be specified in advance or will have the same variables as are included in each mean model.

17. Refer to the discussion of residuals for Poisson models on page 288. Demonstrate how well the standard normal approximation holds for large residuals from a Poisson model with different means. Specifically, for counts y <- c(0:10), and for mean yhat <- 2:

    (a) Compute Pearson residuals pear <- (y-yhat)/sqrt(yhat).

    (b) Compute both the Poisson and the standard normal cumulative probabilities for a *larger* value than the Pearson residual, pp <- 1-ppois(y,yhat) and pn <- 1-pnorm(pear), respectively.

    (c) Use cbind(y,pear,pp,pn) to print the results. Compare these probabilities for any Pearson residual above 2, 3, and 4. Are the probabilities similar? Are there any other Pearson residuals below 2 that have probability < 0.05? Are there any other Pearson residuals above 2 that have probability > 0.05?

    (d) From parts (a) – (c), what can you conclude about interpreting Pearson residuals from models where $\hat{y} = 2$?

    (e) Repeat these steps for $\hat{y} = 1, 0.5, 0.25, 0.1$. Comment on whether you are comfortable using the above 2, 3, and 4 guidelines to identify large Pearson residuals in these cases.

18. Continuing Exercise 4, use the variables Distance and Grass as suggested by the BMA. Include the interaction of these two variables in the model and perform diagnostics as follows:

    (a) Plot standardized Pearson residuals against distance, the estimated probabilities, and the linear predictor. Interpret the results.

    (b) Perform the Hosmer-Lemeshow test using 10 groups. State hypotheses, test statistic, p-value, and conclusions. Also, examine the Pearson residuals from the groupings and indicate whether they show any particular pattern.

    (c) Perform the Osius-Rojek and Stukel tests and draw conclusions.

    (d) Do an influence analysis. Identify any influential observations and indicate how they are influential. Interpret the results.

    (e) Write a brief summary of the results of the diagnostics and conclude whether the model is reasonably good for the data.

19. Margolin et al. (1981) provide data showing the number of revertant colonies of *Salmonella* in response to varying doses of quinoline. There are six different dose levels, and three independent observations per dose. The data are provided in the data frame salmonella from the aod package.

    (a) Notice that the spacings among the dose levels are approximately constant on a log scale. Plot the response (y) against both dose and base-10 log-dose (log10(y+1)) and comment on any apparent trends. Note that we use base 10 here because the dose levels 10, 100, and 1000 are simply 1, 2, and 3 in base-10 log. The +1 is used in the log here because $\log_{10} 0$ is not defined, but $\log_{10} 1 = 0$. In this problem all logs should be taken this way.

    (b) Estimate a Poisson regression model for y with either dose, log-dose, or a categorical version of dose (treat dose as a six-level categorical factor by using factor(dose) in the formula) as an explanatory variable. What conclusions can you draw based on tests for a dose effect?

(c) Examine the deviance/df statistics for each model. Do the model fits appear to be good? Explain.

(d) Plot the standardized residuals against fitted means for the two models that use dose in a numerical manner. Comment on what these plots tell you.

(e) Plot standardized residuals against fitted means for the model treating dose as categorical. There is no question about the linearity of the response or the appropriateness of the link function here. What does this plot suggest?

20. Refer to the Challenger data from Exercise 4 in Chapter 2. Fit a model including only temperature as an explanatory variable.

    (a) Examine the deviance/df statistic. Also conduct a graphical assessment of the residuals. Discuss the fit of the model.

    (b) Examine the goodness of fit for the model and draw conclusions.

    (c) Conduct an influence analysis and draw conclusions.

        i. Explain why h_{14} is so large.
        ii. Re-estimate the logistic regression without observation 14 in the data set. For both the model with and without this observation, compare confidence intervals for the parameters, and plot the estimated probability of O-ring failure against temperature on one plot. Do the results change in a meaningful way?

21. Refer to the death penalty data with a binary interpretation of the response for presence or absence of death penalty (see Exercises 10 and 13). Consider a model consisting of the variables LEGAL, HDI, and GNI. Assess the model fit and identify any concerns.

22. Refer to Exercise 12 in Chapter 4. Plot standardized residuals vs. the 1-2-3-4-5 scores for political ideology. Which models examined in this previous exercise fit the data reasonably well?

23. Agresti (2007) provides data on the social behavior of horseshoe crabs. These data are contained in the HorseshoeCrabs.csv file available on our website. Each observation corresponds to one female crab. The response variable is Sat, the number of "satellite" males in her vicinity. Physical measurements of the female—Color (4-level ordinal), Spine (3-level ordinal), Width (cm), and Weight (kg)—are explanatory variables.

    (a) Fit a Poisson regression model with a log link using all four explanatory variables in a linear form. Test their significance and summarize results.

    (b) Compute deviance/df and interpret its value.

    (c) Examine residual diagnostics and identify any potential problems with the model.

    (d) Carry out the GOF test described on page 296 using the PostFitGOFTest() function available in the PostFitGOFTest.R program of the same name from our website. Use the default number of groups. ($M/5$ when $M \leq 100$)

        i. State the hypotheses, test statistic and p-value, and interpret the results.
        ii. Plot the Pearson residuals for the groups against the interval centers (available in the pear.res and centers components, respectively, of the list returned by the function). Use this plot and the residual plots from part (c) to explain the results.

24. Continuing Exercise 23, conduct an influence analysis. Interpret the results.

25. Continuing Exercise 23, notice that there is one crab with a weight that is substantially different from the rest. This can be seen, for example, in a histogram of the weights. Remove this crab from the data and repeat the steps from Exercise 23. Has this fixed any problems with the model? Are there any other problems with the model, and what could be done to solve these problems?

26. Refer to Exercise 18 from Chapter 4 on the salamander count data.

    (a) Compute deviance/df. What does this value suggest about the model?
    (b) Examine the standardized residuals and perform an influence analysis. Interpret the results.

27. Continuing Exercise 26:

    (a) Carry out the GOF test described on page 296 using the `PostFitGOFTest()` function from PostFitGOFTest.R on the book's website. Use the default number of groups, which is $n/5$ when $n \leq 100$. State the hypotheses, test statistic and p-value, and interpret the results. Plot the Pearson residuals for the groups (`$pear.res`) and against the interval centers (`$centers`). Use this plot and the residual plots from Exercise 26 to explain the results.
    (b) Perform a Monte Carlo simulation to investigate whether the GOF test holds the correct size for this model and data. Specifically, assuming that the model is contained in a `glm` object called `mod.fit`, use `set.seed(1348765911)`, and repeat the following steps 1000 times:

        i. Simulate Poisson response data from the estimated model. For example, use `predict()` to save the estimated means to an object called `mean` and simulate new responses using

        ```
        y <- rpois(n = length(mean), lambda = mean)
        ```

        ii. Estimate the Poisson regression model for the simulated data and obtain the predicted values from the model for each observation.
        iii. Apply `PostFitLOFTest()` to the simulated data and predicted values; obtain the p-value (`pval` component of the returned object)

    The fraction of p-values that lie below 0.05 is the estimated size of the test when a type I error level of this same value is used. Is this fraction close to 5% as one would hope for if the test holds the correct size? Use a confidence interval to account for simulation variability in answering this question.

28. Refer to Exercise 26 from Chapter 4 on the Alcohol Consumption data, in which a Poisson regression model was fit using first Saturday drink consumption as the response and `prel`, `nrel`, `posother`, `negother`, `age`, `rosn`, and `state` as the explanatory variables. Examine the fit of this model and draw conclusions.

29. Refer to the wheat kernels example from Section 3.3. Examine the fit of the model with the six explanatory variables. Use the `multinomDiag()` function contained in our R program multinomDiagnostics.R to help with your calculations.

30. Refer to Exercise 16 from Chapter 4. Fit the ZIP model with all listed variables as linear terms.

(a) Carry out the goodness-of-fit test described on page 296 using the `PostFitGOFTest()` function, which is available on the website for this book. Use the default number of groups. Plot the Pearson residuals for the groups (`$pear.res`) against the interval centers (`$centers`). Use this plot to help explain the results from the test.

(b) Plot the Pearson residuals against each explanatory variable and interpret the results.

(c) Add a quadratic term for `numchron` to the model. Is this term significant? How well does the model fit now? Explain.

(d) Suggest how the model might be improved further.

31. Repeat the Hosmer-Lemeshow test on the placekick data with `distance` as the explanatory variable and using $g = 5, 6, \ldots, 12$ groups. Report the results and comment on the sensitivity of the test to the number of groups.

32. Refer to the example on page 303 regarding simulating data with overdispersion. In general, consider a model in which $Y \sim Po(\mu)$, where μ has a normal distribution with mean τ and variance σ^2. Using the fact that $E(Y) = E_N[E_P(Y|\mu)]$ and $Var(Y) = E_N[Var_P(Y|\mu)] + Var_N[E_P(Y|\mu)]$, where E_N, Var_N are taken with respect to the normal distribution for the mean μ and E_P, Var_P are taken with respect to the Poisson distribution for the response Y, show that $E(Y) = \tau$ and $Var(Y) = \tau + \sigma^2$.

33. Refer to the example on "Beetle egg-laying response to crowding" in Section 4.4.

(a) Check the final model used in the example (contained in `zip.mod.t0`) for overdispersion. When plotting the Pearson residuals versus temperature, the `jitter()` function will be useful. This function adds a small amount of random noise to the temperature values which then helps to distinguish between similar Pearson residuals at the same temperature.

(b) Fit a zero-inflated negative binomial model. Compare the parameter estimates and confidence intervals to those from the zero-inflated Poisson model. Check this model for signs of overdispersion.

34. Continuing Exercise 13 from Chapter 4, complete the following.

(a) Find the sample mean and variance for each day of the week. Is there evidence of overdispersion? Explain.

(b) Using a negative binomial regression model instead of a Poisson regression model, complete all of the parts given in Exercise 13(c) from Chapter 4.

(c) Consider the ANOVA model $Y_{ij} = \mu + \alpha_i + \epsilon_{ij}$ where ϵ_{ij} are independent, normally distributed errors with mean 0 and variance $\sigma^2 > 0$. This is the usual linear model found in a one-way analysis of variance. Below is the code that can be used to find the ANOVA table and perform multiple comparisons (the data are contained within `starbucks`):

```
mod.fit.anova <- aov(formula = Count ~ Day, data =
    starbucks)
summary(mod.fit.anova)

# Least significant differences
```

```
pairwise.t.test(x = starbucks$Count, g = starbucks$Day,
    p.adjust.method = "none", alternative = "two.sided")

# Tukey honest significant differences
TukeyHSD(x = mod.fit.anova, conf.level = 0.95)
```

Compare these results to those obtained using the Poisson and negative binomial regression models. Which of the three modeling approaches is the most appropriate? Discuss.

35. For the Salmonella data from Exercise 19, there may be signs of overdispersion. Consider only the model that treats dose as a categorical explanatory variable.

    (a) Repeat the analysis from the example on page 316 to see whether a quasi-Poisson or a negative binomial is a better option. In particular, create a plot like Figure 5.10 and compute the quadratic regression as shown in the example. What do you conclude about the two model options?

    (b) Fit the quasi-Poisson model to the data and test the significance of the dose effect. How do the results compare to those from Exercise 19?

    (c) Plot standardized residuals against the estimated means. Comment on the results.

36. For the horseshoe crab data from Exercise 23, there may be signs of overdispersion.

    (a) Repeat the analysis from the example on page 316 to see whether a quasi-Poisson or a negative binomial is a better option. In particular, create a plot like Figure 5.10 and compute the quadratic regression as shown in the example. What do you conclude about the two model options?

    (b) Fit the quasi-Poisson model to the data and test the model effects again. How do the results compare to those from Exercise 23?

    (c) Plot standardized residuals against the estimated means. Comment on the results.

37. Kupper and Haseman (1978) describe vaguely an experiment in which pregnant female mice were assigned to one of two groups, labeled "treatment" and "control." The data are available in the `mice` data frame of the `aod` package. From each female the number of pups born in the litter (n) and the count of those that were affected in some way (y) were recorded.

    (a) Fit a Logistic regression model to these data. Test the significance of the treatment effect and find a confidence interval for the odds ratio for the effect of "treatment" relative to "control."

    (b) Note that the number of explanatory variable levels is fixed, so it would not increase if the sample size were to grow. This means that we can use the residual deviance to test the fit of the model formally. Perform this test. State the hypotheses, the test statistic, the p-value, and the conclusions.

    (c) Fit a quasi-binomial regression model and repeat the analysis from part (a). Are the results substantially different?

    (d) Use a beta-binomial model to repeat the analysis from part (a). This can be done using the `betabin()` function of the package `aod`. Are the results substantially different?

Chapter 6

Additional topics

6.1 Binary responses and testing error

Measurement processes are often imperfect. Instruments may be improperly calibrated (e.g., bathroom scales), local conditions may not all be identical (e.g., temperatures in a region), or a target may be difficult to measure (e.g., homeless persons in a census). Even binary responses are sometimes measured with error: a "success" may actually be a mismeasured "failure" and vice versa. For example, a referee may mistakenly declare a placekick successful when it was not, and vice versa.

We focus in this section on those binary responses resulting from diagnostic tests for infectious diseases. For example, tests for infections such as HIV, West Nile Virus, and chlamydia usually are not 100% accurate. The reasons for testing error can include (1) levels of the bacteria, virus, or other causal microbe that are below detectable levels (often the case for new infections), (2) inhibitors within a specimen that block a positive outcome from being detected, and (3) laboratory error. Fortunately, diagnostic tests usually have high accuracy levels, as we saw in Exercise 6 of Chapter 1 for the Aptima Combo 2 Assay. However, the possibility of testing error still needs to be taken into account in a statistical analysis to enable correct inferences. The purpose of this section is to discuss basic techniques to account for testing error that can be used in situations examined in Chapters 1 and 2.

6.1.1 Estimating the probability of success

Define Y as a Bernoulli random variable that denotes a diagnostic test's *measured* result, either positive (1) or negative (0). When Y is measured with the possibility of testing error, we need to define another Bernoulli random variable $\tilde{Y}$ that denotes the *true* status, positive (1) or negative (0). This leads to the following measures of accuracy:

- Sensitivity: $S_e = P(Y = 1|\tilde{Y} = 1)$
- Specificity: $S_p = P(Y = 0|\tilde{Y} = 0)$

Thus, sensitivity is the probability that a measured result is positive given that it is truly positive, and specificity is the probability that a measured result is negative given that it is truly negative. Ideally, we would like these conditional probabilities to equal 1, which means a perfect diagnostic test. Unfortunately, this is often not the case. Thus, the testing error rate is $1 - S_e = P(Y = 0|\tilde{Y} = 1)$ for truly positive individuals and $1 - S_p = P(Y = 1|\tilde{Y} = 0)$ for truly negative individuals.

In the evaluation of a diagnostic test, values for S_e and S_p are usually determined by performing tests on known true positive and negative specimens. For example, infectious disease assays often go through a clinical trial in this manner, and the resulting outcomes are given as part of a product insert for the assay. In fact, the data used for Exercise 6 of

Chapter 1 were obtained from Gen-Probe's product insert. Interestingly, the estimates of sensitivity and specificity from these trials are typically taken to be the true values without the acknowledgment of variability that inevitably occurs. We will do the same here, but we address alternative analysis methods in Section 6.1.3.

There is a simple relationship between Y and $\tilde{Y}$ that can be expressed in terms of probabilities. Using the definition of marginal and conditional probabilities, the probability of a positive test response is

$$\begin{aligned} P(Y=1) &= P(Y=1 \text{ and } \tilde{Y}=1) + P(Y=1 \text{ and } \tilde{Y}=0) \\ &= P(Y=1|\tilde{Y}=1)P(\tilde{Y}=1) + P(Y=1|\tilde{Y}=0)P(\tilde{Y}=0) \\ &= S_e P(\tilde{Y}=1) + (1-S_p)P(\tilde{Y}=0). \end{aligned}$$

By letting $P(Y=1) = \pi$ and $P(\tilde{Y}=1) = \tilde{\pi}$ and then solving for $\tilde{\pi}$ in the above expression, we can compactly write the true probability of positivity as

$$\tilde{\pi} = \frac{\pi + S_p - 1}{S_e + S_p - 1}. \tag{6.1}$$

To estimate $\tilde{\pi}$, suppose that a sample of n individuals is tested for some disease, yielding responses $y_1, \ldots, y_n$. For example, $y_1, \ldots, y_n$ may represent the measured positive and negative HIV test results for blood donations submitted to a blood bank. The observed responses can be modeled using a Bernoulli distribution with probability π. From this model, the maximum likelihood estimate of $\tilde{\pi}$, say $\hat{\tilde{\pi}}$, can be found by the usual methods. The likelihood function is

$$\begin{aligned} L(\tilde{\pi}|y_1,\ldots,y_n) &= \pi^w(1-\pi)^{n-w} \\ &= [S_e\tilde{\pi} + (1-S_p)(1-\tilde{\pi})]^w [1 - S_e\tilde{\pi} - (1-S_p)(1-\tilde{\pi})]^{n-w} \end{aligned} \tag{6.2}$$

where $w = \sum_{i=1}^n y_i$ is the observed number of positive test responses in a sample of size n. Maximizing Equation 6.2 results in the MLE of $\tilde{\pi}$:

$$\hat{\tilde{\pi}} = \frac{\hat{\pi} + S_p - 1}{S_e + S_p - 1}, \tag{6.3}$$

where $\hat{\pi} = w/n$ (see Exercise 3). Alternatively, one can use the invariance property of maximum likelihood estimators[1] and simply substitute $\hat{\pi}$ in for π in Equation 6.1. Notice that when there is no testing error ($S_e = S_p = 1$), $\hat{\tilde{\pi}} = \hat{\pi}$ as would be expected.

There is one unfortunate problem with Equation 6.3. Typically, S_p will be a large number close to 1. However, some diseases have such low infection rates, and hence small values of $\hat{\pi}$, that the value of S_p may not be large enough to prevent the numerator from being negative. Because the denominator is positive in all realistic applications, this leads to a negative value for $\hat{\tilde{\pi}}$, which does not make sense for a probability. A simple solution, although not necessarily ideal, is to set the estimate of $\tilde{\pi}$ to 0 if the MLE is negative. On the other hand, there may be a larger issue to solve when the MLE is negative; that is, perhaps the diagnostic test is just not accurate enough to obtain a meaningful estimate of $\tilde{\pi}$.

The estimated variance of $\hat{\tilde{\pi}}$ can also be found in a similar manner as in Section 1.1.2. Exercise 3 shows that the variance is

$$\widehat{Var}(\hat{\tilde{\pi}}) = \frac{\hat{\pi}(1-\hat{\pi})}{n(S_e + S_p - 1)^2}.$$

---
[1] See Appendix B.4 or p. 319 of Casella and Berger (2002).

This variance is the same as given in Equation 1.3 for the case of no testing error, except for $(S_e + S_p - 1)^2$ in the denominator. Thus, because $1 < S_e + S_p < 2$ in all realistic applications, the presence of testing error essentially reduces the effective sample size from n to $n(S_e + S_p - 1)^2$. The end result is larger variability for the estimator when there is testing error present. This result is intuitive because variability is a measurement of uncertainty. If we are uncertain about a true response, this should be reflected by a larger variance!

The $(1-\alpha)100\%$ Wald confidence interval for $\tilde{\pi}$ is

$$\hat{\tilde{\pi}} - Z_{1-\alpha/2}\sqrt{\frac{\hat{\pi}(1-\hat{\pi})}{n(S_e + S_p - 1)^2}} < \tilde{\pi} < \hat{\tilde{\pi}} + Z_{1-\alpha/2}\sqrt{\frac{\hat{\pi}(1-\hat{\pi})}{n(S_e + S_p - 1)^2}}.$$

The performance of the interval is similar to that for the Wald interval for π. Exercise 2 explores this behavior further. A likelihood ratio interval for $\tilde{\pi}$ is presented in Exercise 5 as an alternative to the Wald interval.

Example: Hepatitis C prevalence among blood donors when testing error is included (HepCPrevSeSp.R)

Section 1.1.2 gives an example where 42 out of 1875 blood donors tested positive for hepatitis C in Xuzhou City, China. While no values of S_e and S_p are given in the corresponding paper by Liu et al. (1997), it is common for diagnostic tests to not be perfectly accurate in similar settings. For example, Wilkins et al. (2010) list values of $S_e = 0.96$ and $S_p = 0.99$ for RNA polymerase chain reaction (PCR) assays. These authors also list both higher and lower values for S_e and S_p with other types of hepatitis C diagnostic tests.

The proportion of individuals testing positive for hepatitis C was $\hat{\pi} = 42/1875 = 0.0224$. Assuming the same accuracy levels as for the PCR assay, the MLE for the overall prevalence becomes

$$\hat{\tilde{\pi}} = \frac{0.0224 + 0.99 - 1}{0.96 + 0.99 - 1} = 0.0131.$$

The estimated variance of $\hat{\pi}$ is $0.0224(1-0.0224)/1875 = 0.00001168$, and the estimated variance of $\hat{\tilde{\pi}}$ is $0.0224(1-0.0224)/[1875(0.96+0.99-1)^2] = 0.00001294$. When testing error is included in the analysis, we see then that the estimated prevalence decreases while the estimated variance increases. The 95% Wald intervals for π and $\tilde{\pi}$ are $0.01570 < \pi < 0.02910$ and $0.006002 < \tilde{\pi} < 0.02010$, respectively.

To examine the effects of testing error further, additional calculations are shown in Table 6.1 for values of S_e and S_p given in Wilkins et al. (2010) for other diagnostic tests. The rows of the table are ordered by the overall uncertainty in the accuracy of the test. We see that as the overall uncertainty increases, the estimated variance increases as well. Also, nonsensical estimates for $\tilde{\pi}$ occur for smaller values of S_p–including a value of -0.301 in the last row which uses the S_e and S_p from a recombinant immunoblot assay. We would like to point out that this comparison of estimates may be somewhat unfair, because different assays will produce different estimates of π. In particular, we would not actually expect an estimate as extreme as -0.301 because a low S_p will lead to more false positive individuals, which then increases $\hat{\pi}$. Still, these comparisons serve the purpose of showing that more uncertainty leads to more variability and perhaps even to nonsensical estimates.

Table 6.1: Prevalence estimates for various values of S_e and S_p.

S_e	S_p	$\hat{\tilde{\pi}}$	$\widehat{Var}(\hat{\tilde{\pi}})$	95% Wald confidence interval
1.00	0.98	0.002	1.22×10^{-5}	(-0.004, 0.009)
0.94	0.97	-0.008	1.41×10^{-5}	(-0.016, -0.001)
0.87	0.99	0.014	1.58×10^{-5}	(0.007, 0.022)
0.79	0.80	-0.301	3.36×10^{-5}	(-0.312, -0.290)

6.1.2 Binary regression models

The presence of testing error can be incorporated into the estimation of a logistic regression model (or other binary regression models) in a similar manner as was shown in Section 6.1.1. The general approach is to model the true probability based on responses $y_1, ..., y_n$ that include possible testing error. Then we apply Equation 6.2 to work back and forth between the two different probabilities.

The logistic regression model now has the form

$$\text{logit}(\tilde{\pi}_i) = \beta_0 + \beta_1 x_{i1} + \cdots + \beta_p x_{ip},$$

where $\tilde{\pi}_i$ is the true probability of success for observation $i = 1, \ldots, n$. The model uses a likelihood function in a familiar form to what was given in Section 2.2.1:

$$L(\boldsymbol{\beta}|\mathbf{y}) = \prod_{i=1}^{n} \pi_i^{y_i} (1 - \pi_i)^{1-y_i}$$

$$= \prod_{i=1}^{n} [S_e \tilde{\pi}_i + (1 - S_p)(1 - \tilde{\pi}_i)]^{y_i} [1 - (S_e \tilde{\pi}_i - (1 - S_p)(1 - \tilde{\pi}_i))]^{1-y_i},$$

where π_i is the success probability for observation y_i, $\boldsymbol{\beta} = (\beta_0, \ldots, \beta_p)$, and $\mathbf{y} = (y_1, ..., y_n)$. Iterative numerical procedures again need to be used to estimate the regression parameters. The estimated covariance matrix can be found by inverting the negative of the Hessian matrix, as discussed in Appendix B.3.4 and in Section 2.2.1.

Example: Prenatal infectious disease screening (HIVKenya.R, HIVkenya.csv)

Verstraeten et al. (1998) screened a sample of pregnant women in Kenya for HIV with the purpose of evaluating a new screening method. While not discussed in this paper, subsequent work by Vansteelandt et al. (2000) and others examined the data in the context of binary regression models. A portion of their data set is available in the data set corresponding to this example. The response variable hiv is a 0 for a negative test and a 1 for a positive test. The explanatory variables are parity (number of times the woman has previously given birth), age (in years), marital.status (1 = single, 2 = married polygamous, 3 = married monogamous, 4 = divorced), and education (1 = none, 2 = primary, 3 = secondary, 4 = higher). We focus only on the age explanatory variable here and leave an analysis involving the other variables as part of Exercise 4. Below are the results from fitting a logistic regression model to the data assuming no testing error:

```
> set1 <- read.csv(file = "c:\\data\\HIVKenya.csv")
> head(set1)
  parity age marital.status education hiv
```

```
1      1    16           3            2    0
2      0    17           3            1    0
3      2    26           3            2    0
4      3    20           3            2    0
5      1    18           3            1    0
6      5    35           3            2    0
```

```
> mod.fit <- glm(formula = hiv ~ age, data = set1, family =
    binomial(link = logit))
> round(summary(mod.fit)$coefficients, 4)
            Estimate Std. Error z value Pr(>|z|)
(Intercept)  -1.8618     0.6917 -2.6917   0.0071
age          -0.0273     0.0287 -0.9528   0.3407
```

The actual testing involved a series of up to three tests per individual in order to reduce the probability of a misdiagnosis. Unfortunately, the data from each test is not available, and only the final test diagnoses are given by hiv in the data set. For illustration purposes, we examine next how to estimate the model with $S_e = 0.98$ and $S_p = 0.98$.

We present two ways to estimate the model using R. First, we follow a similar process as illustrated in Section 2.2.1 when we used our own function to evaluate the log-likelihood and then optim() to maximize it. Below is our code and output:

```
> Se <- 0.98
> Sp <- 0.98
> X <- model.matrix(mod.fit)
> logL <- function(beta, X, Y, Se, Sp) {
    pi.tilde <- plogis(X%*%beta)   # Find exp(X*beta)/(1 +
        exp(X*beta))
    pi <- Se*pi.tilde + (1 - Sp)*(1 - pi.tilde)
    sum(Y*log(pi) + (1-Y)*log(1-pi))
  }

> mod.fit.opt <- optim(par = mod.fit$coefficients, fn = logL,
    hessian = TRUE, X = X, Y = set1$hiv, control = list(fnscale =
    -1), Se = Se, Sp = Sp, method = "BFGS")
> mod.fit.opt$par    # beta.hats
(Intercept)         age
 -2.0068169  -0.0334271
> mod.fit.opt$value  # log(L)
[1] -177.27
> mod.fit.opt$convergence   # 0 means converged
[1] 0

> # Estimated covariance matrix (multiply by -1 because of
    fnscale)
> cov.mat <- -solve(mod.fit.opt$hessian)
> cov.mat
             (Intercept)           age
(Intercept)    0.7884776  -0.032263002
age           -0.0322630   0.001388222
> sqrt(diag(cov.mat))    # SEs
(Intercept)         age
 0.88796260  0.03725886
```

```
> z <- mod.fit.opt$par[2]/sqrt(diag(cov.mat))[2]   # Wald statistic
> 2*(1-pnorm(q = abs(z)))   # p-value
      age
0.3696344
```

The logL() function evaluates the log-likelihood function using matrix algebra to calculate $\tilde{\pi}_i$ for $i = 1, \ldots, n$. This matrix algebra representation allows the code to be easily generalized to more than one explanatory variable. A non-matrix-algebra alternative for one explanatory variable is given in the corresponding program. The estimated logistic regression model is

$$\text{logit}(\hat{\tilde{\pi}}) = -2.0068 - 0.03343\text{age}.$$

A numerical estimate of the Hessian matrix is produced by specifying hessian = TRUE in optim() and leads to the estimated covariance matrix for $\hat{\beta}_0$ and $\hat{\beta}_1$ as given by Equation B.6. A Wald test of $H_0 : \beta_1 = 0$ vs. $H_a : \beta_1 \neq 0$ has a p-value of 0.3696, which indicates little evidence that a linear age term is needed.

The model can also be estimated by writing a new link function for the family argument of glm(). Specifically, the link function form can be seen by rewriting our model as

$$\text{logit}\left(\frac{\pi + S_p - 1}{S_e + S_p - 1}\right) = \beta_0 + \beta_1 \text{age},$$

where Equation 6.1 is used to express the model in terms of $\hat{\pi}$. Solving for $\hat{\pi}$ on the left side results in the inverse link function:

$$\pi = \frac{S_e \exp(\beta_0 + \beta_1 \text{age}) - S_p + 1}{1 + \exp(\beta_0 + \beta_1 \text{age})}.$$

We code these equations into our own function below:

```
> my.link <- function(Se, Sp) {
    # mu = E(Y) = pi
    linkfun <- function(mu) {
      pi.tilde <- (mu + Sp - 1)/(Se + Sp - 1)
      log(pi.tilde/(1-pi.tilde))
    }
    linkinv <- function(eta) {
      (exp(eta)*Se - Sp + 1)/(1 + exp(eta))
    }
    mu.eta <- function(eta) {
      exp(eta)*(Se + Sp - 1)/(1 + exp(eta))^2
    }
    save.it <- list(linkfun = linkfun, linkinv = linkinv, mu.eta
        = mu.eta)
    class(save.it) <- "link-glm"
    save.it
  }

> mod.fit2 <- glm(formula = hiv ~ age, data = set1, family =
    binomial(link = my.link(Se, Sp)))
> round(summary(mod.fit2)$coefficients, 4)
            Estimate Std. Error z value Pr(>|z|)
(Intercept)  -2.0097     0.9398 -2.1383   0.0325
```

```
age            -0.0333      0.0397 -0.8396   0.4011
> vcov(mod.fit2)
            (Intercept)         age
(Intercept)  0.88329971 -0.036451992
age         -0.03645199  0.001573203
```

The `mu.eta()` function within `my.link()` gives the partial derivative of π with respect to the systematic component, $\eta = \beta_0 + \beta_1 \texttt{age}$:

$$\frac{\partial \pi}{\partial \eta} = \frac{\exp(\eta)(S_e + S_p - 1)}{[1 + \exp(\eta)]^2}$$

(see Section 2.3). The `glm()` function is used then in the same manner as with logistic regression except for the value of `link` inside of `binomial()`. The regression parameter estimates are practically the same as those obtained from using `optim()`. The small differences between the estimated covariance matrices from the two methods arise because `glm()` uses Equation B.5 in its calculation, whereas `optim()` uses Equation B.6.

The square root of the estimated variance for $\hat{\beta}_1$ is 0.0287 when there is no testing error and 0.0397 when there is testing error. This increase occurs because there is more uncertainty in the response when testing error is present, which then is exhibited in the variances for the regression parameter estimates. Exercise 6 examines this behavior further.

6.1.3 Other methods

The methods described in Section 6.1.2 are meant to be used in situations when the values of S_e and S_p are known or when the variability from estimating S_e and S_p is unknown or cannot be easily obtained. When the original data used to estimate S_e and S_p are available, these data can then be used when constructing an interval estimate for $\tilde{\pi}$ to account for the additional variability.

In other situations, there are more than two response categories of interest for a given problem as we saw in Chapter 3. These additional response categories subsequently lead to more types of misclassification that need to be addressed. Chapter 2 of the book by Buonaccorsi (2010) discusses both of the above situations, and we refer the reader to this book for detailed discussion.

Küchenhoff et al. (2006) provide an innovative way to account for testing error in categorical response and/or explanatory variable settings. Their "misclassification, simulation, and extrapolation" (MC-SIMEX) method is implemented by the `simex` package of R (Lederer and Küchenhoff, 2006), and we show how to use this package in Exercise 7. In situations when only a binary response variable is subject to testing error, simulations in Küchenhoff et al. (2006) show that the maximum likelihood methodology we described in Sections 6.1.1 and 6.1.2 provides nearly unbiased estimators, while MC-SIMEX produces somewhat biased estimators. Therefore, our presented analysis methods generally would be preferred when S_e and S_p are known.

The analysis of data measured with error, either in response or explanatory variables, is a large and active research area. Books such as Buonaccorsi (2010), Carroll et al. (2010), and Gustafson (2004) provide thorough accounts.

6.2 Exact inference

Most of the inference procedures examined so far rely on a statement similar to the following:

> As the sample size goes to infinity, the statistic's distribution approaches a χ^2 (or normal) distribution.

Unfortunately, infinity is the one sample size that we can never have. There are many situations where a "named" distribution, such as normal or chi-square, serves as a good approximation to the actual distribution for the statistic. However, there are some situations where a named distribution does not, particularly when the sample size is small. The purpose of this section is to develop inference procedures that do not rely on large-sample approximations. These inference procedures are often referred to as *exact* in the sense that they use the actual distribution for the statistic of interest under minimal assumptions about the structure of the data. These exact inference procedures often also provide us ways to assess whether or not a large-sample distributional approximation for a statistic is appropriate.

Exact inference methods were first discussed in Section 1.1.2 with respect to the Clopper-Pearson interval for the probability-of-success parameter in a binomial distribution. We showed that the interval always had a true confidence level at least at the stated level, but it could be much more. This led us to labeling the interval as conservative. Unfortunately, most other exact inference methods are conservative as well. For hypothesis testing, exact tests tend to reject the null hypothesis less often than the stated type I error level when the null hypothesis is true (it will never reject at a higher rate). This behavior persists when the null hypothesis is false, so that exact tests may have lower power than other tests. Still, a conservative test is generally better than a liberal test when error-rate control is important, and especially in settings with small sample sizes when non-exact tests may become excessively liberal.

We do not have space to cover all possible exact inference procedures, so we begin in Section 6.2.1 with one of the most commonly used exact procedures, Fisher's exact test. We generalize the ideas of exact inference in Section 6.2.2, where permutation methods allow us to test for independence. In Section 6.2.3 we introduce exact inference in the context of logistic regression. Finally, we conclude with a survey of other exact inference procedures in Section 6.2.4 for those readers who would like to consider them further.

6.2.1 Fisher's exact test for independence

We begin by examining a test for independence in a 2 × 2 contingency table, a structure studied previously in Section 1.2. Let two groups be represented by the rows of the table, and the responses "success" and "failure" by the columns. Recall that an odds ratio (OR) with a value of 1 means that the odds of success for group 1 are the same as the odds of success for group 2; i.e., the odds are independent of group designation. Because of this result, we will simply write $H_0 : OR = 1$ vs. $H_a : OR \neq 1$ as our hypotheses in a test for independence between the column responses and the row responses.

Hypergeometric distribution

The hypergeometric probability distribution plays an important role in constructing an exact test for independence, so we provide a brief review here. The distribution is typically introduced using an example such as the following:

Table 6.2: Two representations of a 2 × 2 table. The representation on the left uses the notation from Section 1.2.1, and the representation on the right uses the notation from the hypergeometric distribution in the context of the urn example.

		Response				Urn	Drawn out	Remaining	
		1	2						
Group	1	w_1	$n_1 - w_1$	n_1	Color	Red	m	$a - m$	a
	2	w_2	$n_2 - w_2$	n_2		Blue	$k - m$	$b - k + m$	b
		w_+	$n_+ - w_+$	n_+			k	$n - k$	n

Suppose an urn has a red balls and b blue balls with $n = a + b$. Suppose $k \leq n$ balls are randomly drawn from the urn without replacement. Let M be the number of red balls drawn out.

The random variable M has a hypergeometric distribution. The PMF is

$$P(M = m) = \frac{\binom{a}{m}\binom{b}{k-m}}{\binom{n}{k}}$$

for $m = 0, \ldots, k$ subject to $m \leq a$ and $k - m \leq b$. Note that a, b, n, and k are all fixed quantities. The only variable on the right side is m.

The hypergeometric distribution can be used to find the probability of observing a particular 2 × 2 table under independence. Table 6.2 shows a 2 × 2 table from two perspectives. The left side shows the notation for a 2 × 2 table under the independent binomial model that was given in Section 1.2.1. The right side shows the same table written in the notation for a hypergeometric distribution. The big difference between the two is that all of the marginal counts are fixed (i.e., known) before any sampling takes place for the hypergeometric distribution, whereas in the binomial model the column margins—representing the total number of observed successes and failures—are considered random. However, under certain assumptions, models for the two tables can be made equivalent, as we show next.

In the independent binomial model, we observe two random variables, W_1 and W_2. We use the observed values w_1 and w_2 to compare the respective probabilities of success, π_1 and π_2. The total number of successes, w_+, tells us nothing about how different π_1 and π_2 might be, and hence it is not important to the comparison. Thus, there is no harm in "conditioning on" w_+ (i.e., assuming it is known instead of random). In fact, there are mathematical benefits to making this assumption. When $H_0 : \pi_1 = \pi_2$ is true, then it can be shown that conditioning on w_+ in the independent binomial model leads to the hypergeometric model. See Exercise 1 for details of this derivation.

Because all of the marginal totals are known in the hypergeometric model, once we observe one cell count, the remaining three counts are obtained by subtraction and the whole table is known. Thus, the hypergeometric distribution gives us the probability of observing a particular contingency table under independence. This is useful for constructing a test for independence, because it provides a method for calculating the exact p-value for a test. If there is low probability of a table as extreme as the one observed, this casts doubt on the independence assumption. Fisher's exact test, which we discuss next, is based on this idea.

Fisher's exact test

Salsburg (2001) describes one of the most frequently discussed events in statistical history as follows. One day in Cambridge, England, during the late 1920s, a number of individuals

Table 6.3: Hypothetical outcome of Fisher's experiment.

		Lady's response		
		Milk	Tea	
Actual	Milk	3	1	4
	Tea	1	3	4
		4	4	8

Table 6.4: Probability of possible outcomes for Fisher's experiment. Note that the last odds ratio is $4 \times 4/(0 \times 0)$, which is undefined. This has been labeled as > 9.

M	$P(M = m)$	$\widehat{OR}$	X^2
0	0.0143	0	8
1	0.2286	1/9	2
2	0.5143	1	0
3	0.2286	9	2
4	0.0143	>9	8

gathered together for afternoon tea. A lady in the group said that she could differentiate between whether the milk or the tea was poured first into a cup. Sir Ronald Fisher, who contributed immensely to the early foundations of statistics, was in the group and proposed a simple experiment to test the lady's claim. He suggested to pour tea first into four cups and to pour milk first into four cups. The lady was blinded to which cups had milk or tea poured first, but she knew there were four of each. Thus, it would be expected that she would actually choose four milk and four tea, leading to fixed values for both the row and column totals within a 2 × 2 contingency table summarizing her choices. Table 6.3 shows a hypothetical outcome of the experiment where two cups were identified incorrectly.

If the lady could not actually differentiate which was added to a cup first, then her selection of milk responses would in reality be four random cups, and the odds of a milk response would be the same regardless of whether milk or tea was truly added first (odds for row 1 equal to odds for row 2). This is the same sampling structure as in the "ball-and-urn" example from earlier, where milk and tea now represent the two colors of the balls. We can therefore calculate the probabilities of all possible 2 × 2 tables under this scenario using the hypergeometric distribution. These probabilities are shown in Table 6.4 where m is the count in the (1,1) cell of the contingency table. We see that the closer to 1 the estimated odds ratio is, the more likely a particular contingency table is observed by random selection. The opposite is true as the estimated odds ratio becomes farther from 1.

Suppose the lady responded as shown in Table 6.3. The estimated odds ratio is 9, indicating that the estimated odds of a milk response are 9 times as large when milk is actually poured first than when tea is poured first. Thus, it might appear that the lady is able to detect what was poured into the cup first. The probabilities given in Table 6.4 measure how likely this or other choices are, if she did not really know the difference. We can use these probabilities then in the context of a standard hypothesis test, where we calculate a p-value as the probability of an event at least as extreme as the one observed. The probability of randomly guessing three or more of the milk cups correctly is $P(M \geq 3) = 0.2286 + 0.0143 = 0.2429$. This probability is fairly large, so we would conclude that the lady's response would not be unusual for someone who was guessing at random. Suppose instead that she had correctly identified all four milk-first cups. Then we would begin to believe that this was a skill on her part, because the probability of doing this by guessing is only $P(M \geq 4) = 0.0143$.

The scenarios described in the last paragraph show how to calculate a p-value for Fisher's exact test of $H_0 : OR \leq 1$ vs. $H_a : OR > 1$. A one-sided alternative hypothesis makes more sense here than a two-sided because only $OR > 1$ means that the lady can correctly detect what was poured into the cup first. In general for a one-sided test, the p-value is calculated as the sum of the table probabilities only in the direction represented by the alternative hypothesis.

For other situations, a two-sided test may be of interest. The p-value for a two-sided test is calculated as the probability of observing an M at least as extreme as the one observed (i.e., observing a contingency table with probability less than or equal to $P(M = m)$). For demonstration purposes, a two-sided test p-value is $0.0143+0.2286+0.2286+0.0143 = 0.4858$ if Table 6.3 is observed.[2]

We remarked earlier that exact inference procedures are often conservative in rejecting the null hypothesis. This is the case for Fisher's exact test as well, due to the highly discrete nature of the hypergeometric distribution. It rarely happens that there is a configuration of table probabilities under H_0 that leads to exactly the stated significance level α. For the one-sided version of the lady tasting tea, there is only a 0.0143 probability of observing a p-value below $\alpha = 0.05$ under independence, so a test at this level actually has a type 1 error rate of 0.0143, and no test with exactly a 0.05 type I error rate is possible.

Example: Tea tasting (Tea.R)

Salsburg (2001) indicates that the lady did correctly respond for all 8 cups, leading to a p-value of 0.0143. To duplicate these calculations in R, we can use the dhyper() function to find the probabilities of the hypergeometric distribution or use fisher.test() to compute the p-value. Below are the calculations:

```
> M <- 0:4
> # Syntax for dhyper(m, a, b, k)
> data.frame(M, prob = round(dhyper(M, 4, 4, 4), 4))
  M prob
1 0 0.0143
2 1 0.2286
3 2 0.5143
4 3 0.2286
5 4 0.0143

> c.table <- array(data = c(4, 0, 0, 4), dim = c(2,2), dimnames =
    list(Actual = c("Milk", "Tea"), Response = c("Milk", "Tea")))
> c.table
       Response
Actual Milk Tea
  Milk    4   0
  Tea     0   4

> fisher.test(x = c.table, alternative = "greater")
```

---

[2] An alternative method sometimes used for calculating two-sided test p-values is simply to take twice the minimum of the two one-sided p-values. For the data in Table 6.3, we calculate
$$2\min\{0.5, P(M \leq 3), P(M \geq 3)\} = 2\min\{0.5, 0.9857, 0.2429\}$$
$$= 0.4858,$$
where 0.5 is included in the brackets to make sure the calculated p-value does not exceed 1. The two presented p-value calculation methods are equivalent when the corresponding distribution is symmetric.

```
        Fisher's Exact Test for Count Data

data:  c.table
p-value = 0.01429
alternative hypothesis: true odds ratio is greater than 1
95 percent confidence interval:
 2.003768      Inf
sample estimates:
odds ratio
       Inf
```

We use the `alternative = "greater"` argument value in `fisher.test()` to specify $H_a : OR > 1$. The default value for the argument is `"two.sided"`. The p-value given matches what we calculated before using the hypergeometric distribution. Note that there is also an estimate of the odds ratio along with a confidence interval. The confidence interval endpoints are calculated using a non-central hypergeometric distribution that does not assume independence (details are available on page 99 of Agresti, 2002).

Multiple hypergeometric distribution for $I \times J$ tables

Fisher's exact test can be extended to tables larger than 2×2 by using the multiple hypergeometric distribution. The row and column totals are still fixed in this setting, but now there are $(I-1)(J-1)$ different random variables. Probabilities for each possible contingency table can still be calculated leading to a test for independence in the same manner as before. The probability of observing a specific set of cell counts, n_{ij}, $i = 1, \ldots, I;\ j = 1, \ldots, J$ is

$$\frac{\left(\prod_{i=1}^{I} n_{i+}!\right) \left(\prod_{j=1}^{J} n_{+j}!\right)}{n! \left(\prod_{i=1}^{I} \prod_{j=1}^{J} n_{ij}!\right)},$$

where we use the notation of Section 3.2 to describe the table counts. Because there may be a very large number of contingency tables, efficient algorithms have been developed to perform the calculations. A related algorithm is discussed in Section 6.2.2.

Example: Fiber-enriched crackers (FiberExact.R, Fiber.csv)

An example in Section 3.2 examined testing for independence between fiber source for crackers and bloating severity in subjects eating the crackers. A Pearson chi-square and LR test for independence resulted in p-values of 0.0496 and 0.0262, respectively. There were a number of very low expected cell counts, so we were somewhat concerned that the large-sample χ_9^2 approximation used with the test statistics might not be close to their actual distributions. Applying Fisher's exact test allows us to avoid this concern. Below is code continuing from that example:

```
> fisher.test(x = diet.table)

        Fisher's Exact Test for Count Data

data:  diet.table
p-value = 0.06636
alternative hypothesis: two.sided
```

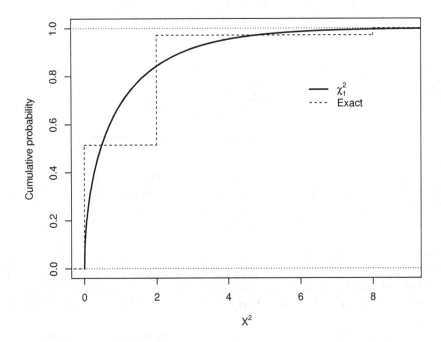

Figure 6.1: Exact and chi-square CDFs for lady-tasting-tea example.

The p-value is 0.0664 which again indicates moderate evidence against independence. It is interesting to note that this p-value is a little larger than the p-values for the Pearson chi-square and LR tests. This could be due to the conservative nature of the Fisher's exact test and/or the χ_9^2 approximation not working as well as we would like. We will perform a more detailed examination of this approximation in the next section.

6.2.2 Permutation test for independence

The actual probability distribution for the Pearson chi-square test statistic X^2 is not exactly $\chi^2_{(I-1)(J-1)}$ in a test for independence. Rather, it is closely related to the hypergeometric (or multiple hypergeometric) distribution when the row and column totals for a contingency table are fixed. Table 6.4 gives each possible value of X^2 that could be observed for the lady-tasting-tea experiment. By summing probabilities over duplicate values of X^2, we obtain the exact PMF under independence as $P(X^2 = 0) = 0.5143$, $P(X^2 = 2) = 0.4571$, and $P(X^2 = 8) = 0.0286$. Figure 6.1 plots the exact cumulative distribution function (CDF) for X^2 along with a χ_1^2 CDF approximation (please see Tea.R for the code). While the χ_1^2 distribution roughly goes through the middle of the exact function, we see there are definitely some differences between the two. For example, the p-value for $X^2 = 2$ is $P(X^2 \geq 2) = 0.4857$ using the exact distribution, while the χ_1^2 approximation yields 0.1572. For $X^2 = 8$, the exact p-value is 0.0286, while the approximate is 0.0047.

Using the hypergeometric distribution is easy for 2×2 contingency tables; however,

Table 6.5: Raw form of the data in Table 6.3. The left table contains the original data. The center and right tables show possible permutations.

Row	Column	Row	Column	Row	Column
1	$z_1 = 1$	1	$z_2 = 1$	1	$z_1 = 1$
1	$z_2 = 1$	1	$z_1 = 1$	1	$z_2 = 1$
1	$z_3 = 1$	1	$z_3 = 1$	1	$z_7 = 2$
1	$z_4 = 2$	1	$z_4 = 2$	1	$z_4 = 2$
2	$z_5 = 1$	2	$z_5 = 1$	2	$z_5 = 1$
2	$z_6 = 2$	2	$z_6 = 2$	2	$z_8 = 2$
2	$z_7 = 2$	2	$z_7 = 2$	2	$z_3 = 1$
2	$z_8 = 2$	2	$z_8 = 2$	2	$z_6 = 2$

when either the sample size or the size of the table is large, the number of tables whose probabilities comprise the p-value may be so great that it is very difficult to compute the p-value exactly. Instead, we can use a general procedure known as a *permutation test* to estimate the exact distribution of a test statistic. A permutation test randomly permutes (re-orders) the observed data a large number of times. Each permutation is done in a way that allows the null hypothesis to be true and the marginal counts in the contingency table remain unchanged.[3] For each permutation, the statistic of interest is calculated. The estimated PMF under the null hypothesis is calculated from these statistics by simply forming a relative frequency table or histogram. P-values are then calculated directly from this PMF.

To test for independence with X^2, we can first decompose the contingency table into its raw data form (see Section 1.2.1); i.e, we create a data set where each row represents an observation and two variables represent the row category and column category of an observation in the contingency table. Table 6.5 (left) shows this format for the data in Table 6.3. We identify each column response in Table 6.5 with its original observation number using $z_j, j = 1, \ldots, 8$. This allows us to see that many distinct permutations of the column responses may lead to the same contingency table. For example, the original data and the center permutation in Table 6.5 lead to the same contingency table, for which $X^2 = 2$. The permutation on the right causes a different contingency table to be formed, for which $X^2 = 0$.

Under independence between the row and column variables, and conditioning on the observed marginal row and column counts, any random permutation of the column observations is equally likely to be observed with the given row observations. There are $8! = 40,320$ different possible permutations of the original data in Table 6.5 with each having a probability of $1/40,320$ of occurring under independence. For this setting, one can show mathematically that there are 20,736 permutations that lead to $X^2 = 0$, 18,432 permutations that lead to $X^2 = 2$, and 1,152 permutations that lead to $X^2 = 8$. Using these permutations, we can calculate the exact distribution for X^2 as $P(X^2 = 0) = 20,736/40,320 = 0.5143$, $P(X^2 = 2) = 18,432/40,320 = 0.4571$, and $P(X^2 = 8) = 1,152/40,320 = 0.0286$. These probabilities are the same as those provided by the hypergeometric distribution!

Frequently, the number of permutations and the number of possible X^2 values are so large that it is difficult to calculate the PMF by evaluating every possible permutation. Instead, we can estimate the p-value similar to what was done in Section 3.2.3 for the multinomial distribution: randomly select a large number of permutations, say B, calculate X^2 on each permutation, and use the empirical distribution of these calculated values as

---

[3] Remember that for hypothesis testing in general, we assume the null hypothesis to be true and find a probability distribution for the test statistic of interest.

an estimate of the exact distribution. This estimate is often referred to as the *permutation distribution* of the statistic X^2. Using this distribution to do a hypothesis test is referred to as a *permutation test*.

Below is an algorithm for finding the permutation estimate of the p-value using Monte Carlo simulation:

1. Randomly permute the column observations while keeping the row observations unchanged.[4]

2. Calculate the Pearson chi-square statistic on the newly formed data. Denote this statistic by X^{2*} to distinguish it from the value calculated for the original sample.

3. Repeat steps 1 and 2 B times, where B is a large number (e.g., 1,000 or more).

4. (Optional) Plot a histogram of the X^{2*} values. This serves as a visual estimate of the exact distribution of X^2.

5. Calculate the p-value as the proportion of the X^{2*} values greater than or equal to the observed X^2; i.e., calculate (# of $X^{2*} \geq X^2$)/B.

Note that the p-value may change from one implementation of these steps to the next. However, the p-values will be very similar as long as a large B is used. Also, note that we could replace the Pearson chi-square statistic with the LRT statistic to perform a different permutation test for independence.

Example: Fiber-enriched crackers (FiberExact.R, Fiber.csv)

The chisq.test() function performs a permutation test using the X^2 statistic when the simulate.p.value = TRUE argument value is specified:

```
> set.seed(8912)
> chisq.test(x = diet.table, correct = FALSE, simulate.p.value =
    TRUE, B = 1000)

        Pearson's Chi-squared test with simulated p-value (based
        on 1000 replicates)

data:   diet.table
X-squared = 16.9427, df = NA, p-value = 0.03896
```

We use the set.seed() function at the beginning so that we can reproduce the same results whenever we run the code. The B = 1000 argument value in chisq.test() specifies the number of permuted data sets. We obtain a p-value of 0.03896, which indicates moderate evidence against independence.[5]

Running the previous code again, but without set.seed(8912), will likely result in a slightly different p-value. Because this p-value is a sample proportion, we can use a confidence interval procedure from Section 1.1.3 to estimate the true probability that

---

[4] Permuting the row observations in addition to the column observations is not necessary. Only one set needs to be permuted to achieve an equal probability for any possible set of pairings.

[5] Note that chisq.test() calculates its p-value as $(1 + \text{\# of } X^{2*} \geq X^2)/(1+B)$, which is slightly different than our p-value formula. This alternative formula treats the original sample as one of the random permutations and ensures that no p-value is ever exactly 0, which otherwise can happen if $X^{2*} < X^2$ for all permutations. Both formulas are widely used to calculate p-values.

would result from using all possible permutations. A 95% Wilson confidence interval is (0.029, 0.053), which does not change our original conclusion about independence. A much larger number of permutations can be used provided that it does not take too long to complete the calculations. Using $B = 100,000$ permutations and the same seed number as before, we obtain a p-value of 0.0455 with a 95% Wilson confidence interval (0.04438, 0.04697).

Unfortunately, the `chisq.test()` function does not provide any information about the permutation distribution other than the relative position of X^2. To obtain the permutation distribution and determine whether a χ_9^2 approximation is really appropriate for X^2, we need to perform some calculations ourselves. We begin doing this by arranging the contingency table counts into the raw form of the data. This is accomplished by first converting `diet.table` to a data frame. Next, we repeat each row of the data frame, using the `rep()` function, corresponding to the number of times a particular `fiber` and `bloat` combination is observed. The raw data are contained in the `set2` data frame as shown below:

```
> set1 <- as.data.frame(as.table(diet.table))
> tail(set1)  # Notice 2 obs. for fiber = both and bloat = high
    fiber  bloat Freq
11    gum medium    3
12   both medium    3
13   none   high    0
14   bran   high    0
15    gum   high    5
16   both   high    2

> set2 <- set1[rep(1:nrow(set1), times = set1$Freq), -3]
> tail(set2)  # Notice 2 rows for fiber = both and bloat = high
     fiber bloat
15.1   gum  high
15.2   gum  high
15.3   gum  high
15.4   gum  high
16    both  high
16.1  both  high

> # Verify data is correct
> xtabs(formula = ~ set2[,1] + set2[,2])
        set2[, 2]
set2[, 1] none low medium high
    none     6   4      2    0
    bran     7   4      1    0
    gum      2   2      3    5
    both     2   5      3    2
> X.sq <- chisq.test(set2[,1], set2[,2], correct = FALSE)
Warning message:
In chisq.test(set2[, 1], set2[, 2], correct = FALSE) :
  Chi-squared approximation may be incorrect
> X.sq$statistic
X-squared
 16.94267
```

The `xtabs()` and `chisq.test()` functions confirm that the raw data were created correctly and the same X^2 value as before was obtained.

To illustrate one permutation of the data, we use the following code:

```
> set.seed(4088)
> set2.star <- data.frame(row = set2[,1], column =
    sample(set2[,2], replace = FALSE))
> xtabs(formula = ~ set2.star[,1] + set2.star[,2])
              set2.star[, 2]
set2.star[, 1] none low medium high
          none    4   3      2    3
          bran    6   4      1    1
          gum     3   3      5    1
          both    4   5      1    2
> X.sq.star <- chisq.test(set2.star[,1], set2.star[,2], correct
    = FALSE)
Warning message:
In chisq.test(set2.star[, 1], set2.star[, 2], correct = FALSE) :
    Chi-squared approximation may be incorrect
> X.sq.star$statistic
X-squared
8.200187
```

The sample() function randomly permutes the column observations, and data.frame() combines the row observations with these permuted column observations to form a permuted data set. Notice that $X^{2\star} = 8.20$. This process is repeated $B = 1,000$ times using the for() function:

```
> B <- 1000
> X.sq.star.save <- matrix(data = NA, nrow = B, ncol = 1)

> set.seed(1938)
> for(i in 1:B) {
    set2.star <- data.frame(row = set2[,1], column =
        sample(set2[,2], replace = FALSE))
    X.sq.star <- chisq.test(set2.star[,1], set2.star[,2],
        correct = FALSE)
    X.sq.star.save[i,1] <- X.sq.star$statistic
  }
There were 50 or more warnings (use warnings() to see the first
    50)

> mean(X.sq.star.save >= X.sq$statistic)
[1] 0.05
> summarize(result.set = X.sq.star.save, statistic =
    X.sq$statistic, df = (nrow(diet.table) - 1) *
    (ncol(diet.table) - 1), B = B)
[1] 0.05
```

The p-value is 0.050 indicating moderate evidence against independence, which agrees with our results from before. Note that the warning messages given by R are not a cause for concern here because these are generated when chisq.test() encounters small expected cell counts, which would be an issue only if we were comparing the

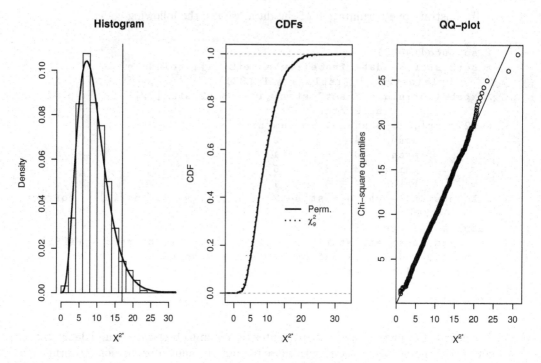

Figure 6.2: Permutation and chi-square distributions. The vertical line in the histogram is drawn at $X^2 = 16.94$.

computed X^{2*} values to the large-sample χ_9^2 distribution.[6] We have written a function named `summarize()` (see corresponding program) to produce the plots in Figure 6.2 and calculate the p-value. We see that the χ_9^2 distribution approximates the permutation distribution very well, leaving little reason to doubt the χ_9^2 approximation for X^2 under independence. For example, the QQ-plot gives the quantiles of a χ_9^2 distribution plotted against corresponding ordered X^{2*} values. These plotted values are generally very similar as evidenced by the vast majority of points lying on 1-to-1 line drawn out from the origin (for example, $\chi^2_{9,0.5005} = 8.348$ is plotted against the 501st ordered X^{2*} value of 8.492).

There are other ways to obtain the permutation distribution. For example, the `boot()` function in the `boot` package provides a convenient way to perform permutation tests in general, where the `sim = "permutation"` argument value needs to be specified in the function call. We provide an example of how to use `boot()` in our corresponding program.

6.2.3 Exact logistic regression

Exact inference methods for logistic regression provide an alternative to the large-sample methods discussed in Chapter 2. In that chapter, we made inferences about a regression

---

[6] If desired, using the `options(warn = -1)` before the `for()` function will prevent the warnings from being printed. This should be done only when one is sure that the warnings are not of concern. Changing the `warn` argument value to 0 allows the warnings to be printed again.

parameter β_j using an approximate normal distribution for the parameter's corresponding estimator $\hat{\beta}_j$. The use of the normal distribution was justified by the fact that MLEs are approximately normally distributed in large samples. Exact methods are useful in situations when small sample sizes may not be large enough for such approximations to work well. Furthermore, exact methods allow estimation and inference when the convergence of maximum likelihood estimates may not occur, such as in complete separation situations (see Section 2.2.7).

The general approach to exact logistic regression is similar to that used in Fisher's exact test. For each parameter we study, we first identify which statistics do or do not contribute to that parameter's estimation. Those statistics that do contribute are called the sufficient statistics for the parameter.[7] As shown in the previous section, there may be many ways to permute the observed data that leave the values of certain statistics unchanged. We apply this approach to inference on regression parameters by holding the explanatory variables fixed and permuting the observed responses. We then focus on the subset of these permutations in which the values of certain sufficient statistics are held constant and look at the distribution of other statistics within that subset. This is referred to as the exact conditional distribution of the latter statistics.

To formalize these ideas, write the joint probability distribution (also the likelihood function) for logistic regression as

$$P(Y_1 = y_1, \ldots, Y_n = y_n) = \prod_{i=1}^{n} P(Y_i = y_i)$$

$$= \prod_{i=1}^{n} \pi_i^{y_i} (1 - \pi_i)^{1-y_i}.$$

Because

$$\pi_i = \frac{\exp(\beta_0 + \beta_1 x_{i1} + \cdots + \beta_p x_{ip})}{1 + \exp(\beta_0 + \beta_1 x_{i1} + \cdots + \beta_p x_{ip})}$$

and

$$1 - \pi_i = \frac{1}{1 + \exp(\beta_0 + \beta_1 x_{i1} + \cdots + \beta_p x_{ip})},$$

we can write the joint probability compactly as

$$P(Y_1 = y_1, \ldots, Y_n = y_n) = \frac{\exp[\sum_{i=1}^{n} y_i(\beta_0 + \beta_1 x_{i1} + \cdots + \beta_p x_{ip})]}{\prod_{i=1}^{n} [1 + \exp(\beta_0 + \beta_1 x_{i1} + \cdots + \beta_p x_{ip})]}. \quad (6.4)$$

Applying Theorem 6.2.10 of Casella and Berger (2002) to Equation 6.4, the sufficient statistic for the regression parameter β_j is $\sum_{i=1}^{n} y_i x_{ij}$, $j = 0, \ldots, p$, where $x_{i0} = 1$ for $i = 1, \ldots, n$ corresponds to the intercept parameter β_0.

Without loss of generality, suppose our interest is in β_p. Inference on β_p is carried out using the exact conditional distribution of its sufficient statistic, say $T = \sum_{i=1}^{n} Y_i x_{ip}$, holding constant the values of the other sufficient statistics at their observed values. Let $t = \sum_{i=1}^{n} y_i x_{ip}$ be the observed value of T and let I be a vector denoting values of the sufficient statistics for all other parameters, $I = \{\sum_{i=1}^{n} y_i x_{ij}$ for $j = 0, \ldots, p-1\}$. Denote a permutation of $y_1, \ldots, y_n$ by $y_1^*, \ldots, y_n^*$. The joint probability for $Y_1, \ldots, Y_n$ conditional on I is shown in Exercise 9 to be

$$P(Y_1 = y_1, \ldots, Y_n = y_n | I) = \frac{\exp(\beta_p \sum_{i=1}^{n} y_i x_{ip})}{\sum_R \exp(\beta_p \sum_{i=1}^{n} y_i^* x_{ip})}, \quad (6.5)$$

---

[7] Further details on sufficiency are available in Section 6.2 of Casella and Berger (2002).

where R is the set of all permutations of $y_1, \ldots, y_n$ such that the values in I remain unchanged.

Next, define U as the number of possible values that $\sum_{i=1}^{n} y_i^* x_{ip}$ can take on as formed by different permutations of $y_1, \ldots, y_n$ in R. Let these distinct values be denoted as $t_1, t_2, \ldots, t_U$. Also, define $c(t_u)$ as a count for the number of permutations in R such that $\sum_{i=1}^{n} y_i^* x_{ip} = t_u$. Then by using Equation 6.5, Exercise 9 also shows that the exact conditional PMF of T given I is

$$P(T = t_u | I) = \frac{c(t_u) \exp(\beta_p t_u)}{\sum_{v=1}^{U} c(t_v) \exp[\beta_p t_v]}, \tag{6.6}$$

for $u = 1, \ldots, U$.

Because finding the set R is not trivial, specific algorithms have been developed to identify the permutations that comprise R (e.g., see Hirji et al., 1987) and to compute the PMF.[8] Once found, this PMF is used in a similar manner as permutation distributions were used in Section 6.2.2. Simply, we can calculate a p-value based on how extreme the observed value t is relative to the PMF. Small probabilities indicate evidence against the null hypothesis. We will demonstrate this process shortly in an example.

To estimate β_p, a conditional maximum likelihood estimate (CMLE) can be obtained by maximizing Equation 6.5. However, when the random variable $\sum_{i=1}^{n} Y_i x_{ip}$ takes on its minimum or maximum possible observed value, then a finite estimate does not exist. In these cases, a *median unbiased estimate* (MUE) can be found instead, and this estimate will always be finite. The estimate is calculated essentially by finding the median of the values $t_1, \ldots, t_u$ using the conditional distribution in Equation 6.6. Specific calculation details are available in Mehta and Patel (1995).

Exact logistic regression is performed in R by the logistiX() function of the logistiX package. This package is currently limited in several ways: (1) only binary explanatory variables can be used, (2) joint tests for regression parameters are not possible, and (3) out-of-memory problems can occur even when using the 64-bit version of R with a large amount of memory available. Alternatively, the elrm() function of the elrm package (Zamar et al., 2007) provides approximations to exact logistic regression using Markov Chain Monte Carlo (MCMC) methods. We will not discuss the details on how the MCMC methods work, but rather refer the reader to the corresponding paper on the package and also to Section 6.6 for an introduction to MCMC methods. The elrm() function avoids the limitations of logistiX() by allowing for quantitative explanatory variables, although multiple trials are still needed for each explanatory variable pattern (EVP; see Section 5.2). The function can also perform specified joint tests of regression parameters, and it is less prone to out-of-memory problems. We discuss how to use both packages in the next example.

Example: Cancer free (MP5.1.R)

Mehta and Patel (1995) provide a series of examples demonstrating exact logistic regression. In particular, Section 5.1 of their paper examines the proportion of individuals (w/n) who were in remission from osteosarcoma over a three-year period. The explanatory variables included within a logistic regression model were lymphocytic infiltration (LI, 1 = present and 0 = not present), gender (gender, 1 = male and 0 = female), and any osteiod pathology (AOP, 1 = yes and 0 = no). Below are the data given first in EVP form (binomial response format needed for elrm()) and then in a

---

[8]Mehta et al. (2000) propose Monte Carlo simulation algorithms where a large number of randomly selected permutations of the data are chosen while holding I fixed. These algorithms are available in some software products, including the same authors' LogXact software, but they are not available in R at this time.

format where each sampled individual is a row of a data frame (Bernoulli response format needed for logistiX()).

```
> # Number disease free (w) out of total (n) per EVP
> set1 <- data.frame(LI = c(0,0,0,0,1,1,1,1), gender =
    c(0,0,1,1,0,0,1,1), AOP = c(0,1,0,1,0,1,0,1), w =
    c(3,2,4,1,5,3,5,6), n = c(3,2,4,1,5,5,9,17))
> sum(set1$n)   # Sample size
[1] 46

> # Transform data to one person per row
> set1.y1 <- set1[rep(1:nrow(set1), times = set1$w), -c(4:5)]
> set1.y1$y <- 1
> set1.y0 <- set1[rep(1:nrow(set1), times = set1$n-set1$w),
    -c(4:5)]
> set1.y0$y <- 0
> set1.long <- data.frame(rbind(set1.y1, set1.y0), row.names =
    NULL)
> head(set1.long)
  LI gender AOP y
1  0      0   0 1
2  0      0   0 1
3  0      0   0 1
4  0      0   1 1
5  0      0   1 1
6  0      1   0 1
> nrow(set1.long)   # Sample size
[1] 46

> ftable(formula = y ~ LI + gender + AOP, data = set1.long)
                    y  0  1
LI gender AOP
0  0      0           0  3
          1           0  2
   1      0           0  4
          1           0  1
1  0      0           0  5
          1           2  3
   1      0           4  5
          1          11  6
```

The contingency table summary given by ftable() shows that everyone who does not have lymphocytic infiltration present is disease free for three years. Thus, there is complete separation, and this causes problems when fitting a logistic regression model through maximum likelihood estimation and glm():

```
> mod.fit <- glm(formula = w/n ~ LI + gender + AOP, data = set1,
    family = binomial(link = logit), weights = n, trace = TRUE,
    epsilon = 1e-8)  # Default epsilon value specified
Deviance = 2.943682 Iterations - 1
Deviance = 2.039803 Iterations - 2

<OUTPUT EDITED>
```

```
Deviance = 1.627772 Iterations - 19
Deviance = 1.627772 Iterations - 20
> round(summary(mod.fit)$coefficients, 4)
             Estimate Std. Error z value Pr(>|z|)
(Intercept)   23.4920 11084.3781  0.0021   0.9983
LI           -21.3842 11084.3781 -0.0019   0.9985
gender        -1.6362     0.9123 -1.7935   0.0729
AOP           -1.2204     0.7712 -1.5825   0.1135
```

Interestingly, glm() concludes its iterative numerical procedure after 20 iterations without any warning messages, because the relative change in the residual deviance is below the default value for epsilon. This can lead one to believe that convergence of the regression estimates has actually occurred. However, one should have strong concerns about the regression estimates due to the extremely large values of $\widehat{Var}(\hat{\beta}_0)^{1/2}$ and $\widehat{Var}(\hat{\beta}_1)^{1/2}$. In fact, when we make the convergence criterion more strict by decreasing the value of epsilon from its default, the estimates of β_0 and β_1 continue to diverge from zero without bound. Therefore, the output given by glm() should not be used.

As an alternative, we first perform exact logistic regression using the logistiX() function with the required Bernoulli response format for the data. There is no formula argument within the function. Instead, the function has an x argument for a data frame that contains the values of the explanatory variables and a y argument for a data frame that contains the values of the response variable. Below is how we estimate the model:

```
> library(package = logistiX)
> mod.fit.logistiX <- logistiX(x = set1.long[,1:3], y =
    set1.long[,4], alpha = 0.05)
> summary(mod.fit.logistiX)
Exact logistic regression

Call: logistiX(x = set1.long[, 1:3], y = set1.long[, 4], alpha =
    0.05)

Estimation method:       LX
CI method:               exact
Test method:             TST

Summary of estimates, confidence intervals and parameter
    hypotheses tests:

  estimates     2.5 %     97.5 % statistic    pvalue cardinality
1 -1.886007      -Inf  0.1614947        19 0.07418173           8
2 -1.547928 -4.023839  0.3627064        16 0.13915271          10
3 -1.156158 -2.997241  0.5113281        12 0.21815408          12

> names(mod.fit.logistiX)
[1] "estout"  "ciout"   "distout" "tobs"    "call"

> mod.fit.logistiX$estout    # beta estimates
  varnum method    estimate
1      1    MUE   -1.886007
2      1    MLE -999.000000
```

```
3         1       LX      -1.886007
4         1       CCFL    -2.414586
5         2       MUE     -1.506908
6         2       MLE     -1.547928
7         2       LX      -1.547928
8         2       CCFL    -1.416228
9         3       MUE     -1.140495
10        3       MLE     -1.156158
11        3       LX      -1.156158
12        3       CCFL    -1.106643
```

For the logistic regression model logit$(\pi) = \beta_0 + \beta_1 \text{LI} + \beta_2 \text{gender} + \beta_3 \text{AOP}$, the estimates are $\hat{\beta}_1 = -1.8860$ (MUE), $\hat{\beta}_2 = -1.5480$ (CMLE), and $\hat{\beta}_3 = -1.1562$ (CMLE). The estout component of mod.fit.logistiX returns the MUEs and CMLEs (shown as method = MLE in the output) for each slope parameter. In addition, estimates given in the method = CCFL rows correspond to Firth's modified likelihood approach, and estimates given in the method = LX rows are those reported in the model summary (CMLEs when the estimate is finite, MUEs otherwise). The latter estimates are what the LogXact software would also report. No estimates of β_0 are given by logistiX()—likely because inference is usually focused on the parameters representing effects of the explanatory variables.

The exact conditional distributions for the sufficient statistics of β_1, β_2 and β_3 are given within the distout component of mod.fit.logistiX. For demonstration purposes, below is the conditional distribution corresponding to β_1:

```
> just.for.beta1 <- mod.fit.logistiX$distout$varnum == 1
> distL1 <- mod.fit.logistiX$distout[just.for.beta1, ]
> distL1$rel.freq <- round(distL1$counts/sum(distL1$counts), 4)
> distL1
  varnum t.stat     counts rel.freq
8      1     19   29445360   0.0371
7      1     20  147312480   0.1856
6      1     21  271271448   0.3417
5      1     22  231819344   0.2920
4      1     23   95325664   0.1201
3      1     24   17473144   0.0220
2      1     25    1204008   0.0015
1      1     26      19448   0.0000
> mod.fit.logistiX$tobs[2]    # Sufficient stat for beta1
[1] 19
> distL1$count[1]/sum(distL1$counts)   # Left-tail test
[1] 0.03709087
> sum(distL1$counts[c(1,6:8)])/sum(distL1$counts)   # Two-tail test
[1] 0.06064205
```

Define $T = \sum_{i=1}^{n} Y_i x_{i1}$. The possible values of T are given in the t.stat column as $19, 20, \ldots, 26$, the counts are the $c(t_u)$ values, and the corresponding conditional probabilities from Equation 6.6 are given in the rel.freq column. The observed value for the test of $H_0 : \beta_1 = 0$ vs. $H_a : \beta_1 \neq 0$ is $t = 19$, and the 2-sided test p-value is the sum of probabilities for outcomes whose conditional probability is no larger than the one for $t = 19$:

$$P(T = 19 | I) + P(T \geq 24 | I) = 0.0371 + 0.0220 + 0.0015 + 0.0000$$
$$= 0.0606.$$

Alternatively, the output from summary() calculates the p-value as

$$2\min\{0.5, P(T = 19|I), P(T \geq 24|I)\},$$

which results in $2P(T = 19|I) = 0.0742$. From these p-values, we see that there is moderate evidence that lymphocytic infiltration has an effect on whether or not there is a three-year, disease-free period. For this specific case, a left-tail test of $H_0 : \beta_1 \geq 0$ vs. $H_a : \beta_1 < 0$ would likely be preferred because lymphocytic infiltration would naturally be expected to lower the disease-free probability. The p-value for this test is 0.0371.

We next perform the MCMC approximation to exact logistic regression using the elrm() function with the required EVP format for the data. As with glm(), a formula argument specifies the model; however, a weights argument is not needed even though the data are in EVP form. Below is the code and output:

```
> library(package = elrm)
> mod.fit.elrm1 <- elrm(formula = w/n ~ LI + gender + AOP,
    interest = ~ LI, iter = 101000, dataset = set1, burnIn = 1000,
    alpha = 0.05)
Generating the Markov chain ...
Progress: 100%
Generation of the Markov Chain required 4 secs
Conducting inference ...
Inference required 0 secs

> summary(mod.fit.elrm1)

Call: [[1]]
elrm(formula = w/n ~ LI + gender + AOP, interest = ~ LI, iter =
    101000, dataset = set1, burnIn = 1000, alpha = 0.05)

Results:
   estimate p-value p-value_se mc_size
LI -1.88707 0.06177    0.00169   1e+05

95% Confidence Intervals for Parameters
      lower      upper
LI    -Inf   0.1668935

> names(mod.fit.elrm1)
 [1] "coeffs"          "coeffs.ci"       "p.values"
 [4] "p.values.se"     "mc"              "mc.size"
 [7] "obs.suff.stat"   "distribution"    "call.history"
[10] "dataset"         "last"            "r"
[13] "ci.level"
```

The interest argument specifies what specific regression parameters are of interest for estimation and inference. Unlike with the logistiX() function, only those explanatory variables identified in this argument have information returned on them. More than one explanatory variable can be specified in the interest argument by using the usual syntax for a formula argument. In this case, a joint test involving the corresponding regression parameters is also performed.

The MCMC methods implemented by elrm() produce a dependent random sequence of possible sufficient statistic values for the regression parameter of interest, where the values in the sample are approximately at the same relative frequency as the probabilities given by Equation 6.6. The `iter` argument within `elrm()` specifies the total sample size, and the `burnIn` argument specifies the size of the initial sample to discard (this helps to ensure that the sample is representative of the corresponding distribution). From the `summary(mod.fit.elrm1)` output, we see that this results in a sample of size 100,000 for inference purposes. The `plot()` method function that can be used with `mod.fit.elrm1` displays the history of the sampling process and the frequency histogram for the distribution of the sufficient statistic values. While we do not show these plots here, we do display the relative frequency distribution below.

```
> mod.fit.elrm1$distribution    # Sorted by second column
$LI
     LI   freq
[1,] 25 0.00152
[2,] 24 0.02270
[3,] 19 0.03755
[4,] 23 0.11882
[5,] 20 0.18875
[6,] 22 0.28704
[7,] 21 0.34362

> sum(mod.fit.elrm1$distribution$LI[1:3,2])    # Two tail test
[1] 0.06177
```

For example, the estimated exact probability of observing $t = 19$ is 0.03755. Using this distribution, we obtain an estimate of the exact p-value for a test of $H_0 : \beta_1 = 0$ vs. $H_a : \beta_1 \neq 0$ as $\hat{P}(T = 19|I) + \hat{P}(T \geq 24|I) = 0.06177$, which is also given in the output produced by the `summary()` function.

Surprisingly, each time that `elrm()` is run, a different sample is taken even if the `set.seed()` function is used just prior to running it. This prevents one from being able to reproduce the same results with multiple invocations, which we find quite disappointing. However, as long as a large sample size is taken, the variability in the results from one invocation to another of `elrm()` should be small.

Within the corresponding program for this example, we provide code implementing the modified likelihood methods proposed by Firth (1993) using the `logistf()` function discussed in Section 2.2.7. The estimated logistic regression model is $\text{logit}(\hat{\pi}) = 4.2905 - 2.4611\text{LI} - 1.4153\text{gender} - 1.1039\text{AOP}$, which leads to similar conclusions as with the exact methods. Also, the hypothesis test involving $H_0 : \beta_1 = 0$ vs. $H_a : \beta_1 \neq 0$ results in a p-value of 0.0309, which again leads to similar conclusions as with exact logistic regression. For this example, the p-values obtained by testing the significance of each explanatory variable are a little smaller for the modified likelihood approach than for exact logistic regression.

Both exact logistic regression and the modified likelihood approach of Firth (1993) can replace standard maximum likelihood estimation when complete separation occurs. Heinze (2006) discusses the relative merits of the two procedures. Heinze notes that the modified likelihood approach can be applied in situations when explanatory variables are truly continuous or when there are few observations per EVP. Exact methods generally cannot be used in these situations due to problems with evaluating the conditional distribution of a sufficient statistic (there may be only one permutation of $y_1, \ldots, y_n$ that satisfies I). In a very limited simulation study, Heinze (2006) also showed that both the modified likelihood and exact methods lead to conservative tests. However, the modified likelihood approach can produce more powerful tests. Heinze goes on to show that calculating p-values for exact tests using the mid p-value method (often used to limit the conservativeness of a test involving a discrete distribution, see Exercise 10) allows the two tests to have comparable power. However, tests based on a mid-p-value are no longer guaranteed to reject at a rate less than or equal to the type I error level. This mid p-value is available in the `ciout` component of a model fit object produced by `logistiX()`.

It is also important to remember that inference using the modified likelihood approach requires large samples in order for distributional approximations to be accurate. The exact inference methods do not have this requirement. While Heinze (2006) shows that the inferences based on the modified likelihood approach generally work as intended, these conclusions were based on a limited, although realistic, set of simulations where moderately sized samples were used.

6.2.4 Additional exact inference procedures

Exact inference methods are used in many other places involving categorical data, and a large number of R packages include functions to perform these methods. A general survey of current R packages can be found simply by doing a search on "exact" at the CRAN web page that lists all R packages: http://cran.r-project.org/web/packages/available_packages_by_name.html. For example, the `exact2x2` package provides a function named `exact2x2()` which not only performs Fisher's exact test but also provides a Blaker-based alternative that is less conservative (the Blaker interval for a single binomial probability was discussed in Section 1.1.2), and the `mcnemar.exact()` function that provides an exact version of McNemar's test discussed in Section 1.2.6. Also, the `exactLogLinTest` package can perform exact inference for Poisson regression models.

6.3 Categorical data analysis in complex survey designs

All of the statistical methods that we have examined so far assume that we have a simple random sample of units (animals, free throws, etc.) from an infinite population. This assumption allows us to form our usual models: the binomial, the Poisson, and the multinomial. These models are convenient for the mathematical development of analysis methods and sampling distributions, but they may not reflect how data are actually gathered. More complex methods of sampling are often used in surveys, particularly when they are done on human populations. For example, large national surveys, like the Center for Disease Control and Prevention's National Health Interview Survey and Statistics Canada's General Social Survey, feature complexities such as stratification, clustering, and unequal-probability sampling.

When complex sampling designs are used, the resulting observations are no longer independent, and they may have other features that further invalidate our usual models. The purpose of this section is to show how to perform many of the same types of statistical analyses from past chapters when sampled units are selected using a complex sampling design. We present methods for computing confidence intervals for a proportion, performing tests for independence, and fitting regression models.

6.3.1 The survey sampling paradigm

Sampling designs are often constructed to increase the convenience of the sampling process and to reduce variability of particular statistics of interest. We briefly discuss below the specific features of a sampling design that enable us to accomplish these goals. See Lohr (2010), Thompson (2002), or other similar books on survey sampling for more details on the uses of these features.

Stratification and Clustering

Two key features in complex survey designs are *stratification* and *clustering*. Stratification is the labeling of population units into groups, called strata, according to some similar, quantifiable feature(s) prior to sample selection. For example people in a population could be stratified by gender, income, marital status, neighborhood, or any number of possible features. Sampling is then done separately within each stratum according to some probability-based design. The ability to control sample sizes within each stratum can ensure that estimated quantities from each stratum have reasonable precision.

Clustering also represents a grouping of units, but for a completely different purpose. Clusters are generally formed for convenience: units are grouped in such a way that those within a cluster are easily sampled together. For example, in door-to-door surveys it is much easier to sample ten households in a particular neighborhood than to select one household from each of ten different neighborhoods. The drawback with clustering is that the units within a cluster tend to be similar in some ways, and hence responses within a cluster tend to be similar, and hence positively correlated. This means that a sample consisting of only one cluster is not an adequate representation of a whole population, while a sample consisting of multiple clusters contains groups of data that are not independent.

Thus, stratification makes sampling a bit less convenient but improves precision, while clustering makes sampling more convenient but diminishes precision. In practice, these tools are often used together, and often in ways that make it very difficult to determine a good distributional model to represent the data. The usual binomial, multinomial, and Poisson models that assume independent observations are no longer valid and may, in fact, very poorly represent the actual structure of the sample.

In order to summarize the impact that a sampling design has on the variance of an estimated parameter, *design effects* are often calculated for a survey. The design effect, or *deff*, is the ratio of the variance of the estimate under the survey design to the variance that the same sample size would provide under simple random sampling. For example, under simple random sampling from a Bernoulli distribution, we showed in Section 1.1.2 that the variance for the sample proportion is $\widehat{Var}_{SRS,n}(\hat{\pi}) = \widehat{Var}(\hat{\pi}) = \hat{\pi}(1-\hat{\pi})/n$. Under a complex survey design, the variance for the sample proportion, $\widehat{Var}_{DES,n}(\hat{\pi})$, could be something quite different. The deff measures $\widehat{Var}_{DES,n}/\widehat{Var}_{SRS,n}$. In most surveys deffs are greater than 1.

Survey weights

Analysis of complex survey data is complicated by the presence of *survey weights*. Surveys—especially those with stratification—are often conducted with unequal probability sampling, meaning that not all units have the same probability of being included in the sample. For example, suppose that a regional government wants to build a new prison in a small town and wants to gauge support for the project by surveying the population. Further, suppose that 1000 people live in the town and that there are 1,000,000 people in the government's region outside the town. Naturally, they want to be certain to include both town residents and other regional residents in the sample. If a sample of total size 50 were to be drawn by simple random sampling, there is approximately a 95% chance that *no* town people would be selected. Instead, it might make sense to sample, say, 10 people from the town and 40 from the rest of the region, in order to ensure an adequate sample size to estimate parameters for both groups. Then each town person in the sample represents 100 members of the town population, while each outsider in the sample represents 25,000 members of the rest of the population. For constructing estimates of population totals, multiplying town responses by 100 and outsider responses by 25,000 will tend to give unbiased estimates of population totals for the entire region. This is the basis for survey weights: for each sampled unit, its weight estimates how many members of the population that unit represents.[9] Weights can be constant if all members of the overall population have equal probability of being included in the sample, or they can vary considerably according to stratum sizes and other features of the survey design. Weights are often scaled so that their sum across the entire sample is approximately equal to the population size. Sometimes they are instead scaled so that their sum is equal to the sample size, in which case they vary around 1. The scales are important only for the interpretation of count estimates that they may produce. One can convert weights or counts from one scale to the other by multiplying the weights by a constant that depends on the population and sample sizes. We generally assume in this section that weights are scaled to produce population counts.

6.3.2 Overview of analysis approaches

Two main approaches have emerged for analyzing data from complex surveys. They are called *design-based* and *model-based* analysis. They have different uses, different assumptions, and can give substantially different results, so it is important to assess first which method is most appropriate for a given problem. Binder and Roberts (2003, 2009) give excellent discussions of the differences between the two approaches. Design-based inference assumes that there exists a fixed, finite population from which the sample was drawn and about which inferences are to be made. Model-based inference assumes that the population is a temporary thing that is constantly changing and itself is sampled from a "super-population" according to some model.

As an example to highlight the differences between these two approaches, suppose one does a survey on a sample selected from first-year students at a particular university. If the interest is specifically in *that* year's first-year students at *that* university, then a design-based approach would be appropriate. If, instead, the interest is to learn about first-year students more generally, and the results will be applied to other years or other universities, then the first-year students at the selected university are themselves a sample from this larger super-population. A model-based approach would likely be more appropriate in this case. Using the wrong approach can lead to estimates that are biased and inconsistent, meaning that they estimate the wrong population parameter. Also, standard errors calculated under the wrong approach can differ drastically from the correct ones, giving inappropriate confidence intervals or test results. Selecting the appropriate approach is therefore a crucial first step.

Design-based analysis

If units are selected from a fixed population with known probabilities, then these *inclusion probabilities* can be used to construct the survey weights. The methods for constructing weights are covered in standard texts on survey sampling (e.g., Lohr, 2010 or Heeringa

---

[9]Survey weights may also include adjustments for missing data and/or nonresponse. See Lohr (2010) for details.

et al., 2010) and are beyond our present scope. Given the weights, the general approach to design-based inference with categorical data proceeds as follows.

First, the weights are used to form weighted counts of responses in each response category by simply summing the weights for all units in that category. These weighted counts estimate the corresponding population total counts. The variance matrix of the weighted counts is found using one of many possible methods that are standard in survey analysis (see, e.g., Korn and Graubard, 1999 or Lohr, 2010). Next, any statistics or formulas that are normally written as functions of sample counts—including most statistics covered in Chapters 1–4—are instead written in terms of the weighted counts. The delta method is used as described in Appendix B or in Korn and Graubard (1999) to "linearize" these functions with respect to the weighted counts. Such linear approximations make finding the variance of any statistic relatively simple by using Equation B.7. Asymptotic normality of many statistics is established using forms of the central limit theorem that are appropriate for finite population problems (see, e.g., Binder, 1983). Wald-type tests and confidence intervals are usually constructed from the resulting estimates. Additional analyses based on Pearson and likelihood-ratio chi-square test statistics can be replaced by similar calculations on the weighted estimates. Approximate sampling distributions for the statistics are derived (Imrey et al., 1982; Rao and Scott, 1981, 1984) that account for both the variability of weights and the correlations among observations that are created by the survey design.

Alternatively, *resampling methods*—in particular, the jackknife, the bootstrap, and balanced repeated replication—are often used to generate replicate sets of sampling weights, especially in large government-based surveys. The statistic of interest is calculated on each replicate set, so that the variability among these replicate statistics can be used to estimate the standard error of the statistic. This avoids the need to derive approximations using the delta method. See texts such as Korn and Graubard (1999) or Lohr (2010) for details.

It should be noted that hypothesis tests are not commonly performed in strict design-based analysis. This is because it is unreasonable to expect null hypotheses to be exactly true for the finite population to which they are to apply. For example, a null hypothesis that a population proportion is 0.5 cannot possibly be true if the total size of the population is odd! When we do hypothesis tests in a design-based analysis, then, we are generally implying that there is a superpopulation that created the current finite population (e.g., there is a constant flow of people into and out of various age groups), and the hypotheses are meant to apply to this superpopulation. The decision to use design-based inference in this case can still be justified when we do not believe that a model can adequately account for all features of the design. See Binder and Roberts (2009) for an excellent discussion.

Model-based analysis

Using methods from Section 6.5, it is possible to explicitly account for clustering by including a corresponding random effect factor in a model. Furthermore, strata can be included as fixed effects in a model to allow separate estimates to be calculated for each stratum. Thus, it may be possible to account for many of the features of a complex sampling design by augmenting a model appropriately. If the augmented model is correctly specified, then the usual analysis procedures (e.g., generalized linear (mixed) models) meant for simple random samples can be used, and the fact that the data came from a different design is ignored. Estimates and standard errors obtained from the model are then appropriate estimates of their corresponding population quantities. Binder and Roberts (2009) suggest comparing the sizes of design-based and model-based standard errors. If they are similar, then the model seems to account for features of the design successfully and resulting inferences are more likely to be reliable than if the standard errors differ substantially.

Comparison

Design-based estimates of population counts are unbiased, and subsequent estimates based on these counts can be expected to be unbiased, or nearly so, depending on the structure of the estimate. Model-based estimates that ignore weights can be biased—sometimes very badly—as estimates of parameters for the finite population from which a sample is drawn. However, design-based estimates tend to have higher variability than do model-based estimates, and the difference becomes worse the more variability there is in the weights. There is no universal preference for one approach over the other, as the bias vs. variance trade-off is very much case-dependent. Nonetheless, Korn and Graubard (1999, p. 143) note: "In summary, weighted estimation used with variance estimation based on the survey design will yield appropriate analysis in most situations."

Because model-based analyses use existing techniques that we cover elsewhere, we discuss only design-based analysis methods here. We begin our discussion by introducing an example and the R package that is used for design-based analyses in this section.

Example: NHANES 1999-2000 (SurveySmoke.R, SmokeRespAge.txt)

The National Health and Nutrition Examination Survey (NHANES) from the U.S. Centers for Disease Control and Prevention is a program that has run a series of surveys on various health issues since the 1960s. Among the many questions asked in 1999–2000 are questions about lifetime tobacco use and about respiratory symptoms. Specifically, subjects are asked whether they have used at least 100 cigarettes, or smoked a pipe, smoked a cigar, used snuff, or used chewing tobacco each at least 20 times. The responses are recorded as a binary variable for each of the five tobacco items. Subjects are also asked about respiratory symptoms: a persistent cough, bringing up phlegm, wheezing/whistling in chest, and dry cough at night. Presence of each of the four symptoms is also recorded as a separate binary response for each symptom. Finally, there are weights associated with each subject, and an additional set of 52 jackknife replicate weights that are to be used for variance estimation (Section 6.3.4 gives an example of how these are used). The population about which inferences can be drawn is all adults aged 20 or older in the United States at the time the survey was conducted. See http://www.cdc.gov/nchs/nhanes.htm for a full description of the NHANES program, questionnaires, and sampling methods used. We will use these data to demonstrate various forms of analyses by examining relationships among tobacco-use and respiratory-symptom variables that might naturally be of interest to a researcher in public health.

The code and output below shows a small portion of the data. Age is listed first. Next come the five smoking indicators (prefixed by sm_), followed by the four respiratory symptom indicators (prefixed by re_). The survey weights supplied with the data are next (wtint2yr), and last are the 52 sets of jackknife replicate weights, jrep01 – jrep52 (only two are shown below). Notice that these replicate weights are already on the same scale as the original survey weights; in some surveys, these replicate weights are merely "adjustment factors" (numbers varying around 1) that need to be multiplied by the original weights to create new combined weights. It is apparent from the weights that some subjects represent substantially more members of the population than others. For example, observation #5 represents 92,603 individuals in the population, and observation #6 represents just 1,647.

```
> smoke.resp <- read.table(file =
    "C:\\data\\SmokeRespAge.txt", header = TRUE, sep = " ")
> smoke.resp[1:6, 1:13]
```

```
  age sm_cigs sm_pipe sm_cigar sm_snuff sm_chew re_cough
1  77       0       1        1        0       0        0
2  49       1       1        1        0       1        0
3  59       1       0        0        0       0        0
4  43       1       0        0        0       0        0
5  37       0       1        1        0       0        0
6  70       1       0        0        0       0        0
  re_phlegm re_wheez re_night   wtint2yr    jrep01    jrep02
1         0        0        0  26678.636  26923.62  26824.718
2         0        0        0  91050.847  92772.79  91672.817
3         0        0        0  22352.089  22495.11  24238.166
4         0        0        0  21071.164  21209.55  22343.413
5         0        1        0  92603.187  94354.49  93235.761
6         0        0        0   1647.446   1681.59   1648.714
> nrow(smoke.resp)   # Sample size
[1] 4852
```

The survey package in R, authored by Thomas Lumley, performs a wide range of statistical analyses for data from a complex sampling design. The package contains functions that perform several common analyses for categorical data. In particular, there are capabilities for the analysis of single proportions, contingency tables, and logistic, Poisson, and certain multinomial regression models for ordinal responses. Multinomial regression models for nominal responses and automated model selection techniques are not available, and there is very limited capacity for model diagnostics. Lumley (2011) describes the package in much more detail than we can cover here. Also, the package is frequently updated, so one should check the documentation for this package before using the programs below in case functions have been added that supersede those in our programs. Version 3.29.5 was used for the calculations in this section.

After loading the survey package, we need to convert our data into an object that contains information about the design so that the analysis functions can properly calculate statistics and their variances. We use svrepdesign() here because our design has replicate weights; other functions allow the user to instead specify stratification and clustering identifiers that can be used in variance estimation. Arguments identify the columns of the data frame that contain the analysis variables (data), the original survey weights (weights), and the replicate weights (repweights). The description of the survey design (http://www.cdc.gov/nchs/data/nhanes/guidelines1.pdf) recommends using the leave-one-out version of jackknife, which is selected here using the argument value type = "JK1". As noted above, the replicate weights are on the same scale as the survey weights, so combined.weights = TRUE. The scale = 51/52 relates the presence of 52 replicate weights to the formula for leave-one-out jackknife estimates of variance (Lohr, 2010, p. 380). The code below creates an object of the class svyrep.design, which can be used for all analyses within survey.

```
> library(survey)
> jdesign <- svrepdesign(data = smoke.resp[, c(1:10)],
    weights = smoke.resp[,11], repweights =
    smoke.resp[,12:63], type = "JK1", combined.weights =
    TRUE, scale = 51/52)
> class(jdesign)
[1] "svyrep.design"
```

6.3.3 Weighted cell counts

The first and most fundamental step in an analysis of categorical data from a complex survey is estimating the population count of units in each category. Not only might estimating these counts be one of the main goals of the analysis, but also any further calculations, such as those described in Sections 6.3.4–6.3.6, are based upon these counts.

Suppose that the population consists of N units and that the response Y has categories $1, \ldots, I$, containing $N_1, N_2, \ldots, N_I$ members of the population, respectively, where $\sum_{i=1}^{I} N_i = N$. Let the sample consist of n units selected from the population according to some design. For each sampled unit, define its survey weight to be w_s, $s = 1, \ldots, n$ and its observed response to be y_s. For example, on p. 384 we show information on sampled units $s = 1, \ldots, 6$ from a sample of $n = 4852$ for the NHANES data set. The wtint2yr variable in that example corresponds to w_s.

Because the survey weight w_s is the number of population units represented by sampled unit s, the estimate of N_i is simply the sum of weights for all sampled units in category i:

$$\hat{N}_i = \sum_{s=1}^{n} w_s I(y_s = i), \ i = 1, \ldots, I \qquad (6.7)$$

where $I(y_s = i)$ indicates whether or not unit s is in category i (1 for yes, 0 for no). We can similarly estimate the population total as $\hat{N} = \sum_{s=1}^{n} w_s = \sum_{i=1}^{I} \hat{N}_i$.

For convenience, collect the weighted cell counts into a vector, $\hat{\mathbf{N}} = (\hat{N}_1, \hat{N}_2, \ldots, \hat{N}_I)'$. The covariance matrix of $\hat{\mathbf{N}}$, $\widehat{Var}(\hat{\mathbf{N}})$, not only measures the variability of $\hat{\mathbf{N}}$, but also serves as the foundation for variances, confidence intervals, and test-statistic calculations for functions of $\hat{\mathbf{N}}$. This covariance matrix can be found using linearization, jackknife, and bootstrap methods as described earlier and discussed in detail in standard texts on analysis of surveys (e.g., Lohr, 2010). We generally do not provide the details for these methods here, but instead rely on functions from the survey package to complete the calculations when possible.

Degrees of freedom for $\widehat{Var}(\hat{\mathbf{N}})$, say κ, are not always straightforward to find. The default is to use (# of clusters)−(# of strata), although Rust and Rao (1996) indicate that this is generally an overestimate. In the NHANES example, the 52 replicate weights were formed from 52 random subgroups ignoring the design's clustering and stratification. This leads to $\kappa = 52 - 1 = 51$ degrees of freedom for $\widehat{Var}(\hat{\mathbf{N}})$.

Example: NHANES 1999-2000 (SurveySmoke.R, SmokeRespAge.txt)

> For each of the 9 binary responses for tobacco use and respiratory symptoms, we can estimate the population total with the weighted count of positive responses. The code below demonstrates the calculations for the first tobacco item, cigarette use. The result is the estimated number of members of the population who have smoked at least 100 cigarettes in their lifetimes.

```
> wt_cig <- smokerespj$sm_cigs * smokerespj$wtint2yr
> totcigwt <- sum(wt_cig)
> totcigwt
[1] 94480990
```

The variance of this estimate can be found by repeating the calculation on each of the 52 sets of jackknife replicate weights. Let $\hat{N}_i^{(r)}$ be the estimated population count for category i from replicate weight set r. Then the variance is found from the formula for leave-one-out jackknife estimates of variance (Lohr, 2010, p. 381),

$$\widehat{Var}(\hat{N}_i) = [(R-1)/R] \sum_{r=1}^{R} (\hat{N}_i^{(r)} - \hat{N}_i)^2. \tag{6.8}$$

More generally, the variance of a full vector of counts is $\widehat{Var}(\hat{\mathbf{N}}) = [(R-1)/R] \sum_{r=1}^{R} (\hat{\mathbf{N}}^{(r)} - \hat{\mathbf{N}})(\hat{\mathbf{N}}^{(r)} - \hat{\mathbf{N}})'$. Equation 6.8 can be calculated in R as follows:

```
> Nhat.reps <- numeric(length = 52)
> for(r in 1:52){
    Nhat.reps[r] <- sum(smoke.resp$sm_cigs *
       smoke.resp[,j+11])
  }
> sum.sq <- var(Nhat.reps) * 51
> var.tot <- sum.sq * (51/52)
> sqrt(var.tot)   # Estimated standard deviation
[1] 2483739
```

The function svytotal() performs these calculations automatically:

```
> svytotal(x = ~ sm_cigs, design = jdesign)
            total       SE
sm_cigs  94480990  2483739
```

Documentation for NHANES 1999-2000 indicates that the jackknife procedure used to estimate variances may produce an underestimate of the true sampling variance for some statistics. It cautions to be careful when interpreting "marginally significant" results when these variances are used in hypothesis tests or confidence intervals.

6.3.4 Inference on population proportions

Define the population proportions for each category (also called *cell probabilities*) as $\pi_i = N_i/N$, $i = 1,\ldots, I$. The population proportions are estimated as

$$\hat{\pi}_i = \hat{N}_i/\hat{N}, \ i = 1,\ldots, I.$$

We use $\hat{N}$ in the denominator even when N is known, so that the estimated proportions sum to 1 across all categories.

Note that $\hat{\pi}_i$ is a ratio of two weighted estimates based on correlated observations. Its variance estimate is not simply the usual formula for the variance of a binomial proportion, $\hat{\pi}(1-\hat{\pi})/n$, as given in Equation 1.3. Instead, using the delta method from Appendix B.4.2, one can show that

$$\widehat{Var}(\hat{\pi}_i) = \frac{\widehat{Var}(\hat{N}_i) + \hat{\pi}_i^2 \widehat{Var}(\hat{N}) - 2\hat{\pi}_i \widehat{Cov}(\hat{N}_i, \hat{N})}{\hat{N}^2}, \tag{6.9}$$

where the needed design-based estimates of variances and covariances are all found from $\widehat{Var}(\hat{\mathbf{N}})$. Alternatively, this variance can be calculated directly using a replication method. For example, the leave-one-out jackknife with R replicate weights results in the formula

$$\widehat{Var}(\hat{\pi}_i) = [(R-1)/R] \sum_{r=1}^{R} (\hat{\pi}_i^{(r)} - \hat{\pi}_i)^2, \tag{6.10}$$

where $\hat{\pi}_i^{(r)}$ is estimated from the rth set of replicate weights.

Methods for forming confidence intervals for the true proportions are analogous to those from Section 1.1.2. The Wald interval provides a simple calculation,

$$\hat{\pi}_i \pm Z_{1-\alpha/2} \sqrt{\widehat{Var}(\hat{\pi}_i)}$$

but with deficiencies noted in Section 1.1.2. Because $\widehat{Var}(\mathbf{N})$ is based on a sum-of-squares calculation, a $t_{\kappa,1-\alpha/2}$ critical value can be used in place of $Z_{1-\alpha/2}$. However, this makes little practical difference unless $\kappa < 30$.

Kott and Carr (1997) propose a more accurate confidence interval for $\hat{\pi}_i$ by using a modification of the Wilson score interval given by Equation 1.4. Making an analogy to the binomial variance, $\hat{\pi}(1-\hat{\pi})/n$, they define the *effective sample size* for estimating proportions π_i, $i = 1, \ldots, I$, to be the value $n_i^* = \hat{\pi}_i(1-\hat{\pi}_i)/\widehat{Var}(\hat{\pi}_i)$. Then a $100(1-\alpha)\%$ confidence interval for π_i is

$$\frac{2n_i^*\hat{\pi}_i + t^2_{\kappa,1-\alpha/2} \pm t_{\kappa,1-\alpha/2} \sqrt{t^2_{\kappa,1-\alpha/2} + 4n_i^*\hat{\pi}_i(1-\hat{\pi}_i)}}{2(n_i^* + t^2_{\kappa,1-\alpha/2})} \tag{6.11}$$

Surveys are often more concerned with estimating population quantities rather than with testing hypotheses, but should a test of $H_0 : \pi_i = \pi_{i0}$ be needed, the usual Wald tests are used (see Section 1.1.2). The test statistic is

$$Z = \frac{\hat{\pi}_i - \pi_{i0}}{\sqrt{\widehat{Var}(\hat{\pi}_i)}}$$

where the quantities $\hat{\pi}_i$ and $\widehat{Var}(\hat{\pi}_i)$ are estimated from the survey using methods described above. The decision rule is found by comparing Z to the appropriate critical value from the standard normal (or t_κ) distribution according to the alternative hypothesis being tested.

Example: NHANES 1999-2000 (SurveySmoke.R, SmokeRespAge.txt)

We estimate the proportion of the population who have smoked at least 100 cigarettes in their lifetimes using a 95% confidence interval. The svymean() function estimates means and standard errors while accounting for the survey design. Since proportions are means of binary variables, we can use this function to estimate π_i.

```
> cigprop <- svymean(x = ~ sm_cigs, design = jdesign)
> cigprop
           mean     SE
sm_cigs 0.49434 0.0131
```

Additional topics 389

We find $\hat{\pi}_i = 0.49$, so nearly half of the population had used at least 100 cigarettes. The estimated standard error is $\widehat{Var}(\hat{\pi}_i)^{1/2} = 0.013$. The svymean() function recognizes the contents of the design jdesign and uses Equation 6.10 rather than Equation 6.9 for this calculation. A t-based confidence interval is found from

```
> confint(object = cigprob, level = 0.95, df = 51)
             2.5 %     97.5 %
sm_cigs  0.4679677  0.5207125
```

A Wald interval could be found from the code above by omitting the df argument. A Wilson score interval requires a few more calculations. We must first save out the estimated proportion and standard error, and then calculate the effective sample size. Then the interval is calculated according to Equation 6.11.

```
> pihat <- coef(cigprop)

> # Effective sample size for Wilson Score interval
> eff_sample <- pihat*(1-pihat)/vcov(cigprop)
> round(eff_sample, 3)
         sm_cigs
sm_cigs 1448.545

> tcrit <- qt(p = c(0.025, 0.975), df = 51)
> # Wilson Score interval
> j.Wilsonci_wt <- (((2*eff_sample*pihat + tcrit[2]^2) +
     tcrit*sqrt(tcrit[2]^2 + 4*eff_sample*pihat*(1-pihat)))
     / (2*(eff_sample + tcrit[2]^2)))
> round(j.Wilsonci_wt, digits = 3)
[1] 0.468 0.521
```

The confidence intervals are essentially the same, roughly 0.468 to 0.521, due to the large sample size for this survey.

6.3.5 Contingency tables and loglinear models

Methods for analyzing contingency tables are introduced in Sections 1.2, 3.2, and 4.2.4. Simpler analyses for two-way contingency tables focus on tests for independence between the row and column variables. Loglinear models for two-way and larger tables allow for more detailed analyses, including tests for various forms of association or other model features and estimates of proportions or odds ratios.

The same analyses are often done when the data arise from a complex survey. The main difference is in how the calculations are performed. First, the table is constructed from survey-weighted cell counts $\hat{\mathbf{N}}$ or corresponding proportions rather than the sample counts or proportions. For example, a loglinear model for two factors under independence is

$$\log(N_{ij}) = \beta_0 + \beta_i^X + \beta_j^Z, \ i = 1, \ldots, I, \ j = 1, \ldots, J,$$

which is simply Equation 4.3, except modeling $\log(N_{ij})$ instead of $\log(\mu_{ij})$. Second, $\widehat{Var}(\hat{\mathbf{N}})$ is calculated based on the survey design rather than a Poisson model. Finally, the error degrees of freedom for any test or confidence interval from a model must be adjusted to $\kappa - p - 1$, where p is the number of parameters estimated in the model. Note that this last condition limits the size of models that can be considered without using special estimation techniques (Korn and Graubard, 1999, Section 5.2).

We start by considering a two-way contingency table formed from variables X with I categories and Y with J categories. Using the same notation as in Section 3.2.1, let π_{ij} be the proportion of the population satisfying $X = i, Y = j$. Testing independence of X and Y implies $H_0 : \pi_{ij} = \pi_{i+}\pi_{+j}$, where π_{i+} and π_{+j} are the population marginal proportions for $X = i$ and for $Y = j$, respectively. Ordinarily, a Pearson or likelihood ratio (LR) test for independence would be used to compare the observed counts to those expected under the null hypothesis, as discussed in Sections 1.2.3, 3.2.3, and 4.2.4. In the context of a loglinear model for survey data, these two sets of counts are obtained from separate model fits to weighted data. The "observed" counts come from a saturated loglinear model, while the estimated expected counts are from a model fit that satisfies the null hypothesis. While it is straightforward to compute these statistics with weighted survey data, finding the probability distribution of the resulting statistics, which must account for the survey design, is not as simple.

Tests of independence: Rao-Scott methods

The use of a χ^2 distribution for Pearson and LRT statistics is typically justified by the appropriateness of the underlying Poisson or multinomial sampling model for the data. However, these sampling models may be poor representations of the data from complex survey designs, so the resulting χ^2 distribution should not be expected to accurately approximate the true probability distribution of any test statistic. Indeed, there is much evidence to indicate that naive use of χ^2 distributions with Pearson statistics calculated on survey data can result in very poor analyses (e.g., Rao and Scott, 1981; Scott and Rao, 1981; Thomas and Rao, 1987; Thomas et al., 1996). In particular, in many designs the test tends to reject the null hypothesis much more often than the specified α level.

Because Pearson statistics are used so broadly in categorical data analysis, Rao and Scott (1981, 1984) developed corrections to improve their performance in complex survey designs. The corrections are similar to the well-known Satterthwaite (1946) corrections that are common in linear models analysis. Let X^2 be the Pearson test statistic for a particular test, and suppose that X^2 would have ν degrees of freedom under simple random sampling (e.g., in a two-way contingency table, $\nu = (I-1)(J-1)$). Rao-Scott methods compare the mean and variance of X^2 under the actual sampling design to the means and variances from members of the χ^2-distribution family. Using the fact that the mean and variance of the χ^2_ν distribution are ν and 2ν, respectively, relatively simple adjustments are then made to the test statistic and to the degrees of freedom that improve how well a χ^2 distribution matches the actual distribution of X^2.

First-order correction The simplest form of a Rao-Scott correction ensures that the mean of the distribution for the test statistic and the mean of the chosen sampling distribution are the same. This is called a *first-order correction*, because it is based on matching the means (first moments) of the two distributions. For ease of notation and to allow extension to larger loglinear models, let K be the number of cells in the table and relabel the probabilities π_{ij}, $i = 1, \ldots, I$; $j = 1, \ldots, J$ as π_k, $k = 1, \ldots, K$. For each π_k let its value under the null hypothesis be denoted by π_{0k}. Recalling the definition of a deff from Section 6.3.1, let d_k^0 be the deff corresponding to $\hat{\pi}_k$ under the null hypothesis:

$$d_k^0 = nVar_0(\hat{\pi}_k)/[\pi_{0k}(1-\pi_{0k})]$$

where $Var_0(\hat{\pi}_k)$ is the variance of $\hat{\pi}_k$ under the null hypothesis. Suppose that each probability estimate has the same deff, $d_k^0 = d^0$, and, further, that the survey-estimated covariances follow $\widehat{Cov}(\hat{\pi}_{k_1}, \hat{\pi}_{k_2}) = d^0(-\pi_{0k_1}\pi_{0k_2})$ for all pairs (k_1, k_2). Under these conditions, X^2 has a distribution that is asymptotically equivalent to that of $d^0 W$, where $W \sim \chi_\nu^2$. So the distribution of X^2/d^0 is approximately χ_ν^2. Note that under simple random sampling, $d^0 = 1$, so the test statistic remains X^2.

In practice $Var_0(\hat{\pi}_k)$ is rarely known in advance, and not all cells k have exactly the same deff due to chance, so this simple adjustment is not generally available. Instead, one can estimate an average deff across all of the cells using matrix techniques outlined in Rao and Scott (1981) and Scott (2007). Call the resulting estimate $\bar{d}$. Then the first-order Rao-Scott test approximates p-values and critical values for tests involving X^2 by comparing

$$X_{RS1}^2 = X^2/\bar{d}$$

to a χ_ν^2 distribution. This procedure works very well as long as there is not much variability among the d_k^0s. This test is generally available as the "Chisq" argument value in testing procedures within the survey package.

Second-order correction More generally when the deffs are not all approximately equal, X^2 has a distribution that is asymptotically equivalent to $\sum_{l=1}^\nu \delta_l W_l$, where W_l, $l = 1, \ldots, \nu$, are independent χ_1^2 random variables and δ_l, $l = 1, \ldots, \nu$, are quantities called *generalized deffs*. The δ_l's are found using matrix techniques described in Rao and Scott (1981) and Scott (2007). Let c be the coefficient of variation among the δ_l's: $c^2 = \sum_{l=1}^\nu (\delta_l - \bar{\delta})^2/(\nu \bar{\delta}^2)$, where $\bar{\delta} = \bar{d}$ is the average generalized deff. Then the second-order-corrected Rao-Scott test compares

$$X_{RS2}^2 = X^2/[\bar{\delta}(1+c^2)]$$

to a $\chi_{\nu/(1+c^2)}^2$ distribution.

Additional approaches Thomas and Rao (1987) propose modified versions of the Rao-Scott tests that use F distributions rather than χ^2. In particular they compare the statistic

$$F_{TR} = X_{RS1}^2/\nu = X^2/[\nu \bar{\delta}] \tag{6.12}$$

to an $F_{\nu/(1+c^2),\kappa\nu/(1+c^2)}$ distribution, where κ is again the degrees of freedom associated with the variance estimate $\widehat{Var}(\hat{\mathbf{N}})$. Simulations in Thomas and Rao (1987), Thomas et al. (1996), and Rao and Thomas (2003) show that this F-test works well enough under most conditions to be used as a primary procedure for Pearson tests with complex survey data. In fact, this is the default procedure for functions that perform Pearson-type tests in survey, and it can also be specified as the "F" argument value for testing arguments in these functions.

Other methods also exist for computing p-values and critical values from X^2. In particular, survey can compute a "saddlepoint approximation" to the linear combination of chi-squares with estimated deffs, $\sum_{l=1}^\nu \hat{\delta}_j W_j$ (the saddlepoint argument value in tests, Kuonen, 1999). We have not investigated the use of this method for approximating the distribution of Pearson statistics in this context, but we expect that it would be effective when the sample size is large, so that the generalized deff estimates are fairly accurate.

The difference between model deviances—which is a likelihood ratio test when standard sampling models are used—can be corrected in the same manner as the Pearson statistic to provide a viable model-comparison test for comparing any two nested models. Thus, the mainstays of categorical modeling and contingency table analyses have useful analogs that can be applied in the complex-survey setting. These procedures are demonstrated in the examples that follow.

Inference on odds ratios

Odds ratios are introduced in Section 1.2.5 and discussed further in the context of a loglinear model in Section 4.2.4. For ease of notation, suppose X and Y are binary, although they could alternatively represent any two rows or columns from a larger table. Create the resulting 2×2 table of estimated population totals using the same approach as given in Equation 6.7. The population odds ratio is $OR = [\pi_{11}/(1-\pi_{11})]/[\pi_{21}/(1-\pi_{21})] = (N_{11}N_{22})/(N_{21}N_{12})$. Define $\hat{N}_{ij}$ to be the weighted count for $X = i, Y = j, i = 1, 2; j = 1, 2$. The population odds ratio is then estimated by

$$\widehat{OR} = \frac{\hat{N}_{11}\hat{N}_{22}}{\hat{N}_{21}\hat{N}_{12}}.$$

As was the case in Section 1.2.5, it is generally more accurate to base inference on $\log(\widehat{OR})$ than on $\widehat{OR}$ itself. Using the delta method to estimate the variance of $\log(\widehat{OR})$ leads to the formula

$$\widehat{Var}(\log(\widehat{OR})) = \mathbf{a}'\widehat{Var}(\hat{\mathbf{N}})\mathbf{a},$$

where $\hat{\mathbf{N}} = (\hat{N}_{11}, \hat{N}_{12}, \hat{N}_{21}, \hat{N}_{22})'$, $\widehat{Var}(\hat{\mathbf{N}})$ is the 4×4 variance matrix computed using some method that accounts for the survey design, and $\mathbf{a} = (1/\hat{N}_{11}, -1/\hat{N}_{12}, -1/\hat{N}_{21}, 1/\hat{N}_{22})'$ is the vector of derivatives of $\log(OR)$ with respect to $N_{11}, N_{12}, N_{21}, N_{22}$ evaluated at $\hat{\mathbf{N}}$. Inference proceeds by computing Wald intervals for $\log(OR)$ as in Section 1.2.5, leading to the confidence interval

$$\exp\left[\log\left(\widehat{OR}\right) \pm Z_{1-\alpha/2}\sqrt{\widehat{Var}(\log(\widehat{OR}))}\right].$$

As was the case for a confidence interval for π_i, a $t_{\kappa,1-\alpha/2}$ critical value can be used in place of $Z_{1-\alpha/2}$. Also, $\widehat{Var}(\log(\widehat{OR}))$ can be estimated directly using a replication method where available. Hypothesis tests can be conducted using a Wald test statistic and a rejection region or p-value based on the standard normal (or t_κ) distribution. This is equivalent to determining whether the confidence interval contains the hypothesized value.

The calculations for odds ratios and their confidence intervals or hypothesis tests are most easily carried out easily within the context of a loglinear model. As noted earlier in this section, the models are parameterized exactly the same as in Section 4.2.4, although they are applied to the population totals N_{ij} in an $I \times J$ table instead of the population means μ_{ij}. A model with only main effects represents independence between X and Y. A saturated model allows odds ratios between each pair of rows and columns to be estimated by the model through the XY interaction parameters.

Example: NHANES 1999-2000 (SurveySmokeLoglinear.R, SmokeRespAge.txt)

We examine the association between $Y = any$ respiratory symptoms and $X = any$ tobacco use. We define "any respiratory symptoms" (anyresp) as TRUE (1) if any of the binaries for persistent cough, bringing up phlegm, wheezing/whistling in chest, or dry cough at night are 1, and FALSE (0) if all of the binaries are zero. Similarly, "any tobacco use" (anysmoke) is TRUE (1) if any of the binaries for cigarettes, pipe, cigar, snuff, or chewing tobacco are 1, and FALSE (0) if all are 0. We first create the new binaries by applying update() to the existing object created by svyrepdesign(). We then compute the table totals using svytable(). The summary of this object gives the estimated totals and a test for independence using the default Thomas-Rao F-test. Tests for independence can also be conducted using svychisq(), where the statistic argument controls the type of test used.

```
> jdesign <- update(object = jdesign, anytob = sm_cigs +
    sm_pipe + sm_cigar + sm_snuff + sm_chew > 0, anyresp =
    re_cough + re_phlegm + re_wheez + re_night > 0)
> head(jdesign$variables, n = 2)
  age sm_cigs sm_pipe sm_cigar sm_snuff sm_chew re_cough
1  77       0       1        1        0       0        0
2  49       1       1        1        0       1        0
  re_phlegm re_wheez re_night anytob anyresp
1         0        0        0   TRUE   FALSE
2         0        0        0   TRUE   FALSE

> # Table of weighted counts
> wt.table <- svytable(formula = ~ anytob + anyresp, design
    = jdesign)
> summary(wt.table)
       anyresp
anytob      FALSE       TRUE
  FALSE  73873587   13006040
  TRUE   75870428   28375419

        Pearson's X^2: Rao & Scott adjustment

data:  svychisq(~anytob + anyresp, design = jdesign,
    statistic = "F")
F = 40.1833, ndf = 1, ddf = 51, p-value = 6.016e-08

> svychisq(formula = ~ anytob + anyresp, design = jdesign,
    statistic = "F")
        Pearson's X^2: Rao & Scott adjustment
data:  svychisq(~anytob + anyresp, design = jdesign,
    statistic = "F")
F = 40.1833, ndf = 1, ddf = 51, p-value = 6.016e-08
```

The totals show that there are similar numbers of people who experience no respiratory symptoms (anyresp = FALSE) regardless of tobacco use. However, there is a somewhat greater number of people who have used tobacco and experienced respiratory symptoms than who have *not* used tobacco and have experienced respiratory symptoms. This suggests that there may be an association between tobacco use and respiratory symptoms. The Rao-Scott tests confirms this with $F_{TR} = 40.1$ and a p-value of 6.0×10^{-8}.

In the program for this example, we show that the cell totals can be converted to proportions by dividing by the estimated total population size. Here, we present an alternative method for computing these proportions and the odds ratio directly using svymean(). This function also has the added feature to save out the proportion calculations on each jackknife replicate, allowing us to estimate the variance of the log odds ratio values so that we can form a confidence interval. We use the replicate proportions to compute the odds ratio estimate for each replicate, and then apply a variant of Equation 6.9 to estimate the variance of $\log(\widehat{OR})$. Finally, a confidence interval is created using the t_{51}-distribution approximation to the sampling distribution of $\log(\widehat{OR})$. Notice that the input to svymean() uses the interaction() function. This function creates a new factor from all combinations of factors that are supplied as its argument values, and must be used here instead of other interaction-forming operators, ":" or "*".

```
> props22 <- svymean(x = ~ interaction(anytob, anyresp),
    design = jdesign, return.replicates = TRUE)
> props22
                                                  mean      SE
interaction(anytob, anyresp)FALSE.FALSE  0.38652  0.0113
interaction(anytob, anyresp)TRUE.FALSE   0.39697  0.0114
interaction(anytob, anyresp)FALSE.TRUE   0.06805  0.0055
interaction(anytob, anyresp)TRUE.TRUE    0.14847  0.0089

> names(props22)
[1] "mean"         "replicates"

> # Odds Ratio; order of terms is available from output.
> ORhat <- props22$mean[1] * props22$mean[4]
        / (props22$mean[2] * props22$mean[3])
> ORhat
interaction(anytob, anyresp)FALSE.FALSE
                               2.12429

> OR.reps <- props22$replicates[,1]*props22$replicates[,4]
         / (props22$replicates[,2]*props22$replicates[,3])
> reps <- length(OR.reps)
> var.logOR = (reps - 1)*var(log(OR.reps))*((reps - 1)/reps)

> # t-based confidence interval
> exp(log(ORhat) + qt(p = c(0.025, 0.975), df =
    reps-1)*sqrt(var.logOR))
[1] 1.667768 2.705778
```

The proportions reflect the same pattern noted above. We calculate $\widehat{OR} = 2.1$ with a t-based confidence interval from 1.7 to 2.7. This is strong evidence of a positive association between use of any tobacco product and presence of any respiratory symptoms.

This analysis can instead be conducted by fitting a saturated 2×2 loglinear model, finding a confidence interval for the association parameter—which is equivalent to $\log(\widehat{OR})/4$—and rescaling and exponentiating the results. We show how to perform this analysis in the program for this example and provide a similar type of analysis in the next example.

Multiway tables

The loglinear-modeling approach to the analysis of tables constructed from more than two variables was described in Section 4.2.5. The same analyses can be applied to data from a complex survey with modifications as described above using the svyloglin() function from the survey package. Information criteria are *not* generally available for use in model selection, because design-based model fitting does not use true likelihoods. Instead, combinations of forward and backward selection procedures are used. The example below demonstrates one possible approach to model selection, first choosing the complexity of possible interactions in the model and then determining which interactions of that level are needed.

Odds ratios can be computed for any comparison that can be expressed as a combination of model parameters. As always, the key step is to identify which model parameters are related to the odds ratios in question. Odds ratios are estimated primarily from parameters representing two-way interactions, although these can be impacted by higher-order interactions. Reiterating Equation 4.5, the odds ratio between levels i and i' of variable X and levels j and j' of variable Y is

$$OR_{ii',jj'} = \exp(\beta_{ij'}^{XY} + \beta_{i'j'}^{XY} - \beta_{i'j}^{XY} - \beta_{ij'}^{XY}).$$

where each β_{ij}^{XY} is a parameter from the X:Y interaction in a model from svyloglin().[10] Some care is needed when there are also higher-order interactions in a model. Recall the basic rules that govern the identification of terms involved in odds ratios:

1. Odds ratios are always 1 for any pair of variables that are not involved together in two-factor or higher interactions.

2. If a pair of variables is involved in a two-way interaction but does not appear together in higher-order interactions, then odds ratios between them are estimated from the two-way interaction parameters, and their values are constant across all levels of other variables.

3. If a pair of variables appears together in both a two-way interaction *and* three-factor or higher-order interactions, then the odds ratios between them vary depending on the levels of the other factors with which they appear in model terms. One must then estimate odds ratios separately for each level of these other factors (but not for variables with which they are not involved in three-way or higher-order interactions).

See Table 4.4 for details.

Example: NHANES 1999-2000 (SurveySmoke.R, SmokeRespAge.txt)

To what extent does use of one tobacco product relate to use of others? Do people who use one product do so essentially independently of other products, or are there certain products that tend to be used together or used separately?

To address these questions, a loglinear model is sought that describes the associations among the five different tobacco-use binary variables. The counts for certain cells of the full 2^5 table are given below:

```
> alltob <- svytable(formula = ~ interaction(sm_cigs,
    sm_pipe, sm_cigar, sm_snuff, sm_chew), design = jdesign)
> cbind(alltob)
                alltob
0.0.0.0.0   86879626.67
1.0.0.0.0   64660110.33
0.1.0.0.0    1248006.32
```

---

[10] Note that svyloglin() uses a different arrangement of the factors prior to estimation than most other model-fitting functions use. Specifically, it uses a "sum-to-zero" set of contrasts to represent factors rather than the usual "set-first-to-zero" set as described in Section 2.2.6. See the program SurveySmokeLoglinear.R for an important note on the computational implications of this parameterization.

<OUTPUT EDITED>

0.1.0.0.1   0.00

<OUTPUT EDITED>

0.1.0.1.1   0.00

<OUTPUT EDITED>

For example, we estimate that over 86 million people have used none of the tobacco products in the required amounts (combination 0.0.0.0.0), while over 64 million have used cigarettes only (combination 1.0.0.0.0). Notice that there are two zero counts, both involving positive pipe and chew response combined with negative cigarette and cigar response. We will see shortly that this causes problems with comparisons involving models containing the 4-way interaction of these four variables.

Model selection is based initially on comparing models of increasing orders, as these are conveniently constructed in svyloglin(). We start with main effects only, then add interactions by updating the model object's formula to include all two-way interactions, then all three-way interactions, all four-way interactions, and finally, the 5-way interaction (the saturated model). We use the update() function to build the new models based on the original model, as recommended by the survey package author (Lumley, 2011, p. 124). It makes constructing complicated models easier and is computationally more efficient than rewriting entire models in multiple svyloglin() fits. Comparisons between models are done using deviance statistics or Pearson (called "Score" in the output) statistics, which all use the default Thomas-Rao F-corrections given in the anova() method function. The output shows all of the terms of the two models being compared, so we have edited the results for brevity.

```
> ll.tob1 <- svyloglin(formula = ~ sm_cigs + sm_pipe +
    sm_cigar + sm_snuff + sm_chew, design = jdesign)
> ll.tob2 <- update(object = ll.tob1, formula = ~ .^2)
> ll.tob3 <- update(object = ll.tob1, formula = ~ .^3)
> ll.tob4 <- update(object = ll.tob1, formula = ~ .^4)
> ll.tob5 <- update(object = ll.tob1, formula = ~ .^5)
> anova(ll.tob4, ll.tob5)
Analysis of Deviance Table

<OUTPUT EDITED>

Deviance= 3.447026e-12 p= 0.01923078
Score= 5.584675e-12 p= 0.01923078

> anova(ll.tob3, ll.tob5)
Error in solve.default(t(wX2) %*% Psat %*% wX2, t(wX2) %*%
    V %*% wX2) :
  system is computationally singular: reciprocal condition
      number = 3.4043e-17
```

```
> anova(ll.tob2, ll.tob5)
Error in solve.default(t(wX2) %*% Psat %*% wX2, t(wX2) %*%
    V %*% wX2) :
  system is computationally singular: reciprocal condition
      number = 5.96634e-18

> anova(ll.tob1, ll.tob5)
Error in solve.default(t(wX2) %*% Psat %*% wX2, t(wX2) %*%
    V %*% wX2) :
  system is computationally singular: reciprocal condition
      number = 1.23203e-18

> anova(ll.tob3, ll.tob4)
Analysis of Deviance Table

<OUTPUT EDITED>

Deviance= 8.033536 p= 0.01923077
Score= 7.317034 p= 0.01923077

> anova(ll.tob2, ll.tob3)
Analysis of Deviance Table

<OUTPUT EDITED>

Deviance= 45.02365 p= 0.04722619
Score= 55.94341 p= 0.01569002

> anova(ll.tob1, ll.tob2)
Analysis of Deviance Table

<OUTPUT EDITED>

Deviance= 2227.541 p= 7.465893e-58
Score= 29924.8 p= 3.462698e-177
```

Several tests comparing lower-order models to the saturated model report computational errors. In addition, two other tests cause some alarm. In one, the comparison of models ll.tob3 and ll.tob4 results in "Deviance" and "Score" statistics that have different values but *the exact same p-values to 8 digits*. In the other, the comparison of the fourth- and fifth-order models is also not believable—it yields test statistics of essentially zero, but with p-values that are essentially the same as those reported in the previous test with much larger test statistics. This is not possible, and it indicates that there is a problem with the reporting of p-values in the anova.svyloglin() method function, perhaps induced by the presence of zero counts in these models. Thus, it is important to always check fitted models to ensure that no cells have estimated counts that are essentially 0.[11] If they do occur, subsequent calculations should be viewed with caution. We show next how to avoid using models that fit zero counts.

---

[11] Model-estimated counts can be viewed using fitted(object$model), where object is the model-fit object.

Starting instead from the simplest model and working forward, the comparison of the first-order and second-order models, ll.tob1 and ll.tob2, is unambiguous in determining that some kind of association exists among uses of different tobacco products. The statistic $X^2 = 29924.8$ with p-value of essentially 0 clearly rejects the simpler null model of mutual independence in favor of a model with pairwise associations. Further, the comparison of the second-order model with the third-order model, ll.tob3, provides $X^2 = 55.97$ with a p-value= 0.047, suggesting that the levels of at least some associations among pairs of tobacco products may depend on whether a third product is used. As noted above, a comparison of the third- and fourth-order models results in an error. The reason for this becomes evident upon looking at the parameter estimates and confidence intervals for the two models (not shown, see program for code). For ll.tob4, all effects that can be formed from combinations of sm_cigs, sm_pipe, sm_cigar, and sm_chew have confidence intervals that run from at least -200 to 200. These values are nonsensical, considering that they would be exponentiated to form cell counts and odds ratios. They are another reflection of the adverse impact of zero counts on the loglinear model. In the program, we show that excluding the 4-way interaction term for these variables results in a model that is sensibly fitted, but results in very little improvement over the third-order model ($X^2 = 2.11$, p-value= 0.83). We therefore begin working with the third-order model.

Several possibilities exist for developing a final model. We could simply use the full third-order model as our final model, particularly if we were not interested in a parsimonious explanation for the associations in the data. Or we could apply backward elimination or forward selection to the third-order terms from the model, keeping in mind that the choice of significance level for adding or deleting a term has nothing to do with type I error rates in any tests (see discussion on p. 276). We show the results of backward elimination in the program for this example. This shows that there is clearly a sm_pipe:sm_snuff:sm_chew interaction, and that there could be two additional three-way interactions if a threshold level of 0.10 or higher is used. For the sake of simplicity in our demonstration, we choose the model with just the one added three-way interaction. This is named ll.tob39 in the program.

Model-estimated odds ratios can be computed between each pair of factors either by manual manipulation of selected model-parameter estimates, or by the more automated calculations available in svycontrast(). The ordering of the parameter estimates is determined by using either summary(ll.tob39) or coef(ll.tob39). This reveals that the parameter estimates are in the order 1, 2, 3, 4, 5, (12), (13), (14), (15), (23), (24), (25), (34), (35), (45), (245), where 1=sm_cigs, 2=sm_pipe, 3=sm_cigar, 4=sm_snuff, 5=sm_chew, and two or more numbers in parentheses means an interaction among the corresponding variables. Coefficients on the parameter estimates must be assigned to match this ordering. A detailed comment in the program for this example explains that the coefficient should be 4 for odds ratios between variables with two levels. For example, cigarette-pipe odds ratio is determined by the coefficient vector (0, 0, 0, 0, 0, 4, 0, 0, 0, 0, 0). These coefficients are entered in svycontrast() as shown below. The abbreviations used are ct=cigarettes, pi=pipes, cr=cigars, sn=snuff, and ch=chew.

```
> logORs <- svycontrast(stat = ll.tob39, contrasts = list(
    ct.pi = c(0,0,0,0,0,4,0,0,0,0,0,0,0,0,0,0),
    ct.cr = c(0,0,0,0,0,0,4,0,0,0,0,0,0,0,0,0),
    ct.sn = c(0,0,0,0,0,0,0,4,0,0,0,0,0,0,0,0),
    ct.ch = c(0,0,0,0,0,0,0,0,4,0,0,0,0,0,0,0),
    sn.ch.pi0 = c(0,0,0,0,0,0,0,0,0,0,0,0,0,0,4, 4),
```

```
            sn.ch.pi1 = c(0,0,0,0,0,0,0,0,0,0,0,0,0,0,4,-4)
        ))
> ORs   <- as.data.frame(logORs)
> ORs$OR <- exp(ORs$contrast)
> # Calculate degrees of freedom for t distribution
> df <- ll.tob39$df.null - length(coef(ll.tob39)) - 1
> # Confidence intervals
> ORs$lower.CI <- exp((ORs$contrast + qt(0.025, df)*ORs$SE))
> ORs$upper.CI <- exp((ORs$contrast + qt(0.975, df)*ORs$SE))
> round(ORs[,3:5], 2)
              OR lower.CI upper.CI
ct.pi       3.37     2.39     4.74
ct.cr       2.70     2.01     3.62
ct.sn       2.01     1.20     3.37
ct.ch       1.32     0.85     2.05
sn.ch.pi0  52.61    22.80   121.41
sn.ch.pi1   7.43     3.63    15.24
```

For pairs of factors not involved in the three-way interaction, a single odds ratio is computed. All of these associations are positive (although the confidence interval for the cigarette-chew odds ratio does include 1), indicating that people who have used cigarettes are generally more likely to have used pipes, cigars, or snuff than are people who have not used cigarettes. For associations involving pipes, snuff, and chew, a separate odds ratio is computed for each level of the third variable. We denote these with a third variable and the level 0 or 1. The results of this analysis tell a little different story. For instance, snuff use and chew use—the two forms that are not smoked—are very strongly associated, both for pipe users (pi1) and for non-pipe users (pi0). However, they are much more strongly associated among the non-pipe users, where the odds of snuff use are estimated to be 52 times as high among chew users than among non-chew users (95% confidence interval of 23 to 121).

Alternative model estimation and inference methods

An alternative method has been proposed by Clogg and Eliason (1987) to account for survey weights in a loglinear model analysis. The idea is to pool together the survey weights corresponding to subjects in each cell of the table being analyzed. The mean weight is used as an offset for a Poisson rate regression model (see Section 4.3), and the analysis is carried out as if the data had arisen from a simple random sample. This suggestion has been recommended by Agresti (2002, p. 391) and used in some research (e.g., Schwartz and Mare, 2005, Vermunt and Magidson, 2007, and Beller, 2009).

This approach has been scrutinized recently in Skinner and Vallet (2010) and Loughin and Bilder (2010), who find that it is inappropriate for the use for which it is intended. The approach has two main flaws:

1. Although the survey weights are incorporated into the analysis, other features of the survey design (clustering and stratification) are not. This means that standard errors cannot be estimated properly and hence inferences cannot be trusted.

2. Even in rare cases where the survey design has little effect on the standard errors, the method is based on an approximation that assumes that survey weights are constant

among all members of a given table cell (Loughin and Bilder, 2010). The procedure suffers badly when there is variability among the weights within a cell. Standard errors and model tests are adversely affected by this variability, resulting in inferences that are too liberal, by potentially large amounts.

Considering these problems, we strongly caution against using this procedure as a quick substitute for a full design-based analysis.

6.3.6 Logistic regression

Logistic regression is covered in detail in Chapter 2. The same principles given there apply to an analysis of weighted survey data. We describe here how to implement a logistic regression analysis using the survey package. Further detail is available in Lumley (2010, Chapter 6). A detailed discussion of the theory is given in Heeringa et al. (2010).

To begin, plots of binary responses or estimated proportions can be made against chosen explanatory variables. The important thing to remember is that each sampled unit represents different numbers of members of the population according to the survey weights. Bubble plots like Figure 2.5 in Section 2.2.4 can be formed using total weights rather than sample sizes to define the bubble widths. The survey package has a plotting function svyplot() that can do this automatically.

A logistic model is fit using the svyglm() function. This function mimics the glm() function described in Section 2.2.1, except that model fitting does not use maximum likelihood estimation, because the binomial likelihood is inappropriate for weighted, correlated survey data. Instead, a *pseudo-likelihood* is used that resembles the binomial likelihood applied to weighted counts rather than to sample counts (see Section 6.4.3 for a brief description of pseudo-likelihood). Wald tests for the model coefficients are available in the summary() of the model, while confidence intervals for the coefficients are found using confint(). The standard errors are computed internally either by linearization or by replication methods. Tests are versions of Rao-Scott second-order-corrected t-tests: the t statistic is found in the usual manner as the ratio of the estimate to its standard error. The p-value is calculated by comparing the square of this statistic to an F distribution as in Equation 6.12. The default confidence interval is a Wald-type interval; t-intervals can also be created.

There is an anova() method that can perform LRT-like or Wald-based model-comparison tests on objects resulting from svyglm(). This requires that both the full and reduced (null-hypothesis) model are fit to the data. Alternatively, these tests can be performed directly on the full model using the regTermTest() function. This is especially useful for testing the significance of groups of coefficients representing a single factor, or for testing whether all coefficients for terms of a certain order or associated with a particular variable are different from 0.

As with loglinear models, odds-ratio and probability-of-success computation requires additional work. The parameterization of svyglm() is exactly the same as in glm(), so the process of estimating an odds ratio or a probability of success from model parameters is exactly the same as was covered in Section 2.2.3–2.2.5. Log odds ratios can be estimated using the svycontrast() function to compute the proper linear combinations of the coefficients and corresponding standard errors. From these quantities the odds ratios and corresponding confidence intervals are found, again using a t distribution with $\kappa - p + 1$ degrees of freedom, where p is the number of parameters in the model (Korn and Graubard, 1999). Predicted probabilities and confidence intervals can be found in a similar manner by using svycontrast() to compute the logits for the desired combinations of explanatory variables and their corresponding standard errors. The inverse logit link, $\exp(\cdot)/(1 + \exp(\cdot))$, is applied to these logits and also to confidence interval endpoints computed in the logit scale.

Alternatively, if interest is restricted to viewing fitted probabilities for combinations of explanatory variables that are present in the data, these can be obtained using the predict() method.

Example: NHANES 1999-2000 (SurveySmokeLogistic.R, SmokeRespAge.txt)

Here we model presence of any respiratory symptoms (anyresp) against age, which is a continuous covariate, and the binary factors for each tobacco product. We begin by fitting a series of models, starting with age by itself (model M1), then adding all five tobacco-use factors as main effects (M2), and finally including interactions between age and each of the tobacco binaries (M3).[12] Sequential comparisons of these three models gives us a starting point for further backward elimination until a satisfactory reduced model is obtained. We use $\alpha = 0.15$ in our backward elimination because this has been shown to be more akin to what measures like AIC might achieve if they were available with these models (Lee and Koval, 1997, Shtatland et al., 2003). Below are the initial model comparisons of M1 versus M2 and M2 versus M3. We show only the anova() function, but additional calculations from regTermTest() are available in the program for this example.

```
> m1 <- svyglm(formula = anyresp ~ age, design = jdesign,
    family = quasibinomial(link = "logit"))
> m2 <- svyglm(formula = anyresp ~ age + sm_cigs + sm_pipe
    + sm_cigar + sm_snuff + sm_chew, design = jdesign,
    family = quasibinomial(link = "logit"))
> anova(m1, m2)
Working (Rao-Scott+F) LRT for sm_cigs sm_pipe sm_cigar
    sm_snuff sm_chew
 in svyglm(formula = anyresp ~ age + sm_cigs + sm_pipe +
    sm_cigar + sm_snuff + sm_chew, design = jdesign,
    family = quasibinomial(link = "logit"))
Working 2logLR =   60.88791 p= 2.9552e-06
(scale factors:   1.8 1.1 1 0.73 0.33 );  denominator df= 45

> m3 <- svyglm(anyresp ~ age * (sm_cigs + sm_pipe +
    sm_cigar + sm_snuff + sm_chew), design = jdesign,
    family = quasibinomial())
> anova(m2, m3)
Working (Rao-Scott+F) LRT for age:sm_cigs age:sm_pipe
    age:sm_cigar age:sm_snuff age:sm_chew
 in svyglm(formula = anyresp ~ age * (sm_cigs + sm_pipe +
    sm_cigar + sm_snuff + sm_chew), design = jdesign,
    family = quasibinomial())
Working 2logLR =   10.75155 p= 0.08842
(scale factors:   2.1 1.1 0.87 0.56 0.41 );  denominator df=
    40
```

---

[12]Note the use of a quasi-binomial family. This has to do with the fact that we are modeling survey-weighted counts of successes. The help for svyglm() explains: "For binomial and Poisson families use family = quasibinomial() and family = quasipoisson() to avoid a warning about non-integer numbers of successes. The 'quasi' versions of the family objects give the same point estimates and standard errors and do not give the warning."

The evidence from the comparison of models m1 and m2 strongly indicates that at least one of the tobacco use factors contributes significantly to the model already containing age (LRT statistic = 60.9, p-value ≈ 0). The t-tests associated with parameter estimates (not shown) point to cigarettes as the strongest contributor by far ($t = 6.4$, p-value ≈ 0, next largest has $t = 1.47$, p-value = 0.15). The inclusion of age-by-tobacco interactions results in a more modest improvement over the model with main effects only (LRT statistic= 10.8, p-value= 0.09), so we consider the larger of these models for backward elimination of individual terms. We maintain hierarchy in the process, allowing the main effect of a tobacco-use factor to be considered for elimination only after its interaction with age has already been eliminated. This results in the following final model:

```
> m.final <- svyglm(anyresp ~ age + sm_cigs + age *
    (sm_pipe + sm_chew), design = jdesign.age, family =
    quasibinomial(link = "logit"))
> summary(m.final)

<OUTPUT EDITED>

Coefficients:
              Estimate Std. Error t value Pr(>|t|)
(Intercept)  -2.039511   0.148635 -13.722  < 2e-16 ***
age           0.006693   0.002360   2.836  0.00682 **
sm_cigs       0.705353   0.110126   6.405  7.8e-08 ***
sm_pipe       1.654550   0.611060   2.708  0.00954 **
sm_chew       1.461494   0.829967   1.761  0.08505 .
age:sm_pipe  -0.025002   0.009924  -2.519  0.01538 *
age:sm_chew  -0.022896   0.014350  -1.596  0.11760
```

This model indicates that the relationship between presence of any respiratory symptoms and age is influenced by use of pipe and chew, but not other products. Cigarette use has a very significant relationship with presence of respiratory symptoms, independent of age. Notice that the signs of the two interaction coefficients are negative and have larger magnitudes than the age main effect. This implies that the model is estimating that there is actually a *decreasing* relationship between presence of respiratory symptoms and age for users of pipes and chew, which is counterintuitive.

Exploring these features further, we calculate odds ratios relating the change in presence of respiratory symptoms to a 10-year increase in age separately for (a) people whose tobacco use does not include pipes or chew, and people who have used (b) pipe tobacco only, and (c) chew only. Based on the ordering of the parameters estimates in the previous output, the estimated odds ratios for these three cases are (a) $\exp(10\hat{\beta}_1)$, (b) $\exp(10\hat{\beta}_1 + 10\hat{\beta}_5)$, and (c) $\exp(10\hat{\beta}_1 + 10\hat{\beta}_6)$. The estimated regression parameters needed for the odds ratios are entered into svycontrast() as shown below, which enables us to find the estimates and confidence intervals for the odds ratios.

```
> logOR.age <- svycontrast(stat = m.final, contrasts = list(
    tenyr = c(0, 10, 0, 0, 0, 0, 0),
    tenyr_pipe = c(0, 10, 0, 0, 0, 10, 0),
    tenyr_chew = c(0, 10, 0, 0, 0, 0, 10)))
> # Convert object into data frame for further computations
```

```
> ORs.age <- as.data.frame(logOR.age)

> ORs.age$OR <- exp(ORs.age$contrast)
> ORs.age$lower.CI <- exp((ORs.age$contrast + qt(0.025,
    m.final$df.residual) * ORs.age$SE))
> ORs.age$upper.CI <- exp((ORs.age$contrast + qt(0.975,
    m.final$df.residual) * ORs.age$SE))
> round(ORs.age[,3:5],2)
              OR  lower.CI  upper.CI
tenyr       1.07     1.02      1.12
tenyr_pipe  0.83     0.68      1.02
tenyr_chew  0.85     0.64      1.13
```

Each odds ratio is interpreted as the multiplicative change in odds of respiratory symptoms for a 10-year increase in age for a person who has used the specified tobacco product. The estimated odds ratios are above 1 for non-pipe or chew users, although the confidence interval for a person with no tobacco-use history is only 1.02-1.12, suggesting that there is a statistically significant, but very small increase in probability of symptoms with age. The overall effect of age for pipe and chew users is not clear, as both estimated odds ratios are below 1, but their respective confidence intervals contain 1. Additional odds ratios focusing on tobacco use at specific ages are shown in the program for this example.

Below, we also compute confidence intervals for probabilities of respiratory symptoms for 20-year old non-users or cigarette users, and 50-year-old non-users, cigarette users, and users of all three products. Clearly, tobacco use is associated with greater probabilities of respiratory symptoms, although no causal relationship can be inferred from this analysis.

```
> logits.reduced <- svycontrast(m.final,list(
    twenty_none=c(1, 20, 0, 0, 0,  0,  0),
    twenty_cig =c(1, 20, 1, 0, 0,  0,  0),
    fifty_none =c(1, 50, 0, 0, 0,  0,  0),
    fifty_cig  =c(1, 50, 1, 0, 0,  0,  0),
    fifty_all  =c(1, 50, 1, 1, 1, 50, 50))))

> preds.reduced <- as.data.frame(logits.reduced)
> preds.reduced$prob <- plogis(preds.reduced$contrast)
> preds.reduced$lower.CI <- plogis(preds.reduced$contrast +
    qt(p = 0.025, df = m.final$df.residual) *
    preds.reduced$SE)
> preds.reduced$upper.CI <- plogis(preds.reduced$contrast +
    qt(p = 0.975, df = m.final$df.residual) *
    preds.reduced$SE)
> round(preds.reduced[,3:5], digits = 3)
              prob  lower.CI  upper.CI
twenty_none  0.129    0.106    0.157
twenty_cig   0.231    0.196    0.271
fifty_none   0.154    0.134    0.176
fifty_cig    0.269    0.242    0.298
fifty_all    0.431    0.334    0.533
```

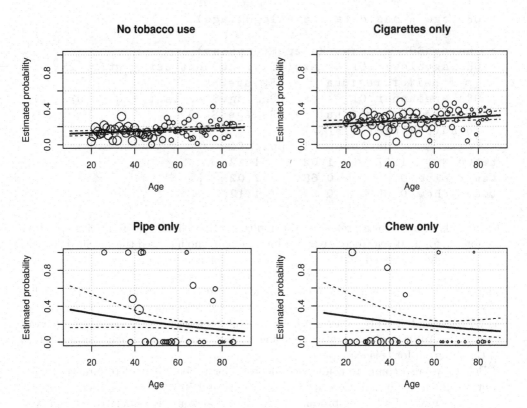

Figure 6.3: Final model fitted to data for four tobacco-use cases. The solid line is the fitted model and the dashed lines are pointwise 95% confidence intervals for the probability of respiratory symptoms at each age. Bubble sizes reflect relative survey weight for each year. Bubbles sizes in different panels are not comparable.

Plots of the model-predicted probabilities and the estimated proportions are given in Figure 6.3 (code is available in our corresponding program). Each bubble represents the weighted sample estimate for the proportion of the population with respiratory symptoms at each age (one bubble per year). Clearly there is not much data with which to estimate the trends for people who have used chew or pipes, so the confidence intervals here are much wider and appear not to preclude a line of zero slope.

6.4 "Choose all that apply" data

Survey questions often include the phrase "choose all that apply" or "pick any" when requesting individuals to select from a pre-defined list of response options or *items*. Individuals may select any number of these items in their responses. For this reason, categorical variables summarizing these responses are referred to as *multiple response categorical variables*

Table 6.6: Joint positive responses for the swine management data.

	Lagoon	Pit	Natural drainage	Holding tank
Nitrogen	27	16	2	2
Phosphorus	22	12	1	1
Salt	19	6	1	0

(MRCVs; Bilder and Loughin, 2004).[13] Allowing individuals to provide multiple responses to a single question creates correlation among these responses. For this reason, analyzing MRCV data is more challenging than analyzing typical single-response categorical variables (SRCVs). The purpose of this section is to show how the Pearson chi-square test can be extended to these data situations and to examine how regression models can be used to explore relationships among MRCVs.

6.4.1 Item response table

Table 6.6 summarizes the responses of 279 Kansas swine farmers who were asked questions in a survey about their swine waste management practices (Richert et al., 1995; Bilder and Loughin, 2007). One question asked farmers to choose all swine waste storage methods that they use among the items: lagoon, pit, natural drainage, and holding tank. Another question asked farmers to choose all contaminants that they test for among the items: nitrogen, phosphorus, and salt. Although the summary of these counts in Table 6.6 looks like an ordinary contingency table, there are several important distinctions. First, because farmers could choose all responses that applied to them, an individual farmer may contribute to more than one cell count in the table. Second, negative answers are not adequately represented in Table 6.6. Farmers who used none of the listed storage methods are not counted, regardless of what contaminants they test for, and vice versa. Thus, there is important information missing in this representation.

A better way to represent these data is to consider both positive and negative responses to each item. Specifically, let W_i denote the binary response (1=positive, 0=negative) to row item i for $i = 1, \ldots, I$, and let Y_j denote the binary response to column item j for $j = 1, \ldots, J$. For example, in the swine waste management data, $Y_1 = 1$ means that a farmer uses a lagoon for waste storage. Then 2×2 contingency tables can be formed from the binary variables for each combination of a row item and a column item. Table 6.7 shows two of the $IJ = 12$ possible tables for the swine waste management example. It is important to note that each farmer appears in exactly one cell in each of the tables; thus, the counts in each table sum to 279, the number of farmers participating in the survey. Also, notice that the counts in Table 6.6 are merely the cell counts for the $W_i = 1$ and $Y_j = 1$ response combination from these 12 tables. We refer to the full set of IJ 2×2 tables as the *item response table* for the data. The item response table is the summary upon which inferences are constructed in the next two sections.

6.4.2 Testing for marginal independence

Procedures to test for independence between two MRCVs were developed in a series of papers by Loughin and Scherer (1998), Agresti and Liu (1999), Thomas and Decady (2004), and Bilder and Loughin (2004). These papers identified two main complications that need

---

[13]In the social sciences literature, the phrase "pick any/n" (said as "pick any out of n") is often used to describe the resulting data when there are n items in the list. Coombs (1964) is likely the first one to coin this terminology.

Table 6.7: Two 2 × 2 contingency tables for the swine waste management data.

		Lagoon				Pit	
		0	1			0	1
Nitrogen	0	123	116	Nitrogen	0	175	64
	1	13	27		1	24	16

to be addressed when testing for independence. First, the usual notion of independence needs to be redefined in the context of an MRCV. Second, constructing and finding the probability distribution of any test statistic under a null hypothesis is complicated by the fact that many responses in the cells of either Table 6.6 or 6.7 are provided by the same individuals. None of the usual sampling models are appropriate for these counts—they are not multinomial counts, and they are not independent Poisson counts—so the theory that leads to using Pearson statistics and chi-square distributional approximations no longer applies. We show below how these issues can be overcome.

Marginal independence hypotheses

The primary interest in many problems involving "choose all that apply" questions is determining whether the probability of a positive response to each item changes depending on the responses to other questions. For example, does the probability of testing for a particular contaminant depend on what waste-storage methods are used? In other words, are each of the twelve 2×2 tables in the item-response table consistent with an independence hypothesis, or does at least one indicate an association between testing a contaminant and using a waste storage method? Each of these 2 × 2 tables is a *marginal* table, because it summarizes data without regard to the responses to all other items. The hypothesis that independence holds in each of these tables is therefore termed *simultaneous pairwise marginal independence* (SPMI), as first named by Agresti and Liu (1999).

Define $\gamma_{ij} = P(W_i = 1, Y_j = 1)$ as a pairwise joint probability for $i = 1, \ldots, I$ and $j = 1, \ldots, J$. Also, define $\gamma_{i+} = P(W_i = 1)$ and $\gamma_{+j} = P(Y_j = 1)$ as marginal probabilities of a positive response to the corresponding items. Then the hypotheses for SPMI are

$$H_0 : \gamma_{ij} = \gamma_{i+}\gamma_{+j} \text{ for } i = 1, \ldots, I \text{ and } j = 1, \ldots, J$$
$$H_a : \gamma_{ij} \neq \gamma_{i+}\gamma_{+j} \text{ for a least one } (i, j) \text{ pair.}$$

One can show that $\gamma_{ij} = \gamma_{i+}\gamma_{+j}$ implies that $P(W_i = w_i, Y_j = y_j) = P(W_i = w_i)P(Y_j = y_j)$ for all combinations of $w_i = 0$ or 1 and $y_j = 0$ or 1, so this is why the null hypothesis can be written using just one cell of each 2 × 2 table.

Instead of writing hypotheses about the pairwise joint probabilities γ_{ij}, one could instead examine the full joint probabilities $P(W_1 = w_1, \ldots, W_I = w_I, Y_1 = y_1, \ldots, Y_J = y_J)$ for each of the 2^{I+J} binary response combinations. A different form of independence involving the joint probabilities, known as *joint independence*, occurs when the combination of responses to one question is independent of the combination of responses to the other: $P(W_1 = w_1, \ldots, W_I = w_I, Y_1 = y_1, \ldots, Y_J = y_J) = P(W_1 = w_1, \ldots, W_I = w_I)P(Y_1 = y_1, \ldots, Y_J = y_J)$. One can show that if joint independence holds, then SPMI does as well, but it is possible for SPMI to be true when joint independence is not. Berry and Mielke (2003) provide details on testing for joint independence. However, we have found testing for joint independence to be of limited use, unless there is interest in the specific response combinations of $w_1, \ldots, w_I$ and $y_1, \ldots, y_J$.

Test Statistics

We next develop tests for SPMI. Treating each 2×2 table separately, maximum likelihood estimates of the pairwise joint probabilities are $\hat{\gamma}_{ij} = m_{ij}/n$, $\hat{\gamma}_{i+} = m_{i+}/n$, and $\hat{\gamma}_{+j} = m_{+j}/n$, where m_{ij} represents the number of positive responses for $W_i = 1$ and $Y_j = 1$, and n is the sample size. Let $X^2_{S,ij}$ be the usual Pearson statistic for testing independence within the $(i,j)^{th}$ 2×2 contingency table. Then the sum of these Pearson statistics is a natural statistic to use for testing simultaneous independence in all tables. We can write this as

$$\begin{aligned}
X^2_S &= \sum_{i=1}^{I}\sum_{j=1}^{J} X^2_{S,ij} \\
&= \sum_{i=1}^{I}\sum_{j=1}^{J} \left[\frac{(m_{ij} - m_{i+}m_{+j}/n)^2}{m_{i+}m_{+j}/n} + \frac{(m_{i+} - m_{ij} - m_{i+}(n-m_{+j})/n)^2}{m_{i+}(n-m_{+j})/n} \right. \\
&\qquad + \frac{(m_{+j} - m_{ij} - m_{+j}(n-m_{i+})/n)^2}{m_{+j}(n-m_{i+})/n} \\
&\qquad \left. + \frac{(n - m_{i+} - m_{+j} + m_{ij} - (n-m_{i+})(n-m_{+j})/n)^2}{(n-m_{i+})(n-m_{+j})/n} \right] \\
&= n\sum_{i=1}^{I}\sum_{j=1}^{J} \frac{(\hat{\gamma}_{ij} - \hat{\gamma}_{i+}\hat{\gamma}_{+j})^2}{\hat{\gamma}_{i+}\hat{\gamma}_{+j}(1-\hat{\gamma}_{i+})(1-\hat{\gamma}_{+j})}.
\end{aligned} \qquad (6.13)$$

If the collection of $X^2_{S,ij}$ statistics were independent, then X^2_S would have a large sample χ^2_{IJ} distribution and SPMI would be rejected if $X^2_S > \chi^2_{IJ,1-\alpha}$. Unfortunately, the individual test statistics are likely to be correlated, because the tables upon which they are based are merely different summaries of the same set of responses. For example, each farmer is represented in each of the IJ 2×2 contingency tables in the swine management data example. Bilder and Loughin (2004) show that ignoring the correlation and performing the test using a χ^2_{IJ} distribution leads to a liberal test—one that rejects H_0 too often when SPMI is true—when there is moderate-to-strong association among an individual's responses within one MRCV (e.g., if farmers who test for one contaminant also tend to test for another). Alternative testing methods proposed by Bilder and Loughin (2004) are described next.

Bonferroni correction

Rather than combine the individual 2×2 table statistics into a sum, one can compute a p-value to test for independence in each table and combine these results into a single test. The simplest way to do this is to use a Bonferroni correction on the individual tests. For each $X^2_{S,ij}$, calculate a p-value, say p_{ij}, using the usual χ^2_1 approximation or an exact method from Section 6.2, and reject SPMI if any p_{ij} is less than α/IJ. Equivalently, an adjusted p-value (Westfall and Young, 1993) for the overall SPMI test can be calculated as $\tilde{p} = IJ \times \min_{ij}(p_{ij})$, where $\tilde{p}$ is truncated to 1 if this product is greater than 1. Then SPMI is rejected if $\tilde{p} < \alpha$. As is often the case with a Bonferroni correction, the overall SPMI test can be conservative—it does not reject often enough when SPMI is true—if the number of 2×2 tables is more than a few.

Bootstrap approximation

A second approach proposed by Bilder and Loughin (2004) involves the bootstrap. The bootstrap is a commonly used technique to estimate a statistic's probability distribution when its distribution may be difficult to obtain mathematically or when large-sample approximations may be poor. We focus here on the application of the bootstrap rather than on details for why it works well in this situation. Interested readers can examine Davison and Hinkley (1997) for more details on the bootstrap.

To implement the bootstrap, data are "resampled"[14] by randomly selecting an observed row-response combination $(w_1, \ldots, w_I)$ and matching it with an independently chosen column-response combination $(y_1, \ldots, y_J)$. For example, one farmer's waste-storage responses may be randomly matched to another farmer's contaminant-testing responses. Repeating this process n times forms a "resample" with very similar characteristics to the original sample—e.g., sample size, proportions of positive responses to each item, and associations among the items within each MRCV—while ensuring that the items from the two different MRCVs are independent of each other.[15] A large number of resamples, say B, are created in this way. For each resample, $b = 1, \ldots, B$, we calculate the test statistic, denoted now by $X_{S,b}^{2*}$, where we use the asterisk in the superscript to differentiate it from the original statistic's observed value. The p-value for the test is the proportion of the $X_{S,b}^{2*}$ values greater than or equal to the observed X_S^2; i.e., calculate $B^{-1}(\# \text{ of } X_{S,b}^{2*} \geq X_S^2)$. This p-value measures how extreme the original test statistic is relative to the distribution of the same statistic under the null hypothesis. Small p-values indicate evidence against SPMI. Bilder and Loughin (2004) show that this test holds the correct size; i.e., the test rejects at the stated α-level when SPMI is true.

Rao-Scott corrections

The test statistic X_S^2 has an asymptotic distribution which is a linear combination of independent χ_1^2 random variables (see Appendix A of Bilder and Loughin (2004)), where the coefficients of the linear combination are unknown. Approximations to this distribution can be found using the Rao-Scott methods described in Section 6.3.5. A first-order correction adjusts X_S^2 so that it has the same mean as a χ_{IJ}^2 random variable. Interestingly, Thomas and Decady (2004) and Bilder and Loughin (2004) show that this adjustment factor is simply 1. Thus, X_S^2 can be compared to a χ_{IJ}^2 distribution for the first-order correction, which results in the same testing method as from naively treating each $X_{S,ij}^2$ as independent. However, as mentioned earlier, performing the test in this manner leads to a liberal test.

A second-order correction adjusts X_S^2 so that it has the same mean and variance as a χ_{IJ}^2 random variable. For $p = 1, \ldots, IJ$, define λ_p to be the coefficients in the large-sample distributional approximation for X_S^2 described above. Then the distribution of the statistic $X_{RS2}^2 = IJ X_S^2 / \sum_{p=1}^{IJ} \lambda_p^2$ can be approximated by a χ^2 random variable with $I^2 J^2 / \sum_{p=1}^{IJ} \lambda_p^2$ degrees of freedom (this may not be an integer). Values of X_{RS2}^2 greater than the $1 - \alpha$ quantile of this χ^2 distribution indicate evidence against SPMI. Bilder and Loughin (2004) show that this testing procedure is generally pretty good but can be a little conservative.

The key to implementing the second-order correction is to estimate λ_p for $p = 1, \ldots, IJ$. These coefficients are functions of the large-sample covariance matrix for the quantities $\hat{\gamma}_{ij} - \hat{\gamma}_{i+}\hat{\gamma}_{+j}$. Details of their calculation are given in Thomas and Decady (2004) and Bilder

---

[14]Resampling is very similar to permuting data, which was discussed in Section 6.2. Permutations of observations result from sampling "without replacement" from an original set of data. Thus, the same observations are retained, but their order is randomized. Resamples of observations result from sampling "with replacement" from an original set of data. Thus, the observations in a data set constructed by resampling take the same values as those in the original data, but these values occur with different frequencies. Some observations may occur multiple times, while others may not appear at all within a resample of n observations.

[15]This form of independence is actually joint independence. Because joint independence implies SPMI, the resampling is also performed under the SPMI null hypothesis.

and Loughin (2004). We implement the calculations here using the function `MI.test()` from the MRCV package.

Example: Swine waste management data (Swine.R)

We now formally examine the swine waste management data first described in Section 6.4.1. Both the data and the corresponding analysis functions are contained in the MRCV package. The `farmer2` data frame within the package contains the response combinations $(w_1, w_2, w_3, y_1, y_2, y_3, y_4)$ for each of the $n = 279$ farmers, where the indexes on the items correspond to the ordering shown in Table 6.6. The first few observations are shown below:

```
> library(package = MRCV)
> head(farmer2)
  w1 w2 w3 y1 y2 y3 y4
1  0  0  0  0  0  0  0
2  0  0  0  0  0  0  1
3  0  0  0  0  0  0  1
4  0  0  0  0  0  0  1
5  0  0  0  0  0  0  1
6  0  0  0  0  0  0  1
```

For example, the second farmer did not test for any of the three contaminants and used only a holding tank. The summarization of these data given by Tables 6.6 and 6.7 is obtained using the `marginal.table()` and `item.response.table()` functions, respectively:

```
> marginal.table(data = farmer2, I = 3, J = 4)
       y1           y2          y3          y4
    count %      count %     count %     count %
w1    27  9.68    16  5.73    2  0.72    2   0.72
w2    22  7.89    12  4.30    1  0.36    1   0.36
w3    19  6.81     6  2.15    1  0.36    0   0.00

> item.response.table(data = farmer2, I = 3, J = 4)
          y1        y2        y3         y4
          0    1    0    1    0    1    0    1
w1  0   123  116  175   64  156   83  228   11
    1    13   27   24   16   38    2   38    2
w2  0   128  121  181   68  165   84  237   12
    1     8   22   18   12   29    1   29    1
w3  0   134  124  184   74  174   84  245   13
    1     2   19   15    6   20    1   21    0
```

To perform the Bonferroni correction, we use the `MI.test()` function with the `type = "bon"` argument value. The `add.constant = FALSE` argument value in `MI.test()` specifies that no constants should be added to cells in the item-response table that have zero counts (adding small constants may be helpful when 0 cell counts are present).

```
> MI.test(data = farmer2, I = 3, J = 4, type = "bon",
    add.constant = FALSE)
Test for Simultaneous Pairwise Marginal Independence (SPMI)
```

```
Unadjusted Pearson Chi-Square Tests for Independence:
X^2_S = 64.83
X^2_S.ij =
     y1    y2    y3    y4
w1 4.93  2.93 14.29  0.01
w2 6.56  2.11 11.68  0.13
w3 13.98 0.00  7.08  1.11

Bonferroni Adjusted Results:
p.adj = 0.0019
p.ij.adj =
      y1     y2     y3     y4
w1 0.3163 1.0000 0.0019 1.0000
w2 0.1253 1.0000 0.0076 1.0000
w3 0.0022 1.0000 0.0934 1.0000
```

The output gives $X_S^2 = 64.83$ and the individual $X_{S,ij}^2$ values. For example, $X_{S,13}^2 = 14.29$ for nitrogen testing and natural drainage, which is the largest value among the 2×2 tables. With a χ_1^2 approximation for the largest X_{ij}^2 value, we obtain a Bonferroni-adjusted p-value of $\tilde{p} = 0.00188$, indicating strong evidence against SPMI. The output also provides each of the p-values for the 2×2 tables. These p-values are Bonferroni-corrected using $\min(IJ \times p_{ij}, 1)$. At the $\alpha = 0.05$ level, the combinations with significant association are salt and lagoon, nitrogen and natural drainage, and phosphorus and natural drainage.

For the bootstrap approximation, resamples are taken from the observed data vectors (w_1, w_2, w_3) and (y_1, y_2, y_3, y_4) independently with replacement from the original data and then $X_{S,b}^{2*}$ is calculated for each resample. This process is demonstrated below for one resample of size $n = 279$:

```
> I <- 3
> J <- 4
> n <- nrow(farmer2)

> set.seed(7812)
> iW <- sample(x = 1:n, size = n, replace = TRUE)
> iY <- sample(x = 1:n, size = n, replace = TRUE)

> # Use resampled index values to form resample
> farmer2.star <- cbind(farmer2[iW, 1:I], farmer2[iY,
    (I+1):(I+J)])
> head(farmer2.star)  # The row numbers here are from iW
    w1 w2 w3 y1 y2 y3 y4
234  0  0  0  1  0  0  0
39   0  0  0  1  1  0  0
157  0  0  0  0  1  0  0
207  0  0  0  1  0  0  0
231  0  0  0  1  0  0  0
236  0  0  0  0  0  1  0

> MI.stat(data = farmer2.star, I = I, J = J, add.constant = FALSE)
$X.sq.S
[1] 1.983276
```

```
$X.sq.S.ij
             [,1]         [,2]       [,3]       [,4]
[1,] 0.014713651 0.080432421 0.5614084 0.013000190
[2,] 0.557564403 0.260648453 0.2937971 0.008984192
[3,] 0.005364064 0.007344591 0.1631658 0.016852915

$valid.margins
[1] 12
```

The MI.stat() function is used to calculate $X_{S,b}^{2*} = 1.98$. We repeat this same resampling process by using the function MI.test(), where type = "boot" specifies the bootstrap test and B specifies the number of resamples.

```
> set.seed(7812)
> MI.test(data = farmer2, I = 3, J = 4, B = 5000, type = "boot",
    add.constant = FALSE, plot.hist = TRUE)
Test for Simultaneous Pairwise Marginal Independence (SPMI)

Unadjusted Pearson Chi-Square Tests for Independence:
X^2_S = 64.83
X^2_S.ij =
     y1    y2    y3   y4
w1  4.93  2.93 14.29 0.01
w2  6.56  2.11 11.68 0.13
w3 13.98  0.00  7.08 1.11

Bootstrap Results:
Final results based on 5000 resamples
p.boot = 0.0004
p.combo.prod = 0.0004
p.combo.min = 0.0048
```

Similar to Section 6.2, we specify a seed number before running MI.test() in order to reproduce the same results at a later time if needed. Out of $B = 5000$ resamples, only two resulted in a $X_{S,b}^{2*}$ that was larger than the observed X_S^2, so the p-value is 0.0004. Figure 6.4 (upper-left corner) provides the bootstrap-estimated probability distribution for X_S^2. The location of the test statistic's observed value is indicated by the vertical line. Obviously, X_S^2 is an extreme value relative to this distribution, resulting in rejection of the SPMI hypothesis.

In addition to performing the test with X_S^2, the function also performs a bootstrap test for SPMI using two p-value combination methods (Loughin, 2004). For these tests, the p-values from each $X_{S,ij}^2$ are found using a χ_1^2 approximation. The IJ p-values are "combined" by multiplying them together or by finding their minimum value. For both p-value combination methods, small observed values relative to the estimated distribution indicate evidence against the null hypothesis. In both cases, there is substantial evidence against SPMI.

The type argument in MI.test() can also have a value of "rs2" for the second-order Rao-Scott correction and "all" for all testing procedures. For the second-order Rao-Scott correction, $X_{RS2}^2 = 36.62$ and the degrees of freedom for the χ^2 distribution is 6.78 (results not shown). This leads to a p-value of 4.5×10^{-6}, which again indicates strong evidence against SPMI.

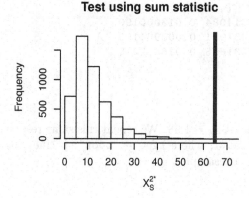
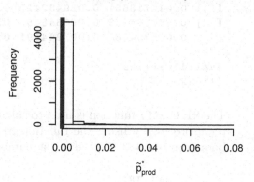
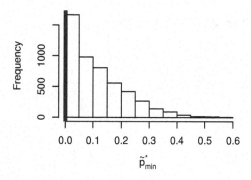

Figure 6.4: Histograms of the estimated probability distributions under SPMI. The vertical lines denote the observed statistic value for the original data.

Other tests for independence involving MRCVs

The testing procedures discussed in this section can be extended to other situations. For example, in the case of one MRCV and one ordinary SRCV, a test for *multiple marginal independence* (MMI) can be performed (see Exercise 5). Suppose that the SRCV is the row variable with I levels and the MRCV is the column variable with J items. Then the item-response table consists of J different $I \times 2$ contingency tables, each of which summarizes counts for one item across the levels of the SRCV. Analogous to X_S^2, Pearson chi-square statistics can be calculated for each of the J contingency tables, and the sum of these individual statistics provides an overall test statistic for MMI (Agresti and Liu, 1999). The Bonferroni, bootstrap, and Rao-Scott approaches can all be applied, and each has properties similar to when they are used for testing for SPMI (Bilder et al., 2000). Additionally, when there is a third SRCV, a test can also be performed for *conditional multiple marginal independence* given this third variable (Bilder and Loughin, 2002).

6.4.3 Regression modeling

As was shown in Sections 4.2.3 and 4.2.4, loglinear Poisson regression models are useful for describing relationships among categorical variables and for estimating the strength of associations using odds ratios. In the context of MRCVs, we can adapt these models both to test for SPMI and to describe associations between MRCVs when SPMI does not hold. This flexibility gives models a distinct advantage over the methods in the previous section.

Consider first the regression model that assumes independence between items W_i and Y_j:

$$\log(\mu_{ab(ij)}) = \beta_{0(ij)} + \beta^W_{a(ij)} + \beta^Y_{b(ij)}, \, a = 1, 2, \, b = 1, 2, \quad (6.14)$$

where $\mu_{ab(ij)}$ is the expected count for row a and column b of the $(i,j)^{th}$ 2×2 contingency table summarizing the pairwise joint responses. This is just Equation 4.3 with extra subscripts (i, j) to identify the two items being modeled. If Equation 6.14 is applied to all IJ 2×2 tables, this model represents SPMI. Odds ratios for each 2×2 table are equal to 1 for Equation 6.14.

Alternatively, we can consider models that allow odds ratios to not be equal to 1. Parameters can be added to Equation 6.14 that relate the odds ratios to the row and column items in much the same way that two-way ANOVA models relate means to levels of factors. These regression models include

1. Homogeneous association: $\log(\mu_{ab(ij)}) = \beta_{0(ij)} + \beta^W_{a(ij)} + \beta^Y_{b(ij)} + \lambda_{ab}$

2. W-main effects: $\log(\mu_{ab(ij)}) = \beta_{0(ij)} + \beta^W_{a(ij)} + \beta^Y_{b(ij)} + \lambda_{ab} + \lambda^W_{ab(i)}$

3. Y-main effects: $\log(\mu_{ab(ij)}) = \beta_{0(ij)} + \beta^W_{a(ij)} + \beta^Y_{b(ij)} + \lambda_{ab} + \lambda^Y_{ab(j)}$

4. W and Y-main effects: $\log(\mu_{ab(ij)}) = \beta_{0(ij)} + \beta^W_{a(ij)} + \beta^Y_{b(ij)} + \lambda_{ab} + \lambda^W_{ab(i)} + \lambda^Y_{ab(j)}$

5. Saturated: $\log(\mu_{ab(ij)}) = \beta_{0(ij)} + \beta^W_{a(ij)} + \beta^Y_{b(ij)} + \lambda^{WY}_{ab(ij)}$

Each of these models allows for the odd ratios, OR_{ij}, say, in each 2×2 contingency table to vary in a particular way. For example, the homogeneous association model uses the association parameter λ_{ab} which is the same for all 2×2 tables. This leads to the equality of each odds ratio, i.e., $OR_{11} = \cdots = OR_{IJ}$, but these odds ratios are not necessarily equal to 1. Also, the Y-main effects model allows the odds ratios to vary across levels of the column MRCV, Y, but they are equal across levels of the row MRCV, W; i.e., $OR_{1j} = \cdots = OR_{Ij}$ for $j = 1, \ldots, J$. Finally, the saturated model is essentially Equation 4.4 with extra subscripts (i, j) to identify the two items being modeled. This model does not assume any type of structure among the odds ratios in relation to the two MRCVs. The previous three models are special cases of the saturated model where we specify an association that lies between independence and complete dependence by allowing odds ratios to vary in a structured way.

A single likelihood function based on the Poisson distribution can be written for any one of the 2×2 tables. However, the parameters for the odds ratios in most of these models apply to several tables at once, so we need a likelihood function that can simultaneously model these tables. A full likelihood function across all 2×2 tables is not as easy to write out, because this would entail specifying a model for the associations among all items within each MRCV. Instead, Bilder and Loughin (2007) propose to use a *pseudo-likelihood* function, which is a construct developed by Rao and Scott (1984) for fitting a loglinear model to contingency table data arising through complex survey sampling (see Section 6.3.5).[16] For the MRCV setting, Bilder and Loughin (2007) form a pseudo-likelihood function simply by multiplying each of the IJ single Poisson likelihood functions together. The parameter

---

[16]This is the same principle used in the generalized estimating equation methods of Zeger and Liang (1986), where a working correlation matrix under independence is used. See also Section 6.5.5.

estimators resulting from maximizing the pseudo-likelihood function are consistent and approximately normally distributed in large samples, just like MLEs (see Appendix B.3.3), as long as the models for the individual tables are correct. The estimates can be found using the `glm()` function with `family = poisson(link = "log")`.

There are some drawbacks to this model-fitting approach. First, the standard errors given by `glm()` are likely to be wrong because the likelihood used in the model does not account for the correlations among the counts in the IJ tables. Fortunately, these standard errors can be corrected using "sandwich" methods similar to those used for generalized estimating equations (see Section 6.5.5). The `genloglin()` function of the `MRCV` package makes the necessary corrections and will be discussed in an example shortly. Second, the model parameter estimates and corresponding odds ratios are likely to be estimated with reduced precision relative to what could be achieved with a full likelihood. This is unavoidable, unless a more complete model for the between-item correlations can be accurately developed and used.

Because we do not use a true likelihood function, information criteria cannot be used to choose between models. Instead, hypothesis tests are used for these model comparisons. Specifically, a Pearson statistic can be calculated to compare two models where one is nested within the other. The statistic has the form

$$X_M^2 = \sum_{a,b,i,j} \frac{\left(\hat{\mu}_{ab(ij)}^{(a)} - \hat{\mu}_{ab(ij)}^{(0)}\right)^2}{\hat{\mu}_{ab(ij)}^{(0)}},$$

where $\hat{\mu}_{ab(ij)}^{(0)}$ and $\hat{\mu}_{ab(ij)}^{(a)}$ are the model predicted counts for the null and alternative hypothesis models, respectively. Once again, the large-sample distribution of X_M^2 is a linear combination of independent χ_1^2 random variables. First- and second-order Rao-Scott corrections can be calculated in order to evaluate the extremeness of X_M^2 under the null hypothesis. Similar to Section 6.4.2, the first-order correction can lead to a liberal test and the second-order correction can be conservative at times (Bilder and Loughin, 2007). Analogous calculations can be applied to statistics created using a likelihood-ratio formulation.

As an alternative to a Rao-Scott correction, the bootstrap can be used to estimate the distribution of X_M^2. Bilder and Loughin (2007) give details of a semi-parametric resampling approach to estimate this distribution that uses the Gange (1995) algorithm for generating correlated binary data. This form of resampling is needed, rather than what was presented in Section 6.4.2, when the test statistics are computed from a model that does not assume SPMI. Both the null- and alternative-hypothesis models are fit to the resampled data sets so that $X_{M,b}^{2\star}$ can be calculated for each resample. The p-value for the test is the proportion of the $X_{M,b}^{2\star}$ values greater than or equal to the observed X_M^2; i.e., calculate $B^{-1}(\# \text{ of } X_{M,b}^{2\star} \geq X_M^2)$. Small p-values indicate evidence against the null-hypothesis model.

Once an adequate model is found, model-estimated odds ratios and their corresponding confidence intervals are calculated to examine associations between MRCVs. Standardized residuals are also calculated to determine if deviations from the specified model exist. We illustrate odds ratio and standardized residual calculation in the next example.

Example: Swine waste management data (Swine.R)

To demonstrate how `genloglin()` works, we summarize the original MRCV binary responses into 2×2 table counts using `item.response.table()`. This time, however, the `create.dataframe = TRUE` argument value is added to `item.response.table()` so that the 2×2 table counts are now given in one column. The columns W and Y contain the names of the items involved in a particular table, and wi and yj identify the four 0-1 combinations of the table. Below is the code and output:

```
> mod.data.format <- item.response.table(data = farmer2, I = 3, J
    = 4, create.dataframe = TRUE)
> head(mod.data.format)
   W   Y  wi  yj  count
1  w1  y1  0   0   123
2  w1  y1  0   1   116
3  w1  y1  1   0    13
4  w1  y1  1   1    27
5  w1  y2  0   0   175
6  w1  y2  0   1    64
```

We apply the `genloglin()` function to fit models to data in this format. The function automatically reformats the data as above so that we can simply use the `farmer2` data frame in the `data` argument. The `genloglin()` function uses `glm()` on the summarized counts to perform the pseudo-likelihood model fitting described earlier, and it corrects the standard errors using a sandwich estimator of the covariance matrix.

The `genloglin()` function includes two ways to specify a model. First, the `model` argument accepts the names `"spmi"`, `"homogeneous"`, `"w.main"`, `"y.main"`, `"wy.main"`, or `"saturated"`, where the "w" and "y" designations correspond to the names assigned by `item.response.table()`. Alternatively, the `model` argument can take on a formulation similar to the `formula` argument of `glm()`. Below we use the first way to specify the model in order to estimate the Y-main-effects model:

```
> set.seed(8922)
> mod.fit1 <- genloglin(data = farmer2, I = 3, J = 4, model =
    "y.main", boot = TRUE, B = 2000)
> summary(mod.fit1)
Call: glm(formula = count ~ -1 + W:Y + wi %in% W:Y + yj %in% W:Y
    + wi:yj + wi:yj %in% Y, family = poisson(link = log), data =
    model.data)

Deviance Residuals:
      Min        1Q    Median        3Q       Max
  -1.58007  -0.13272   0.00043   0.10282   0.79587

Coefficients:
         Estimate   RS SE  z value  Pr(>|z|)
Ww1:Yy1   4.83360  0.06535  73.969  < 2e-16 ***
Ww2:Yy1   4.85571  0.06387  76.023  < 2e-16 ***
Ww3:Yy1   4.87418  0.06314  77.199  < 2e-16 ***
Ww1:Yy2   5.15802  0.04696 109.838  < 2e-16 ***
Ww2:Yy2   5.19427  0.04411 117.750  < 2e-16 ***

<OUTPUT EDITED>

Yy2:wi:yj  -0.70644  0.63025  -1.121  0.26233
Yy3:wi:yj  -3.56978  0.88623  -4.028  5.62e-05 ***
Yy4:wi:yj  -1.39762  0.85852  -1.628  0.10354
---
Signif. codes:  0 '***' 0.001 '**' 0.01 '*' 0.05 '.' 0.1 ' ' 1

(Dispersion parameter for poisson family taken to be 1)

Null deviance: 25401.0663   Residual deviance:     5.8825
```

Number of Fisher Scoring iterations: 4

The output begins with the actual call to `glm()`. The model syntax given in the `formula` argument shows the alternative way that the Y-main-effects model could have been specified, where the variable names are derived from the data frame created by `item.response.table()`. This syntax uses *nested effects* to specify first a loglinear model under independence for each 2 × 2 table using the `W:Y + wi%in%W:Y + yj%in%W:Y` portion of the code. This ensures that the marginal counts for each 2 × 2 table match between the data and the model, as is typically required for loglinear modeling of associations. The `wi:yj + wi:yj%in%Y` syntax then allows the association within each 2 × 2 table to vary in the specified manner across the entire item response table. As an additional example of this syntax, the saturated model can be specified using `model = count ~ -1 + W:Y + wi%in%W:Y + yj%in%W:Y + wi:yj%in%W:Y` in `genloglin()`. This formulation expands the use of nested effects further by essentially estimating Equation 4.4 to each 2 × 2 table.

Continuing in the output, we can relate the estimated parameters to the model specification. For example, `W:Y` in the `formula` argument of `glm()` corresponds to the intercept terms $\beta_{0(ij)}$, which are listed as `Ww1:Yy1` to `Ww3:Yy4` in the output. Also, `wi:yj%in%Y` corresponds to the Y-main effect terms $\lambda^Y_{ab(j)}$ which are listed as `Yy2:wi:yj` to `Yy4:wi:yj` in the output (`Yy1:wi:yj` is not estimated because $\lambda^Y_{ab(1)}$ is set to 0).

We next check the fit of the Y-main-effects model by comparing it to the saturated model to determine if the simpler association structure is adequate. The `boot` argument of `genloglin()` was set to `TRUE` (the default value) previously, so we are able to estimate the distribution of X^2_M using the bootstrap. By specifying `type = "all"` in the `anova()` method function below, we are able to view the results from both the bootstrap and the second-order Rao-Scott approaches.

```
> comp1 <- anova(object = mod.fit1, model.HA = "saturated", type
    = "all")
> comp1
Model comparison statistics for
H0 = y.main
HA = saturated

Pearson chi-square = 5.34
LRT = 5.88

Second-Order Rao-Scott Adjusted Results:
Rao-Scott Pearson chi-square statistic = 10.85, df = 5.23, p =
    0.0624
Rao-Scott LRT statistic = 11.96, df = 5.23, p = 0.0409

Bootstrap Results:
Final results based on 2000 resamples
Pearson chi-square p-value = 0.031
LRT p-value = 0.017
```

From the output, $X^2_M = 5.34$ and the corresponding p-value is 0.031 using the bootstrap approach. The p-value using the second-order Rao-Scott approach is 0.0624. Both p-values indicate that there is marginal evidence of fit problems with the simpler Y-main-effects model.

Observed odds ratios and model-predicted odds ratios for each (W_i, Y_j) combination are found using the `predict()` function:

```
> OR.mod <- predict(object = mod.fit1, alpha = 0.05)
> OR.mod
Observed odds ratios with 95% asymptotic confidence intervals
      y1                  y2                  y3
w1  2.2  (1.08, 4.47)    1.82 (0.91, 3.65)  0.1  (0.02, 0.42)
w2  2.91 (1.25, 6.78)    1.77 (0.81, 3.88)  0.07 (0.01, 0.51)
w3 10.27 (2.34, 44.98)   0.99 (0.37, 2.66)  0.1  (0.01, 0.78)
      y4
w1  1.09 (0.23, 5.12)
w2  0.68 (0.09, 5.43)
w3  0.45 (0.03, 7.83)

Model-predicted odds ratios with 95% asymptotic confidence
    intervals
            y1                  y2                  y3
w1  3.18 (1.54, 6.57)   1.57 (0.78, 3.18)  0.09 (0.02, 0.44)
w2  3.18 (1.54, 6.57)   1.57 (0.78, 3.18)  0.09 (0.02, 0.44)
w3  3.18 (1.54, 6.57)   1.57 (0.78, 3.18)  0.09 (0.02, 0.44)
      y4
w1  0.79 (0.21, 2.98)
w2  0.79 (0.21, 2.98)
w3  0.79 (0.21, 2.98)

Bootstrap Results:
Final results based on 2000 resamples
Model-predicted odds ratios with 95% bootstrap BCa confidence
    intervals
      y1                y2                 y3
w1  3.18 (1.53, 7.3)   1.57 (0.72, 3.16)  0.09 (0.03, 0.33)
w2  3.18 (1.53, 7.3)   1.57 (0.72, 3.16)  0.09 (0.03, 0.33)
w3  3.18 (1.53, 7.3)   1.57 (0.72, 3.16)  0.09 (0.03, 0.33)
      y4
w1  0.79 (0.27, 3.5)
w2  0.79 (0.27, 3.5)
w3  0.79 (0.27, 3.5)
```

The first table gives the observed odds ratios and associated confidence intervals using Equation 1.9. The second and third tables give the model-predicted odds ratios and corresponding confidence intervals. These values are equal within each column because no W effects are estimated in the Y-main-effects model. In the second table, the confidence interval is a Wald interval using model-based estimates of the odds ratio and variance. The last table uses bootstrap-based intervals calculated by the BC_a method; see Davison and Hinkley (1997) for details on the calculation of this type of interval.[17]

---

[17] The bootstrap BC_a confidence interval method is one of two bootstrap-based methods recommended for general practice. The lower and upper bounds of this interval are specific lower and upper quantiles from the B statistics calculated on the resampled data sets. Rather than using the $\alpha/2$ and $1 - \alpha/2$ quantiles, adjusted quantiles are found that take into account possible bias and unstable variance in the statistic being studied.

The fit of the model within each 2×2 table is assessed through standardized Pearson residuals, which are found using the `residuals()` function. Both model-based large-sample and bootstrap-estimated variances are used in their calculation:

```
> resid.mod <- residuals(object = mod.fit1)
> # resid.mod$std.pearson.res.asymp.var  # Excluded to save space
> resid.mod$std.pearson.res.boot.var
       y1         y2          y3         y4
W    1     0    1     0    1     0    1     0
w1 1 -2.39  2.39  1.03 -1.03  0.31 -0.31  1.13 -1.13
   0  2.39 -2.39 -1.03  1.03 -0.31  0.31 -1.13  1.13
w2 1 -0.59  0.59  0.91 -0.91 -0.67  0.67 -0.44  0.44
   0  0.59 -0.59 -0.91  0.91  0.67 -0.67  0.44 -0.44
w3 1  2.83 -2.83 -1.73  1.73  0.24 -0.24 -0.87  0.87
   0 -2.83  2.83  1.73 -1.73 -0.24  0.24  0.87 -0.87
```

Using standard guidelines (See Section 5.2.1), all of the standardized residuals are reasonably small (below 2 in absolute value) except those for (W_1, Y_1) and (W_3, Y_1), which leads us to a similar overall fit conclusion as when using X_M^2. These residuals specifically show that the associations across contaminants for the lagoon waste storage method do not appear to be as homogeneous as those within the other waste storage methods. This suggests that a new model which takes this heterogeneity into account could potentially fit better. We construct this model by adding an additional term that forces a perfect fit to the counts in the (W_3, Y_1) combination table.

```
> set.seed(9912)
> mod.fit.final <- genloglin(data = farmer2, I = 3, J = 4, model
    = count ~ -1 + W:Y + wi%in%W:Y + yj%in%W:Y + wi:yj +
    wi:yj%in%Y + wi:yj%in%W3:Y1, boot = TRUE, B = 2000)
> # summary(mod.fit.final)  # Excluded to save space
> comp.final1 <- anova(object = mod.fit.final, model.HA =
    "saturated", type = "all")
> comp.final1

Model comparison statistics for
H0 = count ~ -1 + W:Y + wi %in% W:Y + yj %in% W:Y + wi:yj + wi:yj
    %in% Y + wi:yj %in% W3:Y1
HA = saturated

Pearson chi-square statistic = 1.81
LRT statistic = 1.86

Second-Order Rao-Scott Adjusted Results:
Rao-Scott Pearson chi-square statistic = 4.29, df = 5.07, p =
    0.5178
Rao-Scott LRT statistic = 4.4, df = 5.07, p = 0.503

Bootstrap Results:
Final results based on 2000 resamples
Pearson chi-square p-value = 0.404
LRT p-value = 0.388
```

This new model fits considerably better than the Y-main-effects model. Comparing the new model to the saturated model, we obtain $X_M^2 = 1.81$ with a bootstrap p-value of 0.404. Also, the standardized residuals are all relatively small in absolute value (not shown). Therefore, this new model appears to reasonably explain the associations between the contaminants and the waste storage methods. The model-estimated odds ratios are 2.48 for (W_1, Y_1) and (W_2, Y_1) with a 95% bootstrap BC$_a$ interval of (1.16, 5.24). Of course, the model estimated odds ratio (W_3, Y_1) and corresponding confidence interval is essentially the same as for the observed values as calculated earlier. All of the other model-estimated odds ratios and standardized residuals are essentially the same as in the Y-main-effects model. In summary, there is moderate to large positive association between lagoon waste storage and testing for each contaminant. Also, there tends to not be contaminant testing when natural drainage is used. For pit and holding tank, there is no clear association with contaminant testing.

Models for three or more MRCVs are possible as well. For example, consider the case of three MRCVs where the third variable is represented by binary responses Z_k for $k = 1, \ldots, K$. Rather than the IJ 2×2 tables as we saw previously, we now have IJK $2 \times 2 \times 2$ tables. The model under complete independence is

$$\log(\mu_{abc(ijk)}) = \beta_{0(ijk)} + \beta_{a(ijk)}^W + \beta_{b(ijk)}^Y + \beta_{c(ijk)}^Z, \qquad (6.15)$$

for $a = 1, 2$, $b = 1, 2$, $c = 1, 2$, $i = 1, \ldots, I$, $j = 1, \ldots, J$, and $k = 1, \ldots, K$. This is the usual Poisson regression model for complete independence between three binary categorical variables, with additional (i, j, k) subscripts added to denote the specific W, Y, Z combination $2 \times 2 \times 2$ table. More complicated models can be formed in which main effects and interactions are added across levels of $X, Y,$ and Z. For example, consider a model that contains the terms included in Equation 6.15 and

$$\lambda_{ab} + \lambda_{ab(i)}^W + \lambda_{ab(j)}^Y + \lambda_{ab(ij)}^{WY} + \delta_{bc} + \delta_{bc(j)}^Y + \omega_{ac}.$$

The λ-parameters allow odds ratios between W and Y to be different for each W, Y combination but remain constant across levels of Z; the δ-parameters allow odds ratios between Y and Z to be different for each Y level, but the same for all X and Z levels; and the ω-parameter allows odds ratios between W and Z to be different from 1, but the same for all levels of X, W, and Z. Model fitting, hypothesis testing, and odds ratio estimation follow in the same manner as described for the two MRCV case. However, as the number of MRCVs increases, finding a suitable model and interpreting its parameters can become more difficult.

6.5 Mixed models and estimating equations for correlated data

All of the models that were introduced in Chapters 2–4 make the critical assumption that the data to which they are fit consist of independent observations. In real applications, data are often collected in clusters, or multiple measurements may be made on the same subject. With such data, it is typical that measurements from the same cluster or unit are more similar to one another than to observations made on different clusters or units. In other words, the data within these groupings are *correlated*.

Performing a statistical analysis on correlated data as if they were independent is not recommended. In particular, when data are positively correlated—i.e., observations from the same group are more similar than those from different groups—a statistical analysis that assumes independence tends to be too liberal. To see why this is true, consider an extreme situation where we have a sample of g groups, each with m observations that all provide identical results, while observations in different groups may vary. Then the multiple measurements in each group are superfluous, and a single measurement from each group is all that is needed. This means that, instead of having a sample of $n = gm$ observations, we effectively have a sample of only $g = n/m$ meaningful observations; the other data tell us nothing new. A statistical analysis should be conducted only on these g numbers, meaning that the sample is effectively much smaller than it appears. If an analysis is done instead on all n observations assuming independence, variances for statistics will be much smaller than they should be. The resulting confidence intervals will be narrower than they should be to provide the stated level of confidence, and p-values smaller than they should be, resulting in excessive type I error rates.

This is a serious issue that should not be ignored in an analysis. There are two basic approaches that account for the effects of analyzing correlated data. The first is to change the statistical model so that it correctly reflects the grouping structure of the data. There are numerous ways to achieve this, but in this section we focus on one of the more commonly used corrective models for counts and proportions, the *generalized linear mixed model*. In Section 6.5.5 we briefly discuss the alternative approach, *generalized estimating equations*, in which models are fit to the data assuming independence, but the inferences resulting from the model are adjusted to account for correlation.

6.5.1 Random effects

Populations of measurements often fall naturally into groups. For example, several different hospitals may participate in a medical study, each with its own set of patients. In a quality-assurance study, we may sample products from several different production runs (or "batches"). In the placekicking example introduced in Chapter 2, placekicks are attempted by different kickers and in different games.

Data that are sampled in groups of this type are called *clustered*. In each of these examples, the grouping variable—hospital, batch, kicker, or game—may be an unavoidable part of the study. Sampling in groups or batches may be needed to ensure that a large enough sample can be collected within a reasonable time frame or budget. Sampling from *multiple* groups helps to ensure that the results of the study apply to units in a larger population of such groups, rather than to just those in one particular group.

In any case, these grouping variables are not of primary interest in the study. They may nonetheless have an impact on the mean or probability of a measured response, because different levels of the variables may have naturally higher or lower response potential (e.g., different kickers may have higher or lower success probabilities than other kickers across all placekicks). This variability imparts a correlation onto the measurements, because all measurements from the same cluster are exposed to that cluster's unique response potential, and will therefore tend to be more similar to one another than they are to measurements from different clusters that have different response potential.

Another way in which measurements can be grouped is when units are independent but multiple observations are made on each unit. These observations are called *repeated measures*, and the units upon which the repeated measurements are taken are generically referred to as "subjects." Repeated measures are quite common in medical studies, where measurements of things like vital signs, disease status, and side effects may be recorded on subjects at specific times. Repeated measures are also used in areas such as studying growth of plants or animals over time, or in measuring soil nutrients and contamination at different prescribed depths. Notice that repeated measures produce correlated data: if a subject's measurement is above average at a particular time or depth interval, it may be likely to remain above average at neighboring times or depths. As with clustered data, correlation among repeated measures taken on the same subject can be alternatively expressed as a different overall response potential for each subject, causing higher- or lower-than-average responses for all measurements on the same subject.

Thus, clustered data and repeated measures are two different versions of the same phenomenon of grouped measurements. However, there are characteristics that distinguish the two structures. First, repeated measures are usually measured in a specific order (in time or space) that may be common to all subjects. For example, measurements of growth might be taken at specific points in time for all subjects. The temporal or spatial measure is typically the only explanatory variable that changes among repeated measures within a subject, although there can be exceptions. All other explanatory variables are usually observed on entire subjects, causing their values to be the same for all measurements within that subject. On the other hand, with clustered data the measurements within a cluster are often *not* ordered in any way. They are said to be *exchangeable* in the sense that one could rearrange the measurements' identifying labels randomly and not lose information. However, different responses within a cluster *may* have different values of some or all explanatory variables.

Consider next the population of all possible levels of a particular grouping variable (e.g., all possible hospitals in a medical study or all possible trees in a growth study). We can imagine that each level has a constant value associated with it that gets added to the overall mean or probability, raising or lowering it by a unique amount. This constant value is called the *random effect* of the level. Because grouping variables like hospitals or tree labels are always categorical, they are often called *random-effects factors*. The levels of a random-effects factor that are used in a particular study are assumed to have been sampled from a population of possible levels.

All of the categorical explanatory variables we have studied prior to this section are *fixed-effects factors*. Their levels are either chosen deliberately or observed among a sample, and comparisons of means or probabilities among those specific levels are of interest. In particular, there is not assumed to be some larger population of levels from which the observed levels were chosen.

For the purposes of modeling, we typically assume that the random effects follow some known distribution with unknown parameters. Most commonly we assume they are normally distributed with mean 0 and unknown variance. The zero mean ensures that the average hospital or tree adds 0 to the measurements taken on units within it, while the unknown variance creates a new parameter in the model, representing the amount by which the measurements from different hospitals or trees may differ from one another. The variance associated with a particular random-effects factor is called its *variance component*.

Understanding the variance component of a random-effects factor can sometimes be the focus of a study. For example, there are many labs that can analyze blood samples, all supposedly using techniques that follow the same protocols and staffed by technicians that have had the same training. A regulatory agency might be interested in assessing whether the protocols and training are adequate. If they are not, then we might expect that there is noticeable variability among labs in the results that they produce on the same samples, and that some labs use methods or technicians that produce consistently higher or lower responses than others. In this case, we are interested in assessing whether there is variability among all labs in the population rather than in comparing a specific lab to another. In other words, we want to know whether the variance component for the lab random-effect factor is zero.

A given study can have more than one random-effects factor, and these factors can be nested or crossed. For example, there may be many labs that can process and analyze blood samples, and many technicians working at each lab. We could gather multiple samples from each of a large number of donors, and send three samples from each donor to a set of randomly selected labs to be analyzed by three different technicians within those labs. Then we have random-effects factors representing labs, technicians nested within the labs, and donors crossed with both labs and technicians.

6.5.2 Mixed-effects models

Models may contain only fixed effects, only random effects, or a combination of fixed and random effects. The latter are called *mixed-effects models*, or "mixed models" for short. Recall from Sections 2.3 and 4.2.1 that logistic, Poisson, and many other regression models for counts or proportions are different types of *generalized linear models* (GLMs). When a mixed model is applied to data whose fixed effects would normally be modeled using a generalized linear model, then we create a *generalized linear mixed model* (GLMM).

Consider first a simple problem in which the only explanatory variable is a random-effects factor with a sampled levels. For example, perhaps we have measured the number of pine beetles on t trees at each of a different random locations in a region. It is possible that trees in different areas have higher or lower mean counts for a variety of reasons that we are not attempting to measure. If we use a Poisson GLM for this problem, the variability in means among locations would likely cause overdispersion (see Section 5.3). Instead, we can fit a Poisson model that allows the mean to vary randomly among different locations. Using a log link, $g(\mu_{ik}) = \log(\mu_{ik})$, where μ_{ik} is the mean beetle count for tree k at location i, we can write the linear predictor as

$$g(\mu_{ik}) = \beta_0 + b_{0i}, \qquad (6.16)$$

where b_{0i}, $i = 1, \ldots, a$ is the random effect of location i; $b_{01}, \ldots, b_{0a}$ are assumed to be a random sample from $N(0, \sigma_{b0}^2)$, and β_0 is the value of the linear predictor (the log-mean) at the average location. The goal in the study might then be to estimate σ_{b0}^2, which would give us an idea regarding how variable beetle counts are throughout the region.

Note that Equation 6.16 is a random-effects model, because the only factor in the model is location, which is random. Notice also that we could write the model a little differently by combining the average log mean with random effect for each location into a single symbol, say $\tau_i = \beta_0 + b_{0i}$, so that it is apparent that the model represents a *random intercept* for each location. In general, random effects combine with their respective fixed "average" parameter to create random parameter values for each subject or cluster.

We can extend this to a generalized linear *mixed* model if we have additional measurements. For example, trees in some locations may be larger than those in other locations, due to recent logging, fires, or other impacts, and larger trees might harbor more beetles simply due to the larger volume in which beetles could reside. Suppose that we measure the circumference of each tree in the study, denoted by $x_{ik}, i = 1, \ldots, a, k = 1, \ldots, t$, and believe that the log-mean beetle count might change linearly with the circumference. A fixed-effects generalized linear model ignoring locations would use $g(\mu_{ik}) = \beta_0 + \beta_1 x_{ik}$, where β_0 is the intercept (technically, the log-mean beetle count on a tree with 0 circumference) and β_1 is the slope of the log-mean (the change in log-mean beetle count per 1-unit increase in circumference).

If we want to allow for the effects of locations in this model, there are several ways to do so. Picture fitting a separate Poisson GLM to the beetle-count/tree-size relationship in each location. We would estimate a different intercept and slope in each case, but the estimates may or may not all be estimating the same underlying population quantity. If we believe that the underlying slopes should all be the same, then the random location effects act only on the intercepts, and the linear predictors are parallel. The corresponding GLMM uses the linear predictor,

$$g(\mu_{ik}) = \beta_0 + \beta_1 x_{ik} + b_{0i}, \qquad (6.17)$$

where $b_{01} \ldots, b_{0a}$ are assumed to be a random sample from $N(0, \sigma_{b0}^2)$. This model specifies that the ratio between mean counts for trees of a given size at two different locations is the same for every size of tree, or equivalently that the ratio of means between two particular sizes of tree is the same at each location. In a logistic regression model, the equivalent assumption would be that the odds ratios between any two levels of the random effect are the same for all values of x or vice versa. Notice that we could rearrange this model as $g(\mu_{ik}) = (\beta_0 + b_{0i}) + \beta_1 x_{ik}$, making it a little more clear that this model represents random intercepts for each location with a common slope for all locations.

Alternatively, if we expect that the true mean ratios (or odds ratios) between levels of x change depending on the level of the random effect, then the random effect is changing the slope as well. For the Poisson regression model this results in the linear predictor

$$g(\mu_{ik}) = \beta_0 + \beta_1 x_{ik} + b_{0i} + b_{1i} x_{ik}, \qquad (6.18)$$

where the additional random effects $b_{11}, \ldots, b_{1a}$ are a random sample from $N(0, \sigma_{b1}^2)$. The random effect b_{1i} measures the amount by which the slope of x in the linear predictor for location i differs from the average slope, β_1. Notice that there are now two variance components, σ_{b0}^2 and σ_{b1}^2, that are estimated when the model is fit to the data. Rearranging this model as we have with other models yields $g(\mu_{ik}) = (\beta_0 + b_{0i}) + (\beta_1 + b_{1i}) x_{ik}$. Thus, it is apparent that this model consists of both random intercepts and random slopes for each location.

One further model feature that can be considered is the relationship between b_{0i} and b_{1i}, $i = 1, \ldots, a$. In the model above, we have specified a separate normal distribution for the random effects on slope and intercept, which implies that we treat them as independent random variables. However, it might be that locations with very low intercepts have very low slopes as well, indicating that the mean beetle count is low for small trees and does not increase by much for larger ones. In other regions where small trees have high counts, it might be that the beetle populations increase even more rapidly in larger trees. Thus, we would have a positive correlation between slope and intercept random effects. This can be specified by adding to Equation 6.18 the assumption that (b_{0i}, b_{1i}), $i = 1, \ldots, a$, have correlation ρ_{01}.

Much more general models can be developed for problems with multiple fixed and random effects. The key in each case is to carefully specify the ways in which each random effect might affect the responses. There are several books on both linear mixed models and generalized linear mixed models that cover this process in detail. See, e.g., Raudenbush and Bryk (2002), Littell et al. (2006), Molenberghs and Verbeke (2005), and Bates (2010).

Example: Falls with Head Impact (FallsGLMM.R, FallsHead.csv)

Falls are a serious problem among the elderly, resulting in injuries, medical expenses, and sometimes death. In particular, head impacts during a fall can have severe consequences. Schonnop et al. (2013) studied video footage of 227 falls among 133 residents at two long-term care facilities in British Columbia, Canada. They recorded numerous attributes of each fall, including the direction of the fall, whether there was impact with any of several body parts during the fall, and the age and gender of the resident. Footage of some falls had obstructed views, and so not all falls have complete data records. We consider a reduced version of this data set[18] consisting of the 215 falls with recorded values for all of the following variables:

- `resident`: a numerical identification code for the resident whose fall was recorded,
- `initial`: a 4-level categorical variable with levels Backward, Down, Forward, and Sideways, and
- `head`: a binary variable indicating whether the fall resulted in the resident's head impacting the floor (1 = yes, 2 = no).

The first few lines of data are shown below, along with the code that shows the number of falls experienced by each subject in the study. These counts are depicted in Figure 6.5.

```
> fall.head <- read.csv("C:\\Data\\FallHead.csv")
> head(fall.head)
  resident  initial head
1       56 Sideways    0
2        9 Backward    0
3       30  Forward    0
4        9     Down    0
5       70 Sideways    0
6       21 Sideways    1

> fallFreq <- table(fall.head$resident)
> plot(x = as.numeric(names(fallFreq)), y = fallFreq, xlab =
    "Resident", ylab = "Fall Frequency", type = "h", lwd = 2)
```

In this example we model the probability of head impact as a function of the initial direction of the fall. The binary variable `head` is our response variable. The direction of the fall, `initial`, is a fixed effect because (a) the four levels—forward, backward, sideways, and straight down—are the only four levels possible, and (b) we are interested in comparing these four levels against one another for the possibility of different effects on the probability of head impact.

From Figure 6.5 it is clear that most residents were observed falling only once, but that some fell numerous times over the course of the study. It is possible that there is a common underlying mechanism behind the multiple falls experienced by a given resident—e.g., a tendency for dizziness, or weakness in a particular leg—that causes the resident to impact the floor in similar ways each time. For this reason, we consider `resident` to be a grouping factor in any analysis we undertake. It is a random-effects factor because the goals of the study are not merely to understand falling behavior among these specific residents. Rather, the observed residents are intended to represent a wider population of residents at similar long-term care facilities. The resulting model has the form,

$$\text{logit}(\pi_{ik}) = \beta_0 + \beta_2 x_{2ik} + \beta_3 x_{3ik} + \beta_4 x_{4ik} + b_i, \tag{6.19}$$

where π_{ik} is the probability that fall k for resident i has a head impact, β_0 is the log-odds of head impact fall for a person with `initial` = "Backward", $x_{2ik}, x_{3ik}, x_{4ik}$

---

[18] Data kindly provided by Dr. Steve Robinovich, Department of Biomedical Physiology and Kinesiology, and School of Engineering Science, Simon Fraser University.

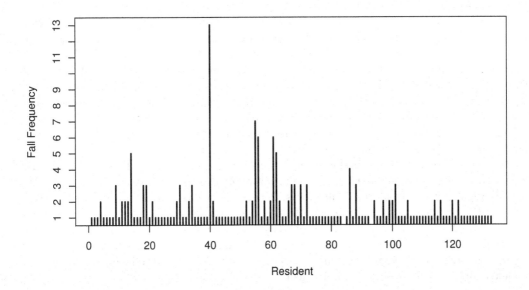

Figure 6.5: Number of falls per resident in the Falls with Head Impact example. Note that falls for two residents had incomplete data and were removed.

are indicator variables for levels "Down", "Forward", and "Sideways" of initial, β_j, $j = 2, 3, 4$ is the difference in log-odds of head impact between level j and level 1 of initial, and b_i is the random effect of resident i upon the log-odds of head impact. We assume that the b_i's are independent and normally distributed with mean 0 and variance σ_{b0}^2.

Although residents were not randomly sampled in any way from the supposed population of residents of long-term care facilities, we make the assumption that their falling experiences are representative of such individuals. This assumption should not be taken lightly. If there is something different about the residents at these facilities—for example, if they have different diets, exercise habits, or ethnicity that somehow relates to their falling experience—then the legitimacy of our assumption is questionable, and the results of the study may not apply to the larger population in the way we hope. We could partially check the representativeness of the sample by comparing relevant demographic characteristics of the sample to those of the population (to the extent that these are known). Any obvious differences would serve as a warning that inferences from the sample may not apply to the desired population. See McLean et al. (1991) for a discussion of the different populations that can be implied by the use of different random-effects structures.

6.5.3 Model fitting

A GLMM is fit to data using maximum likelihood estimation. Writing out a likelihood function turns out to be difficult for a GLMM, because the model contains unobserved random variables—the values of the random effects—in addition to the observed response variable, Y. The mathematical details of the process are provided in numerous books, including Raudenbush and Bryk (2002), Molenberghs and Verbeke (2005), Littell et al. (2006), and Bates (2010). Here, we give only an overview of the process.

Conditional on the random effects (i.e., for a specific cluster or subject), the distribution function of Y has the familiar form of a generalized linear model, typically Poisson or binomial. The mean of the conditional distribution for Y changes for each observation depending on the explanatory variables and on the random effects associated with that observation. Estimating the regression parameters from the conditional distribution for each cluster or subject results in estimates that apply only to the clusters or subjects that were present in the data. However, we want our parameter estimates to apply to the entire populations of random-effects levels, not just to those observed. We need to remove this dependence on the random effects somehow.

There are several ways to do this, but the most common one is to base the maximum likelihood estimates on the marginal distribution of Y; that is, on the average distribution over all possible values of the random effects. To do this, we first form the joint distribution between the response and the random effects. We then obtain the marginal distribution of Y by integrating the joint distribution with respect to the random effects, and form the likelihood from this marginal distribution. The parameters in this likelihood are the regression coefficients and the variance components for the random effects, as well as any modeled correlations between random effects.

For example, the model described in Equation 6.19 specifies that the conditional distribution of the binary response of head impact, given the resident's random effect, follows a binomial distribution with 1 trial and with probability of head impact based on the logit form given in the equation. Denote this distribution by $f(y_i|b_i;\boldsymbol{\beta})$ for resident i, where $\boldsymbol{\beta}$ contains all of the fixed-effect regression parameters. This distribution applies only to resident i's unique random effect value. The joint distribution of both the random effect and the response for resident i is $f(y_i, b_i; \boldsymbol{\beta}, \sigma_{b0}^2) = f(y_i|b_i;\boldsymbol{\beta})g(b_i;\sigma_{b0}^2)$, where $g(b_i;\sigma_{b0}^2)$ represents the normal density function with mean 0 and variance σ_{b0}^2. Because b_i is a random variable, we obtain the marginal distribution of the response, $h(y_i;\boldsymbol{\beta},\sigma_{b0}^2)$, by averaging across all possible values of b_i using an integral:

$$h(y_i;\boldsymbol{\beta},\sigma_{b0}^2) = \int_{-\infty}^{\infty} f(y_i|b_i;\boldsymbol{\beta})g(b_i;\sigma_{b0}^2)\,db_i$$

$$= \int_{-\infty}^{\infty} \pi_{ik}^{y_{ik}}(1-\pi_{ik})^{1-y_{ik}}\frac{1}{\sqrt{2\pi\sigma_{b0}^2}}\exp\left(-\frac{b_i}{2\sigma_{b0}^2}\right)db_i,$$

where π_{ik} is found from Equation 6.19 and is a function of b_i. Finally, the likelihood function for the parameters given all of the data $\mathbf{y}$ is the product $L(\boldsymbol{\beta},\sigma_{b0}^2|\mathbf{y}) = \prod_{i=1}^{a} h(y_i;\boldsymbol{\beta},\sigma_{b0}^2)$.

Unfortunately, the integral in this likelihood cannot generally be evaluated mathematically. We must instead use some alternative method to evaluate the integral, and then maximize this evaluation to find the parameter estimates. There are three general approaches to performing the evaluation[19]:

1. *Penalized quasi-likelihood* or *pseudo-likelihood* ("approximate the model"): Using a Taylor-series approximation, find convenient approximations to the inverse link function that allow the model to be written as "mean + error," the way a normal linear regression model is typically expressed. The mean and error portions are both evaluated

---

[19] The problem can also be framed as a Bayesian estimation problem treating the random effects as parameters with a normal prior. See Section 6.6 for a brief description of Bayesian methods, and Littell et al. (2006) for a more general description of Bayesian approaches to mixed models.

from quantities that are estimated, resulting in "pseudo-data" that are approximately normally distributed. Using the pseudo-data in the approximate model results in an integral that can be evaluated relatively easily using iterative numerical techniques. The pseudo-data are updated with each iteration.

2. *Laplace approximation* ("approximate the integrand"): Because we assume normal distributions for any random effects, the integrand has a form that can be approximated by a simpler function that is easier to integrate mathematically. The resulting function is maximized using iterative numerical techniques.

3. *(Adaptive) Gaussian quadrature* ("approximate the integral"): An integral in one dimension is an area under a curve. This area can be approximated by a sequence of rectangles, very much like approximating a density function with a histogram. Gaussian quadrature is a method for performing this type of computation in any number of dimensions. The rectangles are referred to as the "points of quadrature." Using a large number of quadrature points represents the shape of the integrand better than using fewer wide rectangles, but is also more computationally intensive. Also, using the shape of the function to help select the locations of the quadrature points, a procedure called *adaptive* quadrature, can result in a better approximation than using a preselected grid of points that may be far from any peaks in the function.

The penalized quasi-likelihood (PQL) approach is applicable to fairly general problems, including repeated measures. However, because it is based on an altered version of the data, there are limits to the inferences that can be derived from the model and its estimates. For example, likelihood ratio inference and information criteria are not valid for comparing models with different sets of fixed effects, because these models produce different pseudo-data. Also, PQL produces estimates that are biased, and this bias may be extreme in binomial models with small numbers of trials (Breslow and Lin, 1995). In general, it works best when the models are reasonably approximated by normal distributions, which argues against its use for binomial data involving small numbers of trials or many extreme probabilities and for Poisson data with very small means.

Adaptive Gaussian quadrature (AGQ) with many quadrature points provides the closest approximation to the likelihood, but is also computationally the most difficult. For problems with more than one or two random-effect factors, the computing time of AGQ can become prohibitive with multiple quadrature points. It turns out that AGQ with only one quadrature point is mathematically equivalent to using the Laplace approximation. Therefore, Laplace can be used on problems with more complex random-effects structures than AGQ can handle. However, Laplace does not always provide very accurate estimates of the MLEs. Overall, we recommend using AGQ whenever it is feasible, and Laplace otherwise. If Laplace cannot be performed, then PQL can be used, but inferences should be viewed as very approximate.

Regardless of the fitting method used, the quantities that are returned are essentially the same. Primarily, there are parameter estimates for the regression coefficients and the variance components along with their standard errors, and the deviance corresponding to the log-likelihood function evaluated at the parameter estimates. Secondarily, the estimated values of the random effects for each cluster or subject may be produced. These are sometimes called *conditional modes*.

Fitting GLMMs in R

There are numerous R functions that can estimate the parameters of a GLMM based on various implementations of one of these three methods. The `glmer()` function of the `lme4` package can perform AGQ on binomial and Poisson models with a single random effect. It can also use Laplace approximations on the same distributions with more complicated random-effect structures, including both nested and crossed effects. The `glmmPQL()` function in the `MASS` package applies PQL to any model family available in `glm()`, and can furthermore handle a variety of repeated-measures correlation structures, which `glmer()` cannot do. In this introductory section on GLMMs, we focus on simpler models, and hence use `glmer()`. We recommend consulting more comprehensive resources before attempting to fit more advanced models. Research has shown that failing to properly account for the random-effects structure of a complex model can result in very poor inference, such as type I error rates that are arbitrarily high; see Loughin et al. (2007) and Fang and Loughin (2012) for examples.

Example: Falls with Head Impact (FallsGLMM.R, FallsHead.csv)

The model for our analysis shown in Equation 6.19 contains just one random effect, `resident`. We therefore use the AGQ method available in the function `glmer()` of the `lme4` package. The fixed-effects portion of the `formula` argument specification uses the same syntax as `glm()` and other regression modeling functions. Random effects are incorporated into the `formula` argument value by adding terms of the form `(a|b)`, where `b` is replaced by the name of the random-effects factor and `a` is replaced by one or more terms in `formula` form that indicate the variables whose coefficients are to be taken as random. For example,

- `(1|b)`: random effects are added to the intercept for each level of `b` (e.g., Equation 6.17);
- `(x|b)`: random effects are added to the regression coefficient for `x`;
- `(1|b)+(x|b)`: both the intercept and the regression coefficient for `x` have *independent* random effects (e.g., Equation 6.18);
- `(1+x1+x2|b)`: The intercept and the regression coefficients for `x1` and `x2` have *correlated* random effects.

In our falls example, Equation 6.19 is represented by the syntax, `head ~ initial + (1|resident)`.

To show the effects of the number of quadrature points on the parameter estimates, we fit the model for 1, 2, 3, 5, and 10 points using the `nAGQ` argument. The `glmer()` function produces S4 objects (first discussed in an example on page 184) of class `glmerMod`. Numerous familiar S3 method functions (e.g., `summary()`, `confint()`, `vcov()`, and `predict()`) can be applied to access information that is typically needed in an analysis. We print the variance component estimates ($\hat{\sigma}_{b0}^2$) for each model by accessing the `varcor` element of the model's `summary`. This produces a list of matrices containing the variances and covariances associated with each random effect in the model. The default printout consists of the corresponding standard deviations (square-root of the variance component) and correlations. Here we have only one random effect, so just one number, the square root of the variance component, $\hat{\sigma}_{b0}$, is printed.[20] The results are below.

---

[20] If further access to these quantities is needed, they can be accessed using `summary(obj)$varcor[[q]][i,j]` to produce element (i, j) of the covariance matrix of the qth random effect term given in the `formula` argument.

Additional topics

```
> library(lme4)

> mod.glmm.1 <- glmer(formula = head ~ initial + (1|resident),
    nAGQ = 1, data = fall.head, family = binomial(link = "logit"))
> summary(mod.glmm.1)$varcor
 Groups    Name        Std.Dev.
 resident (Intercept) 0.25193

> mod.glmm.2 <- glmer(formula = head ~ initial + (1|resident),
    nAGQ = 2, data = fall.head, family = binomial(link = "logit"))
> summary(mod.glmm.2)$varcor
 Groups    Name        Std.Dev.
 resident (Intercept) 0.26585

> mod.glmm.3 <- glmer(formula = head ~ initial + (1|resident),
    nAGQ = 3, data = fall.head, family = binomial(link="logit"))
> summary(mod.glmm.3)$varcor
 Groups    Name        Std.Dev.
 resident (Intercept) 0.3034

> mod.glmm.5 <- glmer(formula = head ~ initial + (1|resident),
    nAGQ = 5, data = fall.head, family = binomial(link = "logit"))
> summary(mod.glmm.5)$varcor
 Groups    Name        Std.Dev.
 resident (Intercept) 0.30342

> mod.glmm.10 <- glmer(formula = head ~ initial + (1|resident),
    nAGQ = 10, data = fall.head, family = binomial(link = "logit"))
> summary(mod.glmm.10)$varcor
 Groups    Name        Std.Dev.
 resident (Intercept) 0.30342
```

Notice that using nAGQ = 1 (equivalent to the Laplace approximation) leads to considerably different variance component estimates than does AGQ with three or more points. The last three methods yield very similar estimates, and the computational time in this problem is negligible, so we choose the model with 5 quadrature points to continue our analysis.

We next produce a summary of the results and save the summary as an object for later use before printing it. The relevant portion of the results is below:

```
> summ <- summary(mod.glmm.5)
> summ
Generalized linear mixed model fit by maximum likelihood
    ['glmerMod']
 Family: binomial ( logit )
Formula: head ~ initial + (1 | resident)
    Data: fall.head

      AIC       BIC    logLik  deviance
  279.5190  296.3722 -134.7595  269.5190

Random effects:
 Groups    Name        Variance Std.Dev.
 resident (Intercept) 0.09206  0.3034
```

```
Number of obs: 215, groups: resident, 131

Fixed effects:
                Estimate Std. Error z value Pr(>|z|)
(Intercept)      -0.6447     0.2424  -2.659  0.00784 **
initialDown      -1.1705     0.6782  -1.726  0.08438 .
initialForward    0.9581     0.3648   2.627  0.00862 **
initialSideways  -0.1208     0.3719  -0.325  0.74531
```

The output lists the information criteria and model fit summaries, followed by the variance component estimates, the coefficients for the fixed-effects portion of the regression, and the correlation among the fixed-effect parameter estimates (not shown). The estimated probability of head impact for fall k from resident i is

$$\mathrm{logit}(\hat{\pi}_{ik}) = 0.65 - 1.17 x_{2ik} + 0.96 x_{3ik} - 0.12 x_{4ik} + b_i,$$

where $x_{2ik}, x_{3ik}, x_{4ik}$ are indicator variables for levels 2, 3, and 4 of initial (straight-down, forward, and sideways falls, respectively), and b_i is a random variable drawn from a normal distribution with mean 0 and variance 0.092.

Note that the average value of b_i is zero. Thus, for an average resident with a backward initial direction, $\mathrm{logit}(\hat{\pi}) = -0.6447$, which results in $\hat{\pi} = 0.34$. Similarly for an average resident with a downward initial direction of fall, $\mathrm{logit}(\hat{\pi}) = -0.6447 - 1.1705 = -1.5152$, which translates to $\hat{\pi} = 0.14$. The intercept for different residents varies according to a normal distribution with estimated standard deviation $\hat{\sigma}_{b0} = 0.303$. Thus, we can estimate that approximately 95% of residents who fall backward initially have log-odds of head impact within $\hat{\beta}_0 \pm 2\hat{\sigma}_{b0} = 0.6447 \pm 0.606 = -1.24$ to -0.05. This translates to probabilities between 0.22 and 0.49. Similarly, approximately 95% of residents with downward falls have log-odds of head impact within -1.8152 ± 0.606, which corresponds to probabilities between 0.08 and 0.23. Notice that these intervals are not confidence intervals on a statistic, but rather represent a range of estimated probabilities of head impact for different residents in the population.

The S3 summary() method for glmerMod-class objects contains many useful components. We suggest exploring these using names() and also exploring available method functions using methods(class = "glmerMod"). We show some of the more useful functions below.

```
> # Estimates of fixed-effect parameters
> fixef(mod.glmm.5)
    (Intercept)      initialDown   initialForward
     -0.6446629       -1.1704700        0.9581401
initialSideways
     -0.1207953

> # Conditional Modes (b_i) for random effects listed **BY
    resident ID**, not in data order.
> head(ranef(mod.glmm.5)$resident)
  (Intercept)
1  0.05913708
2 -0.03104619
3 -0.03104619
4 -0.01369721
5  0.05913708
6 -0.02865847
```

```
> # Show that there is one element per resident
> nrow(ranef(mod.glmm.5)$resident)
[1] 131

> # coef = Fixed + Random effects.  Listed **BY resident ID**,
    not in data order.
> head(coef(mod.glmm.5)$resident)
  (Intercept) initialDown initialForward initialSideways
1  -0.5855281    -1.17047      0.9581401      -0.1207953
2  -0.6757081    -1.17047      0.9581401      -0.1207953
3  -0.6757081    -1.17047      0.9581401      -0.1207953
4  -0.6583598    -1.17047      0.9581401      -0.1207953
5  -0.5855281    -1.17047      0.9581401      -0.1207953
6  -0.6733205    -1.17047      0.9581401      -0.1207953
```

The fixed-effect parameter estimates are obtained using fixef(), while the random effect values b_i, $i = 1, \ldots, a$, are obtained with ranef(). Combined fixed and random parameter estimates for each resident are found from coef(). Note that the intercept, which is $\hat{\beta}_0 + b_i$, changes for each resident, but the other three parameter estimates are not associated with random effects and remain constant for all residents.

Next we use various calls to predict() to show the predicted logit, $g(\hat{\pi}_{ik})$, for each resident, the estimated average logit assuming $b_i = 0$, and the respective predicted probabilities of head impact ($\hat{\pi}_{ik}$) and estimates for an average resident.

```
> # Predicted logit for each observation
> logit.i <- round(predict(object = mod.glmm.5, newdata =
    fall.head, REform = NULL, type = "link"), digits = 3)
> # Estimated mean logit for each fall direction, by observation
> logit.avg <- round(predict(object = mod.glmm.5, newdata =
    fall.head, REform = NA, type = "link"), digits = 3)

> # Predicted probability for each observation
> pi.hat.i <- round(predict(object = mod.glmm.5, newdata =
    fall.head, REform = NULL, type = "response"), digits = 3)
> # Estimated average probability for each fall direction, by
    observation
> pi.hat.avg <- round(predict(object = mod.glmm.5, newdata =
    fall.head, REform = NA, type = "response"), digits = 3)

> # Conditional Modes listed in order of the original data (they
    are currently ordered by resident)
> ranefs <-
    round(ranef(mod.glmm.5)$resident[fall.head$resident,], digits
    = 3)

> # All predictions and mean estimates
> head(cbind(fall.head, ranefs, logit.i, logit.avg, pi.hat.i,
    pi.hat.avg))
  resident  initial     head ranefs logit.i logit.avg
1       56 Sideways        0 -0.083  -0.849    -0.765
2        9 Backward        0 -0.073  -0.717    -0.645
3       30  Forward        0 -0.018   0.295     0.313
4        9     Down        0 -0.073  -1.888    -1.815
5       70 Sideways        0 -0.001  -0.766    -0.765
```

```
6         21    Sideways    1    0.098      -0.668        -0.765
  pi.hat.i  pi.hat.avg
1   0.300     0.317
2   0.328     0.344
3   0.573     0.578
4   0.132     0.140
5   0.317     0.317
6   0.339     0.317
```

Note that `logit.i` differs from `logit.avg` by exactly `ranefs`, showing the effect of the random effect values (conditional modes) on the logits, and hence on the probabilities in `pi.hat.i` relative to `pi.hat.avg`. Thus, the "average" quantities are constant for each fall with the same initial fall direction, whereas the predictions vary a bit for each resident. We can determine from this that on average the probability of head impact is highest for forward initial direction (0.58) and smallest for falls that are straight down (0.14).

6.5.4 Inference

The fitted model contains estimates for the fixed effects and variance components, along with the conditional modes. All of these may be desired targets for tests and/or confidence intervals. However, inference in GLMMs must be done carefully. Because we are using ML estimation, parameter estimates should all have approximate normal distributions in large samples, with variances that are not difficult to compute. However, in practice, what constitutes a "large sample" may depend in part on the number of clusters or subjects, the number of observations within clusters or subjects, and the shape of the distribution for the response variables being modeled.

Tests and confidence intervals for fixed-effect parameters

The fixed-effect portion of a GLMM consists of a model like those presented in Chapters 2 and 4. Therefore, the types of analyses that are likely to be needed are the same as what was presented in those chapters. Hypothesis tests for individual parameters or for comparisons of parameters are common, and confidence intervals for these same quantities and for certain means or probabilities are often desired. In those fixed-effects models, likelihood-ratio (LR) methods were generally recommended for large samples, where the requirements for "large" were not too severe. Alternatively, Wald methods based on the large-sample normality of the MLE could be used, but the results were often not as accurate as those from LR methods at comparable sample sizes.

In a GLMM, LR methods can be more difficult to apply than they are in fixed-effect models, because the likelihood can be much more difficult to evaluate, particularly with complicated models or large data sets. Profile LR confidence intervals as described in Appendix B.5.2 are not often used, although improved computational capacity and fitting algorithms are beginning to make them more attractive. Instead, Wald inferences have historically been the standard, even though they tend to be even less accurate in GLMMs than they are with ordinary GLMs.

As an alternative to these methods, our preferred approach is to use a parametric bootstrap. Bootstrap methods are discussed in Sections 3.2.3 and 6.4.2 and are covered more completely in Davison and Hinkley (1997). For a hypothesis test using a parametric bootstrap, data sets (called "resamples") are simulated in a manner similar to that used in

Section 2.2.8, using a version of the GLMM whose parameters are estimated assuming that the null hypothesis is true. For example, we might remove a term from the model and simulate data from the new fitted model to test the significance of the deleted term. The full GLMM is re-fit to each resample, and a test statistic—commonly a Wald statistic or some other easy-to-compute quantity—is computed on each refit. A p-value is computed as the proportion of these simulated test statistic values that are at least as extreme (i.e., that favor the alternative hypothesis at least as much) as the one computed on the original data.

To find a parametric bootstrap confidence interval for a parameter, the simulation model is the original fitted model with all effects intact. The parameter is estimated for each resample, and the set of simulated parameter estimates is manipulated into interval endpoints using some appropriate technique, several of which are described in Davison and Hinkley (1997). Note that the parameter in question may be a regression-model parameter or some other quantity computed from them, such as an odds ratio or a specific mean or probability.

Example: Falls with Head Impact (FallsGLMM.R, FallsHead.csv)

We first show the overall test for equality of probabilities across the four initial fall directions, $H_0 : \beta_2 = \beta_3 = \beta_4 = 0$ in Equation 6.19. The alternative hypothesis is simply that not all initial directions have the same probability, $H_a :$ not all of $\beta_2, \beta_3, \beta_4$ are zero. We conduct this test two ways: using the LRT and using a parametric bootstrap. The LRT for each fixed-effects term in the model is conducted using the drop1() function.

```
> lrt <- drop1(mod.glmm.5, test = "Chisq")
> lrt
Single term deletions

Model:
head ~ initial + (1 | resident)
        Df    AIC     LRT   Pr(Chi)
<none>        279.52
initial  3  289.16  15.638  0.001345 **
```

The LRT gives a test statistic of 15.6, and a p-value of 0.001 based on the large-sample χ_3^2 approximation. We next compare these results to those obtained using the bootstrap by simulating 1000 data sets from the null-hypothesis model using the simulate() function in the lme4 package.[21] Both models are fit to each data set, the LRT statistic is computed each time, and the p-value is estimated by computing the proportion of times the simulated test statistic is at least as large as the observed value.

```
> names(lrt)
[1] "Df"       "AIC"      "LRT"      "Pr(Chi)"
> orig.LRT <- lrt$LRT[2]    # Saves LR Test statistic
```

---

[21] The version of lme4 that we used, 1.0.5, contains a function called bootMer() that can perform the parametric bootstrap more automatically. Unfortunately, this function was not fully operational at the time of this writing and failed to produce results on any model with nAGQ> 1. We recommend checking the documentation for future versions to see if these bugs are fixed. In the example program on the book website, we give code using bootMer() that should work once the function is fully operational if the syntax is not changed.

```
> # Fit Null model
> mod.glmm0 <- glmer(formula = head ~ (1|resident), nAGQ = 5,
    data = fall.head, family = "binomial")
> sims <- 1000
> simfix.h0 <- simulate(mod.glmm0, nsim = sims, seed = 9245982)
> # Fit Model and compute test statistic
> LRT0 <- numeric(length = sims)
> for (i in 1:sims){
    m1 <- glmer(formula = simfix.h0[,i] ~ initial +
        (1|resident), nAGQ = 5, data = fall.head, family =
        "binomial")
    LRT0[i] <- drop1(m1, test="Chisq")$LRT[2]
  }

> summary(LRT0)
    Min.  1st Qu.   Median     Mean  3rd Qu.     Max.
 0.04826  1.27000  2.32600  3.04800  4.28300 17.17000
> pval <- mean(LRT0 >= orig.LRT)
> pval
[1] 0.003

> # Plot results
> hist(x = LRT0, breaks = 25, freq = FALSE, xlab = "LRT
    Statistics", main = NULL, xlim = c(0, 2+round(max(LRT0,
    orig.LRT))))
> abline(v = orig.LRT, col = "red", lwd = 2)
> curve(expr = dchisq(x = x, df = 3), add = TRUE, from = 0, to =
    18, col = "blue", lwd = 2)
```

The parametric bootstrap code runs for several minutes and produces a p-value of 0.003. Both the simulated and the large-sample p-values lead to the same decision to reject the null hypothesis and conclude that the probabilities of head impact are not the same for all four initial falling directions.

Figure 6.6 shows the histogram of the 1000 simulated test statistic values from the parametric bootstrap and the χ_3^2 distribution that is used to approximate the distribution of the test statistic in the LRT. The two distributions are in reasonable agreement, although this certainly is not always the case for other data sets. The reasonably large sample size and the relatively moderate binomial probabilities likely allow the χ_3^2 approximation to serve well in this problem.

In order to determine which initial fall directions have higher or lower probabilities of head impact, we perform tests and construct confidence intervals for all pairwise comparisons. The mcprofile package for LR inferences has not yet been extended to work with glmerMod-class objects. We therefore use Wald-based procedures that are available from various functions in the multcomp package. Code from this package was previously shown in Exercise 20 of Chapter 2 and on page 216 of Section 4.2.3. This package has the added feature that the Type I error rate for the family of tests or confidence intervals can be controlled using one of several adjustments, including simple Bonferroni adjustments and other more complex adjustments. It can also provide unadjusted inferences that have a fixed error rate for each inference, but an inflated error rate for the family. We show below the code and results for the confidence intervals between pairs of levels of initial fall direction using a 95% simultaneous confidence level with a "single step" adjustment for multiple inferences (see p. 110). The confidence

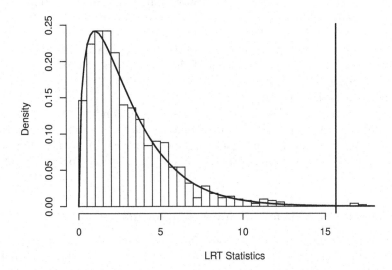

Figure 6.6: Comparison of parametric bootstrap estimate of the LRT statistic's distribution with the large-sample χ_3^2 distribution. The vertical line marks the value of the test statistic in the original data.

intervals are exponentiated so that they can be interpreted as odds ratios. Code for hypothesis tests and for unadjusted inferences is given in the program corresponding to this example.

```
> library(multcomp)

K <- rbind("D-B" = c(0,  1,  0, 0),
           "F-B" = c(0,  0,  1, 0),
           "S-B" = c(0,  0,  0, 1),
           "F-D" = c(0, -1,  1, 0),
           "S-D" = c(0, -1,  0, 1),
           "S-F" = c(0,  0, -1, 1))
> pw.comps <- glht(mod.glmm.5, linfct = K)
> ci.logit.F <- confint(pw.comps, level = 0.95)
> round(exp(ci.logit.F$confint), 2)
    Estimate  lwr   upr
D-B     0.31 0.06  1.74
F-B     2.61 1.03  6.59
S-B     0.89 0.34  2.28
F-D     8.40 1.46 48.52
S-D     2.86 0.49 16.67
S-F     0.34 0.13  0.92
```

The pairwise comparisons are labeled according to the two levels being compared. For example, the estimate 0.31 for "D-B" indicates that the estimated odds of head impact for a downward fall are 0.31 times as large as for a backward fall, or equivalently that the estimated odds of head impact with a backward fall are $1/0.31 = 3.2$ times the odds with a downward fall. The three confidence intervals relating to forward falls all exclude 1. From the direction of the differences we can conclude that head impact is significantly more likely to occur with a forward initial fall direction than with any other direction. All confidence intervals among the remaining three levels contain 1, so that there are no significant differences in probabilities of head impact among these directions. These confidence intervals are all fairly wide, indicating that there remains considerable uncertainty in the estimation of these odds ratios. Finally, the use of a familywise confidence level means that we are 95% confident that *all six* confidence intervals will cover their respective true odds ratios.

Alternatively, we can use the parametric bootstrap to conduct tests and compute confidence intervals. The code for these calculations is included in the program for this example. Westfall and Young (1993) provide additional algorithms that can be used to control familywise confidence levels for resampling-based confidence intervals.

Tests and confidence intervals for variance components

Inference for variance components of random effects is complicated by the fact that we cannot have negative variance, so the variance component parameter is bounded below by zero. In practice, estimates of variance components are often exactly equal to zero, which is a clear indication that the corresponding random effect variance component plays a negligible role in the model and can be eliminated. Even when a non-zero estimate is obtained for one sample, the probability distribution of estimates from all possible samples from the population may still place sizable mass on zero. Furthermore, the sampling distribution for variance components tends to be highly skewed to the right. That is, most samples from a population yield variance component estimates that are slightly below their true values, and may often be zero, but occasionally they are considerably higher. This means that Wald-based inference, which assumes a normal distribution for the variance component estimate, is really not adequate for variance parameters in GLMMs.[22] The lower endpoints of Wald confidence intervals are often negative, and Wald tests are horribly conservative.

At the same time, the lower bound at zero complicates LR inference. In particular, testing the significance of the variance component using $H_0 : \sigma_b^2 = 0$ vs. $H_a : \sigma_b^2 > 0$ involves inference for a boundary value of a parameter, which invalidates the theory that leads to a chi-square distribution for LRT statistics. This does not affect LR confidence intervals the same way it does hypothesis tests, but as noted above, profiling the likelihood to form profile LR confidence intervals can be a difficult computational problem.

This would seem to leave the parametric bootstrap as the best alternative for inference. Tests of the hypothesis $H_0 : \sigma_b^2 = 0$ are relatively straightforward to conduct, provided the run time for repeatedly estimating the model on a large number of resamples is feasible. The process uses the same approach that was used for fixed effects: simulate a large number of new sets of responses from the fitted model that assumes H_0 to be true, fit the full model to each resample, and calculate some test statistic that tests H_0 using the fit of this model. The variance component estimate itself can be used as a test statistic. We then compute the p-value as the proportion of resamples that lead to a variance component estimate at least as large as the one from the original sample, because larger estimates of the variance component favor H_a. This is faster than using a LRT statistic, which requires that two models be fit to each resample. Also, it may be possible that a relatively small number of simulations may determine conclusively whether the null hypothesis should be rejected. For

---

[22]Stroup (2013) p. 182 claims, "No literate data analyst should use the Wald covariance component test, with the possible exception of extremely large studies with tens of thousands or more observations."

example, if 50 simulations are done and 40 of them lead to a more extreme LRT statistic than in the original sample (i.e., a p-value of 0.80), then it is already clear that the null hypothesis should not be rejected and there is little purpose to performing more simulations.

If a problem is too large to allow a parametric bootstrap to be done feasibly, then the usual LRT statistic can be calculated, but the p-value that would normally be found from a chi-square distribution must be evaluated differently. In particular, the actual distribution of the LRT statistic places more mass at 0 than the chi-square approximation allows, so that the standard p-value is at least twice as large as it should be (Littell et al., 2006). Therefore, an approximate p-value can be found by dividing the original p-value by 2.

Confidence intervals represent a much more difficult problem. As long as the profiling can be done reasonably quickly, LR intervals can be formed. However, their properties are not well known when the sampling distribution of the variance component may contain point mass at zero.

Parametric bootstrap confidence intervals for variance components in GLMMs have also not been well researched. Creating resamples from the estimated model and computing variance-component estimates on each resample is reasonably straightforward, assuming that the problem is small enough that a resampling simulation can be applied. However, it is not at all clear how the resulting estimated probability distribution of the variance components should be used. Davison and Hinkley (1997) describe numerous types of confidence intervals that can be calculated from a bootstrap estimate of a sampling distribution—e.g., percentile, bootstrap-t, BC_a, and pivot-based intervals—but all of these require assumptions about the statistic's true probability distribution that are not met when the distribution places considerable point mass on a boundary value. We therefore cannot provide specific guidance for computing parametric bootstrap confidence intervals for variance components when the estimated sampling distribution contains point mass at zero. However, based on general knowledge of how the bootstrap works, we can offer some suggestions. We expect that a percentile or BC_a interval will still provide better true confidence levels than a Wald interval. The percentile interval is particularly easy to compute—simply set the lower confidence limit at the $\alpha/2$ quantile of the simulated variance-component estimates and the upper confidence limit at the $1 - \alpha/2$ quantile of these estimates—although it often provides true confidence levels somewhat below the stated level. The BC_a interval requires some small additional calculation (see Davison and Hinkley, 1997 for details), but in many problems provides confidence intervals whose true confidence levels are close to the stated level. We recommend the BC_a interval when it can be calculated, and the LR interval otherwise.

These intervals can also be used when the variance component estimate is far enough from zero that the boundary problem is not an issue. Indeed, we would anticipate that their ability to provide an appropriate interval would be better in this case, although demonstrating this remains an open research topic.

For model selection, information criteria can be used to select suitable models (see Section 5.1.2). Models with different combinations of fixed and random effects can be compared in this way, as long as the estimation method is not penalized quasi-likelihood. However, Greven and Kneib (2010) warn that the boundary problem affects AIC in linear mixed models in that it tends to favor a model without a random effect over one with the random effect more often than it should. Whether this problem carries over into GLMMs is not yet known.

Example: Falls with Head Impact (FallsGLMM.R, FallsHead.csv)

The estimated variance component for the `resident` effect is $\hat{\sigma}^2_{b0} = 0.092$. Although understanding the variability among residents was not the primary focus of this study, learning about between-resident variability may provide useful information. For example, significant variability among residents may suggest that some residents are able to reduce their probability of head impact by taking preventative measures as they are falling.

To perform the parametric bootstrap test of $H_0 : \sigma_{b0}^2 = 0$ vs. $H_a : \sigma_{b0}^2 > 0$, we first extract the variance component estimate from the original data and save it into an object named orig.vc. Next, we simulate 1000 data sets from the GLM with no random effect (i.e., under H_0) and refit the model for each one so that we can obtain the variance component estimate. These estimates are saved into a 1-column matrix named varcomps0. The p-value is computed and a histogram of the variance component estimates is plotted. Finally, the LRT is conducted manually to compare the null and alternative models.

```
> mod.glm <- glm(formula = head ~ initial, data = fall.head,
    family = binomial(link = "logit"))
> sims <- 1000
> simmod.h0 <- simulate(mod.glm, nsim = sims, seed = 28662819)
> orig.vc <- summary(mod.glmm.5)$varcor[[1]][1,1]
> varcomps0 <- matrix(data = NA, nrow = sims, ncol = 1)

> for (i in c(1:sims)){
    mm <- glmer(formula = simmod.h0[,i] ~ initial + (1|resident),
        nAGQ = 5, data = fall.head, family = binomial(link =
        "logit"))
    varcomps0[i,] <- summary(mm)$varcor[[1]][1,1]
  }

> varcomps0 <- na.omit(varcomps0)
> nrow(varcomps0)
[1] 1000
> pval <- sum(varcomps0 >= orig.vc)/nrow(varcomps0)
> pval
[1] 0.328

> LRstat.vc <- deviance(mod.glm) - deviance(mod.glmm.5)
> # p-value
> (1 - pchisq(LRstat.vc, df = 1))/2
[1] 0.3933802
```

The p-value for the parametric bootstrap test is 0.328. The histogram given in Figure 6.7 shows the highly skewed distribution of the variance component estimates under the null hypothesis. Clearly, no normal approximation should be used with this distribution. The LRT using a χ_1^2 distribution produces a p-value of 0.393, which is in reasonable agreement with the bootstrap result.

The code for computing confidence intervals for σ_{b0}^2 is provided in the program corresponding to this example.

6.5.5 Marginal modeling using generalized estimating equations

As described in Section 6.5.3, estimating parameters in GLMMs can be a complicated and difficult process computationally. The results of a GLMM include predictions for each

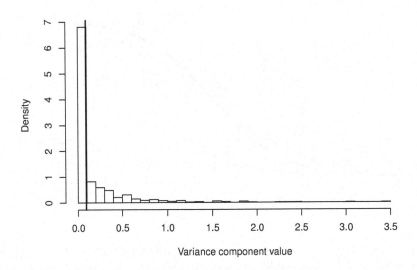

Figure 6.7: Histogram of variance component estimates using parametric resampling under $H_0 : \sigma_{b0}^2 = 0$ for the falls example. The thick vertical line indicates the location of the estimate from the original data.

observed cluster or subject (for ease of discussion, in the remainder of this section we use "subject" to refer to either a subject or a cluster). They also include estimates of various regression parameters for an "average subject" (one with all random effects equal to zero). A GLMM is referred to as a *subject-specific model*, because the addition of random effects allows each subject to have its own different parameter values as described in Section 6.5.2. An alternative approach is to directly model how an explanatory variable relates to the population average response, rather than how it relates to individuals in the population. A direct model for the population average is referred to as a *marginal model*, because the population average is derived from the marginal distribution of the outcome.

The marginal model is fundamentally different from the GLMM as is seen in Figure 6.8. The plot shows the simulated probability of success, $P(Y = 1)$, for a sample of 50 subjects whose probability follows $\text{logit}(\pi_i) = -20 + b_i + 0.2x_i$, $i = 1, \ldots, 50$, where the b_i's are independent $N(0, 1.5^2)$ random variables. The thick solid curve is the probability curve for the "average subject"; i.e., for $b_i = 0$. This is the quantity that is estimated by a GLMM. The dashed curve is the average probability at each value of x, which is the quantity that is estimated by a marginal model. The difference between them occurs because the subject-specific model averages horizontally and the marginal model averages vertically. In linear regression models, this distinction does not matter, but it does matter when the response curve is nonlinear.

In some applications, a marginal model is more relevant than a subject-specific model. For example, in government studies on the association between vaccination frequency and prevalence of a disease, understanding the effects on the population average is more useful for informing national medical policy than understanding the effects on individuals. Nice summaries of the details and interpretations of marginal models are given in Agresti (2002) and Molenberghs and Verbeke (2005).

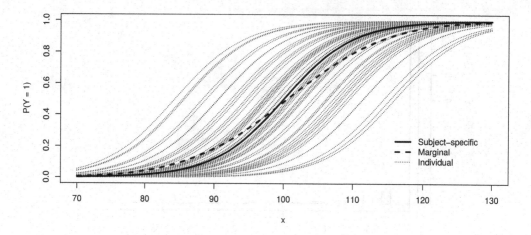

Figure 6.8: Subject-specific model and marginal model for simulated sample of 50 subjects from a population with probability of success following a logistic curve with common slope but different intercepts. The corresponding code for the plot is in LogisticSim.R.

Creating and estimating marginal models

Suppose that we have a subjects, and for convenience, suppose that there are t response measurements per subject (in some settings, the number of responses per subject can vary among subjects). Modeling a marginal distribution involves first specifying a full joint distribution among all t correlated responses within a subject. This requires formulating models for the mean, for 2-way associations among pairs of responses, for 3-way associations, and so forth up to the t-way associations. We may have very little idea how these higher-order associations should be modeled, and we may really be interested only in modeling the mean response. Thus, we consider trying to model the associations as a "nuisance."

As an approximate model, suppose we assume that all associations among responses are zero; that is, the responses within a subject are independent. This means that $n = at$ responses are treated as independent rather than as having come from clusters of responses. It is quite likely that this independence assumption is wrong in many applications, but it allows us to create a "working model" that is easily specified and fit to the data, similar to the pseudo-likelihood approach used in Sections 6.3.6 and 6.4.3. Liang and Zeger (1986) fit this model using a technique closely resembling ML estimation, but not based on a fully specified likelihood. They propose solving a set of *generalized estimating equations* (GEEs) to obtain the parameter estimates.[23] Corresponding estimators have many of the properties that MLEs enjoy. In particular, they are approximately normally distributed in large samples and they are consistent (see Appendix B.3.3) as long as the model for the means is correct, even when the working model for the associations is wrong. However, the variances of the parameter estimates depend on the association structure that we are ignoring in the model. Specifically, variance estimates produced by fitting a standard GLM tend to be too small when responses within subjects are positively correlated. Liang and Zeger (1986) developed a method to correct the variances. The resulting variance estimators

---

[23]"Estimating equations" are functions of parameters that are set to zero to find parameter estimates, like the score function described in Appendix B.3.1 for finding MLEs.

are known as "sandwich" estimators because of their mathematical form as a product of three matrices, where the same matrix is used at each end. They are also called "robust" or "empirically corrected" variances. Further details on their calculation are given in Liang and Zeger (1986), Agresti (2002), and Molenberghs and Verbeke (2005).

The GEE approach can be applied using other assumed association structures for the responses instead of independence. In fact, the efficiency (precision) of the GEE parameter estimates improves when a correlation structure is used that more closely resembles that of the true joint distribution of responses. However, efficiency can also suffer if a correlation model is chosen that requires estimating more parameters than are necessary. Therefore, preference is generally given to simpler models for the association that require fewer parameters.

A popular choice is to assume that every pair of responses has the same non-zero correlation. This is called an *exchangeable* correlation structure. Or, if measurements are made across time, we could assume that the correlation between a pair of responses depends on how far apart the measurements were taken. In particular, if the correlation decreases exponentially with distance, then this is an *auto-regressive* (AR) correlation structure. These two structures are popular because they can both be estimated with a single parameter that is separate from the regression parameters. Alternatively, a completely unspecified correlation structure can be chosen so that there is one correlation parameter estimated for each pairwise combination of the responses. This structure is often referred to as *unstructured*. Note that this results in $t(t-1)/2$ correlation parameters in total, so it is best used only when there is substantial doubt that a simpler structure will suffice. Additional correlation structures can be considered too, but it is typically difficult to discern very finely between different correlation structures, particularly if the number of subjects is not very large.

Inferences from GEEs are performed using Wald methods, so in all cases a "large" sample size is needed. How large is "large" depends on the number of subjects, measurements per subject, explanatory variables, and the sizes of the means or probabilities being estimated. Unlike in other problems, where we could check the quality of the inferences using a simulation based on the estimated model, GEEs are based on an incomplete model for the responses, so it is not entirely clear what model should be used for simulations. When there is concern regarding the adequacy of the Wald inferences and/or variance estimates, a nonparametric bootstrap approach (Davison and Hinkley, 1997) can be used instead. Resampling should be applied to the a subjects (with all t responses included whenever a particular subject is resampled), rather than individual responses, so that the within-subject correlation structure is retained when GEE is applied to the resamples.

The GEE approach is popular because of its relative simplicity. The computations are all straightforward and the variance correction works fairly well in large samples, although it has a tendency to underestimate the true variance of parameter estimates, resulting in liberal inferences. However, the approach is limited primarily to problems with repeated measures or simple clustering. It is not easily applicable to problems with nested or crossed random effects.

Example: Falls with Head Impact (FallsGEE.R, FallsHead.csv)

We re-analyze the falls data with a GEE approach, using a model

$$\text{logit}(\pi_{ik}) = \beta_0 + \beta_2 x_{2ik} + \beta_3 x_{3ik} + \beta_4 x_{4ik}$$

for fall k from resident i. Note that there are no special parameters or effects in this model that distinguish the falls from resident i from those of another resident. That is, any falls for any residents that have the same values of x_2, x_3, and x_4 have the same probability of head impact.

The model is fit using the geeglm() function from the geepack package (Halekoh et al., 2006). This function fits GEEs to model families available in glm(). Also, there is an anova() method function that can be used with the resulting model-fit object. The function computes Wald tests for regression parameters involving numerical explanatory variables, and multiparameter Wald tests for those from categorical explanatory variables (see Appendix B.5.1).[24]

The data for geeglm() must first be arranged so that all data from each cluster are consecutive observations. The function assumes that a new cluster begins with each change in value of the variable listed in the id argument. It automatically estimates a scale parameter as in quasi-likelihood methods (see Section 5.3.3), unless scale.fix = TRUE is specified. Finally, correlation structures are given in the corstr argument, with options including "independence", "exchangeable", and "ar1".

We fit the model for the probability of head impact with initial direction of fall as the explanatory variable. Following the summary of the model fit, we explore the contents of the summary() results and use the anova() function to test whether all levels of initial have equal probability of head impact.

```
> library(geepack)
> # Need to sort data in order of the clusters first.
> fall.head.o <- fall.head[order(fall.head$subject),]
> # Need to specify scale.fix = TRUE, or else quasi-binomial will
    be fit.
> mod.gee.i <- geeglm(formula = head ~ initial, id = subject,
    data = fall.head.o, scale.fix = TRUE, family = binomial(link =
    "logit"), corstr = "independence")
> summ <- summary(mod.gee.i)
> summ
Call: geeglm(formula = head ~ initial, family = binomial(link =
    "logit"), data = fall.head.o, id = subject, corstr =
    "independence", scale.fix = TRUE)
 Coefficients:
               Estimate Std.err  Wald Pr(>|W|)
(Intercept)     -0.6360  0.2475 6.601  0.01019 *
initialDown     -1.1558  0.6510 3.152  0.07583 .
initialForward   0.9544  0.3401 7.875  0.00501 **
initialSideways -0.1085  0.3435 0.100  0.75222
---
Signif. codes:  0 '***' 0.001 '**' 0.01 '*' 0.05 '.' 0.1 ' ' 1
Scale is fixed.
Correlation: Structure = independence
Number of clusters:   131   Maximum cluster size: 13

> names(summ)
[1] "call"          "terms"             "family"
[4] "contrasts"     "deviance.resid"    "coefficients"
[7] "aliased"       "dispersion"        "df"
```

---

[24]Unfortunately, the anova() function computes only "sequential" tests. Each term is tested assuming that only the terms listed before it in the formula are also in the model. This means that the test results may change depending on the order of the terms in the model, which is usually not a good feature. The Wald tests produced by summary() are the more common "partial" tests, which assume that *all* other terms are present before adding the tested term. To get partial tests for a categorical variable in a multivariable model, refit the model without the categorical variable and use anova(mod1, mod2), where mod1 is the full model and mod2 is the smaller model. See Exercise 4.

```
[10] "cov.unscaled"      "cov.scaled"      "corr"
[13] "corstr"            "scale.fix"       "cor.link"
[16] "clusz"             "error"           "geese"
> summ$cov.scaled
         [,1]      [,2]      [,3]      [,4]
[1,]  0.06128  -0.06660  -0.05763  -0.05683
[2,] -0.06660   0.42378   0.09347   0.04674
[3,] -0.05763   0.09347   0.11568   0.05336
[4,] -0.05683   0.04674   0.05336   0.11800

> # anova() method performs Wald chi-square test.
> anova(mod.gee.i)
Analysis of 'Wald statistic'
Table Model: binomial, link: logit
Response: head
Terms added sequentially (first to last)
          Df    X2    P(>|Chi|)
initial    3   21.5   8.2e-05 ***
```

The model fit summary looks much like other regression summaries. Standard errors are computed from the robust sandwich estimates of each parameter estimate's variance. These are also available from the square roots of the diagonal elements of cov.scaled from the resulting object for summary(), which contains the estimated variances and covariances of the regression parameter estimates. The Wald test for equal probability of head impact for each direction is clearly significant, with a test statistic $W = 21.5$ and a very small p-value.

We next do pairwise comparisons among the levels of initial. These are carried out in the same manner as they were for the GLMM analysis and yield the same interpretations with slightly different numbers. An extra step is required to create a vcov() method function that extracts cov.scaled from the model-fit object. The details are given in our corresponding program, along with code to compute confidence intervals for the probability of head impact for each initial fall direction. This program also contains code to re-compute this Wald statistic using matrix methods.

6.6 Bayesian methods for categorical data

Analyses performed in Chapters 1–5 use *frequentist methods* to estimate a fixed, unknown parameter, such as a population prevalence π, an odds ratio OR, or a regression parameter β_r. This involved taking a random sample from a population and assuming a joint probability distribution for the corresponding random variables. Using this information, an MLE was found for the parameter of interest. Confidence intervals were also constructed, usually by taking advantage of likelihood-based properties discussed in Appendix B. The stated confidence level with these intervals was attributed to the hypothetical process of repeating the same sampling and calculations over and over again (from which the name "frequentist" is derived) so that $(1-\alpha)100\%$ of the intervals contain the parameter. Notice that this confidence level does not mean that one specific interval has a $(1-\alpha)$ probability of containing the parameter. Once an interval is constructed, it either contains the parameter or it does not, so the probability, which is unknown to us, is either 1 or 0. Currently, frequentist methods are the most predominant approach to statistical inference.

Bayesian methods present an alternative paradigm for statistical inference. The popularity of these methods has grown immensely over the last 20 years due to the availability of new computational algorithms and faster computers. An advantage to using Bayesian methods is that prior knowledge about an underlying problem can be incorporated into an analysis. This is done by treating a parameter as a random quantity with a known probability distribution that we specify completely. This is called a *prior distribution* (or just the "prior") because it represents what we believe about the parameter prior to considering the data. For example, if a goal is to estimate the prevalence of HIV in a population, one would expect that this prevalence is low even before collecting the data, so a chosen distribution for the probability of HIV in a randomly selected person would concentrate mainly on values near zero. After data are collected, this prior is updated to form a new distribution, known as a *posterior distribution* (or just the "posterior"), to reflect information gained by the data. The specific value of the parameter in the population is treated as a random draw from this posterior. Thus, parameter estimates are summaries of the parameter's posterior, such as the mean or mode, in order to reflect likely values. We can also construct an interval of likely values for the parameter using the posterior. This interval has a $1 - \alpha$ probability of containing the parameter, in contrast to what a confidence interval represents.

The purpose of this section is to discuss the application of Bayesian methods for a number of categorical data settings. We begin with an introduction to the Bayesian paradigm in the context of topics discussed in Section 1.1.1. Next, we extend these ideas to more computationally complex situations involving the regression models discussed in Chapters 2–4. Finally, we provide additional resources for the application of Bayesian methods with R.

6.6.1 Estimating a probability of success

Bayesian methods are based on the application of *Bayes' rule*, which is discussed in most introductory statistics courses. This rule states that for two events A and B, the conditional probability of B given A is

$$P(B|A) = \frac{P(A \cap B)}{P(A)} = \frac{P(A|B)P(B)}{P(A \cap B) + P(A \cap \bar{B})}, \qquad (6.20)$$

where $\bar{B}$ denotes the compliment (or opposite) of B. In our context, one can roughly think of B as representing the parameter and A as representing the data. Then $P(B)$ represents the prior knowledge that one may have about the parameter, and we update this knowledge with information obtained about the data to calculate the posterior probability $P(B|A)$.

Obtaining the posterior distribution

Bayesian inference is based on a more general form of Equation 6.20, stated as

$$p(\theta|y) = \frac{f(y|\theta)p(\theta)}{f(y)}, \qquad (6.21)$$

where θ is the parameter of interest, y represents the observable data, and $p(\cdot)$ and $f(\cdot)$ represent particular probability distributions for θ or Y, respectively. Equation 6.21 is the posterior distribution for θ given information observed about Y. For clarity here, note that not only is Y a random variable, but θ is a random variable as well; we forgo the formality of using a capital version of θ for ease of exposition. We next discuss in detail their probability distributions in the context of estimating a probability-of-success parameter π in a binomial model for the number of successes W out of a fixed number of trials n. Our goal is to obtain the posterior for π given information observed about W.

Throughout Chapters 1 and 2, we used a binomial distribution that was first given in Equation 1.1. We now re-express this distribution as

$$f(w|\pi) = \binom{n}{w} \pi^w (1-\pi)^{n-w},$$

where we use $f(\cdot)$ to represent the PMF for W and add "$|\pi$" to emphasize that probabilities are conditioned on a value for π. This conditional PMF, which is also the likelihood function for w, is used in the numerator of Equation 6.21.

Knowledge about π before a sample is taken can be accounted for by the prior distribution, $p(\pi)$, representing likely values for the parameter. There are many possible priors that could be taken, but the one most often used in this situation is a beta distribution (see Equation 1.5). This distribution allows for $0 < \pi < 1$, is quite flexible in its shape, and leads to a mathematically nice result (to be shown shortly). For example, if we believe that π should be small as in the HIV prevalence example, a beta distribution with $a = 1$ and $b = 10$ may be reasonable because the distribution is right skewed with much of its probability close to 0.[25] The prior distribution for π is also used in the numerator of Equation 6.21.

Finding $f(w)$ corresponding to the denominator in Equation 6.21 is typically the hardest part of a Bayesian analysis. We first need to find the joint probability density function of W and π. Assuming a beta prior with parameters $a > 0$ and $b > 0$, we obtain

$$f(w, \pi) = f(w|\pi) p(\pi)$$
$$= \binom{n}{w} \pi^w (1-\pi)^{n-w} \frac{\Gamma(a+b)}{\Gamma(a)\Gamma(b)} \pi^{a-1} (1-\pi)^{b-1}$$
$$= \frac{\Gamma(n+1)}{\Gamma(w+1)\Gamma(n-w+1)} \frac{\Gamma(a+b)}{\Gamma(a)\Gamma(b)} \pi^{w+a-1} (1-\pi)^{n+b-w-1}$$

for $w = 0, \ldots, n$ and $0 < \pi < 1$. Note that we use the relation $\Gamma(c+1) = c!$ for an integer c. To obtain $f(w)$, we integrate over all possible values of π:

$$f(w) = \int_0^1 \frac{\Gamma(n+1)}{\Gamma(w+1)\Gamma(n-w+1)} \frac{\Gamma(a+b)}{\Gamma(a)\Gamma(b)} \pi^{w+a-1} (1-\pi)^{n+b-w-1} d\pi$$
$$= \frac{\Gamma(n+1)}{\Gamma(w+1)\Gamma(n-w+1)} \frac{\Gamma(a+b)}{\Gamma(a)\Gamma(b)} \frac{\Gamma(w+a)\Gamma(n+b-w)}{\Gamma(n+a+b)}.$$

Putting all the pieces together, we obtain the posterior distribution of π given information observed about W:

$$p(\pi|w) = \frac{f(w|\pi) p(\pi)}{f(w)}$$
$$= \frac{\Gamma(w+a)\Gamma(n-w+b)}{\Gamma(n+a+b)} \pi^{w+a-1} (1-\pi)^{n+b-w-1} \quad (6.22)$$

for $0 < \pi < 1$. Interestingly, Equation 6.22 is a beta distribution just like the prior, but now with parameter values $w + a$ and $n + b - w$ controlling the shape of the distribution. When a posterior is in the same family of distributions as the prior, then these distributions are said to belong to a *conjugate family*. Thus, the beta distribution is conjugate to the binomial distribution.

---

[25] Use `curve(expr = dbeta(x = x, shape1 = 1, shape2 = 10), xlim = c(0,1))` to plot the density function in R.

Bayes estimate

The posterior distribution is used to obtain estimates of π. One reasonable estimate is to use the mean $E(\pi|w)$, which can be shown in this case to be $(w + a)/(n + a + b)$ by using properties about beta distributions (see page 107 of Casella and Berger, 2002). This mean is known as a *Bayes estimate*, and we denote it as $\hat{\pi}_B$. As an alternative, we can use the median or the mode of the posterior as an estimate. Note that finding the mode of the posterior is analogous to finding the MLE from the likelihood function, and it is guaranteed to equal the mean only when the posterior is symmetric and unimodal.

Before observing the data, a sensible estimate for π would have been $E(\pi) = a/(a + b)$ obtained from the prior distribution. Interestingly, the Bayes estimate in this problem can be factored into two components—one that uses $E(\pi)$ and one that uses the MLE $\hat{\pi} = w/n$:

$$\hat{\pi}_B = \left(\frac{n}{n + a + b}\right)\hat{\pi} + \left(\frac{a + b}{n + a + b}\right)E(\pi). \tag{6.23}$$

Thus, the Bayes estimate is a weighted average of the MLE and the mean from the prior distribution. A larger sample size results in more relative weight on the MLE, while a smaller sample size results in more relative weight on the prior mean. In Bayesian models for other parameters, the same principle holds: as the sample size grows, the Bayes estimate relies progressively less on the prior and more on the data. However, showing the relationship in a mathematical closed form is not often as easy as it is in this setting.

Choosing a prior distribution

The choice of a prior distribution obviously plays an important role in any Bayesian analysis. Past experience may suggest possible values of a parameter, which then leads to the use of a particular prior with parameters of its own (parameters of the prior are called *hyperparameters*). For example, our beliefs about π suggest that we can use a beta distribution with particular values of a and b. Selecting a prior in this way is both one of the advantages and one of the criticisms of Bayesian analysis, because different analysts may choose different priors based on different beliefs. These different *subjective* choices could conceivably affect the outcome of the analysis in important ways.

As an alternative to choosing a prior subjectively, a *noninformative prior distribution* is frequently used in practice. This is a distribution that places equal probability or density on all values of the parameter. For our binomial model, a beta distribution with $a = b = 1$ (equivalent to a uniform distribution on the interval $(0, 1)$) is a noninformative prior for estimating π. Of course, this is still specifying a belief about the parameter—specifically that it has an equal chance of being any allowed value—so it is still not the same as a frequentist method that specifies no such belief.

While a uniform distribution is noninformative for estimating π, it is no longer noninformative if the goal is to estimate a monotone transformation of π, like an odds. Thus, a analyst wishing to use the prior may need to choose a different one for each different way that the parameter might be expressed. One way to avoid this problem is to use a Jeffreys' prior, which is a distribution that is invariant to a monotone transformation of the parameter (e.g., see Robert, 2001 and Tanner, 1996 for further details). In general for a parameter θ and data y, the prior is chosen to be proportional to

$$\left[E\left(\frac{\partial^2}{\partial \theta^2}\log[f(y|\theta)]\right)\right]^{1/2}.$$

The quantity within the square root is often referred to as *Fisher information*, and it is very closely related to the variance of a maximum likelihood estimator (see Equation B.5).

Fisher's information is conceptualized as the amount of information about the parameter that is available from the data. Thus, basing the prior on Fisher's information allows for more weight to be assigned to values of θ that have larger Fisher's information. This leads to a different way to think about the notion of "noninformative," in that Jeffreys' prior agrees more with the data than other prior distributions might (Robert, 2001, p. 130).

There are other ways to choose a prior distribution. In particular, probability distributions known as *hyperpriors* can be specified for each hyperparameter of a prior distribution. The resulting analysis is called *hierarchical Bayesian* due to the nested prior specifications. Another way to choose a prior is to estimate it using the observed data through *empirical Bayesian methods*. A problem with this approach is that the prior cannot be fully specified before the data collection, which is somewhat contrary to the Bayesian philosophy of fully stating "prior" beliefs. Nonetheless, some analysts like it for its objectivity. For more information on these and other ways to specify prior distributions, please see full textbooks on Bayesian methods, such as Gelman et al. (2004) and Carlin and Louis (2008).

Credible intervals

In a Bayesian analysis, interval estimates are based on quantiles of the posterior for a parameter. These intervals are called *credible intervals*. The interpretation of a credible interval is that it has $1 - \alpha$ probability of containing the parameter because it is based on the posterior distribution of the parameter. The most popular type of credible interval is an *equal-tail interval* which finds limits such that the probability to the left of the lower limit is $\alpha/2$ and the probability to the right of the upper limit is $\alpha/2$. For example, based on the posterior for π that we derived earlier, we obtain a $(1 - \alpha)100\%$ equal-tail credible interval of

$$\text{beta}\,(\alpha/2;\ w + a,\ n + b - w) < \pi < \text{beta}\,(1 - \alpha/2;\ w + a,\ n + b - w),$$

where $\text{beta}\,(\gamma;\ w + a,\ n + b - w)$ is the γth quantile from the beta distribution with parameter values $w + a$ and $n + b - w$. Equal-tail intervals are relatively easier to calculate than other intervals, and they perform best when the posterior is fairly symmetric.

When the posterior is not symmetric, a better choice is to use a highest posterior density (HPD) credible interval. This interval is calculated by finding the lower and upper limits for an interval that correspond to the region with the highest posterior density, i.e., the most "probable" region. The resulting interval is generally the narrowest possible interval that contains $(1 - \alpha)100\%$ of the posterior density. There usually are not closed-form expressions for the calculation of HPD intervals, so we show how to find this type of interval in the next example. It is important to note that an HPD interval is not invariant to transformations. Thus, simply transforming its limits for π does not necessarily lead to an HPD interval for the same transformation of π.

Example: Bayes estimate and credible intervals (CIpiBayes.R)

Suppose $w = 4$ successes are observed out of $n = 10$ trials. If a noninformative beta distribution with $a = b = 1$ is used as the prior, we obtain a Bayes estimate of $\hat{\pi}_B = (4 + 1)/(10 + 1 + 1) = 0.4167$ and a 95% equal-tail credible interval of $0.1675 < \pi < 0.6921$. The computations in R are straightforward, where one can simply use `qbeta(p = c(0.05/2, 1-0.05/2), shape1 = 4 + 1, shape2 = 10 + 1 - 4)` to compute the interval. If Jeffreys' prior is used instead, one can first show that this prior is proportional to $\pi^{-1/2}(1 - \pi)^{-1/2}$, which leads to the use of a beta distribution with $a = b = 1/2$. The Bayes estimate is 0.4091 and the equal-tail credible interval is $0.1531 < \pi < 0.6963$, providing a lower bound that is a little different than when using $a = b = 1$.

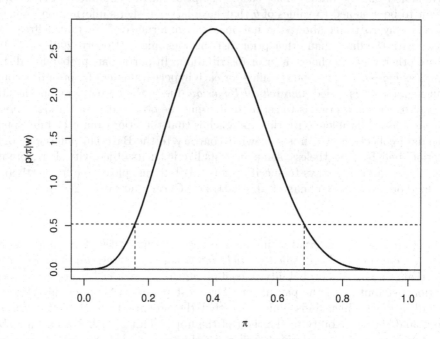

Figure 6.9: Posterior density plot with a horizontal line drawn at $p(\pi|w) = 0.5184$ and vertical lines indicating the limits for the 95% HPD interval.

These estimates and intervals are quite similar to the MLE $\hat{\pi} = 0.4$ and Wilson interval $0.1682 < \pi < 0.6873$ obtained in an example from Section 1.1.1. Part of the reason can be seen by examining the expression for $\tilde{\pi} = (w + Z^2_{1-\alpha/2}/2)/(n + Z^2_{1-\alpha/2})$, the estimator for π used in constructing the Wilson and Agresti-Coull confidence intervals. If $a = b = 1.96/2 = 0.98$, $\hat{\pi}_B$ and $\tilde{\pi}$ are the same.

To find the HPD interval, suppose a horizontal line is drawn across a plot of the posterior distribution $p(\pi|w)$, and suppose the corresponding values of π where the line intersects $p(\pi|w)$ are denoted as π_{lower} and π_{upper}. Iterative numerical procedures are used then to find the one horizontal line such that the area between π_{lower} and π_{upper} is $1 - \alpha$. Figure 6.9 shows the posterior with $a = b = 1$ and a horizontal line at $p(\pi|w) = 0.5184$. The corresponding HPD interval is $0.1586 < \pi < 0.6818$ with a probability of 0.95 between these two limits. This interval is fairly similar to the equal-tail credible interval because there is only slight right-skewness in the posterior.

The computational process for the HPD interval is performed by the `hpd()` function of the `TeachingDemos` package:

```
> w <- 4
> n <- 10
> alpha <- 0.05
> a <- 1
> b <- 1

> library(TeachingDemos)
```

```
> save.hpd <- hpd(posterior.icdf = qbeta, shape1 = w + a, shape2
    = n + b - w, conf = 1 - alpha)
> save.hpd
[1] 0.1585724 0.6817678

> # Verify lower and upper limits are at the same p(pi|w) values
> dbeta(x = save.hpd, shape1 = w + a, shape2 = n + b - w)
[1] 0.5183494 0.5183494

> # Verify area between limits is 1-alpha
> pbeta(q = save.hpd[2], shape1 = w + a, shape2 = n + b - w) -
    pbeta(q = save.hpd[1], shape1 = w + a, shape2 = n + b - w)
[1] 0.95
```

We also provide our own HPD interval code in the corresponding program to give additional insight into the computational process.

The `binom` package used in Chapter 1 to calculate intervals for π can also be used to calculate the equal-tail interval. Both `binom.bayes()` and `binom.confint()` with the `type = "central"`, `prior.shape1 = a`, and `prior.shape2 = b` argument values will calculate the interval. The help documentation for these functions also indicates that the HPD interval can be calculated by specifying `type = "highest"`; however, we have found that the function still returns the equal-tail interval even after this specification (version 1.0-5 of the `binom` package).

Of course, Bayes estimates and credible intervals are affected by the choice of a and b in the prior distribution. For example, we can rewrite the Bayes estimate as

$$\hat{\pi}_B = \left(\frac{n}{n+a+b}\right)\left(\frac{w}{n}\right) + \left(\frac{a+b}{n+a+b}\right)\left(\frac{a}{a+b}\right)$$
$$= \left(\frac{10}{12}\right)0.4 + \left(\frac{2}{12}\right)0.5$$

using Equation 6.23 and $a = b = 1$. This helps to show that this prior leads to a slightly larger estimate of π than does the MLE. Exercise 2 investigates further how Bayes estimates and credible intervals change for different values of a and b.

6.6.2 Regression models

To apply Bayesian methods to regression modeling, one needs to begin with the specification of a prior distribution for the regression parameters $\beta_0, \ldots, \beta_p$. Typically, this involves first assuming that the parameters are independent so that a more complex joint probability distribution is not needed. Also, while a subjective prior could be used for each parameter, more often a noninformative prior or a *weakly informative prior* is chosen that distributes probability rather evenly over a large range. For example, a weakly informative prior could be a normal distribution with a very large variance.

The distribution for the data given the parameters is typically chosen to be some common model, such as those presented in Chapters 2–4. The posterior distribution for $\beta_0, \ldots, \beta_p$

given the data $\mathbf{y} = (y_1, \ldots, y_n)$ has the form

$$p(\beta_0, \ldots, \beta_p | \mathbf{y}) = \frac{f(\mathbf{y}|\beta_0, \ldots, \beta_p) p(\beta_0, \ldots, \beta_p)}{f(\mathbf{y})}$$

$$= \frac{[\prod_{i=1}^n f(y_i|\beta_0, \ldots, \beta_p)] [\prod_{r=0}^p p(\beta_r)]}{\int \cdots \int [\prod_{i=1}^n f(y_i|\beta_0, \ldots, \beta_p)] [\prod_{r=0}^p p(\beta_r)] d\beta_0 \cdots d\beta_p} \quad (6.24)$$

when we take $\beta_0, \ldots, \beta_p$ to be independent. Very often, this expression for the posterior cannot be evaluated in a simple closed form, unlike those given in the previous section.

A variety of general-purpose simulation methods can be used to evaluate Equation 6.24 without giving the complete closed-form expression. These simulation methods are used to obtain samples from the posterior so that the distribution and associated quantities can be estimated. We will focus on one method in particular known as the *Metropolis-Hastings algorithm*. This algorithm is a type of *Markov chain Monte Carlo* (MCMC) simulation method that is implemented by the popular MCMCpack package in R (Martin and Quinn, 2006; Martin et al., 2011). We discuss some details of the algorithm shortly. For a general discussion on the Metropolis-Hastings algorithm, please see Chib and Greenberg (1995) and Chapters 6 and 8 of Robert and Casella (2010).

Example: Placekicking (PlacekickBayes.R, Placekick.csv)

We consider again the placekicking example from Chapter 2, where our purpose is to estimate the model logit$(\pi) = \beta_0 + \beta_1$distance. The MCMClogit() function in MCMCpack fits the logistic regression model using a *random walk* version of the Metropolis-Hastings algorithm (to be discussed later in this section). Familiar arguments within MCMClogit() are formula and data which specify the model and data set, respectively, and the seed argument which gives a seed number used to begin the simulation. New arguments include mcmc which gives the number of samples to be drawn from the posterior and burnin which specifies the number of initial samples to discard. A *burn-in period* is needed because samples toward the beginning of the simulation process may not represent the posterior well (a by-product of the Metropolis-Hastings algorithm). A verbose argument specifies how often to print information about the sampling process, which can be useful for judging how much longer the sampling process will take.

The prior distribution is also specified within the call to MCMClogit(). Single numerical values for the b0 and B0 arguments correspond to the mean and to the *inverse* of the variance, respectively, for independent normal priors for β_0 and β_1. A vector and matrix can instead be used for b0 and B0, respectively, in order to allow different means and variances for each parameter's prior or to incorporate dependence between the parameters through a multivariate normal distribution. Alternatively, the variance can be specified as B0 = 0 leading to noninformative priors for each parameter. Essentially, an infinite variance is used with a normal distribution in this case, which equivalently can be thought of as a uniform distribution from $-\infty$ to ∞. This is referred to as an *improper prior distribution* because it is not really a probability distribution. Finally, to specify something other than a normal distribution, the user.prior.density argument can be given instead of b0 and B0.

Below is our code and output from MCMClogit(). We chose 100,000 samples for this example mainly to guarantee that a very good representation of the posterior distribution occurs (we will show later that a smaller sample size would have been adequate). The priors are specified to have a mean of 0 and variance of 1000, which is weakly informative.

```
> mod.fit.Bayes <- MCMClogit(formula = good ~ distance, data =
    placekick, seed = 8712, b0 = 0, B0 = 0.001, burnin = 10000,
    verbose = 10000, mcmc = 100000)

MCMClogit iteration 1 of 110000
beta =
  5.81208
 -0.11503
Metropolis acceptance rate for beta = 0.00000

<OUTPUT EDITED>

MCMClogit iteration 100001 of 110000
beta =
  5.58192
 -0.10668
Metropolis acceptance rate for beta = 0.51765

@@@@@@@@@@@@@@@@@@@@@@@@@@@@@@@@@@@@@@@@@@@@@@@@@@@@@@@@@
The Metropolis acceptance rate for beta was 0.51719
@@@@@@@@@@@@@@@@@@@@@@@@@@@@@@@@@@@@@@@@@@@@@@@@@@@@@@@@@

> summary(mod.fit.Bayes)

Iterations = 10001:110000
Thinning interval = 1
Number of chains = 1
Sample size per chain = 1e+05

1. Empirical mean and standard deviation for each variable, plus
   standard error of the mean:

              Mean       SD    Naive SE  Time-series SE
(Intercept)  5.8375  0.327389  1.035e-03     3.149e-03
distance    -0.1156  0.008354  2.642e-05     7.976e-05

2. Quantiles for each variable:

               2.5%     25%     50%     75%    97.5%
(Intercept)  5.2094  5.6131  5.8325  6.0532  6.49744
distance    -0.1323 -0.1211 -0.1155 -0.1099 -0.09948

> HPDinterval(obj = mod.fit.Bayes, prob = 0.95)
               lower      upper
(Intercept)  5.2051395  6.49273705
distance    -0.1322877 -0.09945651
```

The first table in the summary() output provides the Bayes estimates for the regression parameters in the Mean column and the standard deviation of the sampled regression parameters in the SD column. The Bayes estimates and the standard deviations are very similar to their respective MLEs and standard errors obtained in Section 2.2.1. Overall, these similarities are not surprising because of the large sample size and the weakly informative priors. In general, the estimates and variances need not be this similar.

The second table in the `summary()` output gives the estimated quantiles for the posterior distributions. For example, the `distance` row gives the 0.025 and 0.975 quantiles of -0.1323 and -0.0995, respectively. These quantiles are the 95% equal-tail interval limits for β_1. Other quantiles can be requested through the `quantiles` argument of `summary()`. The 95% HPD interval for β_1 is given by the `HPDinterval()` function as $-0.1323 < \beta_1 < -0.0995$, which is almost the same as the equal-tail interval. These intervals are quite similar to the 95% profile LR and Wald interval limits found in Section 2.2.3.

The object returned by `MCMClogit()` contains only the posterior samples. We can obtain summaries of these samples similar to `summary()` as follows:

```
> head(mod.fit.Bayes)
Markov Chain Monte Carlo (MCMC) output:
Start = 10001
End = 10007
Thinning interval = 1
     (Intercept)     distance
[1,]    5.919087  -0.11865928
[2,]    5.919087  -0.11865928
[3,]    5.919087  -0.11865928
[4,]    5.919087  -0.11865928
[5,]    5.919087  -0.11865928
[6,]    5.764444  -0.11098132
[7,]    5.215507  -0.09609833

> colMeans(mod.fit.Bayes)   # Same as Mean column in summary()
(Intercept)     distance
  5.8374803   -0.1155836
> apply(X = mod.fit.Bayes, MARGIN = 2, FUN = sd)   # Same as SD
    column in summary()
(Intercept)     distance
0.327389022  0.008354371
```

It is important to note that the corresponding method functions for `summary()` and `HPDinterval()` are found in the `coda` package, which is installed and loaded with `MCMCpack`. The `coda` package contains functions that summarize MCMC simulation samples produced by `MCMCpack` and a number of other packages. We will discuss the `coda` package in more detail shortly when we use it to examine the quality of the samples obtained.

We saw in Chapter 2 that the p-value for a Wald test of $H_0: \beta_1 = 0$ vs. $H_a: \beta_1 \neq 0$ was $< 2 \times 10^{-16}$. Using Bayesian methods, we can evaluate the location of $\beta_1 = 0$ relative to the posterior distribution for β_1:

```
> beta1 <- mod.fit.Bayes[,2]
> min(beta1)
[1] -0.1523133
> max(beta1)
[1] -0.08290305
> mean(beta1 >= 0)   # 0/100000
[1] 0
```

Thus, $P(\beta_1 \geq 0) < 1/100000$ indicating that it is very likely that β_1 is negative.

Functions of the regression parameters, such as odds ratios and probabilities of success, can be evaluated using the samples from the posterior. Bayes estimates and credible intervals are subsequently found from these evaluations. Below is the code showing how to find an odds ratio for a 10-yard decrease in distance and to find the probability of success for a 20-yard placekick:

```
> OR10 <- exp(-10*beta1)   # OR for a 10 yard decrease in distance
> mean(OR10)   # Bayes estimate
[1] 3.187804
> quantile(x = OR10, probs = c(0.025, 0.975))   # 95% equal-tail
    2.5%     97.5%
2.704197 3.755661
> HPDinterval(obj = OR10, prob = 0.95)   # 95% HPD
        lower    upper
var1 2.675545 3.722519
attr(,"Probability")
[1] 0.95

> beta0 <- mod.fit.Bayes[,1]
> pi20 <- plogis(q = beta0 + beta1*20)   # Estimate of pi at
   20-yards
> mean(pi20)   # Bayes estimate
[1] 0.9710167
> quantile(x = pi20, probs = c(0.025, 0.975))   # 95% equal-tail
    2.5%     97.5%
0.9606018 0.9797005
> HPDinterval(obj = pi20, prob = 0.95)   # 95% HPD
        lower    upper
var1 0.9612459 0.9802097
attr(,"Probability")
[1] 0.95
```

Thus, the Bayes estimate for the odds ratio is 3.19 and for the probability of success is 0.9710. Both of these are again very similar to the MLEs found in Sections 2.2.3 and 2.2.4. The corresponding credible intervals are similar to the confidence intervals as well. If desired, plots of the estimated distributions for each quantity can be obtained using the densplot() function from the coda package (see corresponding program for examples).

Metropolis-Hastings algorithm

The Metropolis-Hastings algorithm is used to generate a large sequence of sampled values from a posterior distribution without fully specifying the distribution. This sample then allows us to estimate the posterior itself. For mathematical and computational reasons, it is more convenient to generate a sampled value by basing it on the previously sampled value rather than by generating a new, independent value. This results in a sequence of simulated parameter values that are serially correlated—each depending on the previous value. The corresponding sequential structure is known as a *Markov chain*.

The chain starts at a chosen location, such as the MLE for a parameter (or vector of parameters) of interest. Given this starting value, a new "proposal" is simulated. For instance, suppose the initial parameter value is denoted by d_0. A random walk Metropolis-Hastings algorithm simulates a proposal as $c_1 = d_0 + \epsilon_1$, where ϵ_1 is a random perturbation to d_0 that is simulated from a chosen probability distribution (e.g., a normal distribution with mean 0 and a specified variance). The proposal c_1 is taken to be the next value in the chain, d_1, with a particular probability (discussed shortly) that depends on the prior and data distributions. If the proposal c_1 is not taken, then d_1 is set to d_0 as the next value in the chain. This process continues, using $c_b = d_{b-1} + \epsilon_b$ for $b = 1, \ldots, B$ and a large B. The MCMC algorithm is said to have converged to the posterior distribution when the empirical distribution becomes "stable" in the sense that different sequences of the chain have empirical distributions that closely resemble one another.

The acceptance probability described in the previous paragraph for c_b is the ratio of two posterior distributions evaluated at c_b in the numerator and d_b in the denominator. The more plausible a proposal value is, the larger its evaluated posterior distribution. Thus, this ratio measures the plausibility of c_b relative to d_{b-1}. If c_b is in a higher posterior density region than d_{b-1}, the ratio is greater than 1. Of course, probabilities cannot be greater than 1, so the acceptance probability is set to 1, guaranteeing acceptance of the proposal as d_b. If c_b is in a smaller posterior density region than d_{b-1}, the ratio is less than 1, meaning it is not guaranteed to be accepted. An important aspect of this implementation is that the two posterior distributions have the same denominators, so the ratio can be taken to be the ratio of the two likelihood functions multiplied by their prior distributions. Thus, the potentially multidimensional integral in the posterior distributions' denominators does not need to be found.

For the model-fitting functions of the MCMCpack package, the random walk Metropolis-Hastings algorithm takes c_b and d_{b-1} as $(p+1)$-dimensional vectors containing potential values of $\beta_0, \ldots, \beta_p$. The ϵ_b term is a sampled value from a $(p+1)$-dimensional multivariate normal distribution. The starting value d_0 for the chain is the MLE for $\beta_0, \ldots, \beta_p$, because the MLE is likely to have reasonably high posterior density. This can help to make convergence of the chain faster than starting with values that are far from the center of the posterior. The mean vector for ϵ_1 contains all 0's, and its covariance matrix is taken to be the covariance matrix for the MLE.

Convergence diagnostics

An important aspect to using the Metropolis-Hastings algorithm is knowing whether convergence has been obtained. Model-fitting functions in MCMCpack return output that can be used with the coda package (name comes from "COnvergence Diagnosis and output Analysis") to evaluate this convergence. Both Plummer et al. (2006) and Robert and Casella (2010) provide nice introductory discussions for coda. More in-depth discussions on diagnostic measures given by coda are available in Cowles and Carlin (1996) and Robert and Casella (2004). This subsection provides a summary of these references along with our own insights.

A trace plot is often the first tool used to assess convergence. This plot simply shows the simulated parameter values in the chain joined by a line in the order that they were generated. The traceplot() and plot() functions produce these plots in R, where the plot() function also includes plots of nonparametric density estimates for the parameters. If the algorithm has converged, both the mean and the variability of the simulated values should remain fairly constant throughout the trace plot. Also, the chain should "mix" well, meaning that it visits all regions of the posterior very frequently. These conditions are often characterized in a trace plot by (1) a thick, dark, central band in the middle of the plot, undulating gently within a fairly constant range, and (2) spikes of varying lengths coming out of the band. The trace plots in Figure 6.10 are examples of a chain that appears to have converged.

Sometimes, the constant mean and/or variance will not occur near the beginning of the chain, especially if a chain's starting value does not happen to fall within a likely area of the posterior distribution. These samples can be removed by specifying a sufficiently long burn-in period. Another potential problem is when the chain tends to stay in one region for a long sequence of samples (i.e., it does not "mix" well). This leads to large positive correlations between successive simulated values. On the trace plot, poor mixing is characterized by what appear to be distinctly different bands and spikes which are much narrower than the full range of parameter values. Solutions to poor mixing include taking a larger sample size and/or thinning the chain by using only every k^{th} simulated value as the sample from the posterior, for an integer $k > 1$. Thinning can be specified by the thin argument in the model-fitting functions of MCMCpack, where the default is no thinning (thin = 1). Alternatively, thinning can be done using the thin argument in the window() function.

The rate at which the Metropolis-Hastings algorithm accepts new proposals into the sample is also an important diagnostic tool for convergence. Depending on the situation, a small or large acceptance may be fine. However, in most cases, a rate that is too large or too small indicates the possibility that not all portions of the posterior have been visited during the simulation process. Desirable acceptance rates are generally between 0.2 and 0.5. The tune argument in model-fitting functions of MCMCpack can help to achieve these desired rates. The numerical value given for this argument simply scales the variance associated with ϵ_b. The default value is 1.1, and changes in this value will affect the acceptance rate for the chain. Specifically, smaller values lead to less variability in the proposal c_b and generally increases the acceptance rate, while larger values have the opposite effect.[26] In order for R to print the acceptance rate, a non-zero value needs to be specified for the verbose argument in the model-fitting function call. Alternatively, the rejection rate, which is one minus the acceptance rate, is printed by using the rejectionRate() function of coda.

A number of hypothesis tests are available to help determine convergence. In particular, the geweke.diag() function performs a hypothesis test to compare the means of the first and last halves of the chain. Test statistic values beyond $\pm Z_{1-\alpha/2}$ indicate non-convergence. If the conclusion is non-convergence, the geweke.plot() function can then be used to determine if convergence eventually occurs by plotting the test statistic for successively later portions of the chain. A longer burn-in period can then be chosen to remove the early, unstable portion of the chain with the help of this plot.

Because of the positive serial correlation among elements within the chain, statistics computed from the chain have more variability than they would have if the elements were independent. The *effective sample size* is roughly defined then as the size of a random sample that would provide an estimate of the mean that has the same precision as the chain's mean estimate. The effective sample size is smaller than the chain's sample size. The stronger the correlation between consecutive elements in the chain, the smaller the effective sample size. If the effective sample size is too small, this may again suggest non-convergence and a larger number of samples would be needed to obtain an accurate representation of the posterior distribution. The effective sample size is found by the effectiveSize() function.

As remarked above, if a diagnostic procedure indicates non-convergence, a simple solution may be just to take a larger sample. Convergence will generally occur with the Metropolis-

---

[26] Because the probability of acceptance is based on the ratio of two posterior distributions, the ratio is more likely to be closer to 1 when there is a smaller amount of variability in the proposal c_b. In other words, the value of c_b will likely be closer to d_{b-1}, making the value of the two evaluated posterior distributions more similar.

Hastings algorithm unless the posterior is not an actual probability distribution, which can happen if an improper prior is chosen. To obtain a larger sample size, one simply reruns the same model-fitting function using a larger value for the mcmc argument. Alternatively, additional chains can be started. For example, the MCMClogit() function can be rerun one or more times for the placekicking example, where different starting values for β_0 and β_1 are used each time by changing the beta.start argument value (the default is the MLE; one could choose values at a reasonable number of standard deviations away from the MLE) and using a different seed number with the seed argument. Multiple chains are joined together with the mcmc.list() function in coda. An advantage of running multiple chains is that one obtains additional insight into convergence that may otherwise not be seen through a single chain. As Robert and Casella (2010) eloquently put, "you've only seen where you've been" by using a single chain. In other words, there could be places in the posterior not visited by a single chain. The gelman.diag() and gelman.plot() functions in coda can be used with multiple chains to test for non-convergence of the mean. These functions calculate variances used in standard analysis of variance—a between-chain variance for the mean and a within-chain variance for the mean. Values larger than 1 for the resulting test statistic indicate possible non-convergence.

Example: Placekicking (PlacekickBayes.R, Placekick.csv)

The output given by head(mod.fit.Bayes) beginning on page 452 shows the $b =$ 10001 to 10007 values for β_0 and β_1 in the chain. Proposed values for β_0 and β_1 were rejected at $b = 10002$ to 10005. At $b = 10006$, the proposed values were accepted and included in the sample. Of the 100,000 samples taken, the acceptance rate was 0.5172 (see output that begins on page 450), which is close to the upper bound of the desired acceptance range.

Invoking plot(mod.fit.Bayes) produces trace and density plots for the entire sample. Because we used a very large sample size, we examine these plots over smaller ranges with the help of the window() function. For example, window(x = mod.fit.Bayes, start = 10001, end = 20000) pulls out the first 10,000 samples from mod.fit.Bayes (those samples from the burn-in period are not included in the object), and Figure 6.10 gives the corresponding trace and density plots. The trace plots here (and for other segments of the chain) have a thick, dark central band in the middle with a fairly constant range of parameter values. Also, the values in the trace plot are rapidly changing over short spans of samples. Therefore, these plots do not indicate non-convergence.

The plot() function also provides density plots for β_0 and β_1, as shown in Figure 6.10. Both of these plots show fairly symmetric distributions, which is why the equal-tail and HPD credible intervals gave similar lower and upper bounds for each regression parameter.

Below are the results from geweke.diag():

```
> geweke.diag(x = mod.fit.Bayes)

Fraction in 1st window = 0.1
Fraction in 2nd window = 0.5

(Intercept)    distance
   -0.03532    0.05640
```

Comparing the test statistics produced by geweke.diag() to $\pm Z_{0.975} = \pm 1.96$, we conclude that there is not sufficient evidence to indicate convergence problems.

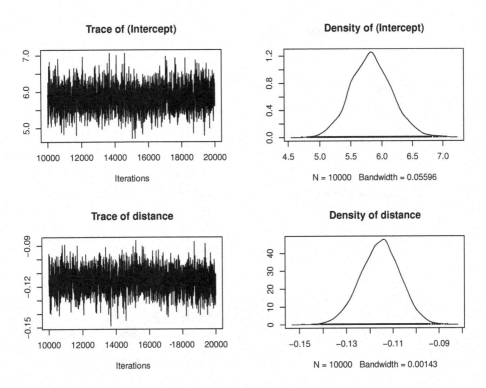

Figure 6.10: Trace and density plots for the first 10,000 samples included in the chain.

The effective sample size given by effectiveSize(mod.fit.Bayes) is 10806.61 for β_0 and 10972.14 for β_1, which is more than enough to obtain an accurate estimate of the mean. Using the same function with the first 10,000 samples of the chain leads to effective sample sizes of 1090.66 for β_0 and 1094.93 for β_1. Again, this should lead to accurate estimates of the mean. However, accurately estimating credible-interval endpoints, which lie in the extreme quantiles of the posterior distribution, typically requires a larger sample size than needed to estimate a posterior mean. To examine quantile estimation, the cumuplot() function calculates the 0.025, 0.5, and 0.975 quantiles of the posterior for each b and plots them against b (other quantiles can be requested using the probs argument). Because our chain is rather long, Figure 6.11 gives a plot for only the first 10,000 samples. We see that the variability in the median quickly dissipates, indicating convergence. The variability in the 0.025 and 0.975 quantiles takes longer to dissipate, but this is not surprising for extreme quantiles estimated by any type of sampling. The upper quantile for distance still seems to be decreasing slightly at the end of the period, so we may need more than the 10,000 post-burn-in samples that are represented in this plot. Therefore, while perhaps a little excessive, we are comfortable with our choice of 100,000 samples. The sampling process took only a few seconds to complete, so little time was lost by using this sample size.

We also ran MCMClogit() twice more to produce two additional chains of size 100,000. These chains used starting values that were one standard deviation away on either side of the MLEs. The specific code is included in our corresponding program. Overall, the Bayes estimates and quantiles of the posterior distributions were very close to those found originally. We also used the mcmc.list() function to com-

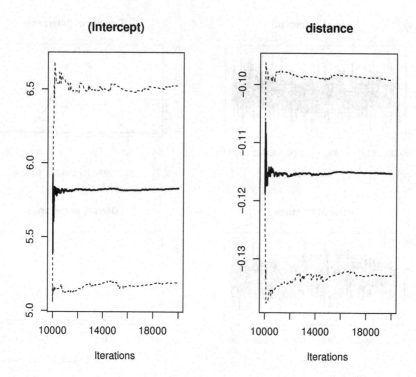

Figure 6.11: Plot of the 0.025, 0.5, and 0.975 quantiles for the first 10,000 samples included in the chain.

bine all three chains into one object for use with `gelman.diag()` to obtain another examination of convergence. The test statistics and the upper interval bounds for tests involving β_0 and β_1 were all equal to 1, indicating that convergence is very likely to have occurred.

A plot of the autocorrelations for a chain can serve as another way to diagnose convergence problems similar to its use in time series analysis to diagnose a nonstationary series. In particular, if the autocorrelations are all large for a long number of lags, this is a sign of non-convergence. The corresponding function in R that produces this plot is `autocorr.plot()`. When we examined the autocorrelation plots for each regression parameter, the autocorrelations quickly decreased toward 0, which supports our previous conclusions regarding convergence.

In addition to convergence of the MCMC methods, another concern when applying Bayesian methods is whether or not inferences may change if some other justifiable prior distribution was chosen instead. This is referred to as *sensitivity analysis*. We investigated this issue by examining different variance specifications with our normal distribution priors. Very similar results as with our original prior distribution were obtained, so this alleviated any of our concerns. This is not surprising because the large sample size for the data reduces the impact of the prior.

The `MCMCpack` package provides the functions needed to estimate a wide variety of models using Bayesian methods. These functions share the same syntax, so it is easy to perform other analyses once you understand how to use a function like MCMClogit(). Some of these functions are investigated in the exercises.

6.6.3 Alternative computational tools

Outside of R, there are a number of software programs available for applying Bayesian methods. Two popular programs that R can interface with are:

1. Bayesian inference using Gibbs sampling (BUGS) and its corresponding variants (http://www.mrc-bsu.cam.ac.uk/bugs); the R2WinBUGS and R2OpenBUGS packages allow R to use this software

2. Just another Gibbs sampler (JAGS; http://mcmc-jags.sourceforge.net); the R2jags package allows R to use this software

Both programs use the "Gibbs sampler," which is an alternative to the Metropolis-Hastings algorithm for sampling from the posterior distribution of the model parameters. There are other R packages than those listed above that can use these external software programs. We recommend that interested readers perform a search on "bugs" or "jags" at http://cran.r-project.org/web/packages/available_packages_by_name.html to obtain a complete listing.

Two more recent additions to the Bayesian capabilities of R include the RStan and INLA packages. The RStan package allows one to use the software "Stan" (http://mc-stan.org). This software uses a different type of MCMC sampling in order to obtain convergence faster than other commonly used MCMC methods. The INLA package (http://www.r-inla.org) implements the integrated nested Laplace approximations (INLA) approach of Rue et al. (2009). This method uses mathematical approximations to integrals rather than simulation. Neither RStan nor INLA are available from CRAN, but information on how to install the packages is available from their respective websites.

There are additional R packages that can perform many similar analyses to those available through MCMCpack. The Bayesian inference task view on CRAN at http://cran.r-project.org/web/views/Bayesian.html provides summaries and links to these packages.

6.7 Exercises

Binary responses and testing error

1. Rogan and Gladen (1978) discuss a survey of 1 million individuals in the United States that was performed to estimate the prevalence of hypertension. For the individuals surveyed, 11.6% had a diastolic blood pressure above 95 mm Hg, which was considered to be hypertensive. Through an additional study, the sensitivity and specificity of the procedure were estimated to be $S_e = 0.930$ and $S_p = 0.911$. Find the MLE for the overall true prevalence of hypertension (using > 95 mm Hg as the cutoff) and a corresponding 95% Wald confidence interval. Describe the effect that accounting for testing error has on understanding the prevalence of hypertension.

2. Section 1.1.3 examined the true confidence levels for confidence intervals constructed for a probability of success π, where $n = 40$, $\alpha = 0.05$, and π ranged from 0.001 to 0.999 by 0.0005. Construct the same types of plots for a Wald confidence interval for $\tilde{\pi}$

ranging from 0.001 to 0.999 by 0.0005. Describe what happens to the true confidence levels as the amount of testing error changes and make comparisons to the case of no testing error. We recommend fixing one of S_e or S_p at 1 first while varying the other accuracy measure.

3. Derive $\hat{\tilde{\pi}}$ and its estimated variance using the following process:

   (a) Find the log of the likelihood function given in Equation 6.2.
   (b) Calculate the derivative of the log-likelihood function with respect to $\tilde{\pi}$.
   (c) Set the derivative found in (b) equal to 0 and solve for $\tilde{\pi}$. The resulting value is the MLE.
   (d) Calculate the second derivative of the log-likelihood function with respect to $\tilde{\pi}$.
   (e) Find the expected value of the result found in (d).
   (f) Invert the result from (e) and multiply by -1. This is $Var(\hat{\tilde{\pi}})$. Substitute $\hat{\tilde{\pi}}$ in for π to obtain $\widehat{Var}(\hat{\tilde{\pi}})$.

4. Consider again the example on prenatal infectious disease screening from Section 6.1.2.

   (a) Estimate the logistic regression model that contains all of the explanatory variables available in the data set as linear terms. Make sure to include `marital.status` as a categorical variable.
   (b) Perform appropriate LRTs on each of the explanatory variables, assuming the remaining variables are in the model.
   (c) Determine the best fitting model for the data. Interpret the explanatory variables using odds ratios.

5. The methods described in Section 6.1.2 provide a convenient way to obtain a LR confidence interval for $\tilde{\pi}$ when there are no explanatory variables. Complete the following to obtain an interval for the hepatitis C prevalence among blood donors example from Section 6.1.1.

   (a) Construct a one-row data frame with variables named `positive` and `blood.donors` to represent the 42 out of 1875 blood donors that tested positive for hepatitis C.
   (b) Estimate the model logit$(\tilde{\pi}) = \beta_0$ using `glm()` with the `my.link()` function given on page 361. Make sure to include `weights = blood.donors` within `glm()` due to the binomial form of the data.
   (c) Calculate the 95% LR confidence interval for β_0 using the `confint()` function.
   (d) Find the 95% LR confidence interval for $\tilde{\pi}$ using the relationship between $\tilde{\pi}$ and β_0. Compare this interval to the Wald interval calculated in Section 6.1.1.
   (e) Discuss why the calculation of a LR confidence interval for $\tilde{\pi}$ leads to limits between 0 and 1.

6. The purpose of this problem is to examine what happens to $\widehat{Var}(\hat{\beta}_1)$ as the amount of testing error increases. For this problem, perform all of your calculations using the data on prenatal infectious disease screening data from Section 6.1.2, where age is used as the explanatory variable to estimate the probability of HIV infection in a logistic regression model.

(a) Estimate the model using $S_e = S_p = 1$ using optim() or glm() with the my.link() function given on page 361. Confirm that you obtain the same answer as when using glm() with family = binomial(link = logit).

(b) While holding S_p fixed at 1, examine what happens to $\widehat{Var}(\hat\beta_1)$ as S_e changes from 0.94 to 1 by 0.01.

(c) While holding S_e fixed at 1, examine what happens to $\widehat{Var}(\hat\beta_1)$ as S_p changes from 0.94 to 1 by 0.01.

(d) Compare your results from (b) and (c) and suggest reasons the variance did or did not substantially change. Discuss any convergence and/or calculation issues that arose.

7. Using the data on prenatal infectious disease screening data from Section 6.1.2 with age as the explanatory variable and HIV as the response variable, estimate the logistic regression model using the MCSIMEX method as follows:

(a) Create a new response variable in set1 that transforms hiv into a variable with the class factor. Call this new variable hiv.factor. This change of class is needed in order to use the mcsimex() function of the simex package.

(b) Estimate a logistic regression model using glm() but do not account for testing error. Store the results from fitting the model in an object named mod.fit.naive. Include the argument value x = TRUE in glm() so that the **X** matrix as described in Section 2.2.1 is in the returned object (this will be needed for mcsimex() in part (d)).

(c) Construct a contingency table that contains S_p, $1 - S_p$, $1 - S_e$, and S_e:

```
test.err <- array(data = c(0.98, 1 - 0.98, 1 - 0.98, 0.98),
    dim = c(2,2), dimnames = list(obs.levels =
    levels(set1$hiv.factor), true.levels =
    levels(set1$hiv.factor)))
```

(d) Estimate the logistic regression model using the MCSIMEX method with the mcsimex() function:

```
library(simex)
mod.fit.mcsimex <- mcsimex(model = mod.fit.naive, mc.matrix
    = test.err, SIMEXvariable = "hiv.factor")
summary(mod.fit.mcsimex)
```

This implementation of the mcsimex() function uses Monte Carlo simulation to approximate the effects of testing error on the regression parameter estimators. Both asymptotic and jackknife-based[27] estimates of standard errors are produced. Compare these estimates and standard errors to those obtained in Section 6.1.2.

---

[27]The jackknife is a procedure similar to the bootstrap in that resampling is used to take samples from the original sample. The difference between the two is that the jackknife uses a "leave-one-out" resampling approach so that n unique resamples are taken where each resample has $n - 1$ observations within it.

Exact inference

1. The hypergeometric distribution for a 2 × 2 table (see Table 6.2) can be found by starting from two independent binomial distributions and applying the following steps:

    (a) Write out the PMFs for W_1 and W_2 with binomial distributions involving n_1 and n_2 trials, respectively, and a common probability of success π. Note that n_1 and n_2 are known constants.

    (b) Find the joint PMF for W_1 and W_2 by multiplying the two PMFs from (a).

    (c) Discuss why the PMF for $W_+ = W_1 + W_2$ is binomial with $n_1 + n_2$ trials and probability of success π.

    (d) Calculate the conditional PMF for W_1 given W_+ as $P(W_1 = w_1 | W_+ = w_+) = P(W_1 = w_1, W_2 = w_2)/P(W_+ = w_+)$ for $w_1 = 0, \ldots, w_+$ subject to $w_1 \leq n_1$ and $w_+ - w_1 \leq n_2$. Relate the PMF found to a hypergeometric distribution.

2. Perform a permutation test for independence with the fiber-enriched-crackers data set as in Section 6.2.2, but now use the transformed LRT statistic $-2\log(\Lambda)$ rather than the Pearson statistic X^2. Compare your p-value to that obtained when using X^2. Is a χ_9^2 approximation appropriate for $-2\log(\Lambda)$?

3. Using the Larry Bird data set introduced in Section 1.2, complete the following:

    (a) Perform Fisher's exact test.

    (b) Perform a permutation test for independence with $B = 10{,}000$ permutations using (i) `chisq.test()` and (ii) code that uses the `for()` function to find each permutation one-by-one. Set a seed number of 1938 at the start of each procedure using `set.seed()`. Compare your p-values to that obtained in part (a). Why are the p-values different?

    (c) Is a χ_1^2 approximation for the Pearson statistic appropriate? Why or why not?

4. Exercise 19 of Chapter 1 examined results from an HIV vaccine clinical trial. Using the data given in Table 1.9, apply the exact inference methods discussed in Sections 6.2.1 and 6.2.2 to determine if there is evidence that the vaccine prevents HIV infection.

5. Section 5.2 of Mehta and Patel (1995) examines a study involving HIV infection rates among infants when the mother was already infected during pregnancy. CD4 and CD8 blood counts were available for infants at six months of age along with their eventual HIV status. The data in EVP form is available in the file ex5.2.csv (w is the number of HIV infections out of n individuals). Note that the CD4 and CD8 counts are coded as ordinal variables with levels of 0, 1, and 2. Mehta and Patel (1995) indicate that the researchers for this study considered these variables to be nominal with three levels each, so we do the same for this exercise.

    (a) Try to fit a logistic regression model using `glm()`, where CD4 and CD8 are used as explanatory variables (without an interaction) to estimate the probability of HIV infection. Does R indicate the model has converged? If so, use a stricter convergence criteria and examine if the parameter estimates change.

    (b) Estimate the same model now using exact logistic regression. Test the importance of CD4 and CD8 individually. Note that `logistiX()` will not be able to perform this test because there are two indicator variables needed to represent each explanatory variable. Also, note that the `factor()` function can be used with the `formula` argument in `elrm()` to create the needed indicator variables.

(c) Provide the estimates of the exact distributions used in (b) and show how the p-values are calculated from these distributions.

6. Use exact logistic regression to analyze the data for the tea-tasting example from Section 6.2.1. Are the conclusions the same as or different from what would be found through Fisher's exact test? Because there is only one binary explanatory variable, small modifications of the code given in Section 6.2.3 are necessary. For example, in order to use `logistiX()`, we recommend entering the data as `tea.long <- data.frame(actual = c(1, 1, 1, 1, 0, 0, 0, 0), y = c(1, 1, 1, 1, 0, 0, 0, 0))` and specifying the x argument in the function as `x = as.matrix(tea.long[,1])`.

7. Foxman et al. (1997) examined the relationship between type of contraceptive used and whether or not a woman had experienced a urinary tract infection (UTI). Data from the study were subsequently used in an advertisement for the LogXact software package and are given in the uti.csv file. The explanatory variables are binary indicating use of an oral pill (`Oral`), condom (`Cond`), lubricated condom (`Lub`), spermicide (`Sperm`), and/or diaphragm (`Dia`) for contraception purposes, where a value of 1 indicates use and a value of 0 indicates non-use. The explanatory variable age is dichotomized into two groups based on < 24 years of age (`Age24` = 1) and ≥ 24 years of age (`Age24` = 0). The binomial response is the number of individuals infected (`Infected`) out of the number of individuals with the same EVP (n). Using these explanatory variables without interactions, complete the following:

   (a) Attempt to fit a logistic regression model using `glm()` to estimate the probability of a UTI and notice that it does not converge. Determine which explanatory variable leads to the non-convergence and describe why non-convergence occurs.

   (b) Try to estimate the logistic regression model using `logistiX()`. Are there any problems that occur during the model fitting process? If so, describe the problems and suggest a reason for their occurrence.

   (c) Estimate the logistic regression model using `elrm()`. Perform hypothesis tests to determine the importance for each explanatory variable, given that the other variables are in the model.

   (d) Compare your results in (c) to fitting the same model using the modified likelihood methods of Firth (1993).

8. Using exact logistic regression methods, analyze the data on hepatitis C prevalence among healthcare workers from Exercise 19 of Chapter 2. Compare your results to those obtained from using MLEs.

9. Let $T_j = \sum_{i=1}^n Y_i x_{ij}$ denote the sufficient statistic for β_j, $j = 0, \ldots, p$. Using this additional notation, complete the following which examines two results from Section 6.2.3.

   (a) Equation 6.5 expresses the conditional probability of $Y_1, \ldots, Y_n$ given I. Denote the p elements of I by $(I_0, I_1, \ldots, I_{p-1})$. Show how Equation 6.5 is found by first noting that
   $$P(Y_1 = y_1, \ldots, Y_n = y_n, T_0 = I_0, \ldots, T_{p-1} = I_{p-1})$$
   $$= \frac{\exp(\beta_0 I_0 + \cdots + \beta_{p-1} I_{p-1} + \beta_p \sum_{i=1}^n y_i x_{ip})}{\prod_{i=1}^n [1 + \exp(\beta_0 + \beta_1 x_{i1} + \cdots + \beta_p x_{ip})]}$$

and

$$P(T_0 = I_0, \ldots, T_{p-1} = I_{p-1}) = \sum_R \frac{\exp(\beta_0 I_0 + \cdots + \beta_{p-1} I_{p-1} + \beta_p \sum_{i=1}^n y_i^* x_{ip})}{\prod_{i=1}^n [1 + \exp(\beta_0 + \beta_1 x_{i1} + \cdots + \beta_p x_{ip})]}.$$

(b) Equation 6.6 expresses the conditional probability of $T_p = t$ given I. Discuss the role that $c(t)$ (and then $c(t_u)$, $u = 1, \ldots, U$) plays in the equation in order to justify the conditional probability.

10. A *mid p-value* is commonly used to lessen the conservativeness of a hypothesis-testing method involving a test statistic with a discrete probability distribution. A p-value is normally calculated as the sum of probabilities for all possible values of the test statistic that are *at least as extreme* as the observed value. The conservative nature of the test arises because in many cases there is no combination of test-statistic values whose probability sums to exactly α. A mid p-value is calculated the same way, except that only half of the probability of the observed value is added to the sum. Thus, a mid p-value is always smaller than an exact p-value. For example, the 2-sided test mid p-value for the lymphocytic infiltration variable in the cancer-free example of Section 6.2.3, where the test statistic is $T = 19$, can be calculated as

$$2\min\{P(T < 19 | I) + 0.5P(T = 19 | I), \, 0.5P(T = 19 | I) + P(T \geq 19 | I)\} = 0.0371.$$

This value is given by the `logistiX()` function in its `ciout` component (TST-Pmid row). Using the exact distributions of $\sum_{i=1}^n Y_i x_{i2}$ and $\sum_{i=1}^n Y_i x_{i3}$, calculate the mid p-values corresponding to tests of significance for gender and for any osteiod pathology. Verify that your answers match those given by `logistiX()`.

Categorical data analysis in complex survey designs

1. The `survey` package contains a data set called `nhanes` that is part of the same large survey that provided the tobacco/respiratory-symptom data for the examples in Section 6.3. The version in the package does not have replicate weights. Instead, it contains variables for *primary sampling units* (PSUs; i.e., clusters), strata, and sampling weights. The `help(nhanes)` page describes the variables in more detail. It also provides the code for creating a `survey.design`-class object from these (also available from `example(nhanes)`), so that the analysis functions in the package can be used.

    (a) According to the documentation for NHANES (http://www.cdc.gov/nchs/data/series/sr_02/sr02_161.pdf) the PSUs are nested within strata. This means, for example, that PSU #1 in one stratum represents a different cluster from PSU #1 in a different stratum. Before starting with survey analysis, determine (i) how many strata there are and (ii) how many PSUs there are in the data. Based on these numbers, how many degrees of freedom should there be initially for tests and confidence intervals (i.e., what is κ)?

    (b) Create a `survey.design` object using `svydesign()` as shown in the help for `nhanes`. Note that there are four categorical analysis variables in the data set: a binary variable for high cholesterol (`HI_CHOL`, the response variable), and demographic variables for race (`race`), age (`agecat`), and gender (`RIAGENDR`). For each variable, use the survey design to obtain estimates of the proportions of the population in each class of the demographic variables, along with confidence intervals for these proportions.

(c) Fit a single logistic-regression model using HI_CHOL as the response variable and the three demographic variables as linear terms.

   i. Report the estimated model.
   ii. Test the effects of each explanatory variable, using $\alpha = 0.05$. Draw conclusions.
   iii. Find a confidence interval for the odds ratio for each pair of age categories. Express all odds ratios so that the older age group is in the numerator. Interpret the results.

2. Refer to the example starting on p. 395. Perform a similar analysis of the associations among the four respiratory symptoms. Which symptoms appear to be related to one another? Use odds ratios to describe the associations.

3. As discussed in Section 6.3.5, one of the advantages of analyzing contingency tables with loglinear models is that it opens up a wider array of tests under the model-comparison framework. Tables with more variables, or with variables having ordinal or other structures, present no difficulty when analyzed as a loglinear model. The anova() function provides a Rao-Scott-type test for any pair of nested loglinear models. Thus, a wider class of hypothesis tests can be considered than merely tests for independence. For example, a linear-by-linear association model for two ordinal variables X and Y can be fit by using categorical versions of X and Y as main effects and numeric versions of the variables in an interaction.[28] The test of goodness-of-fit for the linear-by-linear association compares this model with the model containing the interaction in nominal form.

Refer to the examples from that section and consider a new question: Does the number of respiratory symptoms experienced increase with the number of tobacco products used? This can be addressed using various loglinear models. Recall that there are four binary variables representing different respiratory symptoms and five representing tobacco products used. Summing each of these sets of variables leads to numeric variables num.resp and num.tob. Three models are to be fit: one with no interaction (i.e., independence of number of respiratory symptoms and number of tobacco products used), one with a numeric interaction (i.e., linear-by-linear association model; see Section 4.2.6), and one with a categorical interaction.

(a) Make a 6 × 5 table of the weighted counts and also proportions where the denominator in each row is that row's total weighted count. The following code achieves this:

```
wt.tab <- svytable(formula = ~ num.tob + num.resp, design =
    jdesign2)
sweep(x = wt.tab, MARGIN = 1, STAT = rowSums(wt.tab), FUN =
    "/")
```

Are any trends apparent? Notice that the table has a zero count in it. This will cause problems for model fitting. Because the counts in the last two levels

---

[28] For computational reasons in svyloglin(), this model is obtained by first fitting the independence model to the two factor variables (i.e., with no interaction), and then using update(), where the first argument is the previous model-fit object and the second argument is formula = ~ . + as.numeric(X):as.numeric(Y), with the actual variable names used as X and Y.

of `num.tob` are much smaller than those in other columns, create a new variable (`num.tob2`) that combines the 4 and 5 levels of `num.tob` (e.g., by using the minimum of `num_tob` and 4 for each observation). Fit three models: independence, linear-by-linear association, and nominal association (saturated). Perform the tests of (i) no interaction versus linear-by-linear association, and (ii) linear-by-linear association versus nominal association (the latter two should be constructed using the `update()` function based on the first model). Report the hypotheses, test statistics, p-values, and results of these tests. Interpret the two results.

(b) In the linear-by-linear association model, estimate the association with a point estimate and a 95% confidence interval. Interpret the results.

(c) In the nominal-association model

   i. Estimate the odds ratio for each pair of consecutive rows and columns. That is, estimate $OR_{01,01}, OR_{01,12}, OR_{01,23}, OR_{01,34}, \ldots, OR_{34,23}, OR_{34,34}$, where in each symbol the first pair of numbers stands for the levels of `num.tob2` and the second pair stands for the levels of `num.resp`. Find confidence intervals in each case. (Warning: finding the correct contrast coefficients for `svycontrast()` can be a real challenge due to the parameterization used in `svyloglin()`. Read the note in the SurveySmokeLoglinear.R program on the book's website and work out what the coefficients must be.) Compare results to estimates obtained directly from `wt.tab`.

   ii. Do the confidence intervals suggest that the odds ratios are all similar, or are some clearly different? In particular, is there an overlap to all of the intervals? How does this finding relate to the tests performed in part (b)?

"Choose all that apply" data

1. For $I = 1$ and $J = 1$, show that Equation 6.13 is the Pearson statistic from Section 1.2.3.

2. Suppose the Pearson chi-square statistic,

$$\sum_{i=1}^{I}\sum_{j=1}^{J} \frac{(m_{ij} - m_{i+}m_{+j}/n)^2}{m_{i+}m_{+j}/n},$$

is applied to Table 6.6. Notice that this statistic is essentially one of four components in X_S^2. What would happen to this statistic's value if Table 6.6 instead summarized all of the negative responses rather than positive responses to both MRCVs? Why is this result undesirable?

3. Bilder and Loughin (2004) examine the Kansas farmer survey described in Section 6.4 as well, but they focus on the question about swine waste storage methods along with another question regarding the farmer's sources of veterinary information. Table 6.8 displays the positive responses to both questions. Note that source of veterinary information is a MRCV because farmers could choose up to five responses. The full data set is available in the data file BL2004.csv available on the textbook's website. Using these data, complete the following:

   (a) Construct a table like Table 6.8 for the data. There are examples in the research literature where a Pearson chi-square test for independence has been directly applied to a table like this. Discuss why this is a poor approach.

Table 6.8: Marginal table of swine waste storage and sources of veterinary information responses from 279 Kansas farmers.

	Consultant	Veterinarian	Extension service	Magazines	Feed companies
Lagoon	34	54	50	63	41
Pit	17	33	34	43	37
Natural drainage	6	23	30	49	34
Holding tank	1	4	4	6	2

(b) Construct all of the 20 possible 2 × 2 contingency tables for this example.

(c) Compute $X^2_{S,ij}$ for $i = 1, \ldots, 4$ and $j = 1, \ldots, 5$ and sum them to form X^2_S. What would a naive χ^2_{IJ} approximation to X^2_S conclude in a test for SPMI?

(d) Apply the Bonferroni, bootstrap, and second-order Rao-Scott correction to test for SPMI. Do all testing procedures reach the same conclusion?

(e) In practice, only one of the procedures in (d) should be applied. Discuss the advantages and disadvantages for each of them in general practice.

4. Continuing Exercise 3, estimate the SPMI, homogeneous association, W-main effects, Y-main effects, W and Y-main effects, and saturated models. Which model is best? Interpret the fit of this model using odds ratios.

5. MRCVs are found in areas outside of surveys. For example, wildlife researchers are often interested in determining which types of foods are present in the diet of a particular species of animal. Scats are examined for their content, and the corresponding responses are summarized by MRCVs because each scat may contain more than one type of food.

Riemer et al. (2011) examine the scats of Steller sea lions along the Oregon and California coasts to better understand this species' diet. The `sealion` data frame in the `MRCV` package contains a portion of the data collected at the mouth of the Columbia river in August of 2004 and 2007. For each scat sample, the presence (1) or absence (0) of 11 different prey types are given; thus, prey type is the MRCV. With these data, we would like to determine if there are differences in diet across the two years. Because year is a SRCV, we are interested in testing for MMI as described at the end of Section 6.4.2.

(a) Construct 11 2 × 2 contingency tables to summarize the responses by year (`Date`) and prey presence or absence.

(b) For each 2 × 2 contingency table, compute a Pearson chi-square test statistic (say, X^2_j for $j = 1, \ldots, 11$) and corresponding p-value to test for independence as given in Section 3.2.3.

(c) Perform an overall test for MMI by using a Bonferroni correction with the p-values in (b). State a conclusion about the comparison of diets.

(d) The `MI.test()` function can also perform the test for MMI. The same syntax as used for the SPMI test is also used for the MMI test; however, `I = 1` now needs to be set because `Date` is a SRCV. Use this function to perform the Bonferroni, bootstrap, and Rao-Scott approaches to test for MMI.

6. Wright (2009) examines two different ways to analyze data with one MRCV and one or more SRCVs. Read this paper and comment on the data analysis approaches presented.

Mixed models and estimating equations for correlated data

1. In the example from Section 2.2.1, we ignored the possible effect of the kicker on the probability of success for a placekick. The file PlacekickWithNames.csv contains the kicker's name for each attempted placekick in the placekicking data. We can now assess whether kickers have different overall probabilities of success and/or different relationships between distance and probability of success.

    (a) Determine the number of kickers and the number of attempts for each kicker. These numbers are all reasonably large, so we believe that LR inference should be acceptable here.

    (b) Fit a GLMM for estimating the probability of a successful placekick using the binary variable `good` as the response, `distance` as the numerical explanatory variable, and with a random effect for `kicker`. Use the Laplace method for reasons that will be explained later.

     i. Write out the estimated logit model for this fit. Explain which of the regression parameters are held fixed and which are random according to this model.
     ii. Report the AIC, BIC, and deviance for this model.

    (c) Fit an additional mixed model that allows the `distance` effect to vary randomly among kickers, independently of the intercept variation. This model has multiple random effects, and therefore cannot be fit in `glmer()` using more than one quadrature point.

     i. Write out the estimated logit model for this fit.
     ii. Report the AIC, BIC, and deviance for this model.
     iii. Perform a LRT for the added variance component. What do the results imply about different kickers' responses to increasing distance of a placekick?

    (d) Fit the fixed-effect model with only distance as the explanatory variable.

     i. Report the AIC, BIC, and deviance for this model. Compare these to the other two models. What does this suggest regarding the effect of kicker on the probability of a successful placekick?
     ii. Perform a likelihood ratio test for the intercept variance component relative to the model in part (b). What do the results imply about different kickers' responses to increasing distance of a placekick?

    (e) Were we justified in ignoring the effects of different kickers in our analyses of this data set in Chapter 2 and elsewhere? Explain.

2. Exercise 22 in Chapter 2 examines the effect of picloram herbicide on tall larkspur. A grouping variable, `rep`, was excluded from the analysis done there. In this exercise, we re-analyze the data incorporating the `rep` as a random effect.

    (a) Estimate a mixed-effect logistic regression model for the number of weeds killed using the picloram amount as the fixed-effect variable and `rep` as the random effect. Use 1, 3, 5, and 10 points of quadrature. Report the estimated parameters and the model deviance for each fit.

    (b) Compare the deviances to that from the fixed-effect model with picloram only. Are all of the mixed-model fits reasonable? Explain.

(c) (Use the GLMM with 5 quadrature points here and for the remainder of this exercise.) Use the parametric bootstrap to test whether the rep effect variance component may be zero. Set the seed number to 2939002, and use 500 simulations. Interpret the results in terms of the experiment.

(d) Refer to the FallsGLMM.R example program on the book website. Obtain parametric bootstrap confidence intervals for the fixed- and random-effect parameters. Use the same simulation for both sets of computations. Set the seed number to 37919236, and use 2000 simulations.

3. Continuing 2, refer to Exercise 21 in Chapter 2, where the concept of a *lethal dose level*, x_π, is introduced. In this exercise, we use the parametric bootstrap on the GLMM fit with 5 AGQ points to find a confidence interval for LD90, $x_{0.9} = [\text{logit}(0.9) - \beta_0]/\beta_1$.

   (a) Use the delta method to show that $Var(\hat{x}_\pi) \approx [Var(\hat{\beta}_0) + \hat{x}_\pi^2 Var(\hat{\beta}_1) + 2\hat{x}_\pi Cov(\hat{\beta}_0, \hat{\beta}_1)]/\hat{x}_\pi^2$.

   (b) Compute $\hat{x}_\pi$ from the GLMM fit.

   (c) Compute $\widehat{Var}(\hat{x}_\pi)$ from part (a). All of the needed variance and covariance estimates are available from the matrix computed by vcov() on the model fit.

   (d) Refer to the FallsGLMM.R example program on the book website. Similar to the analysis of pairwise comparisons shown in that program, conduct a parametric bootstrap analysis of x_{90} using the "bootstrap-t" approach described in the program. Base the Z-statistic on the estimate and variance of $\hat{x}_{90}$. Use 2000 simulations and a seed number 37919236 (note that this is the same seed as in the previous problem, so the extra calculations can be added into that simulation loop if desired). Summarize the estimated sampling distribution of the Z-statistic. Does it appear close to normal?

   (e) Find the 95% confidence limits using the critical values of the estimated distribution of Z. Interpret the results.

4. Refer to the alcohol-consumption example studied in Sections 4.2.2 and 5.4.2. The data in the study presented in DeHart et al. (2008) covered a 30-day period for each subject. Our previous analyses considered only the first Saturday for each subject, so that there was only one count per person. The data file DeHartSimplified.csv contains data from the first full week for each subject. We now have multiple counts per subject, requiring a proper treatment of the grouped data.

   (a) Consider a GLMM for these data, using all of the explanatory variables (nrel, prel, negevent, posevent, gender, rosn, age, desired, and state). Add the day of the week, dayweek, at the end of the model formula as a categorical explanatory variable, because we expect different amounts of alcohol consumption on different days of the week. The grouping variable is id, identifying each subject uniquely.

      i. Assume that only the intercept is random and fit this model. What does this assumption imply about the different subjects' drinking habits on different days of the week?

      ii. Try various numbers of quadrature points and choose one that fits best in a reasonable amount of time.

      iii. Test the significance of the variance component using a LRT. Report the results and indicate whether this test is likely to give reasonable results in this problem.

iv. Test the significance of each fixed-effect variable using LRTs. Note that this requires refitting the model to correspond to each different null hypothesis tested. Report the results.

(b) Next consider a GEE approach to the analysis.

i. Describe in terms of the problem what the independence and exchangeable correlation structures mean. Why does exchangeable seem more appropriate?

ii. Fit the fixed-effects portion of the GLMM as a GEE using the exchangeable structure. Report the test results from `summary()` and `anova()`. What differences are there between the results?

iii. Reverse the order of terms in the model and repeat the tests. Do the results change? Explain.

iv. Carry out a test for `dayweek` by fitting the model without it and using `anova()` to compare the resulting model to the full model.

Bayesian methods for categorical data

1. For the Hepatitis B example in Section 1.1.2, complete the following using a beta prior distribution with $a = b = 1$:

    (a) Plot the posterior distribution.

    (b) Compute the Bayes estimate for π.

    (c) Compute the 95% equal-tail and HPD intervals for π. Interpret the intervals.

    (d) Compare your results from (a) and (b) to the frequentist-based estimate and intervals. Provide reasons for their similarities or differences.

    (e) Hepatitis B prevalence would be expected to be small for blood donation settings like this. Propose a prior distribution that would take this into account and complete (a)-(c) again. Do your results change? Discuss.

2. Investigate how the Bayes estimate and 95% credible intervals for π change for different values of a and b in a beta prior distribution. Consider the following observed data scenarios:

    (a) $w = 4, n = 10$

    (b) $w = 1, n = 10$

    (c) $w = 8, n = 20$

    (d) $w = 40, n = 100$

    Include comparisons across (a)-(d) in your responses.

3. Exercise 1 of Chapter 1 examined a number of simple experiments where the purpose was to estimate a probability of success. For each part, discuss how to determine appropriate prior distributions and what they may be. Make sure to include prior distributions that are informative in your discussion. Analyze the available data for each part using Bayesian methods.

4. With the placekicking data example in Section 6.6.2, try using small burn-in periods, small numbers of MCMC samples, different values for the `tune` argument, and different starting values to see the effect these changes have on model estimates and convergence measures.

5. Exercise 7 of Chapter 2 examined the probability of success for a placekick using data from the 2002 and 2003 National Football League seasons. Using `Distance` as the only explanatory variable, complete the following.

   (a) Discuss how the results from the placekicking example in Section 6.6, which used data from a different NFL season, could be used to help choose a prior distribution for the regression parameters.

   (b) Convert `Good` in the data set to a 0-1 variable because `MCMClogit()` does not accept a response variable coded as "Y" or "N".

   (c) Estimate a logistic regression model using Bayesian methods. Comment on the effect that the distance has on the probability of success.

   (d) Find the 95% equal-tail and HPD intervals for the probability of success when the distance is 20-yards. Interpret the intervals.

   (e) Find the 95% equal-tail and HPD intervals for an odds ratio involving a 10-yard decrease in the distance. Interpret the intervals.

   (f) Discuss the measures investigated to ensure that convergence of the MCMC simulation methods occurred.

6. Section 2.2.6 examined a data set involving the Tomato Spotted Wilt Virus. Analyze these data using a logistic regression model and the appropriate Bayesian methods. Include an infestation type and control method interaction within the model. Justify the choice of a prior distribution and provide evidence of convergence for the MCMC sampling. Compare your conclusions to those obtained using maximum likelihood methods.

7. The use of Bayesian methods can provide an alternative solution to the complete separation problem (see Section 2.2.7) that sometimes occurs in logistic regression. The cancer-free example of Section 6.2.3 provides a data set where complete separation occurs. For this data set, complete the following:

   (a) In general, discuss problems that may occur from using the default starting values of `MCMClogit()` when complete separation occurs.

   (b) Estimate the logistic regression model with all three explanatory variables in the model as linear terms. Try a few different normal distribution priors with different amounts of variability.

   (c) Evaluate whether or not convergence to the posterior distribution occurs.

   (d) Compare the resulting model to what was found using the modified likelihood approach.

   The choice of a prior distribution can be more difficult in cases where complete separation exists. Gelman et al. (2008) suggests a solution to the problem that involves rescaling the data and using a Cauchy distribution for the prior. We encourage interested readers to examine the paper and implement their approaches with the `arm` package.

8. For the alcohol consumption example of Section 4.2.2, consider again the model $\log(\mu) = \beta_0 + \beta_1 \text{negevent} + \beta_2 \text{posevent} + \beta_3 \text{negevent} \times \text{posevent}$. Complete the following:

(a) We will use the `MCMCpoisson()` function of MCMCpack to estimate the model for this exercise. Examine the associated help file with this function and compare it to the help file for `MCMClogit()`. Are there any differences in the two functions' syntax?

(b) Estimate the model using `MCMCpoisson()`. Use a noninformative or weakly informative prior distribution.

(c) Provide evidence that shows convergence to the posterior distribution has likely occurred.

(d) Calculate the ratio of means for a 1-unit increase in negative events at each of the three positive-event quartiles. Interpret these values as percent changes, and find the corresponding 95% credible intervals. Compare your answers and their corresponding interpretation to those given in Section 4.2.2.

9. For the wheat-data example of Section 3.3, consider again the multinomial regression model where `class`, `density`, `hardness`, `size`, `weight`, and `moisture` were used to estimate the log-odds for the kernel type. Complete the following:

(a) We will use the `MCMCmnl()` function of MCMCpack to estimate the model for this exercise. Examine the associated help file with this function and compare it to the help file for `MCMClogit()`. Are there any differences in the two functions' syntax?

(b) Estimate the model using `MCMCmnl()` with the argument value `mcmc.method = "IndMH"`. In your implementation of the function, note the following:

   i. We had difficulty running the function when using a weakly informative prior distribution, but a completely noninformative prior worked well.

   ii. A warning message of "`In model.response(mf, "numeric"): using type = "numeric" with a factor response will be ignored`" occurs whenever we use the function. We believe there may be a small problem with an object type, but it does not appear to affect the numerical calculations.

(c) Provide evidence that shows that convergence to the posterior distribution has likely occurred.

(d) Interpret the relationships between the explanatory variables and the wheat types. Compare your interpretations to those given in Section 3.3.

10. With the Larry Bird data set first introduced in Section 1.2, complete the following using Bayesian methods:

(a) Estimate a logistic regression model using the second free throw outcome as the response and the first free throw outcome as the explanatory variable. Compare your model to the estimated model obtained in Exercise 13 of Chapter 2.

(b) Estimate a Poisson regression model (use `MCMCpoisson()`; see Exercise 8) that includes the first and second free throw outcomes as explanatory variables along with their interaction. Compare your model to the estimated model obtained in Section 4.2.4 for the same data example.

(c) After confirming convergence in (a) and (b), judge whether or not the second free throw outcome is dependent on the first.

Appendix A

An introduction to R

Since the introduction of the R statistical software package by Ihaka and Gentleman (1996) and the release of version 1.0.0 in 2000, the use of R has grown immensely. R has quickly become the standard statistical analysis software for those in academics, and its popularity has been growing in industry, especially as new graduates enter the workplace. This popularity has led many commercial statistical software packages, including SAS and SPSS, to incorporate ways to run R directly from their own software. Discussions about R have even made it into mainstream media outlets, including the *New York Times* article "Data Analysts Captivated by R's Power" written by Vance (2009), which is often thought of as a landmark for R's acceptance.

The purpose of this appendix is to help readers begin using R. We also provide details on how to set up program editors that can make reading and writing code easier. Finally, we give a basic regression analysis example using R. All R code used in this appendix is within the AppendixInitialExamples.R and GPA.R programs that are available on the textbook's website.

A.1 Basics

R is a free, open-source software package that is available for download through the Comprehensive R Archive Network (CRAN). Below is the step-by-step process to find the executable file to download:

1. Go to the R website at http://www.r-project.org. Select the CRAN link on the left side.

2. CRAN is mirrored at a large number of locations around the world. Choose the location closest to you.

3. R is available for Linux, Mac, and Windows operating systems. Choose your desired operating system's link. We chose the Windows operating system, and we will use an R version for Windows throughout the book.

4. Select the "base" link for the base distribution of R to download.

5. Select the "Download R 3.*.*" for Windows" link to download the executable file.

After downloading, run the executable file to install R. The default installation settings are satisfactory for most R users. Both the 32-bit and 64-bit versions are installed automatically, but only the 32-bit version is needed for this book.

To begin R, select the icon on the desktop or from the Windows Start or Apps menu. When R is first opened, the graphical user interface (GUI) looks similar to what is shown in

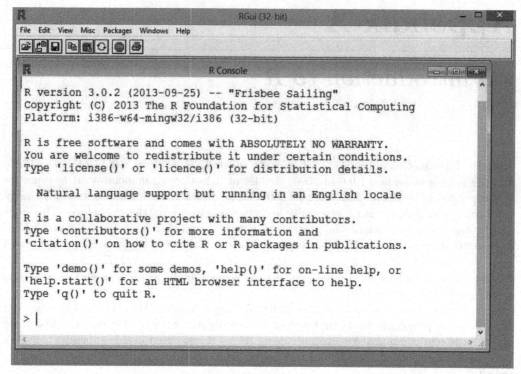

Figure A.1: The R GUI.

Figure A.1. The cursor is positioned in the R Console window at the ">" command prompt waiting for a command. The simplest commands use the R Console as a calculator:

```
> 2+2
[1] 4
> (2-3)/6
[1] -0.1666667
> 2^2
[1] 4
> pnorm(1.96)
[1] 0.9750021
> sin(pi/2)
[1] 1
> log(1)
[1] 0
```

For each example, the <Enter> key is pressed after the line of code is typed. Each numerical result is preceded by a [1] indicating this is the first value in a vector result; we will discuss how multiple values can be produced later.

Results from calculations can be stored in an *object*. Objects are just listings of information, possibly of different types. Simple objects may contain only one number. Complicated objects may contain data, model-fitting information, parameter estimates, residuals, and other relevant output. Below is an example creating an object called **save** to store the results from $2 + 2$:

```
> save <- 2+2
> save
[1] 4
```

The symbol combination <- (less than and a dash) makes the storage assignment; notice that it looks like an arrow pointing the computation results to the object name. This symbol combination is read as "gets." Thus, "save gets the results from 2 + 2." The equal sign can be used to make the storage assignment too, although <- is more prevalent among R users. An object name can be any word or combination of letters, numbers, periods, and underscores provided a letter is first. Periods and underscores are used often to separate multiple descriptive words. With respect to letters, R is case sensitive. Thus, **save** differs from **Save**.

To see a listing of objects in R's working database, either of the following will work:

```
> ls()
[1] "save"
> objects()
[1] "save"
```

To delete an object, use rm() and insert the object name inside the parentheses. All objects are removed automatically from the working database when R is closed (select File > Exit from the menu bar) unless the user chooses to "Save workplace image" on exit.

A.2 Functions

R performs its calculations using *functions*. For example, the pnorm() command used earlier is a function that calculates the probability that a standard normal random variable is less than the specified value. Writing your own function is simple. Suppose you want to write a function to calculate the standard deviation. While the sd() function already exists for this purpose, our new function will provide a basic illustration. Below is our new function's code:

```
> x <- c(1,2,3,4,5)
> sd2 <- function(numbers) {sqrt(var(numbers))}
> sd2(x)
[1] 1.581139
```

In this example, five observations are saved or concatenated into a *vector* object type named x using the c() function. A function called sd2() is then written to calculate the square root of the variance by taking advantage of the functions sqrt() and var(). The function requires a single input, or *argument*, which represents the information that will need to be given to the function when it is run. This argument can be any name, and we call it **numbers** here. The function's body is included within braces, { }, and contains the manipulations of the argument that will lead to the desired result. The sd2() function is stored as an object in R's working database. This function can be run at any time during the R session (i.e., until R is exited). The last line of code runs the function on our object x, which is passed into the function as the value for **numbers**.

When a function has multiple lines of code in its body, the last line corresponds to the returned value. For example, the code

```
> sd2 <- function(numbers) {
+   cat("Print the data \n", numbers, "\n")
+   sqrt(var(numbers))
}
> save <- sd2(x)
Print the data
 1 2 3 4 5 
> save
[1] 1.581139
```

shows a new `sd2()` function with two lines of code in its body, but only the last line is saved into an object. Included within the function is a call to the `cat()` function. This is used to print text and objects from within inside a function (it works outside a function too). Inside `cat()`, the `\n` is a special escape character that moves printed text to the next line. The + signs in the above code were not actually typed. Rather, at the end of each line in the function body, the `<Enter>` key is pressed. When R detects that a complete expression has not been entered, the cursor moves to the next line and shows the + sign to prompt the user for more code.

A.3 Help

Help available within R is accessed by selecting Help from the R menu bar. From the resulting drop down menu, there are a number of help options, including a list of shortcut keystrokes, answers to frequently asked questions, user manuals, and search tools. We are going to focus on the HTML help here. By selecting HTML help, we obtain a list for many of the same help resources, but now within a web page. In particular, the "Packages" link under "Reference" leads to a list of *packages*, which are groupings of R functions and data. For example, the `base` package contains a number of R's basic functions, including `log()` and `ls()` used earlier. Selecting the "base" link displays functions within the `base` package. Other packages include `stats` (contains functions used for the most common statistical calculations) and `graphics` (contains functions for plotting).

For the remainder of this section, we examine the help available for the `pnorm()` function, which was used earlier to find the probability that a standard normal random variable is less than 1.96. Beginning from the web page that lists the available packages, select the `stats` package and then select the "pnorm" link. Under the "Usage" heading in the resulting help web page for the function, the full function syntax is given as

```
pnorm(q, mean = 0, sd = 1, lower.tail = TRUE, log.p = FALSE)
```

The function has arguments `q`, `mean`, `sd`, `lower.tail`, and `log.p` that either need to be specified or will take on the stated default values given after the "=" symbol. The help web page further explains that `q` is a quantile, which we set at 1.96 earlier. Values for arguments can appear in any order in the function call as long as their names are specified, or they can be listed in order without names. Using the optional forms of syntax for `pnorm()`, we produce the same probability as before with the following code:

```
> pnorm(1.96)
[1] 0.9750021
> pnorm(q = 1.96)
[1] 0.9750021
> pnorm(1.96, 0, 1)
[1] 0.9750021
> pnorm(mean = 0, sd = 1, q = 1.96)
[1] 0.9750021
```

The first and third lines of code provide examples where the argument names are excluded, but the numerical values are in the same order as in the function syntax. The second and fourth lines of code include the argument names. When running functions in general, we recommend this latter syntax to avoid possible mistakes and to make code more readable.

The help web page contains other information in addition to syntax for pnorm(). For example, it gives usage examples and syntax information for closely related R functions. Finally, the same help for pnorm() can be opened by typing help(pnorm) and <Enter> at the R Console prompt, or by other methods that depend on the program editor being used (see Appendix A.6).

A.4 Using functions on vectors

Many R functions work directly on vectors of information. For example,

```
> pnorm(q = c(-1.96,1.96))
[1] 0.02499790 0.97500210
> qt(p = c(0.025, 0.975), df = 9)
[1] -2.262157 2.262157
```

produce two results for each function call. The pnorm() function gives the probability of being to the left of -1.96 and 1.96 for a standard normal random variable, and the qt() function gives the 0.025 and 0.975 quantiles from a t-distribution with degrees of freedom specified by the df argument value.

For a more complex example, suppose you would like to find the 95% confidence interval for a population mean using the usual formula based on the t-distribution. Below is a set of code demonstrating how this can be done with vectors and a sample contained within x:

```
> x <- c(3.68, -3.63, 0.80, 3.03, -9.86, -8.66, -2.38, 8.94,
    0.52, 1.25)
> x
[1]  3.68 -3.63  0.80  3.03 -9.86 -8.66 -2.38  8.94  0.52  1.25
> var.xbar <- var(x)/length(x)
> var.xbar
[1] 3.246608
> mean(x) + qt(p = c(0.025, 0.975), df = length(x)-1) *
    sqrt(var.xbar)
[1] -4.707033  3.445033

> t.test(x = x, mu = 2, conf.level = 0.95)
```

```
          One Sample t-test

data:  x
t = -1.4602, df = 9, p-value = 0.1782
alternative hypothesis: true mean is not equal to 2
95 percent confidence interval:
 -4.707033  3.445033
sample estimates:
 mean of x
    -0.631
```

In the code, the `mean()` function calculates the sample mean, the `var()` function calculates the variance, the `length()` finds the length of a vector (sample size), and the `sqrt()` function calculates the square root of the sample size. Because the `qt()` function produces a vector of length 2, it may not be immediately clear how R multiples this vector and `sqrt(var.xbar)`. When R performs arithmetic operations on both a scalar and a vector, it performs the operation separately on the scalar with each element of the vector. Thus, R multiplies both -2.262157 and 2.262157 by $\sqrt{3.246608}$ resulting in a new vector of length 2. When the mean is eventually added to the vector, both the lower and upper confidence interval limits result. While the ability to work with vectors in this manner can simplify code, we recommend caution to make sure the calculations are done correctly.

The last function call to `t.test()` provides a simpler way to calculate the confidence interval. The results from a hypothesis test (population mean = 2 or $\neq$ 2) are also given by the function.

A.5 Packages

Outside of the default installation for R, there are many packages, most of which can be downloaded from CRAN. These packages have been created by R core group members[1], researchers, and everyday users to extend the capabilities of R. Next, we go through a specific installation example involving the `binom` package, which is used in Section 1.1.2.

While in the R Console, select Package > Install package(s) from the R menu bar. A number of CRAN mirror locations from around the world then appear in a new window. After a location is selected, all available packages are listed. We scroll down to select the `binom` package and then select OK. The package is downloaded from the CRAN mirror and installed. In general, if a package depends on additional packages that are not installed already, these packages are automatically installed along with the one selected.

To use the installed package, type `library(package = binom)` and <Enter> in the R Console. This needs to be done only once when R is open. If you exit R and re-open it, the package does not need to be installed again, but `library(package = binom)` does need to be executed.

---

[1] These individuals can modify the R source code; see http://www.r-project.org/contributors.html.

A.6 Program editors

The R Console automatically executes each line as it is entered. It is often more convenient to type multiple lines of code all at once and then execute them together. In order to do this, a *program* (or *script*) needs to be typed into a text editor. Notepad or other basic editors can suffice, but there are several built specifically for use with R that have many convenient features.

A.6.1 R editor

A very limited program editor is available within R. The editor is accessed through File > New script. Code can be typed into the R Editor window and saved as a program using File > Save, while the cursor is within the editor. Note that the file extension used with R programs is .R, and these files contain plain text. A whole program is run by selecting Edit > Run all. Alternatively, a mouse highlighted set of code is run by selecting Edit > Run line or selection.

A.6.2 RStudio

While still fairly new in comparison to other editors, RStudio Desktop (http://www.rstudio.com; hereafter just referred to as "RStudio") is likely the most used among all editors. This software is actually more than an editor because it integrates a program editor with the R Console window, graphics window, R-help window, and other items within one overall window environment. Thus, RStudio is an *integrated development environment* (IDE) for constructing and running R programs. The software is available for free, and it runs on Linux, Mac, and Windows operating systems.

To create a new program using RStudio, select File > New > R Script, or open an existing program by selecting File > Open File. Selecting the corresponding icons on the toolbar work for this purpose as well. Figure A.2 shows a screen capture of RStudio version 0.98.490 where AppendixInitialExamples.R is shown in the program editor.[2] To run a segment of code, highlight it and then select the Run icon in the program editor window (upper right) or just press <Ctrl-Enter>. The code will be automatically transferred to the Console window and executed. Any plots that are created will be sent by default to the Plots tab in the lower-right window. Help can be accessed through the Help or Packages tabs in the lower-right window as well.

One of the most significant benefits to using RStudio rather than R's editor is syntax highlighting. This means that code is colorized according to its purpose. For example, comments are green (these begin with a "#" symbol) and numerical values are blue. This syntax highlighting can make reading and editing R code much easier. Multiple programs can be open at once in different tabs of the Program editor window. Also, the editor can suggest function or package names from any loaded package if <Tab> is pressed at the end of any text string. For example, typing "pn" and pressing <Tab> yields a pop-up window suggesting functions pnbinom(), png(), and pnorm(). Pressing <Tab> where an argument could be given within a function (e.g., after a function name and its opening parenthesis

---

[2]The program editor window of RStudio will only be present if there is a program open. This can be confusing to new RStudio users who open RStudio for the first time without directly including a program. Simply creating a new program or opening a program will open the program editor window.

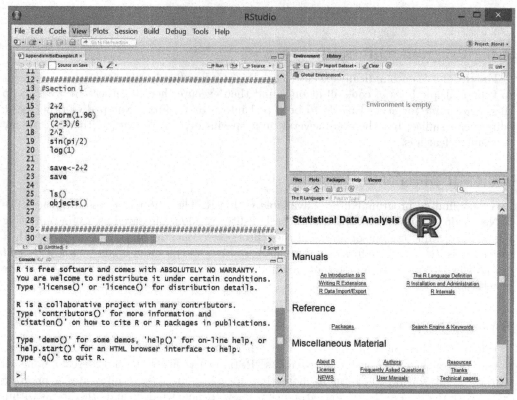

Figure A.2: Screen capture of RStudio.

or after a comma following another argument) gives a list of possible arguments for that function.

A.6.3 Tinn-R

Rather than integrating the program editor and R Console into one window, Tinn-R (http://sourceforge.net/projects/tinn-r and http://nbcgib.uesc.br/lec/software/editores/tinn-r/en) can keep these items separate. This software is available for free, but limited to Windows operating systems.

Figure A.3 shows a screen capture of Tinn-R version 3.0.2.5 with the AppendixInitialExamples.R program. Again, syntax highlighting is automatically incorporated to help distinguish the purpose of the code, although with more distinct coloring than RStudio.[3] In order to run code from the editor, R's GUI needs to be open. This can be opened by selecting the "R control: gui (start/close)" icon from the R toolbar (see #1 in Figure A.3). Tinn-R subsequently opens R in its *SDI* (single-document interface), which is a little different from R's *MDI* (multiple-document interface) that we saw in Figure A.1. The difference between the two interfaces is simply that the MDI uses the R GUI to contain all windows that R opens and the SDI does not. Once R is open in its SDI, program code in Tinn-R can be transferred to R by selecting specific icons on Tinn-R's R toolbar. For example, a highlighted portion of code can be transferred to and then run in R by selecting the "R send:

---

[3]Specific control over what syntax to color and what colors to use is available by selecting **Options > Highlighters** (settings).

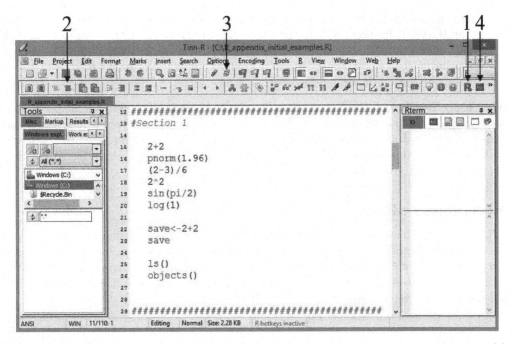

Figure A.3: Screen capture of Tinn-R. Toolbar icons referenced in the text include: (1) R control: gui (start/close), (2) R send: selection (echo = TRUE), (3) Options: return focus after send/control Rgui, and (4) R control: term (start/close).

selection (echo = TRUE)" icon. Note that the transfer of code from Tinn-R to R does not work in the MDI.

Below are some additional important comments and tips for using Tinn-R:

- Upon Tinn-R's first use with R's SDI, the TinnRcom package is automatically installed within R to allow for the communication between the two softwares. This package is subsequently always loaded for later uses.

- When R code is sent from Tinn-R to R, the default behavior is for Tinn-R to return as the window of focus (i.e., the window location of the cursor) after R completes running the code. If Tinn-R and R are sharing the same location on a monitor, this prevents the user from immediately seeing the results in R due to it being hidden behind the Tinn-R window. In order to change this behavior, select Options > Application > R > Rgui and uncheck the Return to Tinn-R box. Alternatively, select the "Options: return focus after send/control Rgui" icon on the Misc toolbar.

- By default, the line containing the cursor is highlighted in yellow. To turn this option off, select Options > Highlighters (settings) and uncheck the Active line (choice) box.

- Long lines of code are wrapped to a new line by default. This behavior can be changed by selecting Options > Application > Editor and then selecting the No radio button for Line wrapping.

- Syntax highlighting can be maintained with code that is copied and pasted into a word processing program. After highlighting the desired code to copy, select Edit > Copy formatted (to export) > RTF. The subsequently pasted code will retain its color.

- Tinn-R can run R within its interface by using a link to a terminal version of R rather than R's GUI. To direct code to the Rterm window (located on the right side of Figure A.3), select the "R control: term (start/close)" icon on the R toolbar. One benefit from using R in this manner is that the syntax highlighting in the program editor is maintained in the R terminal window.

When using Tinn-R and R's GUI, we have found it to be more efficient in a multiple monitor environment. This allows for both to be viewable in different monitors at the same time. Code and output can be side-by-side in large windows without needing to switch back-and-forth between overlaying windows. The same type of environment is achievable with a large, wide-screen monitor as well.

A.6.4 Other editors

RStudio and Tinn-R are the editors that we and our students generally use for R programs. A list of other program editors is available at http://www.sciviews.org/_rgui/projects/Editors.html. Among these, Emacs (http://www.gnu.org/software/emacs) with its Emacs Speaks Statistics (http://ess.r-project.org) add-on are often used with Linux operating systems and both are available for free.

A.7 Regression example

This section provides a simple example showing how to use R in a regression setting. The regression aspect should not be difficult for readers with the needed background for this book. The R aspect will be new to many readers.

A.7.1 Background

Universities often want to be able to predict how well new freshmen will perform given their background. Suppose a university selects a simple random sample of 20 current students at the end of their first semester. The college grade point average (GPA) and the high school GPA are recorded for each student. A hypothetical data set for this scenario is given in the GPA.txt plain text file which is available on the book's website. Using these data, we want to estimate a simple linear regression model with college GPA as the response variable and high school GPA as the explanatory variable.

The GPA.R program contains all of the code used in this subsection. Below is part of the code after running it in R:

```
>###########################################
># NAME: Chris Bilder                       #
># DATE: 3-11-14                            #
># PURPOSE: Regression model for GPA data   #
># NOTES:                                   #
>###########################################
>
> # Read in the data
> gpa <- read.table(file = "C:\\data\\GPA.txt", header = TRUE,
    sep = "")
>
```

```
> # Print data set
> gpa
  HS.GPA College.GPA
1   3.04         3.1
2   2.35         2.3
3   2.70         3.0

<OUTPUT EDITED>

19  2.28         2.20
20  2.88         2.60
```

Lines that begin with a # symbol denote comments that are not executed in R. We use comments like these to inform other users or to make notes for ourselves. The first executable line involves the `read.table()` function, which reads in the GPA.txt file stored at the location given as the `file` argument value. This location needs to be updated to the corresponding location on your computer. Note the use of double back-slashes between folder names. Alternatively, one forward slash can be used. The `header = TRUE` argument value specifies that the first row of the file contains the variable names. The `sep = " "` argument value states that the file is space-delimited. After reading in the data, they are stored as a *data frame* and put into an object named `gpa`. The contents of this object are obtained by just typing its name and then <Enter> at the command prompt.

Data files may come in different formats than we have for GPA.txt. A comma delimited format instead uses commas, rather than spaces, to separate the columns of data. This type of data can be read into R by simply using `sep = ","` in the `read.table()` function code (and changing the file name if needed). Alternatively, a slightly easier way to read in a comma-delimited file is to use the `read.csv()` function which only needs the `file` argument to be specified (it is assumed the first row of the data file contains the variable names). We provide an example of how to use this function in our corresponding program.

Another commonly used data format is an Excel file. The *R Data Import/Export* manual (select Help > Manuals (in PDF)) provides options for how to read in Excel files; however, the manual says "The first piece of advice is to avoid doing so if possible!" Reasons for this recommendation are because of the different Excel file formats (.xls or .xlsx) and 32-bit vs. 64-bit driver issues. The recommended alternative is to save an Excel file in a comma- or space-delimited format instead and use the previously described methods to read in the data. In our corresponding program, we provide an example of how to read in Excel file using the the RODBC package (R Open Database Connectivity).

A.7.2 Data summary

We calculate simple summary statistics for the `gpa` object:

```
> summary(object = gpa)
    HS.GPA          College.GPA
 Min.   :1.980   Min.   :2.000
 1st Qu.:2.380   1st Qu.:2.487
 Median :2.830   Median :2.800
 Mean   :2.899   Mean   :2.862
 3rd Qu.:3.127   3rd Qu.:3.225
 Max.   :4.320   Max.   :3.800
```

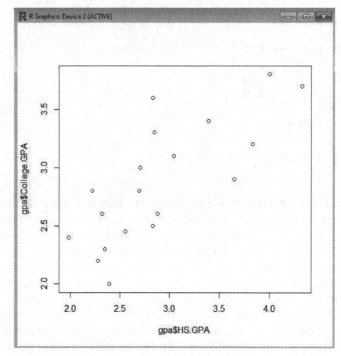

Figure A.4: R Graphics window.

The summary() function summarizes the data for all variables in the object. To create a scatter plot of the data, we use

> plot(x = gpa$HS.GPA, y = gpa$College.GPA)

The plot() function creates plots in an R Graphics window, like the one shown in Figure A.4. In RStudio, the plot will appear in the lower-right window instead. The x and y arguments of plot() state what to plot on the x and y axes, respectively. To specify a variable within gpa, the $ is used to separate the object name from the variable name. If a program editor discussed in Appendix A.6 was used, one can simply highlight gpa$HS.GPA in the plot() function call and then send it to R producing

> gpa$HS.GPA
 [1] 3.04 2.35 2.70 2.55 2.83 4.32 3.39 2.32 2.69 2.83 2.39
[12] 3.65 2.85 3.83 2.22 1.98 2.88 4.00 2.28 2.88

This type of highlighting and running can be useful to new users of R, especially for diagnosing errors in complicated function calls.

There are other ways to access parts of a data frame. R uses a matrix representation for such objects, with variables in columns and observations in rows. Positions within the matrix are stored by [row,column]. For example, gpa[,1] produces the values for all rows in the first column of gpa, giving the same result as gpa$HS.GPA. Other examples include: (1) gpa[1,], which gives all variable values for the first observation, (2) gpa[1,1], which gives the row 1 and column 1 value, and (3) gpa[1:10,1], which gives the first 10 observations of variable 1 (1:10 means 1 to 10 by 1).

Optional arguments for plot() can be used to improve the scatter plot. Figure A.5 displays a plot using the following code:

```
> plot(x = gpa$HS.GPA, y = gpa$College.GPA, xlab = "HS GPA", ylab
    = "College GPA", main = "College GPA vs. HS GPA", xlim =
    c(0,4.5), ylim = c(0,4.0), col = "red", pch = 1, cex = 1.0,
    panel.first = grid(col = "gray", lty = "dotted"))
```

The new arguments are:

- `xlab` and `ylab`: The x- and y-axis labels

- `main`: The plot title

- `xlim` and `ylim`: The x- and y-axis limits, where the lower and upper values are concatenated into a vector

- `col`: The color of the plotting points; colors can be specified by their names (run `colors()` to see all of them or view them at http://research.stowers-institute.org/efg/R/Color/Chart/index.htm) or numbers given in the vector produced by `palette()`

- `pch`: The plotting character, where `pch = 1` denotes an open circle (default); a list of characters can be viewed by running the `pchShow()` function available in the help for `points()`.

- `cex`: The plotting character size relative to the default, where 1.0 is the default

- `panel.first`: The plotting of additional items, where gray colored grid lines with a dotted line type are designated here; the `lty` argument specifies the line type with 1 = "solid", 2 = "dashed", 3 = "dotted", 4 = "dotdash", 5 = "longdash", and 6 = "twodash" (either the number or character string can be used)

The `par()` function, which allows for further control of the graphics parameters, gives additional information on these arguments in its own help.

Plots can be copied into word processing software or exported to a file. When the R Graphics window is the current window, select File > Copy to the clipboard > as a Metafile within R (use Export > Copy Plot to Clipboard in the plot window, and then select Metafile and Copy Plot within RStudio). The plot can be pasted now into a word processing document. Alternatively, select File > Save as within R to save the plot in a specific format (use Export > Save Plot as Image > Save within the plot window of RStudio).

A.7.3 Regression modeling

The `lm()` function

To better understand the relationship between college and high school GPA, we estimate a simple linear regression model using the `lm()` function:

```
> mod.fit <- lm(formula = College.GPA ~ HS.GPA, data = gpa)
> mod.fit
Call: lm(formula = College.GPA ~ HS.GPA, data = gpa)
Coefficients:
(Intercept)  HS.GPA
     1.0869  0.6125
```

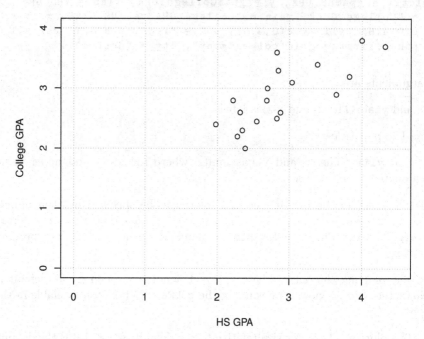

Figure A.5: Scatter plot using optional arguments.

The results from lm() are saved into an object named mod.fit to mean "model fit" (a different object name could have been used). The formula argument of lm() specifies the model in the form response ~ explanatory variable. If there is more than one explanatory variable, a plus symbol is used to separate each explanatory variable. For example, a second explanatory variable z is included using formula = College.GPA ~ HS.GPA + z. The estimated regression model is $\hat{Y} = 1.0869 + 0.6125x$ where $\hat{Y}$ is the estimated College GPA and x is the high school GPA.

The mod.fit object contains much more information than shown so far. Its contents are listed using the names() function:

```
> names(mod.fit)
[1] "coefficients" "residuals"    "effects"  "rank"
[5] "fitted.values" "assign"      "qr"       "df.residual"
[9] "xlevels"      "call"         "terms"    "model"
> mod.fit$coefficients
(Intercept)     HS.GPA
  1.0868795   0.6124941
> round(mod.fit$residuals[1:5],2)
    1     2     3     4     5
 0.15 -0.23  0.26 -0.20 -0.32
```

The mod.fit object is called a *list*. A list has a number of components combined together into one overall object. These components are accessed using the $ syntax that we saw earlier with the gpa data frame. In fact, a data frame is a special type of list where each component is a vector of the same length. Within the mod.fit list, the components include mod.fit$coefficients which contains the estimated regression parameters and mod.fit$residuals which contains the residuals from the model fit. The round() function rounds the residuals to two decimal places at the end of the code.

It is often helpful to view the data, the residuals, and the predicted values together in one data set. In the code below, the data.frame() function creates a new data frame called save.fit containing these items:

```
> save.fit <- data.frame(gpa, C.GPA.hat =
    round(mod.fit$fitted.values, 2), residuals =
    round(mod.fit$residuals, 2))
> head(save.fit)
  HS.GPA College.GPA C.GPA.hat residuals
1   3.04        3.10      2.95       0.15
2   2.35        2.30      2.53      -0.23
3   2.70        3.00      2.74       0.26
4   2.55        2.45      2.65      -0.20
5   2.83        2.50      2.82      -0.32
6   4.32        3.70      3.73      -0.03
```

Inside the data frame, two new variable names (C.GPA.hat and residuals) are created. Also, we round their numerical values to two decimal places. The head() function limits printing to the first six rows of data.

To summarize the information within mod.fit, the summary() function is used:

```
> summary(object = mod.fit)
Call: lm(formula = College.GPA ~ HS.GPA, data = gpa)

Residuals:
     Min       1Q   Median       3Q      Max
-0.55074 -0.25086  0.01633  0.24242  0.77976

Coefficients:
            Estimate Std. Error t value Pr(>|t|)
(Intercept)   1.0869     0.3666   2.965 0.008299 **
HS.GPA        0.6125     0.1237   4.953 0.000103 ***
---
Signif. codes:  0 '***' 0.001 '**' 0.01 '*' 0.05 '.' 0.1 ' ' 1

Residual standard error: 0.3437 on 18 degrees of freedom
Multiple R-squared: 0.5768,     Adjusted R-squared: 0.5533
F-statistic: 24.54 on 1 and 18 DF,  p-value: 0.0001027
```

In the Pr(>|t|) column of the output, we see that the hypothesis test for a linear relationship between high school and college GPA results in a very small p-value of 0.0001.

Object-oriented language

As we have shown, information (i.e., model specifications, estimates, test results) created by functions is stored within an object. Different collections of information are created by functions depending on the types of calculations that are performed. To distinguish objects that contain different collections of information, R assigns each object an attribute called a *class*. We can use the class() function to determine an object's class:

```
> class(mod.fit)
[1] "lm"
> class(gpa)
[1] "data.frame"
```

Thus, `mod.fit` has a class of `lm` and `gpa` has a class of `data.frame` (classes do not need to be function names).

Functions are typically designed to operate on only one or very few classes of objects. However, some functions, like `summary()`, are *generic*, in the sense that essentially different versions of them have been constructed to work with different classes of objects. When a generic function is run with an object, R first checks the object's class type and then looks to find a *method* function with the name format `<generic function>.<class name>`. For example, when `summary(mod.fit)` was run earlier, the function `summary.lm()` is actually used to summarize the fit from a regression model. Also, when `summary(gpa)` was run earlier, the function found `summary.data.frame()`. These two functions produce different output, tailored to suit the type of object on which they are run. If there was no specific method function for an object class, `summary.default()` would be run. For example, `coefficients()` is a generic function that gives the same results as `mod.fit$coefficients`. There is no `coefficients.lm()` function so the `coefficients.default()` function is invoked.

The purpose of generic functions is to use a familiar language set with any object. For example, we frequently want to summarize data or a model fit (`summary()`), plot data (`plot()`), and find predictions (`predict()`), so it is convenient to use the same language set no matter the application. This is why R is referred to as an object-oriented language. The object class type determines the function action. Understanding generic functions may be one of the most difficult topics for new R users. The most important point that readers need to know now is where to find help for these functions. For example, if you want help on the results from `summary(mod.fit)`, examine the help for `summary.lm()` rather than the help for `summary()` itself.

To see a list of all method functions associated with a class, use `methods(class = <class name>)`, where the appropriate class name is substituted for `<class name>`. For our regression example, the method functions associated with the `lm` class are:[4]

```
> methods(class = "lm")
  [1] add1.lm*        alias.lm*         anova.lm         case.names.lm*

<OUTPUT EDITED>

 [33] summary.lm      variable.names.lm* vcov.lm*

    Non-visible functions are asterisked
```

To see a list of all method functions for a generic function, use `methods(generic.function = <generic function name>)`. Below are the method functions associated with `summary()`:

```
> methods(generic.function = "summary")
 [1] summary.aov           summary.aovlist
 [3] summary.aspell*       summary.connection
```

---
[4]See Appendix A.7.4 for more on "non-visible functions."

```
[5] summary.data.frame        summary.Date
```

<OUTPUT EDITED>

One advantage of using RStudio is that you can type "summary" in its help search box to show a list of all functions that start with this word.

Graphics

Now that we have the estimated model, we can plot it on a scatter plot. In our demonstration, we will make two plots side-by-side in one graphics window. First, we use the x11() function[5] to open a new graphics window and then use the par() function to partition it:

```
> x11(width = 8, height = 6, pointsize = 10)
> par(mfrow = c(1,2))
```

The resulting plot window is a 6" × 8" window with a default font size of 10. While opening a new graphics window is not needed (a window will automatically open whenever a plot is constructed), we decided to open a new one here to show how to control the window and font size. The mfrow argument for par() stands for "make frame by row" which results in splitting the plot window into 1 row and 2 columns.

Next, we use a similar call to the plot() function as earlier, but now add the abline() function to plot a line $y = a + bx$ that is solid blue with a thickness twice the default (lwd = 2):

```
> plot(x = gpa$HS.GPA, y = gpa$College.GPA, xlab = "HS GPA", ylab
    = "College GPA", main = "College GPA vs. HS GPA", xlim =
    c(0,4.5), ylim = c(0,4.5), panel.first=grid(col = "gray", lty
    = "dotted"))
> abline(a = mod.fit$coefficients[1], b =
    mod.fit$coefficients[2], lty = solid, col = "blue", lwd = 2)
```

As shown earlier, mod.fit$coefficients is a vector of length 2. To obtain the estimated y-intercept and slope for the model, we use mod.fit$coefficients[1] and mod.fit$coefficients[2], respectively. The resulting plot is given on the left side of Figure A.6.

An alternative to abline() is the curve() function. This function plots mathematical functions that vary over one variable, and it allows one to specify the range over which to draw the mathematical function. The right-side plot in Figure A.6 shows the results from rerunning the previous plot() function call and following it with a new curve() function call as shown below:

```
> curve(expr = mod.fit$coefficients[1] +
    mod.fit$coefficients[2]*x, xlim = c(min(gpa$HS.GPA),
    max(gpa$HS.GPA)), col= "blue", add = TRUE)
```

---

[5]There are a number of functions available to open new graphics windows in R, but these are dependent on the operating system. We have found the x11() to be the most general across operating systems despite it having a Linux-based name. The dev.new() function also works across operating systems, but it does not work for RStudio due to its embedded graphics window. Other functions include windows() and win.graph() for Windows operating systems and quartz() for Mac operating systems.

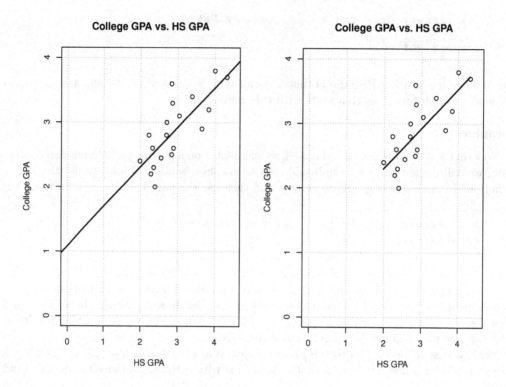

Figure A.6: Scatter plots with estimated model drawn using the `abline()` function (left) and `curve()` function (right).

The `expr` argument specifies the formula for the curve to be plotted, which in our case is the regression model. The letter `x` is required (instead of `HS.GPA` or any other explanatory variable name) for the variable plotted on the x-axis. The `xlim` argument specifies the evaluation limits for the mathematical function, where `min()` and `max()` find the minimum and maximum high school GPAs, respectively. The `add = TRUE` argument value instructs R to add the line to the current plot.

Writing functions

In a simple linear regression analysis, estimation is often followed by plotting. We can partially automate this process by constructing a new R function that does both the estimation and plotting. The code is shown in Figure A.7 from a screen capture of Tinn-R. The function begins by specifying the explanatory variable (`x`), response variable (`y`), and data set (`data`) as its arguments. In the function body, we use almost the same code as earlier; the main difference is the use of `x`, `y`, and `data` rather than `HS.GPA`, `College.GPA`, and `gpa`, respectively. The last line of code in the function body gives `mod.fit` as the returned value from the function. Below are the results from running the code (excluding the plot):

```
> save.it <- my.reg.func(x = gpa$HS.GPA, y = gpa$College.GPA,
    data = gpa)
> names(save.it)
 [1] "coefficients"  "residuals"     "effects"       "rank"
 [5] "fitted.values" "assign"        "qr"            "df.residual"
 [9] "xlevels"       "call"          "terms"         "model"
```

```
my.reg.func<-function(x, y, data) {

    #Fit simple linear regression model and save results in mod.fit
    mod.fit<-lm(formula = y ~ x, data = data)

    #Open a new graphics window - do not need to
    x11(width = 6, height = 6, pointsize = 10)

    #Same scatter plot as before
    plot(x = x, y = y, xlab = "x", ylab = "y", main = "y vs. x",
      panel.first = grid(col = "gray", lty = "dotted"))

    #Include regression model
    curve(expr = mod.fit$coefficients[1] + mod.fit$coefficients[2]*x,
      xlim = c(min(x), max(x)), col = "blue", add = TRUE, lwd = 2)

    #This is the object returned
    mod.fit
}
```

Figure A.7: The my.reg.func() in Tinn-R.

```
> summary(save.it)
Call: lm(formula = College.GPA ~ HS.GPA, data = gpa)

<OUTPUT EDITED>

F-statistic: 24.54 on 1 and 18 DF,  p-value: 0.0001027
```

The function returns the same list as before, and the numerical summary is identical to the previous output. This example shows that it is easy to extend R's capabilities to suit your own needs.

A.7.4 Additional items

There are a number of other plot enhancements that can be used to create better and more descriptive plots. In particular, mathematical expressions can be included on plots using the expression() function. For example, including main = expression(hat(Y) = hat(beta[0]) + hat(beta[1])*x) in plot() prints $\hat{Y} = \hat{\beta}_0 + \hat{\beta}_1 x$ as the plot title. Running demo(plotmath) at the R Console prompt provides syntax examples. Another plot enhancement allows one to have finer control of axis tick marks through using the axis() function. For example, the code below changes the x-axis tick marks from the default to produce a modified version of the scatter plots given in Figure A.6:

```
> plot(x = gpa$HS.GPA, y = gpa$College.GPA, xaxt = "n", xlim =
    c(0, 4.5), ylim = c(0,4.5))
> axis(side = 1, at = seq(from = 0, to = 4.5, by = 0.5))   # Major
    tick marks
> axis(side = 1, at = seq(from = 0, to = 4.5, by = 0.1), tck =
    0.01, labels = FALSE)   # Minor tick marks
```

The xaxt = "n" argument value in plot() prevents the default x-axis tick marks from being plotted. The axis() function is used twice to create a specific sequence of numbers (seq()) for the major and minor tick marks on side = 1 (bottom x-axis). The tck argument sets a specific tick mark length.

One of the reasons R has become popular is it facilitates the sharing of code in an open-source software environment. For example, newly published statistical methodology frequently has corresponding R code available for immediate implementation. This code may be downloadable from an author's website or incorporated into a package. Furthermore, all R code is available for viewing. For example, one can type a function name, such as lm, at an R Console prompt to see its code. This does not work, however, on some "non-visible" functions. Instead, their code may still be seen by using getAnywhere(<function name>) or in the package source code at CRAN.

The open-source environment has also led to the development of a large community for R users to go to when in need of help. In particular, there is an active listserv (http://www.r-project.org/mail.html) where even members of the R core group at times answer submitted questions. Before posting a question to the listserv though, search its archives because the same question may have been asked before. One of the specific search engines for the listserv is available at http://dir.gmane.org/gmane.comp.lang.r.general. Other locations to go to for help include Stack Overflow at http://stackoverflow.com/questions/tagged/r which provides an alternative format than a traditional listserv. This website follows a question and answer format where multiple answers can be submitted and ranked by the community in terms of the answer's usefulness. Also, Stack Overflow's web page at http://stackoverflow.com/tags/r/info provides a number of useful links to find help on R in other locations. Finally, there are a large number of blogs devoted to the use of R. The R-bloggers website at http://www.r-bloggers.com serves as an aggregator for many of these blogs.

Some new R users may be used to a completely point-and-click process when it comes to using a statistical software package. For those individuals, the Rcmdr (R Commander) package may serve as a gentler way to begin using R, because this package provides a GUI for data management and for statistical functions. Once installed, type library(package = Rcmdr) and <Enter> at the R Console prompt to open R Commander. In its first use, additional packages are installed automatically (there can be a significant number of these with a new R installation) to enable all drop down menu options. Once the R Commander window is open, a data set already in the R working database can be selected through the "<No active dataset>" button. Depending on the data type, the drop down menus allow for editing the data, creating data summaries, estimating models, and plotting. For example, suppose the gpa data frame is the active data set within R Commander. We select **Statistics > Fit Models > Linear Regression** from the main menu bar to open a window that allows us to choose **College.GPA** as the response variable and **HS.GPA** as the explanatory variable. We also type mod.fit in the "Enter name for model:" box as the object name for saving the results. Figure A.8 shows what happens after selecting **OK**. The code and output in the script and output windows, respectively, both appear and they are the same as we used earlier in this section. For other applications, R Commander may produce different code from what we might enter. This is because the R Commander package has a number of its own functions built into it to facilitate the computational or graphical processes. However, we believe that this package still serves as a useful tool to help new users learn to write R code.

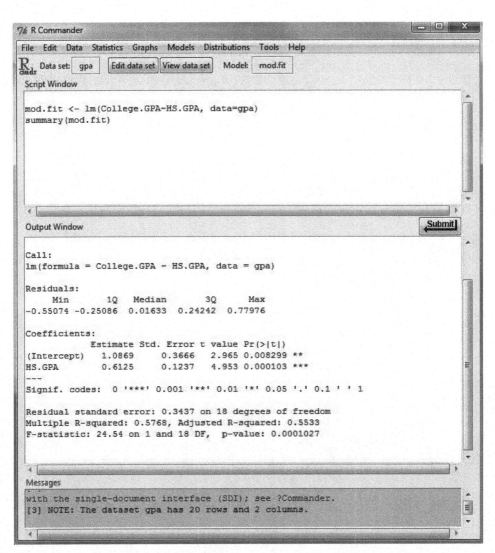

Figure A.8: R Commander window.

Appendix B

Likelihood methods

B.1 Introduction

Likelihood methods are commonly used in statistical analyses. In particular, they provide the basis for most of the categorical data analysis techniques that are covered in this book. The main body of this book is written assuming that readers are already familiar with likelihood-based methods, at least enough to be able to apply them with some confidence. However, we recognize that readers may come from a broad range of disciplines and may not have had any previous exposure to likelihood methods. We therefore offer a brief primer on this important class of procedures. We suggest that readers who have no past exposure to likelihood methods at least look over this appendix before reading the book's main body. We cross-reference this appendix with the sections where likelihood methods are used, so that readers and instructors can also refer to this material as needed.

The scope of this appendix is deliberately limited. An understanding of the theory and asymptotics of likelihood methods is not at all required for this book, although a heuristic appreciation of them is helpful. Accordingly, the descriptions in this appendix are light in mathematics. For a more complete treatment of likelihood methods, please see texts such as Casella and Berger (2002) and Severini (2000).

B.1.1 Model and parameters

The goal of a statistical analysis is to learn something about a population (what we might call the "truth"). To achieve this goal, data are gathered, but data contain variability (or "noise") that prevents us from seeing the truth clearly. In order to extract truth from data, it helps if we start with some idea of what the truth looks like and if we know something about the origin of the noise. For example, rather than just looking at a plot of a response against an explanatory variable, it may help to speculate that the relationship should be a straight line and that the deviations around the straight line should be independent and approximately normally distributed. A *statistical model* is an assumed structure for the truth and the noise (i.e., it is an educated guess). The features of the model are combined into a probability distribution, such as a normal, Bernoulli, or Poisson, that is intended to serve as a useful approximation to reality.

Models generally contain *parameters*. Model parameters relate to the structure of the truth and/or noise and hold the place of unknown or flexible features of the model. They represent population quantities that are often of direct interest to the researcher—such as the mean of a normal distribution, or the probability of success in a binomial distribution—but not always (we often do not really care about the intercept parameter in many linear regression problems). The goal of a statistical analysis is to learn about the model parameters or some functions of the model parameters (like predictions in a regression, which are a function of the slope and intercept parameters). Functions of model parameters are also parameters, since they are also unknown population quantities, so our discussion will not distinguish between parameters in the model and other parameters.

B.1.2 The role of likelihoods

A statistical model serves to relate the data to the parameters. We still need to find values for the parameters and use them to learn about the population. The first step—finding values for the parameters—is called *estimation*. The second step—using them to draw conclusions—is called *inference*.

There are many ways to estimate parameters from a statistical model, but the one that has been adopted almost universally because of its quality and flexibility is maximum likelihood (ML) estimation. This is because the procedure is adaptable to nearly any statistical model, leading to an automatic process for estimation. It also has associated with it a variety of tools that can be used for inference. The procedure and its tools have strengths and weaknesses. We discuss these as they arise in different settings throughout the text.

B.2 Likelihood

B.2.1 Definition

Statistical models are generally described in terms of a *probability mass function* or a *probability density function*. A probability mass function (PMF) for a discrete random variable provides the probabilities for each possible outcome. For a continuous random variable, the corresponding probability density function (PDF) is a little more complicated to interpret, because it assumes that measurements are made to an infinite number of decimal places. Loosely speaking, a PDF describes the relative chances of observing values from different areas of the stated probability distribution. The familiar "normal curve" is an example of a PDF. In both the discrete and continuous cases, higher values of the PMF or PDF are produced by ranges of values that are "more likely" to occur. The *joint* PMF or PDF for a sample of observations can be interpreted as how likely we are to observe the entire sample, given the distribution and its parameters.

In practice, we do not know the values of the parameters, but we do know the data. We therefore cannot know the joint mass or density for our sample exactly. We *can* calculate this quantity if we assume certain values for the parameters. If we change the values of the parameters, we get a different value for the PDF/PMF of the sample, because the data are more or less likely to occur under different parameters. For example, it is very unlikely that we would observe 5 successes in 10 Bernoulli trials when the true probability of success is 0.01. This outcome would be somewhat more likely to happen if the probability of success is 0.30, and even more likely if it is 0.50.

This is the nature of the *likelihood function*: we consider the PMF or PDF in reverse. We observe how the function changes for different values of parameters, while holding the data fixed. We can then make use of this to judge which values of the parameters lead to greater relative chances for the sample to occur. Formally, if we define the joint PMF or PDF of a sample to be $f(\mathbf{y}|\boldsymbol{\theta})$, where $\mathbf{y} = (y_1, \ldots, y_n)$ represents a vector[1] containing the

---

[1] A vector can be thought of as a simple way to express a group of values using one symbol.

n sampled values and $\boldsymbol{\theta} = (\theta_1, \ldots, \theta_p)$ represents a vector for p different parameters[2], then the likelihood function is

$$L(\boldsymbol{\theta}|\mathbf{y}) = f(\mathbf{y}|\boldsymbol{\theta}).$$

Larger values of the likelihood correspond to values of the parameters that are relatively better supported by the data.

A likelihood is not a probability, because the only random part, $\mathbf{y}$, is considered fixed in its construction. In particular, the likelihood is not expected to add to 1 across all values of $\boldsymbol{\theta}$. The actual numerical values of a likelihood are unimportant. The *relative* sizes of the likelihoods for different parameter values are all that matters.

When the observations are drawn independently, the likelihood function is simply the product of the PDFs or PMFs,

$$L(\boldsymbol{\theta}|\mathbf{y}) = \prod_{i=1}^{n} f(y_i|\boldsymbol{\theta}),$$

where we use the $\prod$ symbol to denote multiplying indexed terms together. Thus, likelihoods are very easy to construct for simple random samples, which is the setting for most problems in this book. Also, notice that the value of the likelihood depends on the sample, so that likelihoods—and any features calculated from them—are statistics. This means that such features are random and have probability distributions.

B.2.2 Examples

Below are some simple examples of likelihoods that are commonly used in categorical data analysis.

Example: Bernoulli

Suppose that the random variable Y takes on only two possible values. The Bernoulli PMF for Y is $f(y|\pi) = \pi^y(1-\pi)^{(1-y)}$ with probability-of-success parameter π ($0 < \pi < 1$) and $y = 1$ or 0 denoting a "success" or a "failure," respectively. Let $y_1, \ldots, y_n$ be observations of independent Bernoulli random variables from this PMF. The likelihood for the parameter π is

$$L(\pi|\mathbf{y}) = \prod_{i=1}^{n} \pi^{y_i}(1-\pi)^{(1-y_i)} = \pi^w(1-\pi)^{n-w}, \tag{B.1}$$

where $\mathbf{y} = (y_1, \ldots, y_n)$ and $w = \sum_{i=1}^{n} y_i$. This likelihood is used in Section 1.1.1.

Example: Binomial

An alternative form of the Bernoulli case occurs when the total number of successes, w, is observed instead of the individual trial results. In this case, the joint PMF of $y_1, \ldots, y_n$ cannot be found because we do not know which y_i should be 1 and which should be 0. However, we know that there are $\binom{n}{w} = n!/[w!(n-w)!]$ ways for the w successes to be observed among the n observations, so the PMF of w given π is

$$f(w|\pi) = \frac{n!}{w!(n-w)!}\pi^w(1-\pi)^{n-w}.$$

---

[2] The actual symbols used for the parameters in a given problem vary depending on the context of the problem; see examples later.

If only one set of n trials is run, so that only one total number of successes w is observed, then $L(\pi|w) = f(w|\pi)$. Notice that this is very similar to the Bernoulli likelihood.

Example: Poisson

The Poisson PMF for a random variable Y is $f(y|\mu) = e^{-\mu}\mu^y/y!$ with parameter $\mu > 0$ and y taking on integer values 0, 1, 2, ..., such as for a count of something. Let $y_1, \ldots, y_n$ be observations of independent Poisson random variables. The likelihood for the parameter μ is

$$L(\mu|\mathbf{y}) = \prod_{i=1}^{n} \frac{e^{-\mu}\mu^{y_i}}{y_i!}, \tag{B.2}$$

where $\mathbf{y} = (y_1, \ldots, y_n)$. This likelihood is used in Section 4.1.2.

Example: Multinomial

Consider a random variable Y with responses consisting of one of c categories, labeled 1, 2, ..., c, with respective probabilities of success $\pi_1, \pi_2, \ldots, \pi_c$ ($\sum_{k=1}^{c} \pi_k = 1$). Let $y_1, \ldots, y_n$ be observations of Y measured on independent trials of this type; that is, the possible values for each y_i are the categories $1, 2, \ldots, c$. Likelihoods can be constructed as for the Bernoulli and binomial cases above, depending on whether individual trial results $y_1, \ldots, y_n$ are observed or just the summary counts for each category, $w_1, \ldots, w_c$. The multinomial distribution is based on the summary counts, for which the PMF is

$$f(w_1, \ldots, w_c | \pi_1, \ldots, \pi_c) = \frac{n!}{w_1! w_2! \ldots w_c!} \pi_1^{w_1} \pi_2^{w_2} \ldots \pi_c^{w_c}. \tag{B.3}$$

If only one set of n trials is run, so that only one set of category counts $w_1, \ldots, w_c$ is observed, then we have $L(\pi_1, \ldots, \pi_c | w_1, \ldots, w_c) = f(w_1, \ldots, w_c | \pi_1, \ldots, \pi_c)$. This likelihood is used in Section 3.1.

B.3 Maximum likelihood estimates

We define the maximum likelihood estimate (MLE), $\hat{\boldsymbol{\theta}}$, of a parameter as the value of the parameter at which the likelihood function from the given sample is maximized: $L(\hat{\boldsymbol{\theta}}|\mathbf{y}) \geq L(\tilde{\boldsymbol{\theta}}|\mathbf{y})$ for any possible value $\tilde{\boldsymbol{\theta}}$ of the parameter. The next example illustrates how to find this MLE through simple evaluation of a likelihood function.

Example: MLE for a sample of Bernoulli random variables (LikelihoodFunction.R)

Suppose $\sum_{i=1}^{n} y_i = 4$ successes are observed out of $n = 10$ trials. Given this information, we want to determine the most plausible value for π. The likelihood function

Table B.1: $L(\pi|\mathbf{y})$ evaluated at different values of π.

π	$L(\pi|\mathbf{y})$
0.20	0.000419
0.30	0.000953
0.35	0.001132
0.39	0.001192
0.40	0.001194
0.41	0.001192
0.50	0.000977

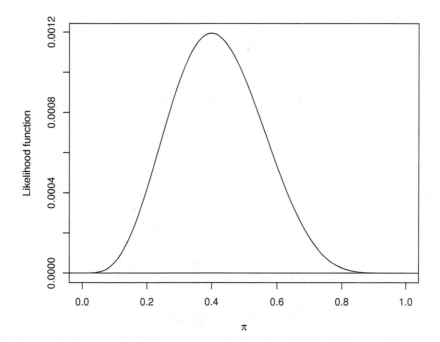

Figure B.1: Bernoulli likelihood function evaluated at $\sum_{i=1}^{n} y_i = 4$ and $n = 10$.

is $L(\pi|\mathbf{y}) = \pi^4(1-\pi)^6$. Table B.1 shows the likelihood function evaluated at a few different values of π, and Figure B.1 plots the function (R code for the table and plot is included in the corresponding program for this example). We can see that the likelihood function reaches its maximum value when $\pi = 0.4$. Therefore, the most plausible value of π, given the observed data, is 0.4, which makes sense because it is the observed proportion of successes. Formally, we say 0.4 is the MLE of π, and we denote it as $\hat{\pi} = 0.4$.

For various mathematical reasons it turns out to be easier to work with the natural log of the likelihood, $\log[L(\hat{\boldsymbol{\theta}}|\mathbf{y})]$, when we attempt to find MLEs. This causes no problem because the log transformation does not change the ordering of likelihood values across different values of $\boldsymbol{\theta}$, so the MLE also maximizes $\log[L(\boldsymbol{\theta}|\mathbf{y})]$.

B.3.1 Mathematical maximization of the log-likelihood function

For simple models with a single parameter, finding the value of θ that maximizes the log-likelihood function is easily done using calculus. The standard technique is to differentiate the log-likelihood function with respect to the parameter, set the result equal to 0, and solve for the parameter. This process is demonstrated on some of the examples from earlier.

Example: Bernoulli

From Equation B.1, we obtain

$$\log[L(\pi|\mathbf{y})] = w \log \pi + (n-w) \log(1-\pi).$$

Differentiating, we find

$$\frac{d}{d\pi} \log[L(\pi|\mathbf{y})] = \frac{w}{\pi} - \frac{n-w}{1-\pi}.$$

Setting this equal to 0 and solving for π leads to $\hat{\pi} = w/n = \sum_{i=1}^{n} y_i/n$. Thus, the MLE for π is the sample proportion of successes, which was shown in the previous example for $\sum_{i=1}^{n} y_i = 4$ and $n = 10$.

Example: Poisson

From Equation B.2, we obtain

$$\log[L(\mu|\mathbf{y})] = -n\mu + \sum_{i=1}^{n} y_i \log \mu - \sum_{i=1}^{n} \log(y_i!).$$

Differentiating, we find

$$\frac{d}{d\mu} l(\mu|\mathbf{y}) = -n + \frac{1}{\mu} \sum_{i=1}^{n} y_i.$$

Setting this equal to 0 and solving for μ leads to $\hat{\mu} = \sum_{i=1}^{n} y_i/n$. Thus, the MLE for μ is the sample mean.

Models with several parameters Many models rely on more than one parameter. The multinomial model in Equation B.3 is an example, because there is a different probability parameter for each category. Other models use regression functions to describe the relationship between population means or probabilities and one or more explanatory variables. In these models, the maximization is carried out exactly as described above, with a separate derivative taken with respect to each parameter. Setting each derivative equal to zero results in a system of equations that needs to be solved simultaneously in order to find the MLE. Theoretically, this presents no difficulty. Practically, the process can be challenging to carry out manually, and the equations may not have closed-form solutions. For example, see the logistic regression likelihood from Section 2.2.

B.3.2 Computational maximization of the log-likelihood function

When the equations have no closed-form solution, they are solved using an educated version of trial-and-error. The general idea is to start with some initial guess at the parameter value, calculate the log-likelihood for that value, and then iteratively find parameter values with larger and larger log-likelihoods until no further improvement can be achieved. Improving the log-likelihood can be done, for example, by calculating the slope of the log-likelihood at the current guess and then moving the next guess some distance in the direction leading to larger log-likelihood values. Alternatively, one can work with the first derivative of the log-likelihood function and seek values of the parameters that cause it to be zero. There are a variety of fast, reliable computational algorithms for carrying out these procedures. One of the most widely implemented is the Newton-Raphson algorithm.

Example: MLE for a sample of Bernoulli random variables (NewtonRaphson.R)

We demonstrate the Newton-Raphson algorithm in a simple setting to find the MLE of π for the previous Bernoulli example where $\sum_{i=1}^{n} y_i = 4$ in $n = 10$ trials. We found earlier that $\hat{\pi} = \sum_{i=1}^{n} y_i / n = 0.4$.

In this context the algorithm uses the following equation to obtain an estimate $\pi^{(i+1)}$ using a previous estimate $\pi^{(i)}$:

$$\pi^{(i+1)} = \pi^{(i)} - \frac{\left.\frac{d}{d\pi}\log[L(\pi|\mathbf{y})]\right|_{\pi=\pi^{(i)}}}{\left.\frac{d^2}{d\pi^2}\log[L(\pi|\mathbf{y})]\right|_{\pi=\pi^{(i)}}}$$

$$= \pi^{(i)} - \frac{\frac{w}{\pi^{(i)}} - \frac{n-w}{1-\pi^{(i)}}}{-\frac{w}{\left(\pi^{(i)}\right)^2} - \frac{n-w}{\left(1-\pi^{(i)}\right)^2}} \tag{B.4}$$

Equation B.4 comes from a first-order Taylor series expansion about $\pi^{(i)}$.[3] To begin using the algorithm, we choose a starting value $\pi^{(0)}$ that we think may be close to $\hat{\pi}$ and substitute this into Equation B.4 for $\pi^{(i)}$ to obtain $\pi^{(1)}$. If $\pi^{(1)}$ is "close enough" to $\pi^{(0)}$, we stop and use $\pi^{(1)}$ as $\hat{\pi}$; otherwise, we substitute $\pi^{(1)}$ into Equation B.4 for $\pi^{(i)}$ to find $\pi^{(2)}$. This process continues until $\left|\pi^{(i+1)} - \pi^{(i)}\right| < \varepsilon$ for some small number $\epsilon > 0$, and we say that *convergence* has been reached at iteration $i+1$.

Using the observed data, suppose we guess $\pi^{(0)} = 0.3$, and we feel that $\varepsilon = 0.0001$ represents "close enough." Table B.2 shows the iteration history where convergence is obtained after 5 iterations. The R code that produced this table is available in the corresponding program for this example. We have included in this program the code to create plots illustrating the Newton-Raphson algorithm.

Generally, readers of this book will not need to implement a Newton-Raphson procedure like this directly. Instead, we will use R functions that take care of these details. Also, note that there are other algorithms besides Newton-Raphson that are used to find maximum likelihood estimates. Many of these are implemented in R's optim() function.

---

[3] Suppose we would like to approximate a function $f(x)$ at a point x_0. The first-order Taylor series expansion approximates $f(x)$ with $f(x_0) + (x - x_0)f'(x_0)$, where $f'(\cdot)$ is the first derivative of $f(\cdot)$ with respect to x.

Table B.2: Iterations for the Newton-Raphson algorithm.

Iteration	$\pi^{(i)}$
1	0.3000000
2	0.3840000
3	0.3997528
4	0.3999999
5	0.4000000

B.3.3 Large-sample properties of the MLE

In order to use an MLE in confidence intervals and tests, we need to know its probability distribution (the probability distribution of a statistic is also known as its "sampling distribution"). It can be shown that *all* of the MLEs we will use share certain properties relating to their sampling distributions that make them very appealing bases for inference procedures. These properties generally hold for large samples; in other words, these properties generally hold *asymptotically*, which means, as the sample size grows toward ∞. Below is a list of the properties:

1. MLEs ARE ASYMPTOTICALLY NORMALLY DISTRIBUTED – This result is analogous to the central limit theorem for sample means. The fact that normality holds asymptotically means that in any given sample, the normal distribution is typically an *approximation* to the correct sampling distribution of $\hat{\boldsymbol{\theta}}$, and the approximation gets better with larger sample sizes.

2. MLEs ARE CONSISTENT – This means essentially that if you sample the whole population (or sample infinitely), the MLE will be exactly the same as the population parameter. In particular, any bias in the estimate (the difference between the expected value of the MLE and the true value of the parameter) vanishes as the sample size grows, and the variance shrinks to 0.

3. MLEs ARE ASYMPTOTICALLY EFFICIENT – This means that as the sample size grows toward ∞ they achieve the smallest variance possible for estimates of their type (e.g., among all asymptotically normal estimates). An important implication of this result is that confidence intervals based on MLEs have the potential to be shorter and tests more powerful than those based on other forms of estimates.

These properties are not guaranteed to hold in samples that are not "large." The normal distribution approximation is generally very good in "large" samples, but may be very poor in "small" samples. Other estimates may have smaller variance than the MLE in finite samples. Unfortunately, there is no uniform way to define "large" or "small." However, it is often not too difficult to simulate data from the chosen model and check whether the MLE has a distribution that appears roughly normal and has an acceptably small bias.

B.3.4 Variance of the MLE

The variance of any MLE is related to the curvature of the log-likelihood function in the neighborhood of the MLE. If the log-likelihood is very flat near the maximum, then there is much uncertainty in the data regarding the location of the parameter (many different values of $\boldsymbol{\theta}$ lead to similarly large likelihoods). Conversely, if the log-likelihood has a sharp peak, then the data show little doubt about the region in which the parameter must lie.

Formally, the variance of the MLE reduces in large samples to a quantity that is estimated directly from the second derivative of the log-likelihood at its peak,

$$\frac{\partial^2}{\partial \theta^2} \log[L(\theta|\mathbf{y})],$$

which is referred to as the *Hessian* matrix. From the Hessian, the estimated asymptotic variance (we will refer to this more simply as the "estimated variance") is computed as

$$\widehat{Var}(\hat{\theta}) = -E\left(\frac{\partial^2}{\partial \theta^2} \log[L(\theta|\mathbf{Y})]\right)^{-1}\bigg|_{\theta=\hat{\theta}}. \tag{B.5}$$

The estimated variance can instead be approximated by

$$\widehat{Var}(\hat{\theta}) \approx -\left(\frac{\partial^2}{\partial \theta^2} \log[L(\theta|\mathbf{y})]\right)^{-1}\bigg|_{\theta=\hat{\theta}}, \tag{B.6}$$

which is often easier to compute. Equation B.6 is *asymptotically equivalent* to Equation B.5, meaning that these estimated variances will essentially be the same in very large samples. Other asymptotically equivalent variants of these formulas are sometimes used as well. The estimated standard deviation of the statistic (i.e., the standard error) is computed by finding the square root of $\widehat{Var}(\hat{\theta})$.

When $p > 1$,

$$\widehat{Var}(\hat{\boldsymbol{\theta}}) = \begin{bmatrix} \widehat{Var}(\hat{\theta}_1) & \widehat{Cov}(\hat{\theta}_1, \hat{\theta}_2) & \cdots & \widehat{Cov}(\hat{\theta}_1, \hat{\theta}_p) \\ \widehat{Cov}(\hat{\theta}_1, \hat{\theta}_2) & \widehat{Var}(\hat{\theta}_2) & \cdots & \widehat{Cov}(\hat{\theta}_2, \hat{\theta}_p) \\ \vdots & \vdots & \ddots & \vdots \\ \widehat{Cov}(\hat{\theta}_1, \hat{\theta}_p) & \widehat{Cov}(\hat{\theta}_2, \hat{\theta}_p) & \cdots & \widehat{Var}(\hat{\theta}_p) \end{bmatrix}.$$

This is known as an estimated *variance-covariance matrix* (sometimes shortened to *covariance matrix* or *variance matrix*). The diagonal elements of the matrix (where the row and column numbers are the same) are the estimated variances of the MLEs. The off-diagonal elements are the covariances between pairs of MLEs. These estimated covariances measure the dependence between the MLEs, and they can be useful for finding the variances of functions of MLEs (see Appendix B.4). Because $\widehat{Cov}(\hat{\theta}_i, \hat{\theta}_j) = \widehat{Cov}(\hat{\theta}_j, \hat{\theta}_i)$, the matrix is symmetric, so that the element in row i, column j equals the element in row j, column i for any $i \neq j$.

Example: Poisson (PoissonLikelihood.R)

The purpose of this example is to find the variance for $\hat{\mu}$ in the Poisson example. Figure B.2 shows the curvature of the log-likelihood function for two different samples, where sample 1 has $\mathbf{y} = (3, 5, 6, 6, 7, 10, 13, 15, 18, 22)$ and sample 2 has $\mathbf{y} = (9, 12)$. For both samples, $\hat{\mu} = 10.5$. We can see that sample 2's log-likelihood function is relatively flat due to its small sample size, while sample 1's log-likelihood function has much more curvature due to its larger sample size. Because the variance for $\hat{\mu}$ is based on this curvature, we would expect the variance for sample 1 to be much less than the variance for sample 2.

Formally, we calculate the variance as

$$\widehat{Var}(\hat{\mu}) = -\left(\frac{\partial^2}{\partial \mu^2} \log[L(\mu|\mathbf{y})]\right)^{-1}\bigg|_{\mu=\hat{\mu}} = \frac{\hat{\mu}^2}{\sum_{i=1}^n y_i} = \frac{\hat{\mu}}{n}.$$

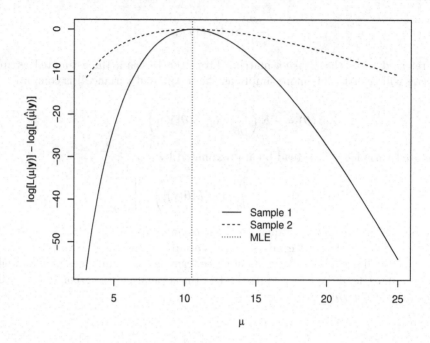

Figure B.2: Poisson log-likelihoods for samples of size $n = 10$ (sample 1) and $n = 2$ (sample 2) with common MLE $\hat{\mu} = 10.5$. Note that the two curves have been shifted vertically so that the log-likelihoods are both 0 at the MLE.

For sample 1, $\widehat{Var}(\hat{\mu}) = 10.5/10 = 1.05$, while for sample 2, $\widehat{Var}(\hat{\mu}) = 10.5/2 = 5.25$. As expected, the variance for $\hat{\mu}$ is larger for sample 2 than for sample 1. The R code for these calculations is included in the corresponding R program for this example.

B.4 Functions of parameters

B.4.1 Invariance property of MLEs

Often our interest is not limited to the model parameters that are directly estimated using the procedures in Appendix B.3.1 or B.3.2. Instead, we may want to estimate parameters that are *functions* of model parameters. As examples, (1) the odds ratio (e.g., Sections 1.2.5, 2.2.3, 3.3.1, and 4.2.3) can be written as a function of probabilities from two binomials, or from multinomial probabilities, or as a function of Poisson means; and (2) predicted values in any regression model can be written as functions of regression coefficients. The *invariance property of MLEs* states that if $\hat{\boldsymbol{\theta}}$ is the MLE for $\boldsymbol{\theta}$ and $g(\boldsymbol{\theta})$ is any function of $\boldsymbol{\theta}$ then $g(\hat{\boldsymbol{\theta}})$ is the MLE for $g(\boldsymbol{\theta})$. In words, the MLE of any function of parameters is just the same function of the parameters' MLE. The important implication of this result is that the large sample properties listed in Appendix B.3.3 hold for functions of MLEs, and all of the

Likelihood methods

inference procedures in Appendix B.5 can be applied to these functions.[4]

B.4.2 Delta method for variances of functions

The delta method is a very general and useful procedure for estimating the variance of a function of a random variable. We present it here to estimate the variance of a function of a MLE, which is not simply the same function of the MLE's variance.

Suppose that $\boldsymbol{\theta}$ consists of p parameters, $\theta_1, \theta_2, \ldots, \theta_p$ and define

$$g'_1(\hat{\boldsymbol{\theta}}) = \frac{\partial}{\partial \theta_1} g(\boldsymbol{\theta})\bigg|_{\boldsymbol{\theta}=\hat{\boldsymbol{\theta}}}, \; g'_2(\hat{\boldsymbol{\theta}}) = \frac{\partial}{\partial \theta_2} g(\boldsymbol{\theta})\bigg|_{\boldsymbol{\theta}=\hat{\boldsymbol{\theta}}}, \ldots, g'_p(\hat{\boldsymbol{\theta}}) = \frac{\partial}{\partial \theta_p} g(\boldsymbol{\theta})\bigg|_{\boldsymbol{\theta}=\hat{\boldsymbol{\theta}}}.$$

Then from a Taylor series approximation to $g(\boldsymbol{\theta})$ centered at $\hat{\boldsymbol{\theta}}$, one can show that

$$\widehat{Var}(g(\hat{\boldsymbol{\theta}})) \approx \sum_{i=1}^{p} [g'_i(\hat{\boldsymbol{\theta}})]^2 \widehat{Var}(\hat{\theta}_i) + 2 \sum \sum_{i>j} [g'_i(\hat{\boldsymbol{\theta}})][g'_j(\hat{\boldsymbol{\theta}})]\widehat{Cov}(\hat{\theta}_i, \hat{\theta}_j), \tag{B.7}$$

where $\widehat{Var}(\hat{\theta}_i)$, $i = 1, \ldots, p$ are the estimated variances of the MLEs and $\widehat{Cov}(\hat{\theta}_i, \hat{\theta}_j)$, $i, j = 1, \ldots, p$, $i \neq j$ the estimated covariances described in Appendix B.3.4.

Example: Variance of an odds ratio

Suppose that there are two independent binomial variables: W_1 with n_1 trials, probability of success π_1 and w_1 observed successes, and W_2 with n_2 trials, probability of success π_2 and w_2 observed successes. As described in Section 1.2.5, inference on the odds ratio $OR = [\pi_1/(1-\pi_1)]/[\pi_2/(1-\pi_2)]$ is done based on the sampling distribution of $\log(\widehat{OR})$. The variance of $\log(\widehat{OR})$ is found using the delta method as follows.

First, note that we have $p = 2$ parameters here, so that $\boldsymbol{\theta} = (\pi_1, \pi_2)'$, and $g(\boldsymbol{\theta}) = [\pi_1/(1-\pi_1)]/[\pi_2/(1-\pi_2)]$. Then

$$g'_1(\hat{\boldsymbol{\theta}}) = \frac{\partial}{\partial \pi_1}[\pi_1/(1-\pi_1)]/[\pi_2/(1-\pi_2)]\bigg|_{(\pi_1,\pi_2)=(\hat{\pi}_1,\hat{\pi}_2)} = \frac{1}{\hat{\pi}_1(1-\hat{\pi}_1)},$$

and similarly $g'_2(\hat{\boldsymbol{\theta}}) = 1/[\hat{\pi}_2(1-\hat{\pi}_2)]$, where $\hat{\pi}_i = w_i/n_i$. Also, we have that $\widehat{Var}(\hat{\pi}_i) = \hat{\pi}_i(1-\hat{\pi}_i)/n_i$, and because W_1 and W_2 are independent, $\widehat{Cov}(\hat{\pi}_1, \hat{\pi}_2) = 0$. Thus,

$$\widehat{Var}(g(\hat{\boldsymbol{\theta}})) \approx \sum_{i=1}^{p} [g'_i(\hat{\boldsymbol{\theta}})]^2 \widehat{Var}(\hat{\theta}_i) + 2\sum\sum_{i>j}[g'_i(\hat{\boldsymbol{\theta}})][g'_j(\hat{\boldsymbol{\theta}})]\widehat{Cov}(\hat{\theta}_i,\hat{\theta}_j)$$

$$= \left[\frac{1}{\hat{\pi}_1(1-\hat{\pi}_1)}\right]^2 \frac{\hat{\pi}_1(1-\hat{\pi}_1)}{n_1} + \left[\frac{1}{\hat{\pi}_2(1-\hat{\pi}_2)}\right]^2 \frac{\hat{\pi}_2(1-\hat{\pi}_2)}{n_2}$$

$$= \frac{1}{n_1\hat{\pi}_1(1-\hat{\pi}_1)} + \frac{1}{n_2\hat{\pi}_2(1-\hat{\pi}_2)}$$

$$= \frac{1}{w_1} + \frac{1}{n_1-w_1} + \frac{1}{w_2} + \frac{1}{n_2-w_2}.$$

This is the formula given in Section 1.2.5.

---

[4]These properties are again approximate in finite samples. The sample sizes required to make them hold satisfactorily for a given function of parameters may be similar to or quite different from those required for the parameters themselves. This can be checked by simulation.

B.5 Inference with MLEs

Throughout this section we consider the problem of doing inference on a single parameter θ, which may be a model parameter or some function of model parameters. Where appropriate, we mention briefly extensions to multiple parameters.

B.5.1 Tests for parameters

Consider the hypotheses
$$H_0 : \theta = \theta_0$$
$$H_a : \theta \neq \theta_0,$$
where θ_0 is some special value of interest. Several different procedures based on likelihood principles can be used to perform this test. All of them are approximate procedures in the sense that they may not achieve the stated type I error rate α exactly. These approximations are very accurate in large samples, but may provide poor results in small samples.

Wald tests

The Wald test (Wald, 1943) for a single parameter is the most familiar likelihood-based test procedure, because it uses the same ideas as the standard normal test that is part of every introductory statistics course. Because the MLE is asymptotically normally distributed with variance estimated as given in Equation B.6 or Equation B.7, we have that $Z_0 = (\hat{\theta} - \theta_0)/\sqrt{\widehat{Var}(\hat{\theta})} \dot\sim N(0,1)$ for a large sample when the null hypothesis is true.[5] Thus, we reject H_0 if

$$|Z_0| = \frac{|(\hat{\theta} - \theta_0)|}{\sqrt{\widehat{Var}(\hat{\theta})}} > Z_{1-\alpha/2}, \tag{B.8}$$

where the critical value $Z_{1-\alpha/2}$ is a $1 - \alpha/2$ quantile from a standard normal distribution. Alternatively, a p-value is calculated as $2P(Z > |Z_0|)$, where Z has a standard normal distribution.[6]

A Wald test is usually simple to conduct, but it is not always very effective. In particular, the $Z_{1-\alpha/2}$ critical value is guaranteed to be close to the correct critical value only in very large samples. In small samples, the normal approximation may be poor and $Z_{1-\alpha/2}$ may not be an appropriate approximation to the true critical value of the test. Thus, this test is recommended only when the sample size is large (in the context of the problem) or when no other test is possible.

A version of the Wald test is available to test hypotheses involving more than one parameter. This can be useful in regression problems with a categorical explanatory variable, which is represented in the regression by several indicator variables, each with a separate parameter. A null hypothesis of no association between the explanatory variable and the

---

[5] $\dot\sim$ means "approximately distributed as."

[6] Note that the critical value is from the normal distribution and not a t-distribution, even though we are estimating the variance in the denominator of Z_0. This is because the t-distribution arises specifically when the variance in the denominator of the test statistic is based on a sum-of-squares calculation on data from a normal distribution. The variance in Z_0 above is based on Equation B.6, which most often is not a sum-of-squares calculation.

response implies that these parameters must all be zero together. Let $\boldsymbol{\theta}_0$ be the hypothesized value of a p-dimensional parameter $\boldsymbol{\theta}$. Let $\hat{\boldsymbol{\theta}}$ be the parameter estimate, and $\widehat{Var}(\hat{\boldsymbol{\theta}})$ be its estimated variance. Then $H_0 : \boldsymbol{\theta} = \boldsymbol{\theta}_0$ is tested using the Wald statistic

$$W = (\hat{\boldsymbol{\theta}} - \boldsymbol{\theta}_0)'[\widehat{Var}(\hat{\boldsymbol{\theta}})]^{-1}(\hat{\boldsymbol{\theta}} - \boldsymbol{\theta}_0),$$

which has an approximate χ_p^2 distribution in large samples.

Likelihood ratio tests

Values of θ that have likelihoods near the maximum are more plausible guesses for the true value of the parameter than those whose likelihoods are much lower than the maximum. A comparison of the value of the likelihood at its peak against the best value of the likelihood when parameters are constrained to follow the null hypothesis is therefore a measure of evidence against the null. It turns out that the best way to make this comparison is through the likelihood ratio (LR) statistic

$$\Lambda = \frac{L(\theta_0|\mathbf{y})}{L(\hat{\theta}|\mathbf{y})}. \tag{B.9}$$

Note that $\Lambda \leq 1$ because $L(\hat{\theta}|\mathbf{y})$ is the maximum value of the likelihood for the given data. The value of Λ is near 1 when $L(\theta_0|\mathbf{y})$ is close to $L(\hat{\theta}|\mathbf{y})$. Closeness is judged by the fact that $-2\log(\Lambda) = 2[\log L(\hat{\theta}|\mathbf{y}) - \log L(\theta_0|\mathbf{y})]$ has an asymptotic χ_1^2 distribution when the null hypothesis is true. Thus, the *likelihood ratio test* (LRT) for $H_0 : \theta = \theta_0$ vs. $H_a : \theta \neq \theta_0$ rejects H_0 when $-2\log(\Lambda) > \chi_{1,1-\alpha}^2$, where $\chi_{1,1-\alpha}^2$ is the $1-\alpha$ quantile from a χ^2 distribution with one degree of freedom.

Example: Poisson (LR-Plots.R)

The left plot in Figure B.3 gives a plot of the log-likelihood function for the previous Poisson example with $n = 10$ (sample 1). Evidence against $\mu = 9$ is fairly light, because its log-likelihood is fairly close to the peak at $\hat{\mu} = 10.5$. However, there is more evidence against $\mu = 5$, whose log-likelihood is much farther from the peak.

A LRT of $H_0 : \mu = 9$ vs. $H_a : \mu \neq 9$ leads to

$$\begin{aligned}-2\log(\Lambda) &= -2\log\left(\frac{L(\mu = 9|\mathbf{y})}{L(\mu = 10.5|\mathbf{y})}\right) \\ &= -2\log\left(L(\mu = 9|\mathbf{y})\right) + 2\log\left(L(\mu = 10.5|\mathbf{y})\right) \\ &= 2.37.\end{aligned}$$

With $\chi_{1,0.95}^2 = 3.84$, we do not reject $H_0 : \mu = 9$. In a similar manner for $H_0 : \mu = 5$ vs. $H_a : \mu \neq 5$, we calculate $-2\log(\Lambda) = 45.81$ leading to a rejection of the null hypothesis.

The right plot in Figure B.3 demonstrates the results from the two hypothesis tests differently. We represent the rejection region here by finding the set of all possible values of μ such that $-2\log(\Lambda) < \chi_{1,1-\alpha}^2$.[7] The figure again shows that $H_0 : \mu = 9$ is not rejected, but $H_0 : \mu = 5$ is rejected.

The LRT generalizes to a broad range of problems involving multiple parameters and hypotheses involving intervals or constraints on the parameters. The general approach is

---

[7] Solve for μ in $-2\log\left(L(\mu|\mathbf{y})\right) + 2\log\left(L(\mu = 10.5|\mathbf{y})\right) = \chi_{1,0.95}^2$ in order to find the rejection regions.

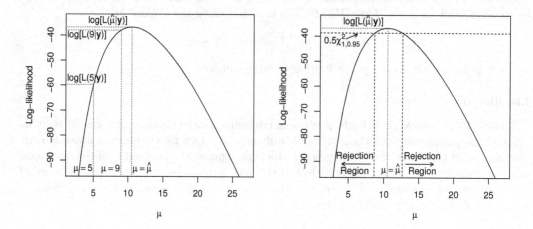

Figure B.3: Left: Poisson log-likelihood for the sample of size 10 with three values of μ indicated. Right: The rejection region for a likelihood ratio test using $\alpha = 0.05$. Any value of μ that lies in the rejection region has its null hypothesis rejected.

that the numerator of Equation B.9 is replaced by the maximum value of the likelihood function across all parameters that satisfy H_0. Additionally, the denominator is replaced by the maximum value of the likelihood function without constraints. The degrees of freedom for the χ^2 distribution are determined by the number of constraints placed on the parameters. This is explained further throughout the book where LRTs are used.

Many problems use models that contain extra parameters that are not involved in the null hypothesis (for example, hypotheses about the mean of a normal model generally do not place constraints on the variance). In these cases, the extra parameters not specified in H_0, say ϕ, are set at their MLEs under the conditions on θ. That is, in the denominator of Equation B.9, the likelihood is maximized with respect to θ and ϕ simultaneously. In the numerator, the likelihood is again maximized for both parameters simultaneously, but subject to the constraint that θ satisfies H_0. Thus, the values for ϕ that yield the best likelihoods in the numerator and denominator may be different.

The LRT is often not simple to do by hand, but there are very good computational techniques that can efficiently maximize likelihoods subject to constraints. This makes LRTs broadly applicable to a wide range of testing problems. Also, the accuracy of the LRT critical value is generally much better for a given sample size than that of the Wald test critical value, so it is generally preferred over Wald tests when both are available.

Score test

An alternative approach to testing hypotheses using likelihoods comes from examining properties of the likelihood function at the null hypothesis. As Figure B.4 shows, the slope of the log-likelihood near the peak should have a smaller magnitude than the slope far from the peak. We can therefore use this slope as a measure of the evidence in the data against H_0. This slope, also called the *score*, is just the first derivative of the log likelihood, evaluated at θ_0.

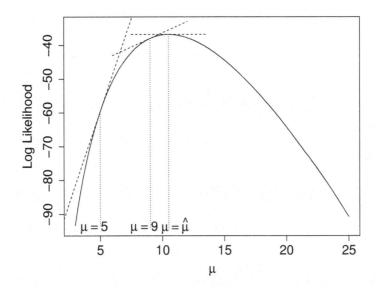

Figure B.4: Poisson log likelihood for artificial sample of size 10 showing the score function (slope) at three values of θ. The score is 0 at the MLE. The magnitude is larger for the θ farther from the MLE ($\theta = 5$) than for the one closer ($\theta = 9$). Code for the plot is in ScorePlot.R.

In particular, for independent random samples, the score is

$$U_0 = \left.\frac{\partial}{\partial \theta} \log[L(\theta|\mathbf{y})]\right|_{\theta=\theta_0}.$$

The central limit theorem ensures that U_0 is asymptotically normally distributed. Under the null hypothesis, the average slope across all possible data sets is zero: $E(U_0) = 0$. It can be shown that the asymptotic variance of the score—measuring how variable the slopes would be at H_0 from likelihood functions calculated based on different data sets—turns out to be estimated analogously to Equation B.5. Specifically,

$$\widehat{Var}(U_0) = -E\left.\left(\frac{\partial^2}{\partial \theta^2} \log[L(\theta|\mathbf{y})]\right)\right|_{\theta=\theta_0}.$$

A score test is then carried out much like the Wald test, comparing $U_0/\sqrt{\widehat{Var}(U_0)}$ to $Z_{1-\alpha/2}$. Multiparameter extensions are carried out just as in the Wald test.

The score test is also based on asymptotics, so that the critical value is approximate for any finite sample. It generally performs better than the Wald test, but not necessarily better than the LRT. Its main advantage is that it uses the likelihood only at the null hypotheses. In some complicated problems, the null hypothesis represents a considerable simplification to the general model—for example, by setting certain parameters to 0—so that calculations are much easier to carry out at θ_0 than anywhere else.

B.5.2 Confidence intervals for parameters

Wald

Like the Wald test, the Wald confidence interval is based on familiar relationships using the normal distribution. We can write $(\hat{\theta} - \theta)/\sqrt{\widehat{Var}(\hat{\theta})} \dot\sim N(0,1)$, where $\hat{\theta}$ and $\widehat{Var}(\hat{\theta})$ are the same as defined in Appendix B.3. Thus,

$$P\left(Z_{\alpha/2} < (\hat{\theta} - \theta)/\sqrt{\widehat{Var}(\hat{\theta})} < Z_{1-\alpha/2}\right) \approx 1 - \alpha,$$

where $Z_{1-\alpha/2}$ is a $1 - \alpha/2$ quantile from a standard normal distribution. After rearranging terms, we obtain

$$P\left(\hat{\theta} - Z_{1-\alpha/2}\sqrt{\widehat{Var}(\hat{\theta})} < \theta < \hat{\theta} - Z_{\alpha/2}\sqrt{\widehat{Var}(\hat{\theta})}\right) \approx 1 - \alpha.$$

Recognizing that $-Z_{\alpha/2} = Z_{1-\alpha/2}$, this leads to the familiar form of a $(1-\alpha)100\%$ confidence interval for θ as

$$\hat{\theta} \pm Z_{1-\alpha/2}\sqrt{\widehat{Var}(\hat{\theta})}. \tag{B.10}$$

An alternative approach to finding a confidence interval is to "invert" a test. That is, we seek the set of values for θ_0 for which $H_0 : \theta = \theta_0$ is *not* rejected. This procedure is easily carried out for the Wald test starting from Equation B.8 and leads to the same interval as Equation B.10.

Likelihood ratio

Likelihood ratio confidence intervals are found by inverting the LRT. A $(1 - \alpha)100\%$ confidence interval for θ is the set of all possible values of θ such that

$$-2[L(\theta|\mathbf{y})/L(\hat{\theta}|\mathbf{y})] \leq \chi^2_{1,1-\alpha}. \tag{B.11}$$

In Figure B.3 this is the interval between the two areas labeled "Rejection Region." As was the case with tests, likelihood ratio confidence intervals tend to be more accurate than Wald in the sense of having a true confidence level closer to the stated $1 - \alpha$ level for a given sample size. However, there is rarely a closed form for the solution, so iterative numerical procedures are needed to locate the endpoints of the interval where equality holds in Equation B.11. These computations can be difficult to carry out in some more complex problems, so not all software packages compute them.

In situations with multiple parameters, such as regression models, we frequently find confidence intervals for individual parameters. For example, suppose that a model contains two parameters, θ_1 and θ_2, and that we want to find a $(1 - \alpha)100\%$ LR confidence interval for θ_1. Because $L(\theta_1, \theta_2|\mathbf{y})$ changes as a function of both parameters, we need to account somehow for θ_2 in the process of computing Equation B.11. One approach is to fix θ_2 at some value, say d, and then find the values of θ_1 that satisfy $-2[L(\theta_1, d|\mathbf{y})/L(\tilde{\theta}_1(d), d|\mathbf{y})] \leq \chi^2_{1,1-\alpha}$, where $\tilde{\theta}_1(d)$ is the MLE of θ_1 when $\theta_2 = d$. However, this may lead to a different confidence interval for each value of d. As an alternative, we can let the value of θ_2 be its MLE for each value of θ_1 that we consider. That is, we fix the denominator of the LR statistic to be the overall maximum, $L(\hat{\theta}_1, \hat{\theta}_2|\mathbf{y})$, and for each different θ_1 we try in the numerator, we set $L(\theta_1, \theta_2|\mathbf{y})$ at the maximum that it achieves across all values θ_2. If we let $\tilde{\theta}_2(c)$ be the MLE of θ_2 when we fix $\theta_1 = c$, then the profile LR confidence interval is the set of values for c that satisfy $-2[L(c, \tilde{\theta}_2(c)|\mathbf{y})/L(\hat{\theta}_1, \hat{\theta}_2|\mathbf{y})] \leq \chi^2_{1,1-\alpha}$.

Score

Score confidence intervals are also found by inverting the score test. This is not as easy to do as it is for the Wald test, however. In the latter case, the standard error in the denominator of the test statistic does not change as one examines different values of θ_0, so that the rearrangement of Equation B.8 is easy. For the score test, the denominator changes with θ_0, and so the rearrangement can result in complicated mathematics unless the form of $\widehat{Var}(U_0)$ is fairly simple. Instead, the interval typically needs to be found iterative numerical procedures similar to those described in Appendix B.3.2. An example where the mathematics *can* be worked out with relative ease is the Wilson score interval in Section 1.1.2.

B.5.3 Tests for models

Many forms of regression are used in categorical analysis. Their use in practice often requires comparing several models involving different subsets of explanatory variables or comparing models with and without groups of explanatory variables (e.g., groups of dummy variables representing a categorical explanatory variable). When model parameters are estimated using maximum likelihood estimation, standard model comparison techniques are available based on likelihood ratio tests.

Consider two models: a *full model*, M_1, consisting of a set of p_1 explanatory variables, and the *reduced model*, M_0, containing a proper subset of p_0 explanatory variables. It is important that the reduced model does not contain any variables that are not also in the full model. Comparing M_0 and M_1 is equivalent to a hypothesis test specifying M_0 as the null hypothesis and M_1 as the alternative. Fit both M_0 and M_1 to the data and let L_{M_1} and L_{M_0} be their respective maximized likelihoods. If we use $\Lambda(M_0, M_1)$ to denote the likelihood ratio L_{M_0}/L_{M_1}, then the LRT for $H_0 : M_0$ vs. $H_a : M_1$ rejects the null hypothesis if

$$-2\log\left(\Lambda(M_0, M_1)\right) > \chi^2_{(p_1-p_0), 1-\alpha}. \tag{B.12}$$

A rejection of H_0 means that at least one of the parameters (thus, one of the variables) that make up the difference between M_0 and M_1 is important to include in a model already containing M_0. Failure to reject H_0 suggests that this simpler model may suffice and that the extra variables do not contribute significantly to the explanatory power of the model. Of course, we can never conclude that the null hypothesis is true for any hypothesis test, so it is not possible to say that M_0 is a "significantly better model" than M_1.

Deviance

In linear regression with normally distributed errors and constant variance, the sum of squared errors (SSE) for any model measures in aggregate how close the predictions are to the observed responses. In more general regression models involving different distributions, the common measure of a model's fit is the *deviance*. The deviance is just $D_M = \Lambda(M, M_{SAT})$ from Equation B.12 where M is the model being fit and M_{SAT} is the *saturated model*, which fits a separate parameter for each observation. The saturated model provides perfect prediction for the observed data, but is often not considered a viable model for predicting new data, because new data is unlikely to have the same random "errors" as the current data. (See Figure B.5 for a simple example of a saturated model.) The reason to use a saturated model in computing deviances is that any other model is guaranteed to be a proper subset of the saturated model, and hence $\Lambda(M, M_{SAT})$ can always be computed. Thus, the deviance is often part of the default output for many modeling procedures in

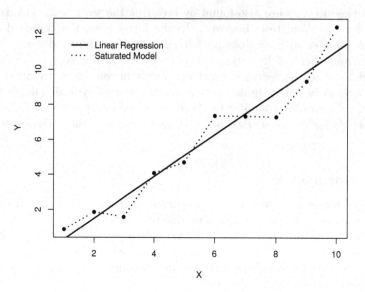

Figure B.5: Comparison of a simple linear regression with the saturated model. Code for the plot is in SaturatedModelExample.R.

categorical data analysis.[8] If we have the deviances from the two models M_0 and M_1 that we want to compare as above, we can compute $\Lambda(M_0, M_1) = D_{M_0} - D_{M_1}$ to make this comparison, and carry out the test as in Equation B.12.

---

[8]For reasons described in Chapter 5, the comparison against $\chi^2_{(p_1-p_0), 1-\alpha}$ is not always done.

Bibliography

Agresti, A. (1996). *An Introduction to Categorical Data Analysis*. Wiley.

Agresti, A. (2002). *Categorical Data Analysis*. Wiley, 2nd edition.

Agresti, A. (2007). *An Introduction to Categorical Data Analysis*. Wiley, 2nd edition.

Agresti, A. and Caffo, B. (2000). Simple and effective confidence intervals for proportions and differences of proportions result from adding two successes and two failures. *The American Statistician*, 54:280–288.

Agresti, A. and Coull, B. (1998). Approximate is better than "exact" for interval estimation of binomial proportions. *The American Statistician*, 52:119–126.

Agresti, A. and Liu, I. (1999). Modeling a categorical variable allowing arbitrarily many category choices. *Biometrics*, 55:936–943.

Agresti, A. and Min, Y. (2001). On small-sample confidence intervals for parameters in discrete distributions. *Biometrics*, 57:963–971.

Agresti, A. and Min, Y. (2005a). Frequentist performance of Bayesian confidence intervals for comparing proportions in 2×2 contingency tables. *Biometrics*, 61(2):515–523.

Agresti, A. and Min, Y. (2005b). Simple improved confidence intervals for comparing matched proportions. *Statistics in Medicine*, 24:729–740.

Aseffa, A., Ishak, A., Stevens, R., Fergussen, E., Giles, M., Yohannes, G., and Kidan, K. (1998). Prevalence of HIV, syphilis and genital chlamydial infection among women in north-west Ethiopia. *Epidemiology and Infection*, 120:171–177.

Bates, D. (2010). *lme4: Mixed-Effects Modeling with R*. Self-published, http://lme4.r-forge.r-project.org/.

Becker, C., Loughin, T., and Santander, T. (2008). Identification of forest-obligate birds by mist netting and strip counts in Andean Ecuador. *Journal of Field Ornithology*, 79:229–244.

Beller, E. (2009). Bringing intergenerational social mobility research into the twenty-first century: Why mothers matter. *American Sociological Review*, 74:507–528.

Belsley, D., Kuh, E., and Welsch, R. (1980). *Regression Diagnostics: Identifying Influential Data and Sources of Collinearity*. Wiley.

Berry, K. and Mielke, P. (2003). Permutation analysis of data with multiple binary category choices. *Psychological Reports*, 92:91–98.

Berry, S. and Wood, C. (2004). The cold-foot effect. *Chance*, 17:47–51.

Bilder, C. (2009). Human or Cylon? Group testing on 'Battlestar Galactica'. *Chance*, 22:46–50.

Bilder, C. and Loughin, T. (1998). "It's Good!" An analysis of the probability of success for placekicks. *Chance*, 11:20–24.

Bilder, C. and Loughin, T. (2002). Testing for conditional multiple marginal independence. *Biometrics*, 58:200–208.

Bilder, C. and Loughin, T. (2004). Testing for marginal independence between two categorical variables with multiple responses. *Biometrics*, 60:241–248.

Bilder, C. and Loughin, T. (2007). Modeling association between two or more categorical variables that allow for multiple category choices. *Communications in Statistics: Theory and Methods*, 36:433–451.

Bilder, C., Loughin, T., and Nettleton, D. (2000). Multiple marginal independence testing for pick any/c variables. *Communications in Statistics: Simulation and Computation*, 29:1285–1316.

Binder, D. (1983). On the variances of asymptotically normal estimators from complex surveys. *International Statistical Review*, 51:279–292.

Binder, D. and Roberts, G. (2003). Statistical inference for survey data analysis. In *ASA Proceedings of the Joint Statistical Meetings*, pages 568–572. American Statistical Association.

Binder, D. and Roberts, G. (2009). Design- and model-based inference for model parameters. In Pfeffermann, D. and Rao, C., editors, *Handbook of Statistics 29B: Sample Surveys: Inference and Analysis*, pages 33–54. Elsevier.

Blaker, H. (2000). Confidence curves and improved exact confidence intervals for discrete distributions. *The Canadian Journal of Statistics*, 28:783–798.

Blaker, H. (2001). Corrigenda: Confidence curves and improved exact confidence intervals for discrete distributions. *The Canadian Journal of Statistics*, 29:681.

Bolker, B. (2009). Dealing with quasi-models in R. http://cran.r-project.org/web/packages/bbmle/vignettes/quasi.pdf.

Bonett, D. and Price, R. (2006). Confidence intervals for a ratio of binomial proportions based on paired data. *Statistics in Medicine*, 25:3039–3047.

Borkowf, C. (2006). Constructing binomial confidence intervals with near nominal coverage by adding a single imaginary failure or success. *Statistics in Medicine*, 25:3679–3695.

Breslow, N. and Lin, X. (1995). Bias correction in generalised linear mixed models with a single component of dispersion. *Biometrika*, 82:81–91.

Bretz, F., Hothorn, T., and Westfall, P. (2011). *Multiple Comparisons Using R*. Chapman & Hall/CRC.

Brown, L., Cai, T., and DasGupta, A. (2001). Interval estimation for a binomial proportion. *Statistical Science*, 16:101–133.

Brown, L., Cai, T., and DasGupta, A. (2002). Confidence intervals for a binomial proportion and asymptotic expansions. *The Annals of Statistics*, 30:160–201.

Brown, P., Stone, J., and Ord-Smith, C. (1983). Toxaemic signs during pregnancy. *Applied Statistics*, 32:69–72.

Brownlee, K. (1955). Statistics of the 1954 polio vaccine trials. *Journal of the American Statistical Association*, 50:1005–1013.

Buonaccorsi, J. (2010). *Measurement Error: Models, Methods, and Applications.* Chapman & Hall/CRC.

Burnham, K. and Anderson, D. (2002). *Model Selection and Multimodel Inference: A Practical Information-Theoretic Approach.* Springer, 2nd edition.

Calcagno, V. and de Mazancourt, C. (2010). glmmulti: An R package for easy automated model selection with (generalized) linear models. *Journal of Statistical Software*, 34.

Carlin, B. and Louis, T. (2008). *Bayesian Methods for Data Analysis.* Chapman & Hall/CRC.

Carroll, R., Ruppert, D., Stefanski, L., and Crainiceanu, C. (2010). *Measurement Error in Nonlinear Models: A Modern Perspective.* Chapman & Hall/CRC.

Casella, G. and Berger, R. (2002). *Statistical Inference.* Duxbury Press, 2nd edition.

Chambers, J. (2010). *Software for Data Analysis: Programming with R.* Springer.

Chib, S. and Greenberg, E. (1995). Understanding the Metropolis-Hastings algorithm. *The American Statistician*, 49:327–335.

Clogg, C. and Eliason, S. (1987). Some common problems in log-linear analysis. *Sociological Methods & Research*, 16:8–44.

Clopper, C. and Pearson, E. (1934). The use of confidence or fiducial limits illustrated in the case of the binomial. *Biometrika*, 26:404–413.

Cognard, C., Gobin, Y., Pierot, L., Bailly, A., Houdart, E., Casasco, A., Chiras, J., and Merland, J. (1995). Cerebral dural arteriovenous fistulas: Clinical and angiographic correlation with a revised classification of venous drainage. *Radiology*, 194:671–680.

Coombs, C. (1964). *A theory of data.* Wiley.

Cowles, M. and Carlin, B. (1996). Markov chain Monte Carlo convergence diagnostics: A comparative review. *Journal of the American Statistical Association*, 91:883–904.

Dalal, S., Fowlkes, E., and Hoadley, B. (1989). Risk analysis of the space shuttle: Pre-Challenger prediction of failure. *Journal of the American Statistical Association*, 84:945–957.

Davison, A. and Hinkley, D. (1997). *Bootstrap Methods and their Application.* Cambridge University Press.

Dawson, L. (2004). The Salk polio vaccine clinical trial of 1954: Risks, randomization and public involvement in research. *Clinical Trials*, 1:122–130.

Deb, P. and Trivedi, P. (1997). Demand for medical care by the elderly: A finite mixture approach. *Journal of Applied Econometrics*, 12:313–336.

DeHart, T., Tennen, H., Armeli, S., Todd, M., and Affleck, G. (2008). Drinking to regulate romantic relationship interactions: The moderating role of self-esteem. *Journal of Experimental Social Psychology*, 44:527–538.

Fang, L. and Loughin, T. (2012). Analyzing binomial data in a split-plot design: Classical approach or modern techniques? *Communications in Statistics: Simulation and Computation*, 42:727–740.

Firth, D. (1993). Bias reduction of maximum likelihood estimates. *Biometrika*, 80:27–38.

Fox, J. (2008). *Applied Regression Analysis and Generalized Linear Models*. Sage Publications, 2nd edition.

Foxman, B., Marsh, J., Gillespie, B., Rubin, N., Koopman, J., and Spear, S. (1997). Condom use and first-time urinary tract infection. *Epidemiology*, 8:637–641.

Francis, T., Korns, R., Voight, R., Boisen, M., Hemphill, F., Napier, J., and Tolchinsky, E. (1955). An evaluation of the 1954 poliomyelitis vaccine trials. *American Journal of Public Health*, 45:1–63.

Friendly, M. (1992). Graphical methods for categorical data. In *SAS User Group International Conference Proceedings*, volume 17, pages 190–200.

Gange, S. (1995). Generating multivariate categorical variates using the iterative proportional fitting algorithm. *The American Statistician*, 49:134–138.

Gelman, A., Carlin, J., Stern, H., and Rubin, D. (2004). *Bayesian Data Analysis*. Chapman & Hall/CRC.

Gelman, A., Jakulin, A., Pittau, M. G., and Su, Y. (2008). A weakly informative default prior distribution for logistic and other regression models. *The Annals of Applied Statistics*, pages 1360–1383.

Gentle, J. (2009). *Computational Statistics*. Springer.

Grechanovsky, E. (1987). Stepwise regression procedures: Overview, problems, results, and suggestions. *Annals of the New York Academy of Sciences*, 491:197–232.

Greven, S. and Kneib, T. (2010). On the behaviour of marginal and conditional AIC in linear mixed models. *Biometrika*, 97:773–789.

Gustafson, P. (2004). *Measurement Error and Misclassificaion in Statistics and Epidemiology: Impacts and Bayesian Adjustments*. Chapman & Hall/CRC.

Halekoh, U., Højsgaard, S., and Yan, J. (2006). The R package `geepack` for generalized estimating equations. *Journal of Statistical Software*, 15.

Hastie, T., Tibshirani, R., and Friedman, J. (2009). *The Elements of Statistical Learning*. Springer, 2nd edition.

Heeringa, S., West, B., and Berglund, P. (2010). *Applied Survey Data Analysis*. Chapman & Hall/CRC.

Heinze, G. (2006). A comparative investigation of methods for logistic regression with separated or nearly separated data. *Statistics in Medicine*, 25:4216–4226.

Heinze, G. and Schemper, M. (2002). A solution to the problem of separation in logistic regression. *Statistics in Medicine*, 21:2409–2419.

Henderson, M. and Meyer, M. (2001). Exploring the confidence interval for a binomial parameter in a first course in statistical computing. *The American Statistician*, 55:337–344.

Hirji, K., Mehta, C., and Patel, N. (1987). Computing distributions for exact logistic regression. *Journal of the American Statistical Association*, 82:1110–1117.

Hoeting, J., Madigan, D., Raftery, A., and Volinsky, C. (1999). Bayesian model averaging: A tutorial. *Statistical Science*, 14:382–417.

Hosmer, D., Hosmer, T., le Cessie, S., and Lemeshow, S. (1997). A comparison of goodness-of-fit tests for the logistic regression model. *Statistics in Medicine*, 16:965–980.

Hosmer, D. and Lemeshow, S. (1980). Goodness-of-fit tests for the multiple logistic regression model. *Communications in Statistics: Theory and Methods*, 9:1043–1069.

Hosmer, D. and Lemeshow, S. (2000). *Applied Logistic Regression*. Wiley, 2nd edition.

Hubert, J. (1992). *Bioassay*. Kendall Hunt Publishing Company, 3rd edition.

Ihaka, R. and Gentleman, R. (1996). R: A language for data analysis and graphics. *Journal of Computational and Graphical Statistics*, 3:299–314.

Imrey, P., Koch, G., Stokes, M., in collaboration with Darroch, J., Freeman, D., and Tolley, H. (1982). Some reflections on the log-linear model and logistic regression. Part II: Data analysis. *International Statistical Review*, 50:35–63.

Korn, E. and Graubard, B. (1999). *Analysis of Health Surveys*. Wiley.

Kott, P. and Carr, D. (1997). Developing an estimation strategy for a pesticide data program. *Journal of Official Statistics*, 13:367–383.

Küchenhoff, H., Mwalili, S., and Lesaffre, E. (2006). A general method for dealing with misclassification in regression: The misclassification SIMEX. *Biometrics*, 62:85–96.

Kuonen, D. (1999). Saddlepoint approximations for distributions of quadratic forms in normal variables. *Biometrika*, 86:929–935.

Kupper, L. and Haseman, J. (1978). The use of a correlated binomial model for the analysis of certain toxicological experiments. *Biometrics*, 34:69–76.

Kutner, M., Nachtsheim, C., and Neter, J. (2004). *Applied Linear Regression Models*. McGraw-Hill/Irwin, 4th edition.

Lambert, D. (1992). Zero-inflated poisson regression, with an application to defects in manufacturing. *Technometrics*, 34:1–14.

Larntz, K. (1978). Small-sample comparisons of exact levels for chi-squared goodness-of-fit statistics. *Journal of the American Statistical Association*, 73:253–263.

Lederer, W. and Küchenhoff, H. (2006). A short introduction to the SIMEX and MCSIMEX. *R News*, 6:26–31.

Lee, K. and Koval, J. (1997). Determination of the best significance level in forward logistic regression. *Communications in Statistics: Simulation and Computation*, 26:559–575.

Lesaffre, E. and Albert, A. (1989). Multi-group logistic regression diagnostics. *Journal of the Royal Statistical Society, Series C*, 38:425–440.

Liang, K. and Zeger, S. (1986). Longitudinal data analysis using generalized linear models. *Biometrika*, 73:13–22.

Littell, R., Milliken, G., Stroup, W., Wolfinger, R., and Schabenberger, O. (2006). *SAS for Mixed Models*. SAS Institute, 2nd edition.

Liu, P., Shi, Z., Zhang, Y., Xu, Z., Shu, H., and Zhang, X. (1997). A prospective study of a serum-pooling strategy in screening blood donors for antibody to hepatitis C virus. *Transfusion*, 37:732–736.

Lohr, S. (2010). *Sampling: Design and Analysis*. Cengage Learning, 2nd edition.

Long, J. (1990). The origins of sex differences in science. *Social Forces*, 68:1297–1316.

Loughin, T. (2004). A systematic comparison of methods for combining p-values from independent tests. *Computational Statistics and Data Analysis*, 47:467–485.

Loughin, T. and Bilder, C. (2010). On the use of a log-rate model for survey-weighted categorical data. *Communications in Statistics: Theory and Methods*, 40:2661–2669.

Loughin, T., Roediger, M., Milliken, G., and Schmidt, J. (2007). On the analysis of long-term experiments. *Journal of the Royal Statistical Society, Series A*, 170:29–42.

Loughin, T. and Scherer, P. (1998). Testing for association in contingency tables with multiple column responses. *Biometrics*, 54:630–637.

Lui, K. and Lin, C. (2003). A revisit on comparing the asymptotic interval estimators of odds ratio in a single 2x2 table. *Biometrical Journal*, 45:226–237.

Lumley, T. (2011). *Complex Surveys: A Guide to Analysis using R*. Wiley.

Margolin, B., Kaplan, N., and Zeiger, E. (1981). Statistical analysis of the Ames *Salmonella*/microsome test. *Proceedings of the National Academy of Sciences USA*, 78:3779–3783.

Martin, A. and Quinn, K. (2006). Applied Bayesian inference in R using `MCMCpack`. *R News*, 6:2–7.

Martin, A., Quinn, K., and Park, J. (2011). `MCMCpack`: Markov chain Monte Carlo in R. *Journal of Statistical Software*, 42.

Martin, C., Herrman, T., Loughin, T., and Oentong, S. (1998). Micropycnometer measurement of single-kernel density of healthy, sprouted, and scab-damaged wheats. *Cereal Chemistry*, 75:177–180.

Maugh, T. (2009). Results of AIDS vaccine trial 'weak' in second analysis. *Los Angeles Times*.

McCullagh, P. and Nelder, J. (1989). *Generalized Linear Models*. Chapman & Hall/CRC, 2nd edition.

McLean, R., Sanders, W., and Stroup, W. (1991). A unified approach to mixed linear models. *The American Statistician*, 45:54–64.

McNemar, Q. (1947). Note on the sampling error of the difference between correlated proportions or percentages. *Pyschometrika*, 12:153–157.

Mebane, W. and Sekhon, J. (2004). Robust estimation and outlier detection for overdispersed multinomial models of count data. *American Journal of Political Science*, 48:392–411.

Mehta, C. and Patel, N. (1995). Exact logistic regression: Theory and examples. *Statistics in Medicine*, 14:2143–2160.

Mehta, C., Patel, N., and Senchaudhuri, P. (2000). Efficient Monte Carlo methods for conditional logistic regression. *Journal of the American Statistical Association*, 95:99–108.

Meinshausen, N. (2007). Relaxed lasso. *Computational Statistics and Data Analysis*, 52:374–393.

Michalewicz, Z. (1996). *Genetic Algorithms + Data Structures = Evolution Programs*. Springer, 3rd edition.

Miller, A. (1984). Selection of subsets of regression variables (with discussion). *Journal of the Royal Statistical Society, Series A*, 147:389–425.

Milliken, G. and Johnson, D. (2001). *Analysis of Messy Data, Volume III: Analysis of Covariance*. Chapman & Hall/CRC.

Milliken, G. and Johnson, D. (2004). *Analysis of Messy Data Volume I: Designed Experiments*. Chapman & Hall/CRC, 2nd edition.

Molenberghs, G. and Verbeke, G. (2005). *Models for Discrete Longitudinal Data*. Springer.

Moore, D. and Notz, W. (2009). *Statistics: Concepts and Controversies*. W.H. Freeman & Company, 7th edition.

Mullahy, J. (1986). Specification and testing of some modified count data models. *Journal of Econometrics*, 33:341–365.

Newcombe, R. (1998). Improved confidence intervals for the difference between binomial proportions based on paired data. *Statistics in Medicine*, 17:2635–2650.

Newcombe, R. (2001). Logit confidence intervals and the inverse sinh transformation. *The American Statistician*, 55:200–202.

Osius, G. and Rojek, D. (1992). Normal goodness-of-fit tests for multinomial models with large degrees of freedom. *Journal of the American Statistical Association*, 87:1145–1152.

Plummer, M., Best, N., Cowles, K., and Vines, K. (2006). CODA: Convergence diagnosis and output analysis for MCMC. *R news*, 6:7–11.

Potter, D. (2005). A permutation test for inference in logistic regression with small- and moderate-sized data sets. *Statistics in Medicine*, 24:693–708.

Pregibon, D. (1981). Logistic regression diagnostics. *The Annals of Statistics*, 9:705–724.

Raftery, A. (1995). Bayesian model selection in social research. *Sociological Methodology*, 25:111–163.

Rao, J. and Scott, A. (1981). The analysis of categorical data from complex sample surveys: Chi-squared tests for goodness of fit and independence in two-way tables. *Journal of the American Statistical Association*, 76:221–230.

Rao, J. and Scott, A. (1984). On chi-squared tests for multiway contingency tables with cell proportions estimated from survey data. *The Annals of Statistics*, 12:46–60.

Rao, J. and Thomas, D. (2003). Analysis of categorical response data from complex surveys: An appraisal and update. In Chambers, R. and Skinner, C., editors, *Analysis of survey data*, pages 85–108. Wiley.

Raudenbush, S. and Bryk, A. (2002). *Hierarchical linear models: Applications and data analysis methods*. Sage Publications, 2nd edition.

Rerks-Ngarm, S., Pitisuttithum, P., Nitayaphan, S., Kaewkungwal, J., Chiu, J., Paris, R., Premsri, N., Namwat, C., de Souza, M., Adams, E., Benenson, M., Gurunathan, S., Tartaglia, J., McNeil, J., Francis, D., Stablein, D., Birx, D., Chunsuttiwat, S.,

Khamboonruang, C., Thongcharoen, P., Robb, M., Michael, N., Kunasol, P., and Kim, J. (2009). Vaccination with ALVAC and AIDSVAX to prevent HIV-1 infection in Thailand. *New England Journal of Medicine*, 361:2209–2220.

Richardson, M. and Haller, S. (2002). What is the probability of a kiss? (It's not what you think). *Journal of Statistics Education*, 10:9–9.

Richert, B., Tokach, M., Goodband, R., and Nelssen, J. (1995). Assessing producer awareness of the impact of swine production on the environment. *Journal of Extension*, 33.

Riemer, S., Wright, B., and Brown, R. (2011). Food habits of Steller sea lions (*Eumetopias jubatus*) off Oregon and northern California, 1986–2007. *Fishery Bulletin*, 109:369–381.

Robert, C. (2001). *The Bayesian Choice: From Decision-Theoretic Foundations to Computational Implementation*. Springer.

Robert, C. and Casella, G. (2004). *Monte Carlo Statistical Methods*. Springer.

Robert, C. and Casella, G. (2010). *Introducing Monte Carlo Methods with R*. Springer.

Rogan, W. and Gladen, B. (1978). Estimating the prevalence from the results of a screening test. *American Journal of Epidemiology*, 107:71–76.

Root, R. (1967). The niche exploitation pattern of the blue-gray gnatcatcher. *Ecological Monographs*, 37:317–350.

Rours, G., Verkooyen, R., Willemse, H., van der Zwaan, E., van Belkum, A., de Groot, R., Verbrugh, H., and Ossewaarde, J. (2005). Use of pooled urine samples and automated DNA isolation to achieve improved sensitivity and cost-effectiveness of large-scale testing for *Chlamydia trachomatis* in pregnant women. *Journal of Clinical Microbiology*, 43:4684–4690.

Rue, H., Martino, S., and Chopin, N. (2009). Approximate Bayesian inference for latent Gaussian models by using integrated nested Laplace approximations. *Journal of the Royal Statistical Society, Series B*, 71:319–392.

Rust, K. and Rao, J. (1996). Variance estimation for complex surveys using replication techniques. *Statistical Methods in Medical Research*, 5:293–310.

Salsburg, D. (2001). *The Lady Tasting Tea: How Statistics Revolutionized Science in the Twentieth Century*. Henry Holt and Company, LLC.

Satterthwaite, F. (1946). An approximate distribution of estimates of variance components. *Biometrics Bulletin*, 2:110–114.

Schonnop, R., Yang, Y., Feldman, F., Robinson, E., Loughin, M., and Robinovitch, S. (2013). Prevalence of and factors associated with head impact during falls in older adults in long-term care. *Canadian Medical Association Journal*, 185:E803–E810.

Schwartz, C. and Mare, R. (2005). Trends in educational assortative marriage from 1940 to 2003. *Demography*, 42:621–646.

Scott, A. (2007). Rao-Scott corrections and their impact. In *Proceedings of the Section on Survey Research Methods*, pages 3514–3518. American Statistical Association.

Scott, A. and Rao, J. (1981). Chi-squared tests for contingency tables with proportions estimated from survey data. In Krewski, J., Platek, R., and Rao, J., editors, *Current Topics in Survey Sampling*, pages 247–266. Academic Press.

Seeber, G. (2005). Poisson Regression. In Armitage, P. and Colton, T., editors, *Encyclopedia of Biostatistics, online.* Wiley.

Severini, T. (2000). *Likelihood Methods in Statistics.* Oxford University Press.

Shtatland, E., Kleinman, K., and Cain, E. (2003). Stepwise methods using SAS Proc Logistic and SAS Enterprise Miner for Prediction. In *SAS Users Group International*, volume 28, paper 258. SAS Institute.

Skinner, C. and Vallet, L. (2010). Fitting log-linear models to contingency tables from surveys with complex sampling designs: An investigation of the Clogg-Eliason approach. *Sociological Methods & Research*, 39:83–108.

Snee, R. (1974). Graphical display of two-way contingency tables. *The American Statistician*, 28:9–12.

Stroup, W. (2013). *Generalized Linear Mixed Models: Modern Concepts, Methods, and Applications.* CRC Press.

Stukel, T. (1988). Generalized logistic models. *Journal of the American Statistical Association*, 83:426–431.

Suess, E., Sultana, D., and Gongwer, G. (2006). How much confidence should you have in binomial confidence intervals? *Stats: The Magazine for Students of Statistics*, 45:3–7.

Swift, M. (2009). Comparison of confidence intervals for a Poisson mean—further considerations. *Communications in Statistics: Theory and Methods*, 238:748–759.

Tango, T. (1998). Equivalence test and confidence interval for the difference in proportions for the paired-sample design. *Statistics in Medicine*, 17:891–908.

Tanner, M. (1996). *Tools for Statistical Inference: Methods for the Exploration of Posterior Distributions and Likelihood Functions.* Springer, 3rd edition.

Tauber, M., Tauber, C., and Nechols, J. (1996). Life history of *Galerucella nymphaeae* and implications of reproductive diapause for rearing univoltine chrysomelids. *Physiological Entomology*, 21:317–324.

Thomas, D. and Decady, Y. (2004). Testing for association using multiple response survey data: Approximate procedures based on the Rao-Scott approach. *International Journal of Testing*, 4:43–59.

Thomas, D. and Rao, J. (1987). Small-sample comparisons of level and power for simple goodness-of-fit statistics under cluster sampling. *Journal of the American Statistical Association*, 82:630–636.

Thomas, D., Singh, A., and Roberts, G. (1996). Tests of independence on two-way tables under cluster sampling: An evaluation. *International Statistical Review*, pages 295–311.

Thompson, S. (2002). *Sampling.* Wiley.

Thorburn, D., Dundas, D., McCruden, E., Cameron, S., Goldberg, D., Symington, I., Kirk, A., and Mills, P. (2001). A study of hepatitis C prevalence in healthcare workers in the west of Scotland. *Gut*, 48:116–120.

Tibshirani, R. (1996). Regression shrinkage and selection via the LASSO. *Journal of the Royal Statistical Society, Series B*, 58:267–288.

Turner, D., Ralphs, M., and Evans, J. (1992). Logistic analysis for monitoring and assessing herbicide efficacy. *Weed Technology*, 6:424–430.

Vance, A. (2009). Data analysts captivated by R's power. *The New York Times*.

Vansteelandt, S., Goetghebeur, E., and Verstraeten, T. (2000). Regression models for disease prevalence with diagnostic tests on pools of serum samples. *Biometrics*, 56:1126–1133.

Ver Hoef, J. and Boveng, P. (2007). Quasi-Poisson vs. negative binomial regression: How should we model overdispersed count data? *Ecology*, 11:2766–2772.

Vermunt, J. and Magidson, J. (2007). Latent class analysis with sampling weights: A maximum-likelihood approach. *Sociological Methods & Research*, 36:87–111.

Verstraeten, T., Farah, B., Duchateau, L., and Matu, R. (1998). Pooling sera to reduce the cost of HIV surveillance: A feasibility study in a rural Kenyan district. *Tropical Medicine and International Health*, 3:747–750.

Vos, P. and Hudson, S. (2005). Evaluation criteria for discrete confidence intervals: Beyond coverage and length. *The American Statistician*, 59:137–142.

Wald, A. (1943). Tests of statistical hypotheses concerning several parameters when the number of observations is large. *Transactions of the American Mathematical Society*, 54:426–482.

Wardrop, R. (1995). Simpson's paradox and the hot hand in basketball. *The American Statistician*, 49:24–28.

Wedderburn, R. (1974). Quasi-likelihood, generalized linear models, and the Gauss-Newton method. *Biometrika*, 61:439–447.

Westfall, P. and Young, S. (1993). *Resampling-Based Multiple Testing: Examples and Methods for P-Value Adjustment*. Wiley.

Wilkins, T., Malcolm, J., Raina, D., and Schade, R. (2010). Hepatitis C: Diagnosis and treatment. *American Family Physician*, 81:1351–1357.

Williams, D. (1975). The analysis of binary responses from toxicological experiments involving reproduction and teratogenicity. *Biometrics*, 31:949–952.

Wilson, E. (1927). Probable inference, the law of succession, and statistical inference. *Journal of the American Statistical Association*, 22:209–212.

Wright, B. (2009). Use of chi-square tests to analyze scat-derived diet composition data. *Marine Mammal Science*, 26:395–401.

Yee, T. (2010). The `VGAM` package for categorical data analysis. *Journal of Statistical Software*, 32.

Yuan, M. and Lin, Y. (2006). Model selection and estimation in regression with grouped variables. *Journal of the Royal Statistical Society, Series B*, 68:49–67.

Zamar, D., McNeney, B., and Graham, J. (2007). `elrm`: Software implementing exact-like inference for logistic regression models. *Journal of Statistical Software*, 21(3):1–18.

Zeger, S. and Liang, K. (1986). Longitudinal data analysis for discrete and continuous outcomes. *Biometrics*, pages 121–130.

Zeileis, A., Kleiber, C., and Jackman, S. (2008). Regression models for count data in R. *Journal of Statistical Software*, 27.

Zhou, X. and Qin, G. (2005). A new confidence interval for the difference between two binomial proportions of paired data. *Journal of Statistical Planning and Inference*, 128:527–542.

Zou, H. (2006). The adaptive LASSO and its oracle properties. *Journal of the American Statistical Association*, 101:1418–1429.

Index

abline(), 3, 489
acat(), 192
Adjacent-categories model, 186, 192
aggregate(), 74
Agresti-Caffo interval for $\pi_1 - \pi_2$, 29
Agresti-Coull interval for π, 12
Agresti-Min interval for $\pi_{+1} - \pi_{1+}$, 45
AIC(), 268
AlcoholDetailed.R, 330
AlcoholPoRegs.R, 205
All-subsets regression, 268
AllGOFTests.R, 294
AllSubsetsPlacekick.R, 269
Alternating stepwise selection, 276
Anova(), 77, 79
anova(), 77, 79, 396, 416, 442, 465
Anova.polr(), 179
aod package, 317
AppendixInitialExamples.R, 473, 479
apply(), 73, 118, 160
arm package, 471
array(), 26, 149
as.factor(), 261
as.numeric(), 85, 89, 465
assocstats(), 37, 149
Asymptotically equivalent, 503
Asymptotics, 502
autocorr.plot(), 458
axis(), 491

Backward elimination, 224, 274
base package, 476
Baseline-category logit model, 153
Bayes estimate, 446
Bayes' rule, 444
Bayesian
    credible interval, 12
    empirical, 447
    methods, 444
    model averaging, 281
BeetleEggCrowding.R, 242
BeetleEggCrowding.ZIP.R, 247
Bernoulli distribution, 1, 62, 497

bestglm package, 271
Beta distribution, 14
Beta-binomial model, 316
Beta.R, 15
betabin(), 317
Bias-variance tradeoff, 266, 345
bic.glm(), 282
Binary regression model, 61
binom package, 14, 23, 449, 478
binom.bayes(), 449
binom.blaker.limits(), 15
binom.confint(), 14, 16, 17, 449
binom.coverage(), 23
binom.plot(), 23
Binomial distribution, 2, 25, 73, 497
Binomial regression model, 61
Binomial.R, 2, 7
Bird.R, 27, 30, 35, 43
BirdCountPoReg.R, 212
BirdOverdisp.R, 311, 314
BirdQuasiPoi.R, 305
Blaker interval for π, 15, 380
BlakerCI package, 15
BMA package, 271, 282
BMAPlacekick.R, 282
Bonferroni, 84, 110, 407
boot package, 130, 372
boot(), 130, 372
boot.ci(), 130
bootMer(), 433
Bootstrap, 151, 407, 414
    confidence interval, 130, 417, 437, 469
    parametric, 130, 432
Bubble plot, 93
Burn-in period, 450

c(), 26, 475
car package, 77, 135, 212, 218, 299, 331
Case-control studies, 41, 58
cat(), 476
Categorical explanatory variables, 100, 163, 168, 211, 218, 421

cbind(), 70
cex, 485
chisq.test(), 36, 149, 369
Choose all that apply, 404
ci.pi(), 93
CIpi.R, 10, 13, 15, 52
CIpiBayes.R, 447
class(), 487
Clopper-Pearson interval for π, 14
Clustered data, 304, 420
Clustering, 381
coda package, 452, 454
coef(), 283, 431
coefficients(), 161, 488
col, 485
colors(), 485
Complete separation, 112
Complex sampling design, 380
Comprehensive R Archive Network (CRAN), 473
Conditional distribution, 426
Conditional modes, 427
Conditional multiple marginal independence, 412
Confidence interval
   bands, 91
   interpretation, 12
confint(), 85, 162, 179
confint.default(), 86
confint.glm(), 212
confint.mcprofile(), 110
confint.multinom(), 162
ConfLevel.R, 19, 20
ConfLevel4Intervals.R, 18, 22, 23
ConfLevelTwoProb.R, 31
ConfLevelWaldOnly.R, 22
Conjugate family, 445
Conservative, 11
Consistency, 120
Contingency table
   comparison of models for, 229
   exact test, 362
   for multiple-response categorical variables, 405
   $I \times J$, 47, 143, 366
   loglinear model, 218
   nominal multinomial regression model, 163
   proportional odds regression model, 179
   survey data, 389
   tests, 34, 146
   three-way, 222
   two-way, 25
   zero counts, 42
Continuity correction, 36
contour(), 73
Contrast coefficients, 228, 398
Contrast matrix, 90, 228
contrasts(), 102
Convergence, 65, 110, 454, 501
Cook's distance, 297
cooks.distance(), 298
Correlated data, 419
Covariance matrix, 69, 503
Coverage, 11, 17
Credible interval, 447
   equal tail, 447
   for π, 447
   for $\pi_1 - \pi_2$, 30
   highest posterior density, 447
Crosstabulations, 28
Crossvalidation, 278
Cumulative distribution function (CDF), 122
   inverse, 123
Cumulative logit, 170
CumulativeLogitModelPlot.R, 171, 185
cumuplot(), 457
curve(), 64, 489
cv.glmnet(), 278

data.frame(), 2, 37, 487
datasets package, 262
dbinom(), 2
Deff
   generalized deff, 391
ΔD_m, 297
Delta method, 38, 41, 383, 505
ΔX_m^2, 297
deltaMethod(), 135, 158
deltaMethod.polr(), 175
deltaMethod.polr2(), 175
densplot(), 453
Design effect (deff), 381
Design-based inference, 382
dev.new(), 489
Deviance, 81, 511
   change in residual, 297
   residual, 81, 110
dhyper(), 365
diag(), 70

DiagnosisCancer.R, 46, 47
diffpropci.mp(), 46
diffpropci.Wald.mp(), 46
diffscoreci(), 57
Disease prevalence, 5
Dispersion parameter, 310
dlogis(), 123
dmultinom(), 142
dose.p(), 137
drop1(), 77, 433

Effect parameters, 211
Effective sample size, 388
effectiveSize(), 455
elrm package, 374
elrm(), 374
Emacs Speaks Statistics, 482
Empirical distribution, 368
    cumulative distribution function, 150
epitools package, 199
Estimation, 496
Evidence weights, 285
Exact
    conditional distribution, 373
    confidence interval, 14, 199
    inference, 14, 362
    test, 362, 363, 367
exact2x2 package, 380
exact2x2(), 380
exactLogLinTest package, 380
Examine.logistic.model.R, 321
examine.logistic.reg(), 321
Exchangeable, 421
exp(), 39
expand.grid(), 31
Expected length, 53
Explanatory variable pattern form, 287
Exposure, 240
expression(), 3, 64, 491
extractAIC(), 348
Extreme value distribution, 124

F-distribution, 15
factor(), 102, 149
FallsGEE.R, 441
FallsGLMM.R, 423, 428, 433, 437
family, 67, 192, 311, 360
Familywise confidence level, 90, 110
Fiber.R, 148, 164, 168, 180, 183
FiberExact.R, 366, 369
Fisher information, 446

Fisher scoring, 68
Fisher's exact test, 363
Fisher, Ronald, 364
fisher.test(), 365
Fitted value, 205
Fixed effects, 421
fixef(), 431
for(), 23
formula, 67, 75, 80, 95, 224, 236, 248
Forward selection, 224, 272
Frequentist inference, 12, 443
ftable(), 191, 375

Gamma function, 14
Gaussian quadrature, 427
geeglm(), 442
geepack package, 442
gelman.diag(), 456
gelman.plot(), 456
Generalized estimating equations, 440
Generalized linear mixed model, 310, 422
    fitting methods, 425
Generalized linear model, 65, 121, 203, 266, 285
    linear predictor, 121
    link, 121, 203, 266, 286, 360
    random component, 121
    systematic component, 121
Genetic algorithm, 269, 334
    mutation, 269
genloglin(), 414
Geometric mean, 216
getAnywhere(), 492
geweke.diag(), 455
geweke.plot(), 455
glht(), 136, 212
GLM (Generalized linear model), 121
glm(), 65, 102, 111, 118, 139, 206, 242, 311, 360, 376, 414
glm.nb(), 314
glmDiagnostics.R, 298, 338
glmer(), 427, 468
glmInflDiag(), 298
glmmPQL(), 428
glmnet package, 278
glmnet(), 278
glmulti package, 269
glmulti(), 269, 282, 318, 334
glmultiFORmultinom.R, 348
glmultiFORpolr.R, 348
glmultiFORzeroinfl.R, 349

Goodness of fit
    deviance/df, 293
    Hosmer and Lemeshow test, 293
    ordinal score set, 238
    Osius-Rojek test, 294
    Pearson statistic, 292
    Poisson model, 296
    residual deviance statistic, 81, 292
    Stukel test, 294
    test, 239
GPA.R, 473, 482
graphics package, 73, 476
grplasso package, 280
Gumbel distribution, 124

Hat matrix, 114, 297
hatvalues(), 298
head(), 28, 487
help(), 477
HepCPrev.R, 16
HepCPrevSeSp.R, 357
Hessian matrix, 69, 72, 360, 503
Heteroscedasticity, 302
hist(), 8
HIVKenya.R, 358
HLtest(), 294
Hosmer and Lemeshow test, 293
hpd(), 448
HPDinterval(), 452
hurdle(), 257
Hypergeometric distribution, 362
Hyperparameters, 446
Hyperprior distribution, 447

I(), 95
ifelse(), 20
Independence, 40, 144, 219
    exact test for, 363, 367
    for MRCVs, 406
    test for (large-sample), 146, 163, 179, 220, 235
    test for (survey), 390
Indicator variables, 100, 211, 218
Inference, 496
Influence, 296
influencePlot(), 299
Information criteria, 239, 267
    Akaike (AIC), 267
    Bayesian (BIC), 267
    corrected Akaike (AIC_c), 267
INLA package, 459

Intensity, 195
Interaction, 94, 103, 132, 168, 187, 209, 219, 238, 271, 395, 419
    high-order, 223, 238, 395
interaction(), 393
italic(), 3
Item response table, 405
item.response.table(), 409, 414
Iterative numerical procedures, 65, 124
Iteratively reweighted least squares, 65

Jackknife, 385, 387, 393, 461
Jeffreys interval for π, 11, 447
jitter(), 353
Joint independence, 406
Joint probabilities, 44

Laplace approximation, 427
LarryBirdPoReg.R, 220
LASSO, 277
LASSOPlacekick.R, 278
Law of large numbers, 7
legend(), 33, 93
length(), 75, 478
Lethal dose level, 137
levels(), 100
Leverage, 297
Liberal, 11
library(), 90, 478
Likelihood
    modified, 114, 377, 379
Likelihood function, 496
    Bernoulli distribution, 8, 356, 497, 500, 501
    binomial distribution, 74
    independent binomial distributions, 27
    logistic regression, 64
    multinomial distribution, 142, 143
    Poisson distribution, 198, 498, 500, 503, 507
Likelihood ratio
    confidence interval, 510
    confidence interval, binomial probability, 17
    confidence interval, Poisson mean, 199
    confidence interval, profile, 84, 87, 96, 212, 510
    confidence interval, variance component, 436

test, 17, 76, 507
test for equal binomial probabilities, 35
test for independence, 147, 220
test for proportional odds, 182
test for variance component, 436
test of fit, 239
test with quasi-likelihood, 310
test, model comparison, 82, 511
test, Rao-Scott, 391
LikelihoodFunction.R, 498
Linear predictor, 63, 87, 203, 266
lines(), 33
link, 126
lm(), 72, 485
lme4 package, 428
lmtest package, 77, 249
Loess, 289
loess(), 290
log(), 71, 476
logistf package, 115
logistf(), 115, 379
logistftest(), 115
Logistic regression, 62
    comparison to loglinear model, 231
    exact, 114, 372
    Firth penalty, 114, 377
    testing error, 358
Logistic.R, 122
LogisticRegBiasVarianceSim.R, 345
LogisticSim.R, 440
logistiX package, 374
logistiX(), 374
Logit
    confidence interval, 52
    link, 121
    transformation, 52, 63
logLik(), 71
logLik.glm(), 71
Loglinear model, 203, 218
    comparison to logistic regression, 231
    for MRCV, 412
    homogeneous association, 223
    linear association, 234
    linear-by-linear association, 238, 465
    mutual independence, 222
    survey data, 394
LR-Plots.R, 507
lrm(), 176
LRT (likelihood ratio test), 17
lrtest(), 77, 249

ls(), 475
lty, 485
lwd, 489

Main effects, 95
Marginal distribution, 426
Marginal model, 439
Marginal probability, 44
marginal.table(), 409
Marginality, 224, 270, 334
Markov chain, 453
    Monte Carlo, 374, 450
MASS package, 137, 154, 172, 314, 428
Matched pairs data, 43
max(), 490
Maximum likelihood estimate, 9, 498
    conditional, 374
    invariance property, 504
    large-sample properties, 9, 502
    variance of, 502
MC-SIMEX, 361
MCMC, 450
    burn-in period, 455
    convergence diagnostics, 454
    effective sample size, 455
mcmc.list(), 456
MCMClogit(), 450, 471
MCMCmnl(), 472
MCMCpack package, 450, 458
MCMCpoisson(), 472
McNemar's test, 47
mcnemar.exact(), 380
mcnemar.test(), 47
mcprofile package, 88, 154, 212, 344, 434
mcprofile(), 90, 109, 210, 216
mean(), 7, 478
Median unbiased estimate, 374
methods(), 68, 488
Metropolis-Hastings algorithm, 450, 453
mfrow, 63, 489
MI.stat(), 411
MI.test(), 409, 467
Mid p-value, 464
Mid-p interval for π, 15
midPci(), 15
min(), 490
Model averaging, 268, 281
Model selection, 265
Model space, 268
Model-based inference, 382

Monte Carlo simulation, 6, 20, 116, 148, 201, 251, 303, 345, 352, 369
MP5.1.R, 374
MRCV (Multiple response categorical variable), 405
MRCV package, 409, 414, 467
multcomp package, 99, 136, 212, 216, 434
multinom(), 153, 159, 165, 190, 301
multinomDiag(), 352
multinomDiagnostics.R, 301, 352
Multinomial distribution, 44, 141, 143, 498
Multinomial regression model, 153, 229
Multinomial.R, 142, 145
multinomRob package, 318
Multiple hypergeometric distribution, 366
Multiple inference, 84, 216, 434
Multiple marginal independence, 412
Multiple response categorical variable (MRCV), 404

names(), 67, 486
Negative binomial distribution, 313
Nested effects, 416
Newton-Raphson algorithm, 501
NewtonRaphson.R, 501
nnet package, 153
Non-convergence.R, 112, 114
Non-proportional odds model, 182
Nonlinear models, 121
Null deviance, 81

o.r.test(), 294
Object-oriented language, 488
objects(), 475
Odds, 40, 152
Odds ratio, 40, 151
    logistic regression, 82, 95, 105
    loglinear model, 219, 223
    MRCV model, 413
    multinomial regression, 160
    ordinal, 234, 238
    probit model, 125
    profile likelihood ratio interval, 84
    proportional odds model, 177
    survey data, 392
    variance, 505
    Wald confidence interval, 83, 96, 105
Offset, 241
optim(), 65, 71, 359, 501
options(), 115, 372

orscoreci(), 57
Osius-Rojek test, 294
Outliers, 289, 300
OverdispAddVariable.R, 307
Overdispersion, 199, 289, 302, 422
OverdispSim.R, 303

P-value combination testing method, 411
palette(), 485
panel.first, 485
par(), 63, 485, 489
Parallel coordinates plot, 154, 341
Parameter, 495
parcoord(), 154
Parsimony, 266
paste(), 3
pch, 485
pchisq(), 81
pchShow(), 485
PDF (probability density function), 496
Pearson chi-square test, 34, 147, 367
Pearson statistic, 292, 301
    change in, 297
    in MRCV, 406, 414
    survey data, 390
Penalized quasi-likelihood, 426
Percentage change (PC), 204
Permutation
    distribution, 369
    test, 368
persp3d(), 33
Pick any/c, 404
PiPlot.R, 63, 124
Placekick-FindBestModel.R, 318
Placekick.R, 65, 70, 71, 74, 77, 82, 85, 88, 91, 96, 111, 125
PlacekickBayes.R, 450, 456
PlacekickDiagnostics.R, 290, 294, 298
plogis(), 123
plot(), 3, 23, 454, 484
PMF (Probability mass function), 496
pnorm(), 122, 475, 476
PoiConfLevels.R, 201
pois.approx(), 199
pois.exact(), 199
Poisson distribution, 195, 498
Poisson regression, 201
    exact, 380
    hurdle model, 246
    MRCV, 413
    with offsets, 241

with rates, 241
zero-inflated (ZIP), 245
`poisson.test()`, 199
PoissonLikelihood.R, 503
PoissonNormalDistPlots.R, 259
PolIdeolNominal.R, 225
PolIdeolOrd.R, 233, 235, 239
`polr()`, 172, 178, 183, 186
PoRegLinks.R, 204
Posterior distribution, 444, 445, 449
`PostFitGOFTest()`, 296
PostFitGOFTest.R, 296, 338, 351, 352
Power, 362
`predict()`, 87, 89, 157, 174, 243, 417, 431, 488
`predict.glm()`, 89
Primary sampling units, 464
`print()`, 23, 270
Prior distribution, 444, 445
  improper, 450
  Jeffreys', 446
  noninformative, 446
  weakly informative, 449
Probability density function (PDF), 14, 122, 496
  joint, 496
Probability mass function (PMF), 496
  exact, 367
  joint, 496
Probability of success, 1
  complementary log-log regression, 124
  logistic regression, 86
  probit regression, 124
Product multinomial model, 144
Product symbol, $\prod$, 497
`prop.table()`, 261
`prop.test()`, 17, 31, 35
PropCIs package, 15, 31, 46, 57
Proportional odds model, 170
pscl package, 247, 263
Pseudo-likelihood, 400

`qbeta()`, 15
`qnorm()`, 11
QQ-plot, 150
`qqnorm()`, 258
`qt()`, 477
`quartz()`, 489
Quasi-binomial model, 311
Quasi-likelihood, 310
Quasi-Poisson model, 311

R
  argument, 475, 476
  buffered output, 23
  class, 487
  Commander, 492
  components of a list, 67
  console window, 474
  data frame, 2, 483, 486
  factor, 100
  function writing, 475, 491
  generic function, 488
  graphics window, 484
  help, 476
  list, 26, 68, 486
  listserv, 492
  method function, 68, 488
  multiple-document interface (MDI), 480
  object, 474
  package, 476, 478
  R editor, 479
  S4, 184
  single-document interface (SDI), 480
  slots, 184
  Stack Overflow, 492
R2jags package, 459
R2OpenBUGS package, 459
R2WinBUGS package, 459
Random effect, 310, 421
`ranef()`, 431
Rao-Scott correction, 390, 408, 414
Rate, 195, 240
Raw data, 28, 368, 370
`rbinom()`, 7
Rcmdr package, 492
`read.csv()`, 483
`read.table()`, 66, 483
Recycles, 70
Regression, linear, 61
  parameters, 61
`regTermTest()`, 400
`rejectionRate()`, 455
Relative risk, 37, 135
`relevel()`, 100
`rep()`, 370
Repeated measures, 420
Representativeness, sample, 425
Resampling, 383, 408
Residual, 286
  deviance, 287
  Pearson, 287

standardized Pearson, 152, 287, 338, 418
studentized, 299
residual
    Pearson, 152
residuals(), 288, 298, 418
residuals.glm(), 288
rev(), 43
rgl package, 33, 73
riskscoreci(), 57
rjava package, 269
rm(), 475
rms package, 176
rmultinom(), 142
RODBC package, 483
round(), 3, 487
rowSums(), 27
RStan package, 459
rstandard(), 288, 298
RStudio, 479
runif(), 116

Salk.R, 38, 42
sample(), 371
Sampling distribution, 502
SamplingDist.R, 116, 118
Sandwich estimator, 441
Saturated model, 62, 81, 219, 511
Score
    confidence interval, 511
    function, 114
    interval for μ, 199
    interval for π, 12
    interval for $\pi_1 - \pi_2$, 29, 57
    test, 11, 16, 508
    test equality of two probabilities, 34
scoreci.mp(), 45, 57
Scores
    ordinal, 232
sd(), 160, 475
segments(), 138
Sensitivity, 50, 355
Sensitivity analysis, Bayesian, 458
seq(), 492
set.seed(), 7
SGL package, 280
sim.all(), 118
SimContingencyTable.R, 26
simex package, 361, 461
simulate(), 433
Simultaneous inference, 84, 137, 229

Simultaneous pairwise marginal independence (SPMI), 406
Single-step error rate control, 110, 229
slotNames(), 184
solve(), 70
Sparsity, 266
Specificity, 50, 355
spm(), 331
SPMI (Simultaneous pairwise marginal independence), 406
sqrt(), 93, 475
Standard error, 70, 503
Statistical model, 1, 495
stats package, 47, 77, 199, 476
step(), 273
StepwisePlacekick.R, 273, 275, 276
Stoplight.R, 196, 199
Stratification, 381
Stukel test, 294
stukel.test(), 294
Subject-specific model, 439
Sufficient statistic, 373
sum(), 26
summarize(), 372
summary(), 68, 70, 149, 484, 487
summary.glm(), 69
summary.polr(), 179
summary.table(), 36
survey package, 385, 464
Survey weights, 381
SurveySmoke.R, 384, 386, 388, 395, 401
SurveySmokeLoglinear.R, 392, 395, 466
svrepdesign(), 385
svychisq(), 392
svycontrast(), 398, 466
svydesign(), 464
svyglm(), 400
svyloglin(), 466
svymean(), 388
svyplot(), 400
svytable(), 392
sweep(), 465
Swine.R, 409, 414
symbols(), 93

t(), 70
t.test(), 478
table(), 8, 28
Tea tasting experiment, 364
Tea.R, 365
TeachingDemos package, 448

Test type
    partial, 442
    sequential, 442
Testing error, 355
Tinn-R, 480
TinnRcom package, 481
TomatoVirus.R, 101, 104, 106
traceplot(), 454
Transformations, 94
True confidence level, 11, 17, 31, 42, 84, 200, 251, 304, 437

uniroot(), 86
update(), 392, 396, 466

var(), 7, 475
Variable selection, 265
    biased parameter estimates due to, 277, 282
Variance component, 421
    inference, 436
Variance-covariance matrix, 69, 503
vcd package, 37, 149
vcov(), 69, 86
vcov.glm(), 69
VGAM package, 160
vglm(), 160, 183, 186

Wald confidence interval, 510
    binomial probability, 10
    binomial probability with testing error, 357
    difference of binomial probabilities, 29
    difference of marginal probabilities, 45
    logit transformation, 52
    odds ratio, 41
    odds ratio in logistic regression, 83, 96, 105
    Poisson mean, 199
    probability in logistic regression, 87
    probability in multinomial regression, 158
    relative risk, 38
Wald test, 506
    equal binomial probabilities, 34
    equal marginal probabilities, 47
    regression parameter, 76
    survey data, 388
    variance component, 436

wald(), 91, 96, 99
wald2ci(), 31
weightable(), 270
Weighted least squares, 65
Weights, replicate, 384
Wheat.R, 154, 160, 173, 178
Wilson interval for π, 11
win.graph(), 489
window(), 455
windows(), 489

x11(), 489
xtabs(), 28, 148

Zero inflation, 245, 336
zeroinfl(), 248
ZeroInflPoiDistPlots.R, 246
ZIPSim.R, 250, 257